Principles and Practices of Engineering Hydrology

Principles and Practices of Engineering Hydrology

Edited by Stacy Keach

MURPHY & MOORE
www.murphy-moorepublishing.com

Published by Murphy & Moore Publishing,
1 Rockefeller Plaza,
New York City, NY 10020, USA
www.murphy-moorepublishing.com

Principles and Practices of Engineering Hydrology
Edited by Stacy Keach

International Standard Book Number: 978-1-63987-452-1 (Hardback)

Cataloging-in-Publication Data

Principles and practices of engineering hydrology / edited by Stacy Keach.
p. cm.
Includes bibliographical references and index.
ISBN 978-1-63987-452-1
1. Hydraulic engineering. 2. Engineering. 3. Hydrology. 4. Fluid mechanics. I. Keach, Stacy.
TC153 .E54 2022
627--dc23

TABLE OF CONTENTS

PREFACE

Engineering hydrology is the applied science of water resources engineering. It is concerned with the study of the hydrological cycle like runoff, precipitation, transpiration, estimation of water resources, etc. It analyzes the problem of floods and drought and aims to formulate preventive measures. This field finds its application in planning and management of water resources, floods forecasting, calculating rainfall, determining water balance, catchment hydrology, creating measures to control erosion and sedimentation and hydropower generation. This field focuses on determining water yield from a basin, studying the groundwater development, and determining the maximum intensity of the storm. This book brings forth some of the most innovative concepts and elucidates the unexplored aspects of engineering hydrology. From theories to research to practical applications, case studies related to all contemporary topics of relevance to this field have been included herein. This book is a vital tool for all researching or studying this domain as it gives incredible insights into emerging trends and concepts.

After months of intensive research and writing, this book is the end result of all who devoted their time and efforts in the initiation and progress of this book. It will surely be a source of reference in enhancing the required knowledge of the new developments in the area. During the course of developing this book, certain measures such as accuracy, authenticity and research focused analytical studies were given preference in order to produce a comprehensive book in the area of study.

This book would not have been possible without the efforts of the authors and the publisher. I extend my sincere thanks to them. Secondly, I express my gratitude to my family and well-wishers. And most importantly, I thank my students for constantly expressing their willingness and curiosity in enhancing their knowledge in the field, which encourages me to take up further research projects for the advancement of the area.

Editor

1

Robust global sensitivity analysis of a river management model to assess nonlinear and interaction effects

L. J. M. Peeters[1], G. M. Podger[2], T. Smith[2], T. Pickett[3], R. H. Bark[4], and S. M. Cuddy[2]

[1]CSIRO Land and Water, Water for a Healthy Country Flagship, Adelaide, Australia
[2]CSIRO Land and Water, Water for a Healthy Country Flagship, Canberra, Australia
[3]CSIRO Land and Water, Water for a Healthy Country Flagship, Brisbane, Australia
[4]CSIRO Ecosystem Sciences, Water for a Healthy Country Flagship, Brisbane, Australia

Correspondence to: L. J. M. Peeters (luk.peeters@csiro.au)

Abstract. The simulation of routing and distribution of water through a regulated river system with a river management model will quickly result in complex and nonlinear model behaviour. A robust sensitivity analysis increases the transparency of the model and provides both the modeller and the system manager with a better understanding and insight on how the model simulates reality and management operations.

In this study, a robust, density-based sensitivity analysis, developed by Plischke et al. (2013), is applied to an eWater Source river management model. This sensitivity analysis methodology is extended to not only account for main effects but also for interaction effects. The combination of sensitivity indices and scatter plots enables the identification of major linear effects as well as subtle minor and nonlinear effects.

The case study is an idealized river management model representing typical conditions of the southern Murray–Darling Basin in Australia for which the sensitivity of a variety of model outcomes to variations in the driving forces, inflow to the system, rainfall and potential evapotranspiration, is examined. The model outcomes are most sensitive to the inflow to the system, but the sensitivity analysis identified minor effects of potential evapotranspiration and nonlinear interaction effects between inflow and potential evapotranspiration.

1 Introduction

Water managers rely heavily on models to predict future water availability, optimize water use and evaluate water management strategies in order to find a balance between environmental, social and economic demands on the system. It is therefore crucial to be aware of the ability of a model to capture the dynamics of the hydrological cycle relevant to the water management question. In recent decades, addressing this issue has been the focus of much research in hydrological model calibration and predictive uncertainty analysis (Gupta et al., 2012).

For a modeller, to arrive at a "well"-calibrated model or to produce sensible and robust prediction intervals, it is essential to have a thorough understanding of how the hydrological system works and how this system is represented in the model – how a variation in parameters, boundary conditions or driving forces will affect the prediction of interest. The knowledge gained from such sensitivity analysis is not only of relevance during model development, it also provides added value to the model as it can focus management and monitoring to those aspects of the system and model that are most important to the management of water resources (Saltelli et al., 2008). Additionally, discussing model sensitivities with stakeholders will remove the notion of the model being a "black box" and can provide stakeholders with a better appreciation of the accuracy of the model, which has proven to be a key aspect of adoption of model results by management (Patt, 2009; Bark et al., 2013).

River management models such as eWater Source (Welsh et al., 2013) are increasingly used, especially in Australia, in the development of basin-wide water allocation plans. As these plans directly affect the livelihood of people and the health of ecosystems, it is essential that the models underpinning these plans have wide support and are robust. It is therefore essential that practitioners have a set of tools for sensitivity analysis available, tailored to the needs of water allocation modelling. The most straightforward sensitivity analysis technique is One-At-a-Time (OAT) sensitivity analysis in which one model aspect is changed while the others are fixed. The sensitivity of the model output to variation of the tested parameter is proportional to the gradient of the response surface. This is formalized in gradient-based calibration routines, such as Levenberg–Marquardt optimization. Examples of such OAT sensitivity analysis are Doherty and Hunt (2009), Foglia et al. (2009), Castaings et al. (2009) and Peeters et al. (2011). This methodology is attractive as it requires a very limited number of model runs, about two or three model runs per parameter evaluated, and, as long as the model behaves linearly, parameter interaction effects can be explored (Hill and Tiedeman, 2007). Saltelli and Annoni (2010) highlight that OAT sensitivity analysis only provides reliable and robust results if it can be shown that the model behaviour is linear. This condition is seldom satisfied for hydrological models or even known before a sensitivity analysis. The Elementary Effects method (Campolongo et al., 2007) is more robust against nonlinearity in the model behaviour, whilst still being frugal in the number of model runs.

Global sensitivity analysis techniques however do not require the model behaviour to be linear (Saltelli et al., 2008). The most straightforward global sensitivity analysis is either random or density-based sampling of parameter space and visualizing scatter plots of the parameter value against the prediction of interest (Wagener and Kollat, 2007; Peeters et al., 2013). Variance-based methods, such as Sobol' sensitivity analysis (Saltelli and Annoni, 2010; Nossent et al., 2011), use a scheme of structured resampling of a random base sampling to decompose the variance of the metric of interest into the main effects of a parameter and interaction effects of other parameters.

The main drawback of variance-based methods is that it assumes that the entire effect of a parameter can be summarized by the variance (Borgonovo, 2007; Borgonovo et al., 2011). Variance-based sensitivity indices will therefore be less reliable if the response to a parameter has a skewed or multi-modal distribution. Density-based sensitivity analysis techniques attempt to account for this by incorporating the entire distribution of the response of a prediction of interest in the metric in a way that does not require any assumptions on the shape of the distribution. The methodology suggested by Plischke et al. (2013) implements such a density-based sensitivity analysis technique which is independent of the parameter sampling scheme. This has the added benefit that as no model runs need to be devoted to the resampling of a base sampling, more computing resources can be directed to exploration of parameter space.

The goal of this study is to apply a density-based sensitivity analysis in a river management modelling context to assess its capability to identify and quantify nonlinear effects and to extend the methodology to account for interaction effects. An idealized, hypothetical river management model implemented in the eWater Source platform (Welsh et al., 2013) serves as testing platform to assess the ability of the sensitivity analysis methodology to quantify the influence of a small number of forcing variables upon a variety of model outcomes.

The next section presents the theoretical background and numerical implementation of the Plischke et al. (2013) global sensitivity analysis method. The river management model is briefly introduced before presenting the results of the sensitivity analysis and summarizing the findings in the discussion and conclusion sections.

2 Methods

The sensitivity analysis introduced in Plischke et al. (2013) provides a robust, global density-based sensitivity analysis, independent of sampling strategy. This section provides a short summary of this methodology. For a detailed overview the interested reader is referred to Plischke et al. (2013).

Consider X and Y the set of variables that comprise the input and output respectively of a river system model. Fixing X to a single realization, the parameter combination x, results in a conditional cumulative distribution of Y equal to $F_{Y|X=x}(y)$ and an equivalent density function $f_{Y|X=x}(y)$. The importance of fixing X to x can be quantified by the separation between the unconditional $F_Y(y)$ and the conditional $F_{Y|X=x}(y)$ or, similarly, the separation between $f_Y(y)$ and $f_{Y|X=x}(y)$. Using the L1-norm, the separation between the two density functions can be written as

$$s(x) = \int\limits_Y |f_Y(y) - f_{Y|X=x}(y)| \mathrm{d}y. \quad (1)$$

The importance of factor X on outcome Y can then be defined as

$$\begin{aligned} \delta(Y, X) &= \frac{1}{2} \mathrm{E}[s(X)] \\ &= \frac{1}{2} \int\limits_X f_X(x) \int\limits_Y |f_Y(y) - f_{Y|X=x}(y)| \mathrm{d}y \mathrm{d}x. \end{aligned} \quad (2)$$

The sensitivity index $\delta(X, Y)$ varies between 0 and 1 and it can be shown that this index is zero when X and Y are completely independent (Plischke et al., 2013).

To compute $\delta(X, Y)$ the integrals in Eq. (2) need to be approximated numerically. This can be achieved by taking n samples of the parameter space X and computing the corresponding values for Y. The method does not impose any restrictions on the sampling strategy of the parameter space.

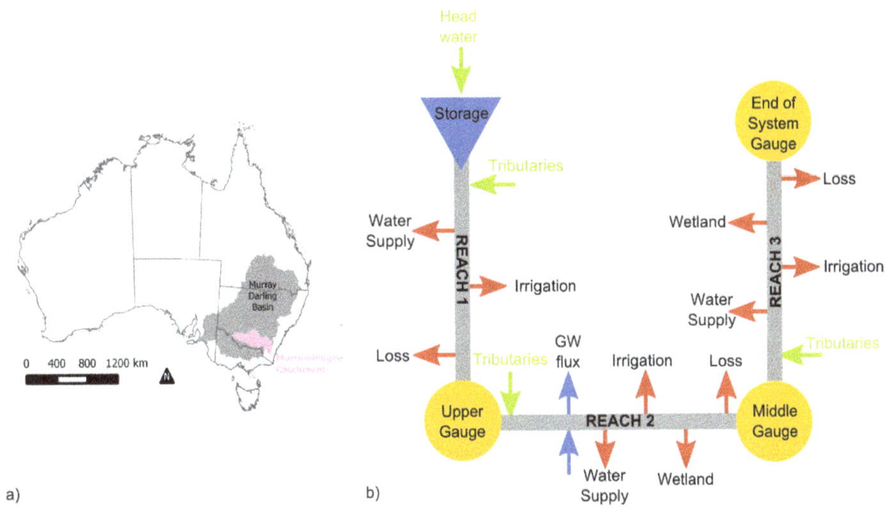

Figure 1. **(a)** Map showing the extent (indicated by pink shading) of the idealized river system model within the Murray–Darling Basin and **(b)** schematic structure of the river management model.

This implies that the methodology can be applied with random sampling, quasi-random sampling (e.g. Latin Hypercube Sampling or Sobol' sequences) or Markov chain Monte Carlo simulation.

The resulting data set is partitioned into M classes C_m with $m = 1, \ldots, M$. For each class C_m, the density function can be approximated with a kernel smoothing function with kernel $K(.)$ and bandwidth α (Devroye and Gyorfi, 1985):

$$\hat{f}_Y(y) = \frac{1}{n}\sum_{i=1}^{n}\frac{1}{\alpha}K\left(\frac{y-y_i}{\alpha}\right)$$
$$\hat{f}_{Y|C_m}(y) = \frac{1}{n_m}\sum_{i:x_i\in C_m}^{n_m}\frac{1}{\alpha_m}K\left(\frac{y-y_i}{\alpha_m}\right), \tag{3}$$

where n_m is the number of samples in class C_m and α_m the corresponding bandwidth for the kernel smoothing function.

The next step is to approximate the L1-norm between the two distributions for each class. Using a predefined number of quadrature points $\{\tilde{y}_j,\ j = 1, \ldots, l\}$, the separation can be computed as

$$s_{m,j} = \hat{f}_Y(\tilde{y}_j) - \hat{f}_{Y|C_m}(\tilde{y}_j)$$
$$\hat{S}_m = \frac{1}{2}\sum_{j=1}^{l-1}\left(|s_{m,j+1}| + |s_{m,j}|\right)\left(\tilde{y}_{j+1} - \tilde{y}_j\right). \tag{4}$$

The sensitivity index δ can then be approximated by

$$\hat{\delta} = \frac{1}{2n}\sum_{m=1}^{M} n_m \hat{S}_m. \tag{5}$$

To avoid bias in the sensitivity index and to assess the robustness of the sensitivity index estimate, it is recommended to perform a bootstrap of the sensitivity index (Efron, 1977) and to adjust $\hat{\delta}$ with the mean of the bootstrap $\bar{\delta}^*$:

$$\hat{\hat{\delta}} = 2\hat{\delta} - \bar{\delta}^*. \tag{6}$$

$\hat{\hat{\delta}}$ provides the sensitivity index of the main effect of a variable. Plischke et al. (2013) however does not provide a method to explore second-order effects, i.e. the interaction between two variables. To estimate second-order effects between variables X_1 and X_2, the samples are subdivided into n groups of equal intervals for X_1. The sensitivity index $\hat{\delta}$ for X_2, $\hat{\delta}_{X_2}$, is computed for each interval. If there is no interaction effect between X_1 and X_2, then $\hat{\delta}_{X_2}$ will not vary with the level of X_1. To quantify this, the variance of $\hat{\delta}_{X_2}$ is computed over all n levels of X_1. Small variances indicate small interaction effects and vice versa.

3 Model description and setup

The case study is a hypothetical river system model (Fig. 1), based on a simplified version of the Murrumbidgee River model in New South Wales, Australia (Dutta et al., 2012; Podger et al., 2014). Using the full version of the Murrumbidgee River model was not warranted, not only because of the complexity of the system and the management rules, but, more importantly, because of legal issues with regard to model licensing and confidentiality. The idealized, hypothetical model retains most of the relevant complexity practitioners encounter when creating water allocation models, which is more than sufficient to illustrate the sensitivity analysis methodology.

In the model, water is routed from a storage reservoir through three river reaches. Routing starts in reach 1 at the storage reservoir with hydropower generators that receive water from a single tributary inflow. In reach 1, water is taken

Table 1. Output variables of the Source river system model.

Name	Description	Units
UpperFlow	Flow rate at the gauge at the end of the first reach	$m^3\,s^{-1}$
MiddleFlow	Flow rate at the gauge at the end of the middle reach	$m^3\,s^{-1}$
EndFlow	Flow rate at the gauge at the end of the final reach	$m^3\,s^{-1}$
$AlgalBloom	Monetary value generated by recreation as function of the risk of algal blooms	10^6 AUD
$Stor	Monetary value generated by recreation on storages	10^6 AUD
$TotalAg	Monetary value generated by irrigated agriculture	10^6 AUD
Hydropower	Electricity generated from the storage reservoir	kWh
GenSec	Percentage of time general security licences receive their full entitlement	%

from the system for town water supply and irrigation and water is received from unregulated rain-fed tributaries. From the Upper Gauge at the end of reach 1, water is routed through reach 2. In this reach, interaction with groundwater is taken into account by an exchange flux. As in reach 1, water is received from unregulated, rain-fed tributaries and water is taken out for irrigation and town water supply. In addition to these offtake, water is diverted into an off-river wetland system. Reach 3 starts at the middle gauge and is similar to reach 2. It also has offtake for town water supply, irrigation and off-river wetlands and receives inflow from rainfed tributaries. Groundwater–surface water interaction is not taken into account in this reach. Each reach has a term representing unaccounted losses. The loss relationships are taken from the more complex model. The total travel time from headwater to end-of-system is 18 days (3 days for reach 1, 6 days for reach 2 and 9 days for reach 3). These values, together with the other parameters influencing routing of water are also taken and aggregated from the more complex model.

Daily time series of rainfall and evaporation from 1895 to 2006 are obtained from SILO (http://www.longpaddock.qld.gov.au/silo/) for sites representative of each of the three reaches. These time series are used to simulate inflow from tributaries and compute irrigation demand. Inflow into the main storage in the model is taken from daily gauged data from 1895 to 2006.

The town water demands are based on a fixed annual pattern (8.8, 3.0 and $1.2 \times 10^6\,m^3\,year^{-1}$ for reaches 1, 2 and 3 respectively). Irrigation demands are based on a reach-based aggregation of irrigation use as well as rationalizing of crop types. There are environmental demands for the wetlands in reach 2 and 3, which are designed to establish and maintain favourable habitat conditions for indigenous fauna and flora (Janssen, 2012).

Two aspects of water management are considered: $347\,m^3\,s^{-1}$ order constraint on storage releases, i.e. the maximum flow that can be requested by water users in the system of the storage, and an annual allocation system. The allocation system comprises high and general security order debit annual accounting schemes. Water is first allocated from the storage to high security entitlement holders and only once these are fulfilled is water allocated to general security entitlement holders. The start of the water year is 1 July with allocations updated continuously throughout the year, where these include allowances for minimum tributary inflows and delivery losses. At the end of the water year accounts are reset to zero. Licence entitlements were aggregated on a reach basis. Two socio-economic indicators have been included to indicate the impacts of storage volumes on recreational usage and mid-river flows on algal blooms and the associated impact on recreational usage. There are three storage volume categories (< 10, < 50 and $> 50\,\%$) for recreational usage based on visitor numbers. Recreational benefits are calculated for periods of time the model is at each threshold, using the Crase and Gillespie (2008) 100 000 visitor estimate to Lake Hume. Estimates of visitor numbers at high and low storage volumes are based on this estimate and the actual Tourism Research Australia (TRA) average, low and high visitor numbers in the Murrumbidgee catchment in the period 2003–2010 (DRET, 2010). Benefit transfer recreation values are taken from the same study (updated to 2012 Australian dollars (AUD) using the Australian Consumer Price Index, CPI). There are three risk of algal bloom categories (no bloom, alert and bloom) – no bloom occurs if there is a flow of at least $11.6\,m^3\,s^{-1}$ in the previous 7 days and alert if this flow occurs within the previous 14 days; if flow does not exceed $11.6\,m^3\,s^{-1}$ in the previous 14 days, algal bloom is simulated to occur. Australian dollars have been associated with loss of amenity in the weeks when there is an alert or bloom using the thresholds, estimated visitor numbers using TRA data and high and low estimates of river recreation based on survey data (DRET, 2010), and benefit transfer of general recreation benefits from Morrison and Hatton MacDonald (2010) (2010 AUD values are updated to 2012 AUD using the CPI and where the full value is used for no bloom, a proportion based on Crase and Gillespie (2008) for an alert and 0 AUD for a bloom).

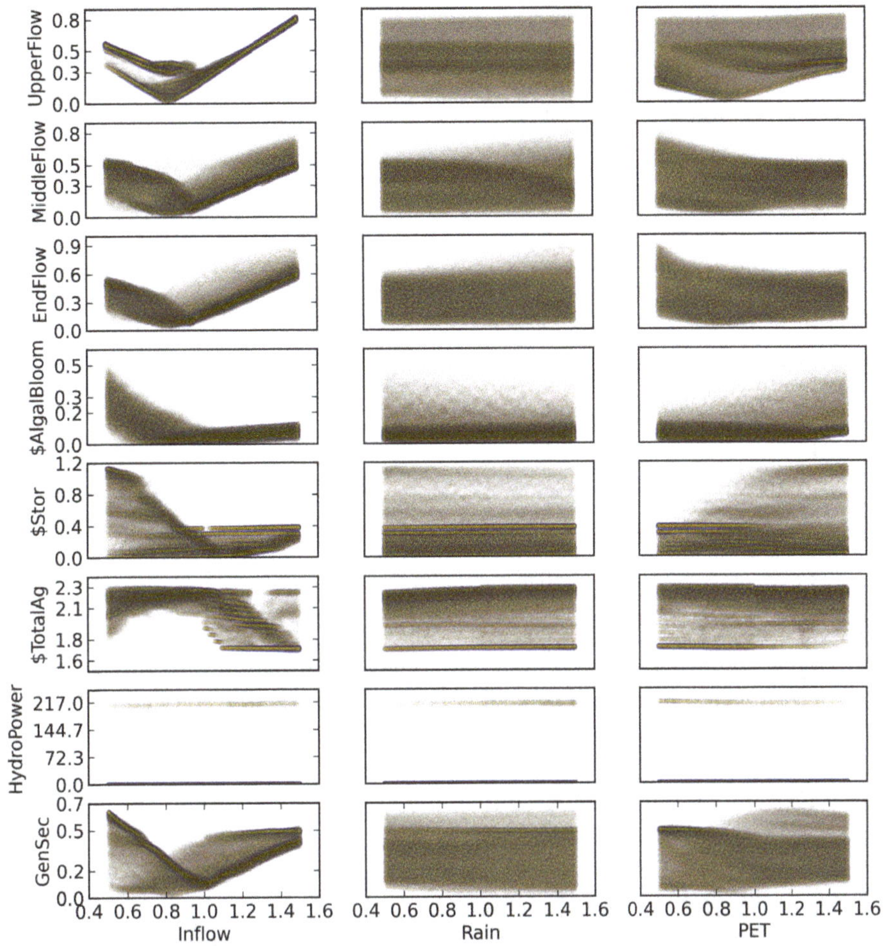

Figure 2. Scatter plots of $\hat{M}$, the difference between kernel density estimates for each simulation and the kernel density estimate of the reference simulation for all forcing data and model output variables for the eWater Source hypothetical river management model.

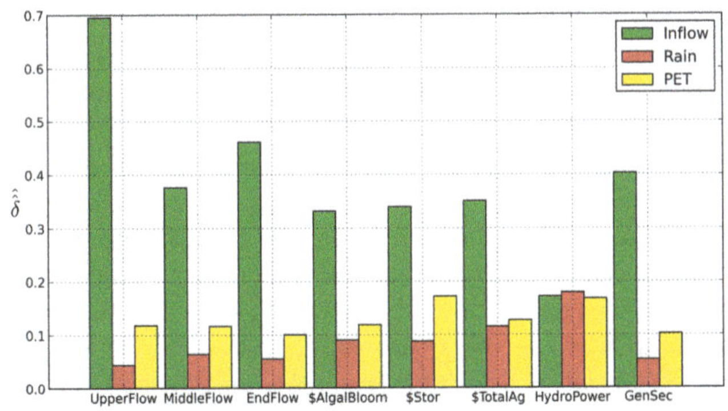

Figure 3. Sensitivity indices, $\hat{\hat{\delta}}$, for all forcing data and model output variables for the eWater Source hypothetical river management model.

4 Results

In the sensitivity analysis, the three main forcing variables are considered: the system inflow (Inflow), the precipitation (Rain) and the potential evapotranspiration (PET). The latter two affect the inflow into the reaches and the irrigation demand. Inspired by the work of Leblanc et al. (2012), the forcing variables are changed through a multiplier to the corresponding input time series with the range of the multiplier for each variable between 0.5 and 1.5. This range encompasses both historical variation in hydrological input and output, as well as the expected change under various climate change models and scenarios. While elaborate schemes are available to perturb hydrological time series, this is not warranted in this study as the focus is on metrics that integrate the entire flow time series. As such, the emphasis of this research is on changes in total flow in or out of the model, rather than in changes of the timing of flow.

Using Sobol' sequences (Sobol, 1976), 100 000 quasi-random samples of the three input variables are generated. For each of these samples a range of output time series is

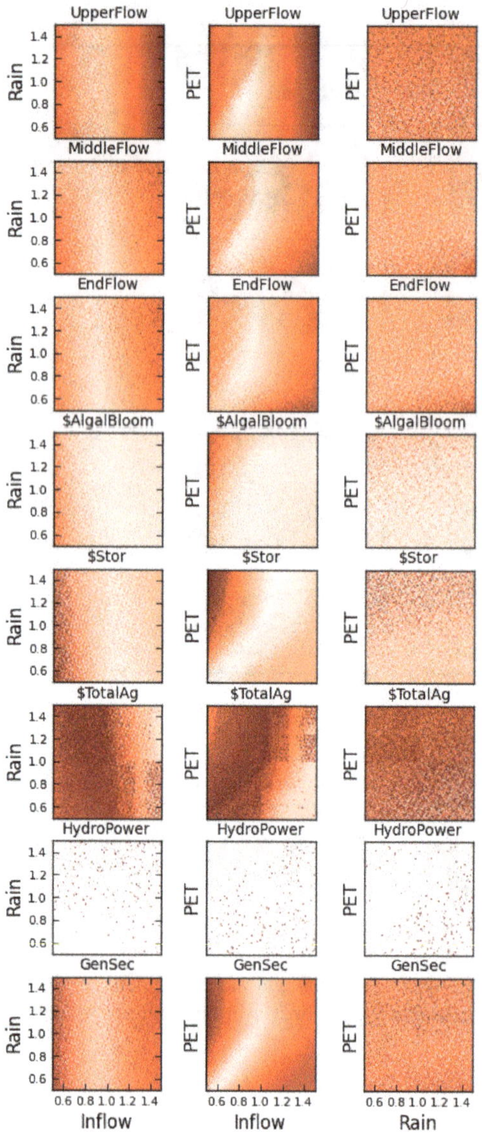

Figure 4. Scatter plots of interaction of the driving forces. The intensity of the colour scale is proportional to the model outcome value, where dark red indicates high values and light red indicates low values.

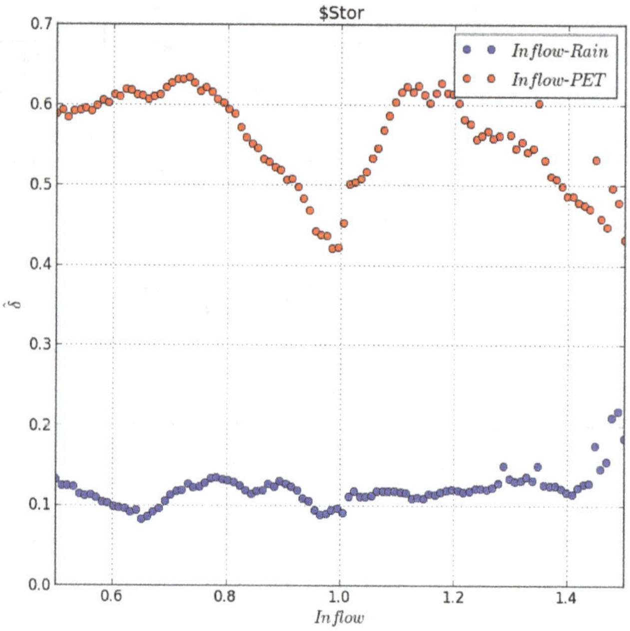

Figure 5. Sensitivity index $\hat{\delta}$ of the effect of Rain (blue) and PET (red) on \$Stor for 100 equal intervals of Inflow.

calculated (Pickett et al., 2013). Table 1 lists the names of the output series and a short description.

Each of the output variables in Table 1 is a daily time series. The metric for the sensitivity for different forcing data ($\hat{M}$) is the difference between the kernel density estimate of the daily times series of a randomly selected reference simulation ($\hat{f}_{\mathrm{Yref}}(y)$) and the kernel density estimate of the daily time series for the changed forcing data ($\hat{f}_{\mathrm{Ysim}}(y)$):

$$\hat{f}_{\mathrm{Yref}}(y) = \frac{1}{n}\sum_{j=1}^{n}\frac{1}{\alpha}K\left(\frac{y_{\mathrm{ref}} - y_{\mathrm{ref},i}}{\alpha}\right)$$

$$\hat{f}_{\mathrm{Ysim}}(y) = \frac{1}{n}\sum_{j=1}^{n}\frac{1}{\alpha}K\left(\frac{y_{\mathrm{sim}} - y_{\mathrm{sim},i}}{\alpha}\right)$$

$$d_j = \hat{f}_{\mathrm{Yref}}(\tilde{y}_j) - \hat{f}_{\mathrm{Ysim}}(\tilde{y}_j)$$

$$\hat{M} = \frac{1}{2}\sum_{j=1}^{l-1}(d_{j+1} + d_j)\left(|\tilde{y}_{j+1} - \tilde{y}_j|\right) \tag{7}$$

The choice of this metric is motivated by the fact that, since the case study is an idealized, hypothetical model, it is not possible to directly compare the results with observations. In addition to this, and more importantly, the variety of model outcomes examined in this study are more than likely to be affected by different aspects of the hydrograph. Similar to choosing an objective function in traditional calibration or a likelihood function in uncertainty analysis, such metric needs to be tailored to be able to capture the relevant aspects of the hydrograph. Choosing an ill-suited metric can have huge consequences for the sensitivity analysis, calibration or uncertainty analysis, as pointed out in Montanari and Koutsoyiannis (2012) and Nearing (2014). The metric presented in Eq. (7) is designed to provide an as general and robust as possible measure of the difference between two time series as not to bias the interpretation of the sensitivity analysis.

4.1 Main effects

Figure 2 shows the scatter plots of sensitivity metric $\hat{M}$ for all combinations of forcing data and output variables. It is clear that the dominant influencing driving variable is Inflow, as

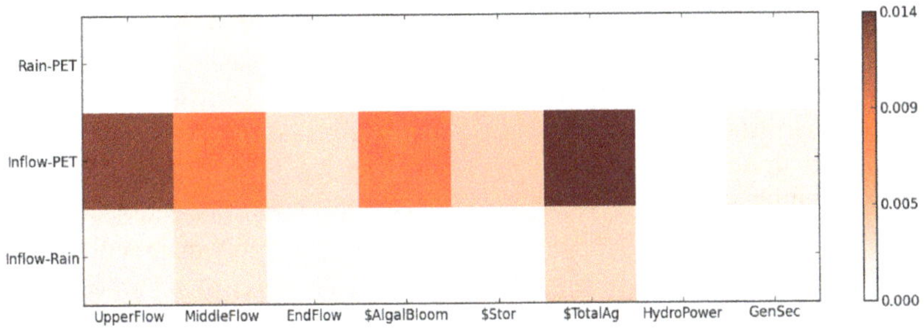

Figure 6. Var($\hat{\delta}_{X1-X2}$) for all combinations of driving forces for all model outcomes. High values indicate potential interaction between $X1$ and $X2$. The values for Hydropower are omitted in order not to distort the visualization.

a strong response is noticeable for variations in this driving variable for all output variables with the exception of HydroPower. The effects of "Rain" and "PET" are less pronounced. A very striking feature are the many nonlinearities in the response surface of the hypothetical model. This is mostly due to a number of threshold values used in the management rules of the river management system. For instance, generation of hydro-power is only possible when the storage level in the dam exceeds a predefined threshold related to the height of the water intake point for the turbines.

Figure 3 shows a barplot of the sensitivity indices $\hat{\hat{\delta}}$ for all main effects. These indices confirm the dominant influence of Inflow on most output variables. They provide a relative ranking of the influence of the input variable Inflow on the various output variables. MiddleFlow, EndFlow and GenSec respond to a similar degree to changes in Inflow and the same is true for the output variables related to monetary value ($AlgalBloom, $Stor and $TotalAg). HydroPower is least influenced by Inflow, which, from Fig. 2, is clearly related to the threshold-induced nonlinear behaviour.

The methodology is also able to quantify the often small and nonlinear effects of the other forcing variables. This is especially noticeable for PET. There is a clear but highly nonlinear effect of PET on $Stor, which is reflected in a higher $\hat{\hat{\delta}}$. The output variable HydroPower has a bimodal distribution where the majority of simulations have an $\hat{M}$ close to zero. Nevertheless, the global sensitivity method is able to distinguish and quantify the subtle trends in the non-zero values for the different input variables.

4.2 Interaction effects

The previous section established the importance of Inflow as the main driving variable. It is however from both a management and modelling perspective interesting to have an understanding of how the interaction between variables affects the model outcome.

Figure 4 shows plots with the factor values on the x- and y-axis, with a colour scale to visualize $\hat{M}$ for the three combinations of interaction of the driving forces (Inflow-Rain, Inflow-PET and Rain-PET) for all eight model outcomes. The first column shows that the effect of Inflow on most of the model outputs does not vary with the value of Rain. There is however a clear interaction between Inflow and PET for most of the model outputs; while the Inflow response is the dominant feature in the plots, the shape of this response depends on the value of PET. HydroPower is a noted exception as it displays very little structure in the scatter plots. This is because hydropower is generated by release of water from the reservoir in function of the demand and the water level in the reservoir. These management rules create a buffer to immediate impact from rainfall and inflow and also result in nonlinear, threshold related behaviour.

Very little structure is noticeable in the third column of Fig. 4, which shows the interaction between Rain and PET, reflecting the limited influence both driving forces have as a main effect.

To quantify the interaction effect for each interaction combination in Fig. 4, the variance of the $\hat{\delta}$ of the variable on the y-axis is computed for 100 equal intervals of the variable on the x-axis. By using Sobol' sequences to generate the 100 000 samples of the parameter space, each equal interval of the x-axis variable has approximately 1000 samples to compute the $\hat{\delta}$.

Figure 5 illustrates this for the interaction effects of Inflow, Rain and PET on $Stor. The sensitivity index values for Rain are low and hardly vary for different levels of Inflow, which is an indication of very limited interaction between Rain and Inflow, as confirmed by the scatter plot (Fig. 4). The $\hat{\delta}$ values for PET vary markedly with the level of Inflow. This sensitivity index reaches a minimum for Inflow values close to 1, while reaching peaks close to values of 0.75 and 1.1. This is reflected in the variance of the $\hat{\delta}$ values which is 4.5×10^{-4} for the Inflow–Rain couple and 3.5×10^{-3} for Inflow–PET. Figure 6 shows the variance of the sensitivity indices for all interaction pairs for all model outcomes. The values for Hydropower are much higher than for the other model outcomes due to the nonlinear behaviour. They were omitted from Fig. 6 as they distorted the visualization.

The most dominant interaction effects are between Inflow and PET for $TotalAg and UpperFlow, followed by $AlgalBloom, $Stor and MiddleFlow.

5 Discussion

The sensitivity analysis of the hypothetical river management model highlights inflow as a crucial variable of the model and how this affects the economic, environmental and sociological functions of the river. This emphasizes the importance of an accurate characterization of the flow rates of upstream areas when modelling flow routing in regulated systems comparable to the case study, i.e. the regulated river systems of the Murray–Darling Basin in Australia. An accurate characterization of flow rates not only entails maintaining a dense river gauge network, it also means adequately describing the measurement uncertainty in the flow rates, not in the least the uncertainty introduced by the rating curve that describes the stage–discharge relationship (Tomkins, 2012). The work of Hughes et al. (2014) illustrates this as they identify the inflow from ungauged catchment as crucial in the calibration of river management models.

Direct precipitation in the storage, wetlands and irrigation areas has a very minor influence on the model outcomes. This is mostly due to the small volume of rainfall (0.633 $km^3\,yr^{-1}$) compared to the inflow volume (4.4 $km^3\,yr^{-1}$) and the correlation between the inflow volume and rainfall. Any effect of rainfall will therefore be dwarfed by the effect of inflow to the system. The interaction effect of Inflow and PET is mostly due to the feedback mechanism as irrigation requirements increase with increasing potential evapotranspiration.

Such parameter interaction is well known in other areas of hydrological modelling, such as in rainfall–runoff modelling (Gallagher and Doherty, 2007; Zhang et al., 2013; Peeters et al., 2013) and in groundwater modelling (Doherty and Hunt, 2009), although it has not received much attention in river system modelling. Letcher et al. (2007) discuss the importance of interacting effects in water allocation models, without however providing a rigorous quantitative framework to evaluate the effects.

The sensitivity analysis in this study was limited to multiplying factors on three driving forces. It would be very insightful to include other model parameters in the sensitivity analysis, especially those controlling storage volumes and irrigation requirements. Along the same lines, including the parameters of the management rules, e.g. rules on allocations, in the sensitivity analysis can yield additional understanding of the operational management of the river system, as shown by Micevski et al. (2011).

6 Conclusions

The density-based sensitivity analysis of Plischke et al. (2013) has been applied to a river management model representing an idealized regulated river system representative of the southern Murray–Darling Basin in Australia to identify the main and interaction effects of three driving forces on several hydrological and socio-economic model outcomes.

The extended sensitivity analysis method presented in this paper provides a quantitative measure of sensitivity of the main and interaction effects and, through a combination with qualitative visual inspection of scatter plots, proved to be able to identify not only major effects but also subtle interactions, even in the presence of strong nonlinearities.

Due to the small dimensionality of the case study, it was possible to visualize all main effects and their interactions through scatter plots for all model outcomes. Although this will be challenging for higher-dimensional problems, the visual inspection of scatter plots is an invaluable complement to the sensitivity indices.

Understanding the dynamics of river system models is often not intuitive, especially in larger or basin-scale models (Johnston and Smakhtin, 2014). A robust and comprehensive sensitivity analysis is an invaluable step in model development to elucidate the often intricate interactions between driving forces, management rules and parameters. Increased understanding of the model will not only lead to improvements in calibration and prediction, it also has enormous potential in establishing the credibility and understanding of models.

Acknowledgements. The authors thank Russell Crosbie and Dave Penton for their constructive comments.

Edited by: H. Cloke

References

Bark, R., Peeters, L., Lester, R., Pollino, C., Crossman, N., and Kandulu, J.: Understanding the sources of uncertainty to reduce the risks of undesirable outcomes in large-scale freshwater ecosystem restoration projects: An example from the Murray-Darling Basin, Australia, Environ. Sci. Pol., 33, 97–108, 2013.

Borgonovo, E.: A new uncertainty importance measure, Reliability Eng. Syst. Safety, 92, 771–784, 2007.

Borgonovo, E., Castaings, W., and Tarantola, S.: Moment Independent Importance Measures: New Results and Analytical Test Cases, Risk Anal., 31, 404–428, 2011.

Campolongo, F., Cariboni, J., and Saltelli, A.: An effective screening design for sensitivity analysis of large models, Environ. Modell. Softw., 22, 1509–1518, 2007.

Castaings, W., Dartus, D., Le Dimet, F.-X., and Saulnier, G.-M.: Sensitivity analysis and parameter estimation for distributed hydrological modeling: potential of variational methods, Hydrol. Earth Syst. Sci., 13, 503–517, doi:10.5194/hess-13-503-2009, 2009.

Crase, L. and Gillespie, R.: The impact of water quality and water level on the recreation values of Lake Hume, Aust. J. Environ. Manage., 15, 21–29, 2008.

Devroye, L. and Gyorfi, L.: Nonparametric Density Estimation: The L1 View, John Wiley & Sons, Ltd New York, NY., 1985.

Doherty, J. and Hunt, R. J.: Two statistics for evaluating parameter identifiability and error reduction, J. Hydrol., 366, 119–127, 2009.

DRET: Destination visitor survey: strategic regional research report – New South Wales, Victoria and South Australia. Impact of the drought on tourism in the Murray River Region., Tech. rep., Department of Resources, Energy and Tourism, Tourism Research Australia, 57 pp., 2010.

Dutta, D., Hughes, J., Vaze, J., Kim, S., Yang, A., and Podger, G.: A daily river system model for the Murray-Darling Basin: development, testing and implementation, in: 34th Hydrology and Water Resources Symposium 2012, Sydney, 19–22 November 2012, 1057–1066, available at: http://www.hwrs2012.org.au/, 2012.

Efron, B.: Bootstrap methods: another look at the jackknife, The Annal. Stat., 7, 1–26, 1977.

Foglia, L., Hill, M. C., Mehl, S. W., and Burlando, P.: Sensitivity analysis, calibration, and testing of a distributed hydrological model using error-based weighting and one objective function, Water Resour. Res., 45, W06427, doi:10.1029/2008WR007255, 2009.

Gallagher, M. R. and Doherty, J.: Parameter interdependence and uncertainty induced by lumping in a hydrologic model, Water Resour. Res., 43, W05421, doi:10.1029/2006WR005347, 2007.

Gupta, H. V., Clark, M. P., Vrugt, J. A., Abramowitz, G., and Ye, M.: Towards a comprehensive assessment of model structural adequacy, Water Resour. Res., 48, W08301, doi:10.1029/2011WR011044, 2012.

Hill, M. C. and Tiedeman, C. R.: Effective groundwater model calibration, Wiley, 2007.

Hughes, J., Dutta, D., Vaze, J., Kim, S., and Podger, G.: An automated multi-step calibration procedure for a river system model, Environ. Modell. Softw., 51, 173–183, 2014.

Janssen, V.: Indirect Tracking of Drop Bears Using GNSS Technology, Aust. Geogr., 43, 445–452, doi:10.1080/00049182.2012.731307, 2012.

Johnston, R. and Smakhtin, V.: Hydrological Modeling of Large river Basins: How Much is Enough?, Water Resour. Manage., 28, 1–36, doi:10.1007/s11269-014-0637-8, 2014.

Leblanc, M., Tweed, S., Van Dijk, A., and Timbal, B.: A review of historic and future hydrological changes in the Murray-Darling Basin, Global Planet. Change, 80–81, 226–246, 2012.

Letcher, R. A., Croke, B. F. W., and Jakeman, A. J.: Integrated assessment modelling for water resource allocation and management: A generalised conceptual framework, Environ. Modell. Softw., 22, 733–742, 2007.

Micevski, T., Lerat, J., Kavetski, D., Thyer, M., and Kuczera, G.: Exploring the utility of multi-response calibration in river system modelling, 19th International Congress On Modelling and Simulation (modsim2011), 12–16 December 2011, Perth, Australia, 3889–3895, 2011.

Montanari, A. and Koutsoyiannis, D.: A blueprint for process-based modeling of uncertain hydrological systems, Water Resour. Res., 48, W09555, doi:10.1029/2011WR011412, 2012.

Morrison, M. and Hatton MacDonald, D.: Economic valuation of environmental benefits in the Murray-Darling Basin, Tech. rep., A report to the Murray Darling Basin Authority, 2010.

Nearing, G.: Comment on "A blueprint for process-based modeling of uncertain hydrological systems" by Montanari and Koutsoyiannis, Water Resour. Res., 50, 6260–6263, doi:10.1002/2013WR014812, 2014.

Nossent, J., Elsen, P., and Bauwens, W.: Sobol' sensitivity analysis of a complex environmental model, Environ. Modell. Softw., 26, 1515–1525, 2011.

Patt, A.: Uncertainties in environmental modelling and consequences for policy making, chap. Communicating uncertainty to policy makers, pp. 231–251, NATO science for peace and security series – C: Environmental Security, Springer, available at: http://www.pik-potsdam.de/news/public-events/archiv/alter-net/former-ss/2009/14.09.2009/patt/literature/bavaye-book.pdf, 2009.

Peeters, L., Lerat, J., and Rassam, D.: Improving parameter estimation in transient groundwater models through temporal differencing, in: MODSIM 2011: Sustaining Our Future, 12–16 December 2011, Perth, Australia, p. 7, 2011.

Peeters, L., Crosbie, R., Doble, R., and Van Dijk, A.: Conceptual evaluation of continental land-surface model behaviour, Environ. Modell. Softw., 43, 49–59, 2013.

Pickett, T., Smith, T., Bulluss, B., Penton, D., Peeters, L., Podger, G., and Cuddy, S.: Approaches to distributed execution of hydrologic models: methods for ensemble Monte Carlo risk modelling with and without workflows, in: MODSIM 2013 20th International Congress on Modelling and Simulation, 1–6 December 2013, Adelaide, Australia, p. 7, 2013.

Plischke, E., Borgonovo, E., and Smith, C. L.: Global sensitivity measures from given data, Eur. J. Operational Res., 226, 536–550, 2013.

Podger, G., Cuddy, S., Peeters, L., Smith, T., Bark, R., and Black, D.: Risk management frameworks: supporting the next generation of Murray-Darling Basin water sharing plans, in: Evolving Water Resources Systems: Understanding, Predicting and Managing Water-Society Interactions, Proceedings of ICWRS2014, Bologna, Italy, June 2014 (IAHS Publ. 364, 2014, 452–457), 364, 452–457, 2014.

Saltelli, A. and Annoni, P.: How to avoid a perfunctory sensitivity analysis, Environ. Modell. Softw., 25, 1508–1517, 2010.

Saltelli, A., Ratto, M., Andres, T., Campolongo, F., Cariboni, J., Gatelli, D., Saisana, M., and Tarantola, S.: Global Sensitivity Analysis. The Primer, John Wiley & Sons, Ltd, doi:10.1002/9780470725184, 2008.

Sobol, I.: Uniformly distributed sequences with an additional uniform property, USSR Comput. Mathem. Mathem. Phys., 16, 236–242, 1976.

Tomkins, K. M.: Uncertainty in streamflow rating curves: methods, controls and consequences, Hydrol. Process., 28, 464–481, doi:10.1002/hyp.9567, 2012.

Wagener, T. and Kollat, J.: Numerical and visual evaluation of hydrological and environmental models using the Monte Carlo analysis toolbox, Environ. Modell. Softw., 22, 1021–1033, 2007.

Welsh, W. D., Vaze, J., Dutta, D., Rassam, D., Rahman, J. M., Jolly, I. D., Wallbrink, P., Podger, G. M., Bethune, M., Hardy, M. J., Teng, J., and Lerat, J.: An integrated modelling framework for regulated river systems, Environ. Modell. Softw., 39, 81–102, 2013.

Zhang, C., Chu, J., and Fu, G.: Sobolś sensitivity analysis for a distributed hydrological model of Yichun River Basin, China, J. Hydrol., 480, 58–68, 2013.

2

Pitfalls in using log-transformed flows within the KGE criterion

Léonard Santos, Guillaume Thirel, and Charles Perrin

Irstea, HYCAR Research Unit, 1 rue Pierre-Gilles de Gennes, 92160 Antony, France

Correspondence: Léonard Santos (leonard.santos@irstea.fr)

Abstract. Log-transformed discharge is often used to calculate performance criteria to better focus on low flows. This prior transformation limits the heteroscedasticity of model residuals and was largely applied in criteria based on squared residuals, like the Nash–Sutcliffe efficiency (NSE). In the recent years, NSE has been shown to have mathematical limitations and the Kling–Gupta efficiency (KGE) was proposed as an alternative to provide more balance between the expected qualities of a model (namely representing the water balance, flow variability and correlation). As in the case of NSE, several authors used the KGE criterion (or its improved version KGE′) with a prior logarithmic transformation on flows. However, we show that the use of this transformation is not adapted to the case of the KGE (or KGE′) criterion and may lead to several numerical issues, potentially resulting in a biased evaluation of model performance. We present the theoretical underpinning aspects of these issues and concrete modelling examples, showing that KGE′ computed on log-transformed flows should be avoided. Alternatives are discussed.

1 Introduction

In the context of rainfall–runoff modelling, evaluating the quality of the models' outputs is essential. Deterministic simulations are commonly evaluated using efficiency criteria such as the Nash–Sutcliffe efficiency (NSE, Nash and Sutcliffe, 1970). The choice of the criteria obviously depends on the modeller's objective. For example, one may wish to focus on the overall water balance evaluation, or more specifically on the simulation of different flow ranges – typically high, intermediate or low flows. For these different objectives, given that the model residuals are generally not homoscedastic and often depend on the flow magnitude, one common option to focus more closely on specific flow ranges is to apply various prior transformations on the simulated and observed discharge time series to distort the range of errors, which consequently changes the relative weight of different flow ranges in the criterion. This is commonly done within the NSE criterion, which has been one of the most popular criteria used in hydrological modelling in the past few decades. NSE is the distance to 1 of the ratio between the mean square error of the model and the variance of observed flows. Compared to the basic criterion computed on untransformed flows, a prior squared transformation on flows would put even more weight on high flows, and a logarithmic or inverse transformation would put more weight on low flows, while a square-root transformation would have an intermediate effect (Krause et al., 2005; Oudin et al., 2006; De Vos and Rientjes, 2010; Pushpalatha et al., 2012).

However, the Nash–Sutcliffe criterion was shown to have limitations. Indeed, using a decomposition of NSE based on the correlation, bias and ratio of variances, Gupta et al. (2009) clearly demonstrated that discharge variability is not correctly taken into account for the evaluation. Therefore, Gupta et al. (2009) proposed a new criterion, the Kling–Gupta efficiency (KGE), which was then improved into a modified criterion called KGE′ (Kling et al., 2012). KGE combines the previous components of NSE (correlation, bias, ratio of variances or coefficients of variation) in a more balanced way. It corrects the underestimation of variability and provides direct assessment of four aspects of discharge time series, namely shape, timing, water balance and variability.

Given that this criterion tends to be sensitive to large errors, some users chose to apply prior transformations on flows before computing KGE, e.g. to put more weight on low flows, as done with NSE. For example, Pechlivanidis et al. (2014) applied the logarithmic transformation to use it as a benchmark for fitting a model on low flows. Seeger and Weiler (2014) used it as an objective function. Beck et al. (2016) used the untransformed and log-transformed flows in NSE, R^2 and KGE as an evaluation of different global models, and Quesada-Montano et al. (2018) also used it as an evaluation criterion of the HBV model outputs.

In this technical note we show that the use of a logarithmic transformation when computing KGE or KGE′, applied in a similar way to with NSE, introduces numerical flaws and should be avoided. After reviewing the mathematical formulation of KGE′, we expose the theoretical aspects explaining these flaws and illustrate them with modelling examples. Then we suggest alternatives to circumvent this issue. The tests will be carried out using KGE′ but they are also valid for the initial KGE formulation.

2 The KGE and KGE′ formulations

The KGE and KGE′ criteria (Gupta et al., 2009; Kling et al., 2012, respectively denoted E_{KG} and E'_{KG} in Eq. 1 and Eq. 2) are written as a linear transformation ($f: x \longmapsto 1-x$) of the Euclidian distance to an ideal value (i.e. [1,1,1]) in a three-dimensional space defined by three components of the modelling error:

$$E_{\mathrm{KG}} = 1 - \sqrt{(r-1)^2 + (\beta-1)^2 + (\alpha-1)^2}, \quad (1)$$

$$E'_{\mathrm{KG}} = 1 - \sqrt{(r-1)^2 + (\beta-1)^2 + (\gamma-1)^2}, \quad (2)$$

in which

- r, the Pearson correlation coefficient, evaluates the error in shape and timing between observed (Q_{o}) and simulated (Q_{s}) flows:

$$r = \frac{\mathrm{cov}(Q_{\mathrm{o}}, Q_{\mathrm{s}})}{\sigma_{\mathrm{o}}^2 \sigma_{\mathrm{s}}^2}, \quad (3)$$

where "cov" is the covariance between observation and simulation and σ is the standard deviation, with subscripts "o" and "s" standing for observed and simulated, respectively.

- β, the bias term, evaluates the bias between observed and simulated flows:

$$\beta = \frac{\mu_{\mathrm{s}}}{\mu_{\mathrm{o}}}, \quad (4)$$

where μ is the mean also with subscripts "o" and "s" standing for observed and simulated, respectively.

- α, the ratio between the simulated and observed standard deviations, evaluates the flow variability error:

$$\alpha = \frac{\sigma_{\mathrm{s}}}{\sigma_{\mathrm{o}}}. \quad (5)$$

- γ, the ratio between the simulated and observed coefficients of variation (CV), also evaluates the flow variability error. These coefficients of variation are used to avoid the impact of bias on the variability indicator (Kling et al., 2012):

$$\gamma = \frac{\mu_{\mathrm{o}} \sigma_{\mathrm{s}}}{\sigma_{\mathrm{o}} \mu_{\mathrm{s}}}. \quad (6)$$

The KGE′ values range between $-\infty$ and 1, as for NSE, and it is positively oriented.

3 Issues associated with the use of a prior logarithmic transformation

3.1 Instability when the moments of log-transformed flows become close to zero

Because the three terms γ, β and r are ratios, they can become overly sensitive to the denominator values (here μ_{o}, μ_{s}, σ_{o} or σ_{s}) if they become close to zero. In this case, a small absolute variation in the moments' values can negatively impact the related ratio and thus produce very negative KGE′ values. It is generally unlikely that values of σ_{o}, σ_{s}, μ_{s} and μ_{o} so close to zero can be obtained to produce numerical instability when using untransformed flows. However, when a prior logarithmic transformation is applied, the values of $\mu_{\mathrm{log,o}}$ or $\mu_{\mathrm{log,s}}$ (more rarely $\sigma_{\mathrm{log,o}}$ or $\sigma_{\mathrm{log,s}}$) computed on transformed values can become equal or close to zero (because $\log(1) = 0$). The corresponding ratios r, β or γ would therefore become very large, leading to strongly negative KGE′ values. Thus a small relative difference can lead to very different conclusions. In this case, the score value does not adequately represent the qualities of the model simulation.

3.2 Dependence on the flow unit chosen

KGE′ and NSE criteria are dimensionless. This means that using discharge values expressed in litres per second or in cubic metres per second has no impact on the criteria values. It can be easily demonstrated that γ, β and r remain identical when flow is expressed in any of these two units, since the division by 1000 necessary for the conversion is eliminated in the ratios. When using a prior logarithmic transformation, the NSE criterion is not affected because the squared differences of flows eliminates the multiplicative conversion coefficients in the mean square error (numerator) or in the variance (denominator). However, the KGE′ calculation is altered through the β ratio. Using the example of the average observed flow calculation, the conversion from cubic metres per second to litres per second gives the following:

$$\mu_{\log,o}[\mathrm{L\,s^{-1}}] = \log(1000) + \mu_{\log,o}[\mathrm{m^3\,s^{-1}}]. \quad (7)$$

Consequently, because the conversion term becomes additive when applying the logarithmic transformation, the β ratio value is modified. Similarly, the γ ratio is also altered. Therefore, if the logarithmic transformation is used, the KGE′ (and also the KGE) is no longer a dimensionless value. This can lead to interpretation problems.

3.3 Dependence on the constant added to avoid the zero-flow issue

When using a logarithmic (or an inverse) transformation, the case of null flows, which may exist in the case of intermittent or ephemeral streams, prevents proper calculation. To avoid this, different techniques may be set up in the case of NSE:

- The first involves discarding the zero-flow values from the series, i.e. considering them as gaps (see for example Nguyen and Dietrich, 2018). The drawback is that parts of the hydrographs become neglected, though they can bring important information on the processes at play.
- The second involves adding a small constant to all flow values (Pushpalatha et al., 2012), typically a fraction of average flow. This option is widely used and Pushpalatha et al. (2012) showed that the NSE value has limited sensitivity to this constant with a logarithmic transformation as long as it is small enough compared to flow values. These authors advise a constant equal to 1/100 of the mean observed flows. But the dependence of KGE′ on this constant has not been investigated so far.
- The third involves using a Box–Cox transformation to reproduce the effects of the logarithmic transformation without the zero-flow issue (Box and Cox, 1964; Hogue et al., 2000; Vázquez et al., 2008).

4 Testing methodology

To illustrate these numerical issues and their potential impacts, several tests were carried out in a wide range of catchments, using the GR4J rainfall–runoff model (Perrin et al., 2003).

4.1 Catchment set and data

A daily data set of 240 catchments across France (Fig. 1), set up by Ficchí et al. (2016), was used. The climate data of the SAFRAN daily reanalysis (Vidal et al., 2010) were used as input data. Precipitation and temperature were spatially aggregated in each catchment since the GR4J model is lumped. Potential evapotranspiration was calculated using a temperature-based formula (Oudin et al., 2005). Full details on this data set are available in Ficchí et al. (2016). Observed flows were retrieved for each catchment outlet from the Banque HYDRO (http://www.hydro.eaufrance.fr/ (last access: 29 August 2018), Leleu et al., 2014). The availability of data covers the 2005–2013 period. To avoid requiring a snow model, the catchments with less than 10 % of precipitation falling as snow were selected.

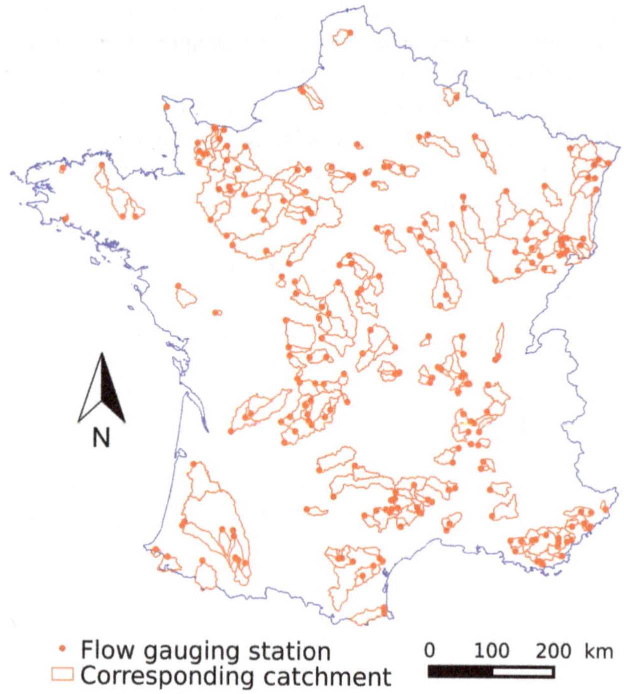

Figure 1. Location of the 240 flow gauging stations in France used for the tests and their associated catchments.

4.2 Model and calibration

The tests were performed with the daily lumped conceptual GR4J model (Perrin et al., 2003). The four parameters of the model are calibrated using the local search optimization algorithm used in Coron et al. (2017). The available records are split into a calibration (from July 2005 to June 2009) and a validation (from July 2009 to July 2013) period following a standard split-sample test procedure (Klemeš, 1986). The calibration procedure was run using the KGE′ on untransformed flows as an objective function. The performance of the model is then evaluated during the validation period using KGE′ on untransformed and log-transformed flows. The performance is also calculated using different transformations that can substitute the logarithmic transformation, namely the square-rooted flows, the inverted flows and the Box–Cox transformed flows. The NSE criterion is also calculated on log-transformed flows to be compared to KGE′ using the same transformation. The zero flows were treated following the conclusions of Pushpalatha et al. (2012), i.e.

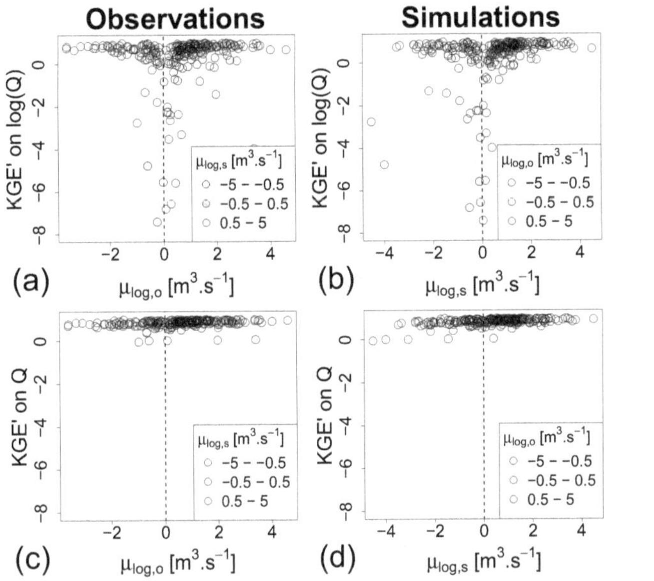

Figure 2. Values of KGE′ on log-transformed flows **(a, b)** versus the mean of the log-transformed observed and simulated flows compared. As a benchmark, the same plots are drawn with untransformed flows **(c, d)**. Each dot represents the performance obtained in validation for one catchment after calibration with the KGE′ on untransformed flows as an objective function. In plots **(a)** and **(c)**, the axis values represent the observed log-transformed flow averages and the color represents the simulated averages, while in plots **(b)** and **(d)** it is the opposite.

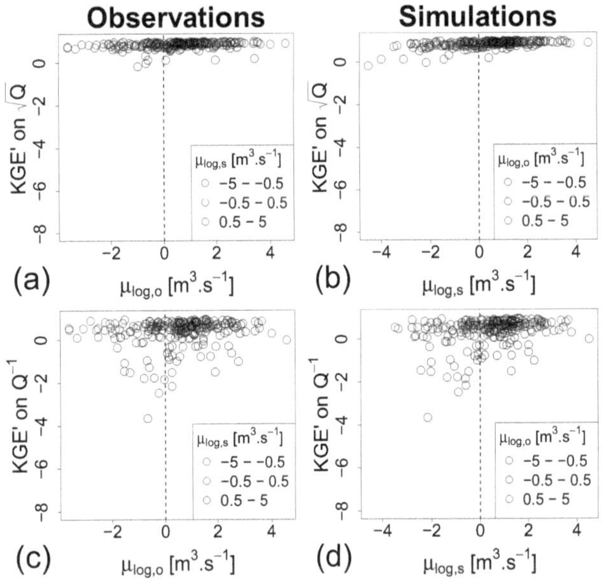

Figure 3. Values of KGE′ on square-root **(a, b)** and inverse **(c, d)** transformed flows versus the mean of the log-transformed observed and simulated flows. Each dot represents the performance obtained in validation for one catchment after calibration with the KGE′ on untransformed flows as an objective function. In plots **(a)** and **(c)**, the axis values represent the observed log-transformed flow averages and the color represents the simulated averages, while in plots **(b)** and **(d)** it is the opposite.

by adding to flows a constant equal to 1/100 of the mean observed flows. The parameter of the Box–Cox transformation is fixed at the value of 0.25, as Vázquez et al. (2008) argue that it is an usual value in hydrological studies.

5 Results

5.1 Instability when the moments of log-transformed flows become close to zero

Figure 2a and b analyse the stability of the KGE′ values with log-transformed flows obtained in the validation period. The KGE′ values were plotted against the mean of the log-transformed observed (a) and simulated (b) flows. When any of these means tends to be close to zero, the KGE′ criterion exhibits unusually low values. This plot illustrates the problem identified in Sect. 3.1. These very negative values may alter model evaluation. When working on a large set of catchments, they may also bias the calculation of the mean performance over the catchment set, by heavily weighting these outlier values. Figure 2c and d shows that the catchments with negative KGE′ values in Fig. 2a and b do not seem to exhibit any specific behaviour when evaluated with the KGE′ values on untransformed flows: the criterion values are not lower in these catchments than in other catchments. Furthermore, this result can be completed by making the same plot for other transformations, giving more weight to low flows. Figure 3 shows that square-root (Fig. 3a and b) and inverse (Fig. 3c and d) transformations do not encounter the same problems as with the logarithm for catchments that have an average log-transformed flow around zero.

The KGE′ on log-transformed flows can also be compared to the NSE using the same transformation. Figure 4 shows that, when KGE′ is significantly lower than NSE, the average of log-transformed flows (observed or simulated) is around zero (red dots in the figure). This tends to confirm that the strongly negative KGE′ values stem more from a numerical issue than an actual problem in simulated values, because the NSE values in these catchments remain positive or around zero.

In this technical note, the impact of a near-zero standard deviation of log-transformed flows is not presented because it is rarer than near-zero mean values. The standard deviations of flows in the catchments studied are indeed all significantly higher than zero.

5.2 Dependence on the flow unit chosen

The dependence of KGE′ on log-transformed flows on the chosen flow units can easily be shown by plotting the KGE′ on log-transformed flows in cubic metres per second versus

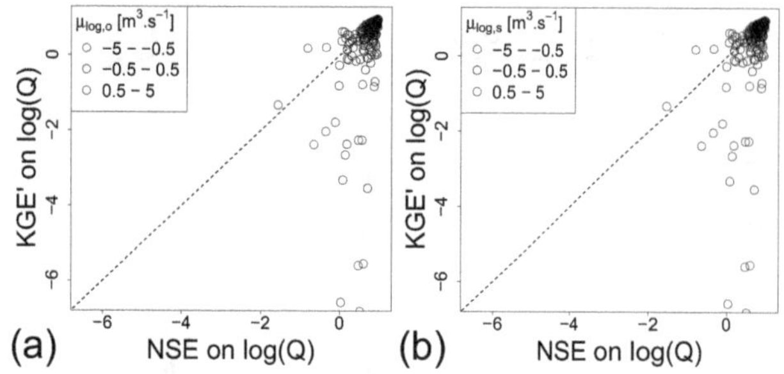

Figure 4. Comparison between KGE′ and NSE values on the validation period using a calibration with KGE′ on untransformed flows as an objective function. The red dots represent the catchments where the average of log-transformed observed **(a)** or simulated **(b)** flows is around 0.

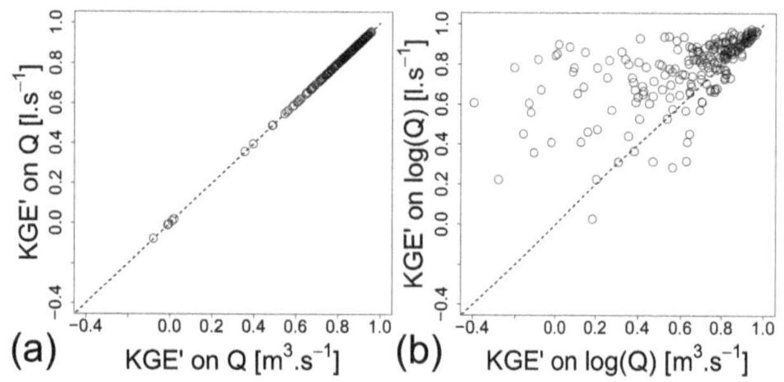

Figure 5. Dependence on flow units of the KGE′ using untransformed flows **(a)** and log-transformed flows **(b)** in the 240 catchments. The parameters used for simulation evaluation were obtained by calibrating GR4J using KGE′ on untransformed flows.

the KGE′ on log-transformed flows in litres per second. Figure 5b shows that, for the catchments tested, the values of KGE′ on log-transformed flows clearly depend on the flow unit used. A more optimistic evaluation of model performance will generally be obtained with the flows in litres per second. As a comparison, Fig. 5a shows that the KGE′ with untransformed flows is not affected by the flow unit change. This dimension dependence makes the KGE′ values based on log-transformed flows very difficult to interpret.

The higher model performance when using litres per second than when using square metres per second can be explained analytically. Considering Eq. (7), the formula of the bias ratio in litres per second regarding the averages in metres per second is as follows:

$$\beta_{\log}[\mathrm{L\,s^{-1}}] = \frac{\log(1000) + \mu_{\log,s}[\mathrm{m^3\,s^{-1}}]}{\log(1000) + \mu_{\log,o}[\mathrm{m^3\,s^{-1}}]}. \tag{8}$$

Because log(1000) is not negligible compared to the averages, adding this constant term would artificially improve β and, by extension, the KGE′ value. The γ ratio is also affected and, due to the interactions between the standard deviation and the averages, modifies the KGE′ value differently.

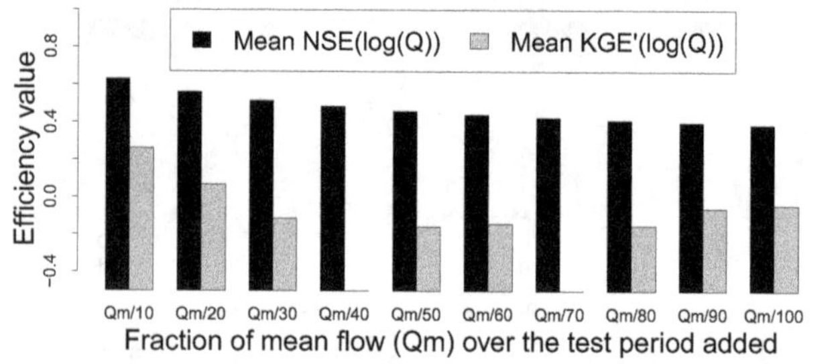

Figure 6. Sensitivity of NSE and KGE′ to the fraction of average flows that is added to flows to avoid zero flows in the logarithmic transformation for 240 catchments over the validation period. This graph is inspired by Fig. 9 in Pushpalatha et al. (2012).

5.3 Dependence on the value added to avoid the zero-flow issue

Pushpalatha et al. (2012) showed that the sensitivity of the NSE criterion on log-transformed flows to the small added constant declines when this constant decreases (from 1/10 to 1/100 of the mean observed flow) and becomes limited for very small values (see Fig. 9 in Pushpalatha et al., 2012). We performed the same test with the KGE′ criterion and we obtained a very different result (Fig. 6). The impact on performance is erratic for different values added to flows and does not show any trend. This may be due to the numerical issues shown in Sect. 5.1. For these reasons, the impact of added values can be major and may alter the model evaluation.

5.4 The case of the Box–Cox transformation

As presented in Sect. 3.3, instead of adding a small value to flows, a Box–Cox transformation can be applied to flows to mimic the logarithm transformation without the zero-flow problem. However, even though it removes the dependence of the KGE′ value to the value added to avoid zero flows, the other issues presented in the previous sections exist as for the logarithm. For catchments in which the log-transformed flows' average is close to zero, the Box–Cox transformed flows exhibit the same behaviour as with the logarithm (Fig. 7). This result is logical because the Box–Cox transformation of 1 is equal to 0, as for the logarithmic transformation.

The Box–Cox transformation is also dependent on the units (Fig. 8a). However, for this last issue, a slight modification of the Box–Cox formula allows one to address this problem. The classical Box–Cox transformation can be written as follows:

$$f_{\mathrm{BC}}(Q) = \frac{Q^{\lambda} - 1}{\lambda}, \tag{9}$$

in which λ is an exponent to be chosen by the user, Q is the flow value for any unit and f_{BC} is the Box–Cox function.

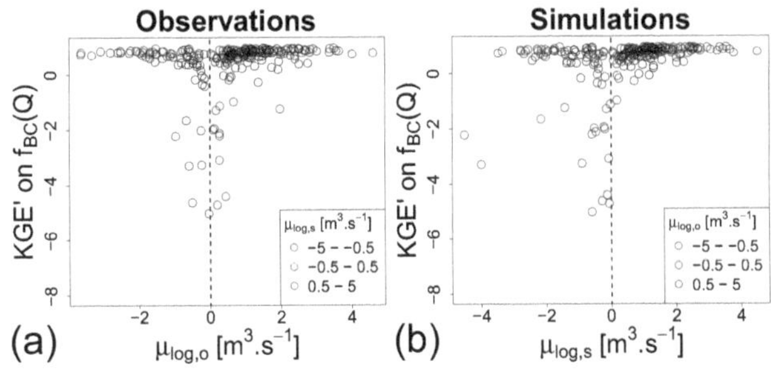

Figure 7. Values of KGE′ on Box–Cox transformed flows versus the mean of the log-transformed observed **(a)** and simulated **(b)** flows. Each dot represents the performance obtained in validation for one catchment after calibration with the KGE′ on untransformed flows as an objective function. In plot **(a)**, the axis values represent the observed log-transformed flow averages and the color represents the simulated averages, while in plot **(b)** it is the opposite.

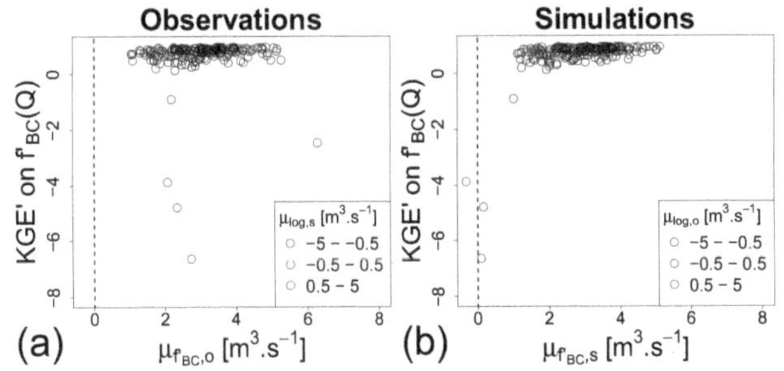

Figure 9. Values of KGE′ on modified Box–Cox transformed flows (Eq. 10) versus the mean of this transformed observed **(a)** and simulated **(b)** flows. Each dot represents the performance obtained in validation for one catchment after calibration with the KGE′ on untransformed flows as an objective function. In plot **(a)**, the axis values represent the observed transformed flow averages and the color represents the simulated averages, while in plot **(b)** it is the opposite.

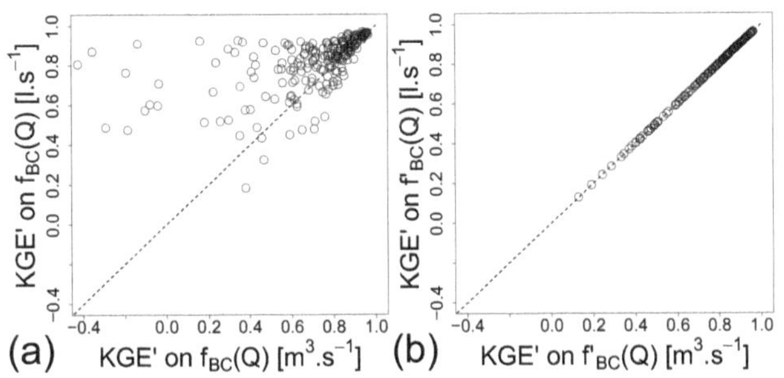

Figure 8. Dependence on flow units of the KGE′ using Box–Cox transformed flows without adaptation (**a**, Eq. 9) and with adaptation (**b**, Eq. 10) in the 240 catchments. The parameters used for simulation evaluation were obtained by calibrating GR4J using KGE′ on untransformed flows.

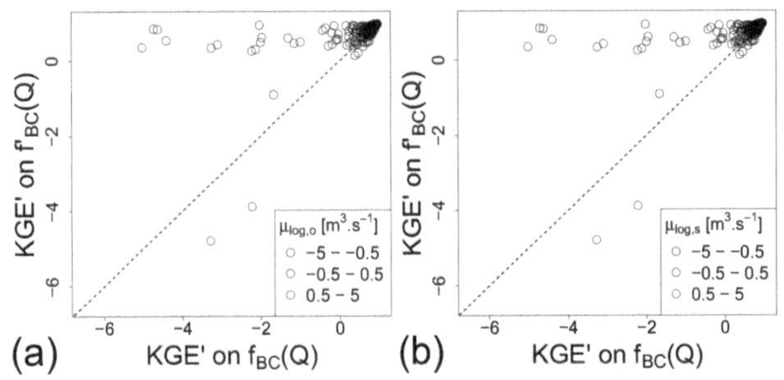

Figure 10. Comparison between KGE′ values on Box–Cox and modified Box–Cox transformed flows on the validation period using a calibration with KGE′ on untransformed flows as an objective function. The red dots represent the catchments where the average of log-transformed observed **(a)** or simulated **(b)** flows is around 0.

Using this equation, the KGE′ on transformed flows will be unit-dependent because of the additive term 1 in the numerator. To avoid this, we can slightly modify the formula, by replacing the term 1 by a constant with a unit dependence (here we propose 1/100 of the mean flow) and by putting it to the power λ:

$$f'_{\mathrm{BC}}(Q) = \frac{Q^{\lambda} - (0.01\mu_{\mathrm{o}})^{\lambda}}{\lambda}. \tag{10}$$

Using Eq. (10), the KGE′ criterion remains dimensionless using the Box–Cox transformation (Fig. 8b).

Furthermore, because the zero of the modified Box–Cox function is not 1 any more, this transformation would reduce the issue of strongly negative values when $\mu_{\mathrm{log,o}}$ or $\mu_{\mathrm{log,s}}$ are around zero. However, there still is an issue if the average of simulated flows is around the zero of the modified Box–Cox function (i.e. if $\mu_{\mathrm{s}} = (0.01\mu_{\mathrm{o}})^{\lambda}$, Fig. 9). This instability occurs more rarely than for the logarithm transformation but can be more frequent if larger percentages of the average of observed flow or different λ value are used. Because this instability is due to μ_{s} (which is only in the denominator of the γ ratio in Eq. 6), it will only affect the KGE′. The KGE is not affected because an α ratio is used instead of the γ ratio (Eqs. 1 and 5).

The modified Box–Cox transformation (Eq. 10) allows unit dependence to be avoided and the instability issues due to the values of average flows to be reduced (especially when using the KGE). The behaviour of this modified transformation also remains similar to the one of the initial Box–Cox transformation except when $\mu_{\mathrm{log,o}}$ or $\mu_{\mathrm{log,s}}$ are around zero (Fig. 10).

6 Summary

6.1 Log transformation should not be used in the KGE or KGE′ criterion

Given the previous results, we can argue that using log-transformed flows to calculate the KGE or the KGE′ criterion

Table 1. Pros (+) and cons (−) of different flow transformations to improve consideration of low flows in KGE′. In the second column, the number of + symbols represents the intensity of the low-flow weight increase. There are parentheses around the last + for inverted root and Box–Cox transformations because the low-flow weight depends on parameters.

Flow transformation	Increased low flow weight	No issue with zero flows	Dimensionless	No issue when flows average around 1
Square root	+	+	+	+
Inverse	+++	−	+	+
Reciprocal of root	++(+)	−	+	+
Logarithm	++	−	−	−
Box–Cox	+(+)	+	+ (if using Eq. 10)	+ (if using Eq. 10)

can lead to difficulties in the interpretation of criterion values. The criterion does not remain dimensionless like NSE with a prior logarithmic transformation. It also becomes overly sensitive when the log-transformed flows' average becomes close to zero, yielding potentially very negative values, or when a small constant is added to flows prior to logarithmic transformation to cope with zero flows. Because of all these issues, logarithmic transformation should be avoided when using KGE′.

6.2 Alternatives

Instead of KGE′ on log-transformed flows, several transformations can be used to calculate KGE′. The pros and cons for several transformations are summarised in Table 1. The reciprocal of root (RoR) is an example of a transformation used in the literature that is not tested in the article but leads to an increase in the weight of low flows (Chapman, 1964; Ishihara and Takagi, 1965; Ding, 1966). As stated in Ding (2018b), it can be parametrized with the value of the power in the root ($Q^{-\frac{1}{N}}$). Depending on the value of N, there will be more or less weight on low flows (Ding, 2018a). The higher N is, the less the weight on low flows is. This N value can also be determined with the recession curves of observed flows. Regarding this table, the modified Box–Cox transformation (Eq. 10) seems to be the best solution but it still faces instabilities for some flow average values (for the KGE′). Thus, there is no ideal solution to avoid all problems. Modellers have to make a choice depending on their specific applications. According to the intensity of low-flow weight increase that is needed, the choice of transformation has to be adapted. Garcia et al. (2016), for example, recommend averaging two KGE′ criteria, computed on untransformed and inverted flows, into a composite criterion.

Note that many studies use NSE on log-transformed flows (see for example Lyon et al., 2017; Nguyen and Dietrich, 2018). Fortunately, the mathematical formulation of NSE avoids all the problematic aspects identified for KGE with the logarithmic transformation. However, this may not be a sufficient argument to continue to use NSE given the issues presented by Gupta et al. (2009) and Schaefli and Gupta (2007):

- the underestimation of variability,
- the low weight of water balance errors for catchments with highly variable flows,
- the poor benchmark represented by the mean flows for catchments with highly variable flows.

6.3 Final remarks

Two additional remarks should be taken into account on this topic. First, as noted by Harald Kling in a personal communication, 2018, prior transformations on flows in KGE (or in NSE) lead to a misinterpretation in the estimation of the water balance. The other components of the KGE also lose their initial physical meaning. KGE on transformed flows can give more information on low flows, but the physical interpretation of the criterion is not as simple as in the case of untransformed flows.

Secondly, even if it did not occur in our experiment, the issue described in this technical note may lead to problems during the calibration process. Indeed, it can create a strongly negative zone in the objective function hyperspace, which may negatively impact the performance of local calibration algorithms.

Data availability. The daily flow data can be downloaded from the Banque HYDRO website (http://www.hydro.eaufrance.fr/, last access 29 August 2018). The climatic data from the SAFRAN reanalysis used in this paper (daily precipitation and temperature) are not freely available. The data was provided to Irstea following a convention between the two institutes. However, the analyses can be reproduced using open data and would lead to similar conclusions.

Author contributions. LS made the technical development and the analysis. The paper was written by him, GT and CP.

Competing interests. The authors declare that they have no conflict of interest.

Acknowledgements. The authors thank Météo France for providing the data used in this work. We also wish to thank Alban De Lavenne, Laure Lebecherel, Maria-Helena Ramos and Cedric Rebolho for the discussions on the different aspects of the issues using the logarithmic transformation with KGE. We thank Andrea Ficchí for his work on the database and Linda Northrup for her correction of the English language of an earlier version of the paper. Finally, we extend our thanks to Harald Kling for discussions on this issue.

We thank the topical editor, Bettina Schaefli, for her careful reading of the paper, her suggestion on the modified Box–Cox transformation and the following discussions. We also thank the two reviewers, Lieke Melsen and Björn Guse, for taking the time to read our paper and for their remarks that helped us to make the paper and the figures more understandable. We thank Sivarajah Mylevaganam for the discussions that helped us to be more precise in the KGE and KGE′ description. Finally, we particularly want to thank John Ding for his suggestion to add the RoR transformation (that we did not know about before) to the article and for the fruitful discussions that followed.

Edited by: Bettina Schaefli

References

Beck, H. E., van Dijk, A. I. J. M., de Roo, A., Miralles, D. G., McVicar, T. R., Schellekens, J., and Bruijnzeel, L. A.: Global-scale regionalization of hydrologic model parameters, Water Resour. Res., 52, 3599–3622, https://doi.org/10.1002/2015WR018247, 2016.

Box, G. E. P. and Cox, D. R.: An Analysis of Transformations, J. Roy. Stat. Soc. B, 26, 211–252, 1964.

Chapman, T. G.: Effects of groud-water storage and flow on the water balance, in: Proceedings of "Water resources, use and management", 291–301, Australian Academy of Science, Melbourne Univ. Press, 1964.

Coron, L., Thirel, G., Delaigue, O., Perrin, C., and Andréassian, V.: The suite of lumped GR hydrological models in an R package, Environ. Model. Softw., 94, 166–177, https://doi.org/10.1016/j.envsoft.2017.05.002, 2017.

De Vos, N. J. and Rientjes, T. H. M.: Multi-objective performance comparison of an artificial neural network and a conceptual rainfall-runoff model, Hydrol. Sci. J., 52, 397–413, https://doi.org/10.1623/hysj.52.3.397, 2010.

Ding, J.: Interactive comment on "Technical note: Pitfalls in using log-transformed flows within the KGE criterion" by Léonard Santos et al., Hydrol. Earth Syst. Sci. Discuss., https://doi.org/10.5194/hess-2018-298-SC2, 2018a.

Ding, J.: Interactive comment on "Technical note: Pitfalls in using log-transformed flows within the KGE criterion" by Léonard Santos et al., Hydrol. Earth Syst. Sci. Discuss., https://doi.org/10.5194/hess-2018-298-SC5, 2018b.

Ding, J. Y.: Discussion of "Inflow hydrograph from large unconfined aquifers" by Ibrahim, H. A. and Brutsaert, W. J., J. Irrig. Drain. Am. Soc. Civ. Eng., 92, 104–107, 1966.

Ficchí, A., Perrin, C., and Andréassian, V.: Impact of temporal resolution of inputs on hydrological model performance: An analysis based on 2400 flood events, J. Hydrol., 538, 454–470, https://doi.org/10.1016/j.jhydrol.2016.04.016, 2016.

Garcia, F., Folton, N., and Oudin, L.: Which objective function to calibrate rainfall–runoff models for low-flow index simulations?, Hydrol. Sci. J., 62, 1149–1166, https://doi.org/10.1080/02626667.2017.1308511, 2016.

Gupta, H. V., Kling, H., Yilmaz, K. K., and Martinez, G. F.: Decomposition of the mean squared error and NSE performance criteria: Implications for improving hydrological modelling, J. Hydrol., 377, 80–91, https://doi.org/10.1016/j.jhydrol.2009.08.003, 2009.

Hogue, T. S., Sorooshian, S., Gupta, H., Holz, A., and Braatz, D.: A Multistep Automatic Calibration Scheme for River Forecasting Models, J. Hydrometeorol., 1, 524–542, https://doi.org/10.1175/1525-7541(2000)001<0524:AMACSF>2.0.CO;2, 2000.

Ishihara, T. and Takagi, F.: A study on the variation of low flow, Bulletin of the Disaster Prevention Research Institute, 15, 75–98, http://hdl.handle.net/2433/124698, 1965.

Klemeš, V.: Operational testing of hydrological simulation models, Hydrol. Sci. J., 31, 13–24, https://doi.org/10.1080/02626668609491024, 1986.

Kling, H., Fuchs, M., and Paulin, M.: Runoff conditions in the upper Danube basin under ensemble of climate change scenarios, J. Hydrol., 424–425, 264–277, https://doi.org/10.1016/j.jhydrol.2012.01.011, 2012.

Krause, P., Boyle, D. P., and Bäse, F.: Comparison of different efficiency criteria for hydrological model assessment, Adv. Geosci., 5, 89–97, https://doi.org/10.5194/adgeo-5-89-2005, 2005.

Leleu, I., Tonnelier, I., Puechberty, R., Gouin, P., Viquendi, I., Cobos, L., Foray, A., Baillon, M., and Ndima, P.-O.: Re-founding the national information system designed to manage and give access to hydrometric data, La Houille Blanche, 1, 25–32, https://doi.org/10.1051/lhb/2014004, 2014 (in French).

Lyon, S. W., King, K., Polpanich, O., and Lacombe, G.: Assessing hydrologic changes across the Lower Mekong Basin, J. Hydrol.: Reg. Stud., 12, 303–314, https://doi.org/10.1016/j.ejrh.2017.06.007, 2017.

Nash, J. E. and Sutcliffe, J. V.: River flow forecasting through conceptual models. Part I – A discussion of principles, J. Hydrol., 10, 282–290, https://doi.org/10.1016/0022-1694(70)90255-6, 1970.

Nguyen, V. T. and Dietrich, J.: Modification of the SWAT model to simulate regional groundwater flow using a multicell aquifer, Hydrol. Process., 32, 939–953, https://doi.org/10.1002/hyp.11466, 2018.

Oudin, L., Hervieu, F., Michel, C., Perrin, C., Andréassian, V., Anctil, F., and Loumagne, C.: Which potential evapotranspiration input for a lumped rainfall–runoff model?, J. Hydrol., 303, 290–306, https://doi.org/10.1016/j.jhydrol.2004.08.026, 2005.

Oudin, L., Andréassian, V., Mathevet, T., Perrin, C., and Michel, C.: Dynamic averaging of rainfall-runoff model simulations from complementary model parameterizations, Water Resour. Res., 42, W07410, https://doi.org/10.1029/2005wr004636, 2006.

Pechlivanidis, I. G., Jackson, B., McMillan, H., and Gupta, H.: Use of an entropy-based metric in multiobjective calibration to improve model performance, Water Resour. Res., 50, 8066–8083, https://doi.org/10.1002/2013WR014537, 2014.

Perrin, C., Michel, C., and Andréassian, V.: Improvement of a parsimonious model for streamflow simulation, J. Hydrol., 279, 275–289, https://doi.org/10.1016/s0022-1694(03)00225-7, 2003.

Pushpalatha, R., Perrin, C., Moine, N. L., and Andréassian, V.: A review of efficiency criteria suitable for evaluating low-flow simulations, J. Hydrol., 420–421, 171–182, https://doi.org/10.1016/j.jhydrol.2011.11.055, 2012.

Quesada-Montano, B., Westerberg, I. K., Fuentes-Andino, D., Hidalgo, H. G., and Halldin, S.: Can climate variability information constrain a hydrological model for an ungauged Costa Rican catchment?, Hydrol. Process., 32, 830–846, https://doi.org/10.1002/hyp.11460, 2018.

Schaefli, B. and Gupta, H. V.: Do Nash values have value?, Hydrol. Process., 21, 2075–2080, https://doi.org/10.1002/hyp.6825, 2007.

Seeger, S. and Weiler, M.: Reevaluation of transit time distributions, mean transit times and their relation to catchment topography, Hydrol. Earth Syst. Sci., 18, 4751–4771, https://doi.org/10.5194/hess-18-4751-2014, 2014.

Vázquez, R. F., Willems, P., and Feyen, J.: Improving the predictions of a MIKE SHE catchment-scale application by using a multi-criteria approach, Hydrol. Process., 22, 2159–2179, https://doi.org/10.1002/hyp.6815, 2008.

Vidal, J.-P., Martin, E., Franchisteguy, L., Baillon, M., and Soubeyroux, J.-M.: A 50-year high-resolution atmospheric reanalysis over France with the Safran system, Int. J. Climatol., 30, 1627–1644, https://doi.org/10.1002/joc.2003, 2010.

3

Accelerated gravity testing of aquitard core permeability and implications at formation and regional scale

W. A. Timms[1,2], R. Crane[2,3], D. J. Anderson[3], S. Bouzalakos[1,2], M. Whelan[1,2], D. McGeeney[2,3], P. F. Rahman[3], and R. I. Acworth[2,3]

[1]School of Mining Engineering, University of New South Wales, Sydney, Australia
[2]UNSW Connected Waters Initiative affiliated with the National Centre for Groundwater Research and Training, Sydney, Australia
[3]Water Research Laboratory, School of Civil and Environmental Engineering, University of New South Wales, Sydney, Australia

Correspondence to: W. A. Timms (w.timms@unsw.edu.au)

Abstract. Evaluating the possibility of leakage through low-permeability geological strata is critically important for sustainable water supplies, the extraction of fuels from coal and other strata, and the confinement of waste within the earth. The current work demonstrates that relatively rapid and realistic vertical hydraulic conductivity (K_v) measurements of aquitard cores using accelerated gravity can constrain and compliment larger-scale assessments of hydraulic connectivity. Steady-state fluid velocity through a low-K porous sample is linearly related to accelerated gravity (g level) in a centrifuge permeameter (CP) unless consolidation or geochemical reactions occur. A CP module was custom designed to fit a standard 2 m diameter geotechnical centrifuge (550 g maximum) with a capacity for sample dimensions up to 100 mm diameter and 200 mm length, and a total stress of ~ 2 MPa at the base of the core. Formation fluids were used as influent to limit any shrink–swell phenomena, which may alter the permeability. K_v results from CP testing of minimally disturbed cores from three sites within a clayey-silt formation varied from 10^{-10} to 10^{-7} m s^{-1} (number of samples, $n = 18$). Additional tests were focussed on the Cattle Lane (CL) site, where K_v within the 99 % confidence interval ($n = 9$) was 1.1×10^{-9} to 2.0×10^{-9} m s^{-1}. These K_v results were very similar to an independent in situ K_v method based on pore pressure propagation though the sequence. However, there was less certainty at two other core sites due to limited and variable K_v data. Blind standard 1 g column tests underestimated K_v compared to CP and in situ K_v data, possibly due to deionised water interactions with clay, and were more time-consuming than CP tests. Our K_v results were compared with the set-up of a flow model for the region, and considered in the context of heterogeneity and preferential flow paths at site and formation scale. Reasonable assessments of leakage and solute transport through aquitards over multi-decadal timescales can be achieved by accelerated core testing together with complimentary hydrogeological monitoring, analysis, and modelling.

1 Introduction

Clay or other low-permeability sediment and rock often dominate sedimentary sequences and can form important aquitards (Potter et al., 1980). These hydraulic barriers often overlie aquifers that yield strategically important fresh water resources and form important cap rocks or seals between shallow aquifers and deeper strata targeted for depressurisation during gas or mineral extraction (Timms et al., 2012). The current work compares the results of steady-state centrifuge permeability testing of semi-consolidated drill core samples with column tests at standard gravity (1 g at earth's surface, 9.8065 m s^{-2}). Results of laboratory tests were also compared with in situ permeability, based on analysis of pore pressure propagation at formation scale.

Thick, low hydraulic conductivity (K), un-oxidised, clay-rich aquitards represent important sites for waste confinement and disposal (including high-level radioactive waste and the sequestration of carbon dioxide and saline effluents) and act as protective covers for regional aquifers (Cherry et al., 2004). Effective shale and claystone flow barriers are required to disconnect shallow aquifer systems from underlying coal seams that are depressurised to produce gas (Timms et al., 2012; APLNG, 2013). Fine-grained geologic media are also commonly used as engineered barriers to limit horizontal seepage of mine water (Bouzalakos et al., 2014), for containment of tailings (Znidarčić et al., 2011), and disposal of municipal refuse and nuclear waste (Rowe et al., 1995). Low-permeability material is defined by K of $< 10^{-8}\,\mathrm{m\,s^{-1}}$ (Neuzil, 1986). The US EPA requires low-permeability waste barriers for hazardous waste landfills with K of $< 10^{-9}\,\mathrm{m\,s^{-1}}$ (US EPA, 1989).

Aquitards volumetrically constitute the bulk of sedimentary geologic deposits (Potter et al., 1980), and are typically assumed saturated if located below a water table (Cherry et al., 2004). Water-saturated K and diffusion coefficients for aquitards are therefore not applicable to variably saturated or non-water-saturated low-permeability strata. Research is lacking for semi-consolidated clayey aquitards (e.g. alluvial, colluvial, and aeolian deposits), compared with aquitard research on glacial tills (Grisak and Cherry, 1975), claystones (Smith et al., 2013; Jougnot et al., 2010), and shale (Neuzil, 1994; Josh et al., 2012). Clay-bearing sediments formed via alluvial, colluvial, and aeolian processes frequently occur in the geosphere. For example clayey-silt aquitards account for 60 % of the ~ 100 m thick alluvial sediment sequences in the Mooki catchment of Australia's Murray–Darling Basin (Farley, 2011). The relative lack of information on the dominant type of sedimentary deposit represents a key gap in the current theoretical understanding of clay mineralogy and geochemistry.

Aquitard research on alluvial sediments is important because recharge by slow seepage provides essential groundwater supplies for municipal water supply and crop irrigation in relatively dry inland settings (Acworth and Timms, 2009). Increased effective stress associated with aquifer drawdown for irrigation, may release saline water stored within shallow aquitards with implications for the continuation of high yields of fresh water. Characterising the effects of variable chemical composition of formation water on the hydraulic conductivity of such sediments is therefore essential to determine the long-term changes to fresh water.

As an example, revised calculation of hydraulic parameters based on water level recovery from a bore pump test in glacial till ($K = 10^{-11}\,\mathrm{m\,s^{-1}}$) has been required to improve the fit with the data emerging over ~ 30 years (van der Kamp, 2011). Various field and laboratory methods are available to directly measure or indirectly calculate hydraulic conductivity along the horizontal (K_h) or vertical (K_v), and saturated and unsaturated or multi-phase flow (e.g. liquid and gas). Obtaining realistic measurements of groundwater flow and solute transport within aquitards is by definition a slow process, requiring relatively time-consuming and expensive field and/or laboratory studies.

Methods for measuring the in situ permeability of clay formations include: slug tests (piezometer tests, falling-head tests), aquifer pumping tests with piezometers in the aquitard, aquifer pumping tests with observation wells in the aquifer only, measurement of seasonal fluctuations of pore pressure, measurement of pore pressure changes and settlement due to surface loading, and numerical analysis of local and regional groundwater flow (van der Kamp, 2001). Neuman and Witherspoon (1968) developed generic analytical solutions for drawdown within an aquiclude, in which vertical flow occurs, but is sufficiently small to have no effect on water levels within an overlying or underlying aquifer. Type curves were presented for analytical solutions applying to an infinitely thick and a finite thickness aquiclude. In contrast, analysis of a leaky aquitard–aquifer system was presented by Neuman and Witherspoon (1972). The ratio method compares drawdown within an aquitard with drawdown in an underlying aquifer from which extraction was occurring. Drawdown data are then used to calculate hydraulic diffusion of pressure transients, and K_v, assuming a uniform, homogeneous aquitard.

Deconvolution of the pressure response to depth through an aquitard can be analysed with a Fourier transform or harmonic analysis (Boldt-Leppin and Hendry, 2003). The hydraulic diffusivity (hydraulic conductivity divided by specific storage) is expressed analytically, either based on the amplitude or phase shift of harmonic signals, assuming that the thickness of the aquitard is semi-infinite. Jiang et al. (2013) further developed the harmonic analysis method for finite aquitards in a multi-layer system in the instance of water level monitoring within aquifers above and below an aquitard, but not monitoring within the aquitard. Coherence analysis of water level fluctuations in bounding aquifers from indeterminate stresses (e.g. pumping, recharge, rainfall or earthquake) was used to derive K_v for deep rock aquitards on the basis of interpolated groundwater level data measured at irregular intervals of at least 10 days over a duration of several decades.

A more direct method of determining in situ hydraulic parameters is possible using fully grouted vibrating wire transducers and high-frequency data recording within deep formations, as recently demonstrated by Smith et al. (2013) for a bedrock claystone at up to 325 m below ground (BG). Pore pressure and barometric pressure were recorded at 30 min intervals and analysed, assuming no leakage in the grouted system, for barometric response, earth tides, and rainfall events. Core samples from the same drill holes were vacuum sealed on site for consolidation testing and triaxial permeameter testing. The in situ compressibility and specific storage calculated from barometric pressure responses were as much as an order of magnitude smaller than laboratory results.

A variety of laboratory testing techniques for low-K samples are also available; however, the reliability of results may

Table 1. Specifications and performance details of the Broadbent GT-18 centrifuge permeameter (CP) system as constructed by Broadbent (2011).

Dimensions/mass	
Diameter (lower rotary stack)	200.0 cm
Radius to top sample chamber	45.0 cm[a]
Radius to base sample chamber	65.0 cm[b]
Total mass	4800 kg
Performance	
Rotational speed	10–875 rpm
Maximum sample length	20.0 cm
Maximum sample diameter	10.0 cm
Maximum sample mass	4.7 kg
Maximum sample density	SG 3.0
Maximum effluent reservoir capacity	1000 mL
Maximum payload	18.11 kg

[a] 385 G at 875 rpm; [b] 556 G at 875 rpm.

depend on factors such as the preparation and size of core samples, configuration of equipment and uncertainties of measurement, the influent water that is used and the stresses that are applied relative to in situ values, and whether permeability is directly measured from steady-state flow, or subject to additional parameters and assumptions with alternative flow regimes. Laboratory testing of clayey-silt cores by standard rigid and flexible wall column techniques requires 1–2 weeks, compared with < 1 week for centrifuge permeameter (CP) methods in unsaturated samples (ASTM, 2010). Constant or falling-head tests in rigid-walled column permeameters at natural gravity require a large water pressure gradient and/or long testing times for low-permeability samples. They are subject to potential leakage, and may not replicate in situ confining stresses. Column testing of core samples is possible for some test conditions in triaxial cells on both K_h and K_v, for example those used in geotechnical and petroleum studies (Wright et al., 2002). However, standard practice for testing ultra-low-permeability cores (e.g. $K_v < 1 \times 10^{-10}\,\mathrm{m\,s^{-1}}$) typically consists of applying a confining pressure to a watertight system and measuring small transient pore pressures with high-resolution pressure transducers (API, 1998).

Geotechnical centrifuges are used to subject porous samples to high artificial gravities in order to characterise their hydraulic and/or consolidation properties (Conca and Wright, 1998; Nakajima and Stadler, 2006; Znidarčić et al., 2011), and for physical modelling as part of geotechnical design (Garnier et al., 2007; Parks et al., 2012). Accelerated gravity acts on both the solid particles and fluids within the porous sample without use of a large fluid pressure gradient to drive flow. The technique can be applied to investigate slow hydrogeological processes over shorter timescales; i.e. flow through low-permeability layers that would take several years under in situ conditions can be reproduced in a geotechnical centrifuge within hours or days, depending on test conditions.

A CP, or a column mounted on a centrifuge strong box, is commonly used for hydraulic characterisation of porous media. Accelerated gravity achieves a steady-state equilibrium for fluid flow through the CP within hours or days of instrument operation (for an unsaturated sample), while simultaneously applying stresses to the solid matrix. A permeameter column, mounted on a geotechnical centrifuge is rotated sufficiently fast to accelerate flow and approximate in situ total stresses, while the target g level is designed to ensure that the matrix is not consolidated and chemical equilibrium is maintained. Steady-state flow can provide more reasonable K results than transient flow techniques. Although transient tests are even more rapid than steady-state tests in the centrifuge, more complex instrumentation is required to ensure reliable results (Zornberg and McCartney, 2010).

The geotechnical centrifuge system described in this paper is moderately sized and relatively economical to operate, whilst able to perform both unsaturated and saturated testing of porous media with real-time measurement of various parameters during flight (Table 1). These attributes mean that CP testing of relatively large diameter cores (up to 100 mm diameter) in this facility is comparable in cost to testing of small cores (38 mm diameter) using alternative methods such as He-gas permeation. The system has been successfully used for testing low-permeability rock cores (Bouzalakos et al., 2013; Jiang et al., 2015). To date, there were no other direct K_v measurements on these deep shales available (APLNG, 2013) and alternative laboratory methods were not successful in obtaining a K_v value from these very low-K rocks (Bouzalakos et al., 2013).

This paper demonstrates novel CP techniques and equipment that have been specifically developed for characterising semi-consolidated clayey-silt cores. K_v results from CP methods are compared with standard 1 g column methods and in situ measurements of permeability, based on harmonic analysis of the high-frequency pore pressure propagation through a thick clayey sequence. The variability, confidence limits and overall reliability of the K_v results to constrain assessments of regional-scale vertical connectivity are considered in the context of sampling and flow and stress conditions within the CP. This paper provides reasonable K_v for at least one local clayey-silt sequence and strategies for future testing that are important contributions towards evaluating flow connectivity at a range of scales. These K_v results can be complimented with hydrogeological data such as pore pressure and tracer data to better constrain numerical flow models.

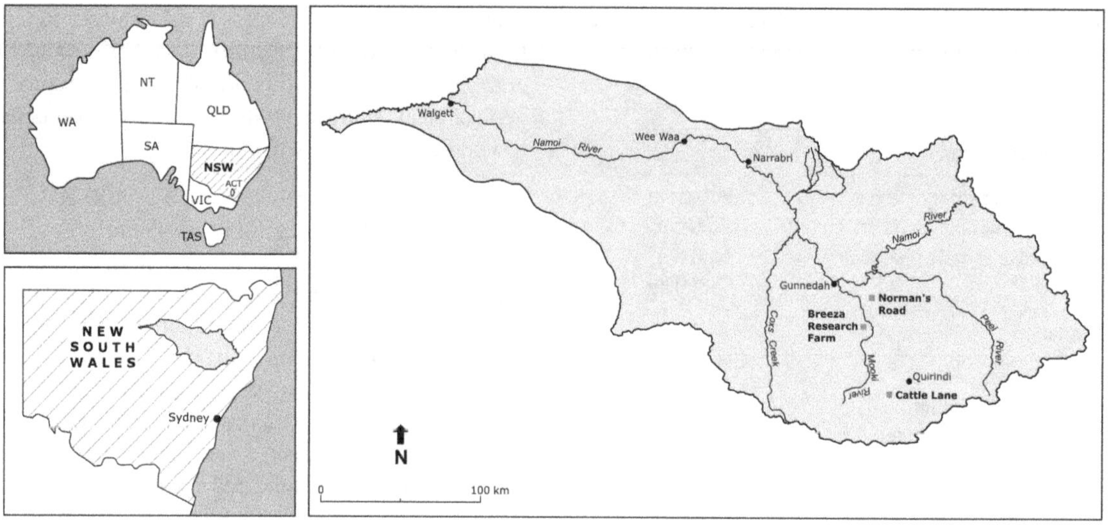

Figure 1. Location of study sites in eastern Australia, state of New South Wales (NSW). The Norman's Road (NR), Breeza Farm (BF) and Cattle Lane (CL) sites are shown within the Namoi catchment.

2 Geology of study sites

Semi-consolidated sediment cores were obtained from three sites in the Australia Murray–Darling Basin, in the Upper Mooki subcatchment of the Namoi catchment (Fig. 1). Groundwater is extracted in this area for irrigation and town water supplies, with drawdowns of more than 10 m over 30 years. It can take years or decades for changing pore pressures to be transmitted through these mixed sediments that are heterogeneous, even though the effects of groundwater extraction were assumed to occur rapidly within homogeneous, high-permeability sediments (Kelly et al., 2013). The alluvial sedimentary geology of the valley features significant heterogeneity but a general fining upwards that reflects climatic drivers of sedimentation (Kelly et al., 2014). This study found that the architectural features and the net (sand and gravel) to gross (total volume) line plot that identifies low-permeability clays and silts of the valley-filling sequence are best represented by a distributive fluvial system. In this type of fluvial system, the avulsion frequency increases at a slower rate than the aggradation rate.

Core drilling was completed at three research sites (Fig. 1) including Cattle Lane (CL), located south of the town of Caroona (31°31′9″ S, 150°28′7″ E), the Breeza farm (BF) operated by the New South Wales (NSW) Department of Primary Industries, south-east of Gunnedah (31°10′32″ S, 150°25′15″ E), and Norman's Road (NR), east-southeast of Gunnedah (31°2′48″ S, 150°26′7″ E).

Clayey-silt sediments at the Cattle Lane site are approximately 30 m thick (Timms and Acworth, 2005) and extend throughout the valley (Wiesner and Acworth, 1999), as shown by numerous CCPT (conductivity cone penetrometer) profiles. The porewater salinity profile at the site, increasing from 10 to 30 m depth through the clay, is consistent with a diffusion dominated transport over thousands of years (Timms and Acworth, 2006). The saturated zone fluctuates in response to rainfall events between the ground surface and approximately 2 m depth, while water levels in the confined gravel aquifer at > 50 m depth display a delayed and dampened response to the same rainfall events. There is no groundwater extraction for irrigation from this aquifer in the vicinity of the site, and the valley has been artificially drained to prevent ponding of surface water and soil salinisation. Detailed geological studies and particle dating have identified that the clayey silt in the top ~ 30 m at this site accumulated gradually at 0.2–0.3 mm year^{-1} by weathering of alkali basalts (Acworth et al., 2015). Flow testing of 100 mm diameter cores from the CL site, reported by Crane et al. (2015) has revealed evidence for dual porosity flow when a hydraulic gradient is imposed on the low-permeability sediments, with further work in progress to identify the nature and significance of these potential flow paths.

Sediments at the Breeza farm and Norman's Road site are relatively heterogeneous, with mixed sandy, clayey-sand, and clayey-silt alluvium overlying a semi-confined aquifer. The saturated zone is approximately 18–20 m below surface and extraction for flood irrigation of crops causes large fluctuations in groundwater levels in the confined aquifers at > 50 m depth. Hydrogeological and hydro-geochemical evidence indicate a leaky aquifer–aquitard system, with the variability in groundwater level responses controlled by a fining upward alluvial sequence (Acworth and Timms, 2009). At the Norman's Road site, highly saline porewater (15 mS cm^{-1}) in the clayey silt in proximity to the surface (< 20 m) appears to have leached into the underlying aquifer, causing a significant increase in salinity of the aquifer (Badenhop and Timms, 2012).

3 Study site characterisation and sampling

3.1 Drilling and core sampling

Equipment and procedures for coring were compliant with ASTM D1587-08 (2008a) to obtain samples that were as undisturbed as possible. A rotary drilling rig equipped with Triefus triple core barrels, lined with seamless clear PET, was used in push coring mode. Local creek water was used as a drilling fluid and casing was used to stabilise the hole behind the push core barrel such that drilling fluid additives were not required. The holes were therefore fully cased to the maximum depth of push core drilling at up to 40 m BG.

The non-rotating core barrel was forced into the formation, whilst a rotating device on the outside of the tube removes the cuttings as the barrel was advanced. The cutting edge of the non-rotating sample tube projects several millimetres beyond the rotary cutters. The thin-walled core barrel complied with the standard for undisturbed sampling, with an area ratio of less than 25 % for an open-drive sampler. The area ratio of 16 % was based on a core barrel design with an external diameter of 110 mm and internal diameter of 101 mm (C size). The 1.5 m length core barrel was a composite open sampling system with a core nose screwed onto the base with a bevelled end to cut the core as the barrel pushed into the formation. After the core was extracted from the ground, an air supply was connected to the top of the core barrel to slide the core out of the barrel whilst it remained in the clear PET liner without rotation, distortion, or compression.

The cores contained within PET liners were transferred directly from the core barrels to a cool room on site, and thence to a laboratory cool room, reducing the potential for moisture loss. Semi-consolidated clay cores were selected from below the saturated zone for CP tests, at depths up to 40 m BG. Sediment core samples of lengths between 50 and 100 mm were prepared for CP testing. The moisture content and bulk density of cores was measured using methods adapted from ASTM D7263-09 (2009). These measurements were completed immediately on the drill site.

The preferred method for preservation of drill core was double plastic bagging of sections of core within their PET liners using a food grade plastic sealing system (with brief application of a vacuum to extract air from the plastic bag). Alternatively, core within PET core barrel liners were trimmed of air or fluid filled excess liner immediately after drilling, and then sealed with plastic tape. All cores were stored at 4 °C in a portable cool room on the drill site and then at the laboratory. Sections of cores, particularly at the nose end, that appeared to be damaged or disturbed were excluded from permeability or bulk density testing. Additional steps that were taken in the laboratory to ensure core testing was representative of in situ conditions are described in Sect. 4.1.

After coring, the holes were completed as monitoring piezometers and the casing was jacked out. The piezometers were constructed of screwed sections of 50 mm PVC casing with O-ring seals, with a 1.5 m machine slotted screen packed with pea-sized washed gravel. The annulus was then filled with a bentonite seal, backfilled to the surface and completed with a steel casing monument and cement monument pad.

3.2 Groundwater sampling for influent

Fluid for K testing (influent) should be taken from the formation at the same depth as the core. Formation water can be synthesised if it is not possible to sample directly from aquitard strata, by estimating the ionic strength, Na / Ca ratio, and pH. In this study, groundwater from piezometers at a similar depth to the core was obtained using standard groundwater quality sampling techniques (Sundaram et al., 2009). A 240 V electric submersible pump (GRUNDFOS MP1) and a surface flow cell were used to obtain representative samples after purging stagnant water to achieve constant field measurements of electrical conductivity and other parameters (Acworth et al., 2015 and unpublished data).

4 Centrifuge permeameter methods and calculations

4.1 Preparation of cores

To ensure that cores were tested under saturated realistic conditions, drill cores were adequately preserved, stored, prepared, and set on a vacuum plate prior to centrifuge testing. Cores from PET drill core liners were trimmed and inserted into an acrylic liner for the CP using a core extruder. The custom made core extruder had five precision cutting blades driven by a motorised piston suitable for a 100 mm diameter core. Cores for CP testing in this study were 100 mm diameter C size core, with a length of 50–100 mm. A close fit between the clay core and the liner was achieved using this extruder.

A vacuum plate system for core samples was designed to ensure fully saturated cores, remove air at the base of the core, and ensure an effective seal between the CP liner prior to testing at accelerated gravity. The vacuum plate device was designed to fit the CP liners containing the cores, drawing ponded influent from the top to the base of the cores using a standard laboratory vacuum pump at 100 kPa of negative pressure. After 12–48 h, or upon effluent flow from the base, the acrylic liners containing the prepared cores were then transferred directly to the CP module without disturbing the sample.

Furthermore, the moisture content and degree of saturation was monitored by measuring weight change of the permeameters during testing, and direct moisture tests of samples before and after CP testing. There was negligible difference observed between the moisture content of the core tests and in situ conditions, and the results were not associated with the time between sampling and testing of the core. Moisture content was not affected by the use of a vacuum to expel air from sealing bags or from the top or base of the cores fitted into the CP liners.

A self-seal was observed forming from material swelling at the interface with the liner within minutes of introducing the influent solution. Prior to the self-seal development, leakage along the liner interface was identified by a flow rate of several orders of magnitude higher than the steady-state flow K_V value. The swelling that occurred to self-seal the core was estimated at less than 0.02 % of the cross sectional area of the core by comparing flow rates through the CP drainage hole (described in Sect. S3 in the Supplement). It was calculated that this area of swelling was sufficient to seal an annulus aperture of ~0.01 mm between the clay core and the acrylic liner.

Given the relatively shallow depth of these cores, and the semi-consolidated status, the maximum g level in the centrifuge was limited to prevent structural changes in the core matrix. To minimise changes in porosity of the core during testing, the g level and the weight of ponded fluid on the cores were therefore designed to ensure that total stress was less than estimated in situ stress at the depth from which the core was drilled.

Blind permeability tests were carried out by an independent laboratory, which adapted a constant/falling-head method (AS 1289 6.7.3/5.1.1, 1991) with methods from Head (1988). For these 1 g column tests, a sample diameter of 45.1 mm and length 61.83 mm was used, and a confining pressure of 150 kPa and back pressure of 50 kPa was applied, providing a vertical uniaxial stress of 100 kPa. The test time was up to 100 h. These standard 1 g column tests used deionised water as the influent.

4.2 Centrifuge permeameter testing

The Broadbent CP module and some unique systems developed as part of this study are described in this section, with further details in Sects. S1 and S3. A conceptual plan of a CP is shown in Fig. 2. The CP contains a cylindrical clay sample with length L and diameter D, and is spinning in a centrifuge around a central axis at an angular velocity ω. The permeameter has an inlet face at a radius r, and a drainage plate at a radius of r_0. The co-ordinate z is defined as positive from the base of the sample towards the central axis of rotation, consistent with definitions in 1 g column testing (McCartney and Zornberg, 2010). This frame of reference is in an opposite direction to that defined by Nimmo and Mello (1991), but is convenient for interpretation and comparison of column flow tests.

Influent was fed from burettes located next to the centrifuge via a pair of custom designed low voltage peristaltic pumps mounted either on the centrifuge beam, or outside the centrifuge and through the low flow rotary union. In this study, the outlet face was a free drainage boundary, and is discussed further in Sects. S2 and S3.

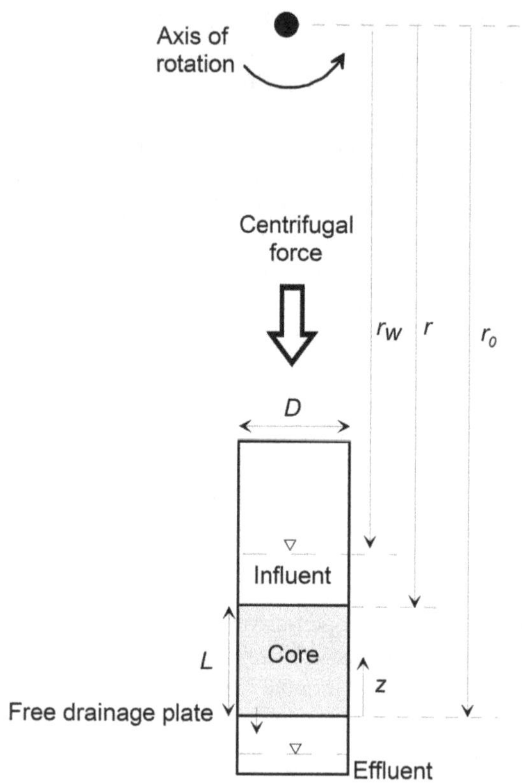

Figure 2. Cross sectional diagram of a core sample subjected to centrifugal force, with a free drainage boundary condition at the base of the core.

The K value is based on flow rate, flow area, radius, and revolutions per minute (RPM), although the method was adapted from a unsaturated/saturated flow apparatus (UFA) centrifuge to this CP system (Sect. 4.3). Importantly, both testing systems are for steady-state flow with free drainage due to zero pressure at the base of the core.

The mass of two core samples were balanced to the nearest 100 g and tested simultaneously at either end of the centrifuge beam. The CP was operated at 10 g for 30 min, and if no rapid flows due to leakage were detected, this was gradually increased to 20, 40 g, and so on, until the maximum total stress on the core approached the estimated in situ stresses of the material at the given depth in the formation. The upper permissible g level was designed to be less than the estimated in situ stress from the depth at which the core was obtained. It was also important to ensure that effective stress (Sect. 4.4) was acceptable, as variable pore fluid pressures during testing could cause consolidation of the core matrix. Influent volume was measured using both a calibrated continuous time record of pump rotations, and manual burette measurements, and effluent volumes were measured by weight. steady-state flow was defined as ±10 % change in discharge over subsequent measurements in time, provided that influent flow rate was within ±10 % of the effluent flow rate. Both of these conditions were required for the testing to be considered as a steady-state flow condition. This protocol provided additional quantitative measures to the ASTM D7664 (2010), which states that steady-state conditions have been attained "if the outflow is approximately equal to the inflow". Section S4 discusses the uncertainty of the measured data in more detail.

4.3 K_{v} calculations and statistical analysis

Hydraulic conductivity calculations for the CP in this study were based on ASTM D6527 (2008b) and ASTM D7664 (2010) with a form of Darcy's Law that incorporates the additional driving force within a centrifuge. The gradient in the centrifuge elevation potential (Nimmo and Mello, 1991), or the gradient in centrifuge "elevation head" (Zornberg and McCartney, 2010) due to the centrifuge inertial force driving was defined as flow away from the centre of rotation (or in the opposite direction to z in Fig. 2). The g level was defined at the mid-point of the core. A ponded influent above the top of the core prevented loss of saturation along the core (Nimmo and Mello, 1991). The centrifuge inertial (elevation) head gradient and hydraulic head gradient (stationary centrifuge at 1 g) were calculated at 0.005 m increments through the core.

Statistical analysis of the data followed a basic small-sampling theory using the Student's t distribution, following the approach of Gill et al. (2005) and extending the approach of Timms and Anderson (2015) for estimating sample numbers required for CP testing. Upper and lower confidence intervals (UCI, LCI) were calculated from the mean $\pm t_{(n-1)} \cdot s_n/n^{1/2}$, where s_n is the sample standard deviation and $t_{(n-1)}$ is the value of the Student's t distribution at the selected confidence limits (CL) of 90 and 99 %. The confidence intervals were calculated for the increasing number (n) of K_{v} data from each core.

4.4 Fluid pressure and total stress calculations

Fluid pressures and hydraulic gradient through the centrifuge core were determined following the approach of Nimmo and Mello (1991). The total fluid pressure P (kPa) was calculated in Eq. (1)

$$P = \rho_w \int_{r_0}^{r'} r\omega^2 \mathrm{d}r\,, \tag{1}$$

assuming a fluid density ρ_w of 1.0 g cm^{-3} and where r is the radius of rotation (cm), and ω is the angular velocity (s^{-1}). The total stress S (kPa) was determined through the centrifuge core, following Eq. (2)

$$S = \rho_s \int_{r_0}^{r'} r\omega^2 \mathrm{d}r\,, \tag{2}$$

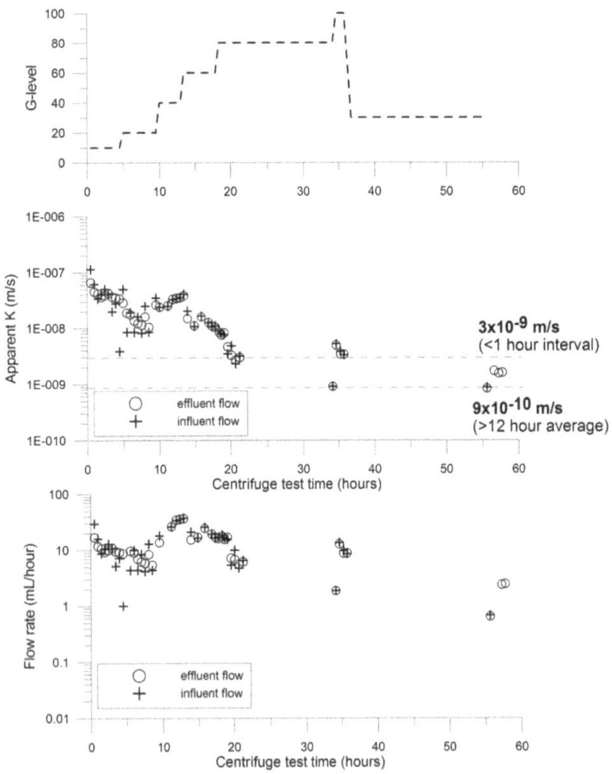

Figure 3. Centrifuge permeameter testing at low stresses of a semi-consolidated clayey-silt core sample (CL 26.1 m depth; Test 39-1) showing variation of g level, K_{V}, and influent and effluent flow rate during the test (after Timms et al., 2014).

assuming core bulk density ρ_s of 1.9 g cm^{-3}. The total stress and fluid pressure were calculated at 0.005 m increments through the core. The effective stress was then calculated as the difference between total stress and fluid pressure. An increase in effective stress associated with decreased fluid pressures near the base of the free draining core may cause consolidation of the core matrix near the boundary.

The total stress applied to the core, relative to stress, may affect the porosity of the core sample, depending on the stress history. In situ stress of the cores (S_i) at the sampling depth below ground (D) was calculated using Eq. (3)

$$S_i = \rho_s D g\,. \tag{3}$$

It was assumed that the overlaying formations were fully saturated and of a similar bulk density to the supplied core samples.

5 Results and discussion

5.1 Core properties and K_{v} results from CP testing

Index properties for five representative cores are provided in Table 2. The cores were typically silty clay (where the clay–

Table 2. Core descriptions and index properties.

Core ID	BF C2.8	BF C2.16 and C2.15	CL C4.8a	CL C4.20a	NR C3.23
Depth (m BG)	11.00–11.35/ 11.35–11.68	22.50–22.90/ 21.93–22.18	11.27–11.47	28.50–28.70	33.00–33.35/ 33.35–33.68
Description	Sandy clay; brown	Clayey silt; brown	Silty clay; brown	Silty clay; pale brown	Clayey silt; brown
Moisture (% wt.)	24.7	28.2	45.7	36.4	–
D_{50} (mm)	0.025	0.0068	–	–	<0.0013
Bulk wet density (g cm^{-3})	1.88	1.81	1.71	1.77	1.72
Particle density (g cm^{-3})	2.52	2.47	2.58	2.50	2.58
Initial void ratio	0.67	0.75	1.20	0.93	0.89
Initial degree of saturation (%)	93	95	96	99	74

Table 3. K_v results from CP tests indicating g level maximum and testing time. The influent source column identifies the site (NR, CL, BF) and depth (P_{20} is piezometer screen at 20 m depth) of groundwater sampling. Calculations were based on Eq. (3) for in situ stress.

Site	Depth (m BG)	K_v (m s^{-1})	g level maximum	Estimated in situ stress (kPa)	Testing time (h)	Influent source
NR	33.8	4×10^{-9}	10	615	~144	NR P_{30}
NR	33.90	2×10^{-9}	10	615	~144	NR P_{30}
NR	34.68	2.4×10^{-7}	10	646	2.6	NR P_{30}
CL	11.75	3.5×10^{-9}	80	219	24	CL P_{15}
CL	14.00	2.2×10^{-9}	80	261	1	CL P_{15}
CL	19.25	2.0×10^{-9}	80	359	24	CL P_{20}
CL	21.70	5.1×10^{-9}	80	404	1	CL P_{20}
CL	26.01	2.4×10^{-9}	80	485	21	CL P_{40}
CL	26.10	1.1×10^{-10}	80	486	21	CL P_{40}
CL	28.33	2.0×10^{-9}	10	526	24	CL P_{40}
CL	28.52	2.7×10^{-9}	80	532	1	CL P_{25}
CL	31.36	1.6×10^{-9}	40	585	24	CL P_{40}
BF	24.07	5.9×10^{-9}	40	449	3	BF CP_{25}
BF	24.14	3.4×10^{-8}	40	450	3	BF CP_{25}
BF	31.40	1.3×10^{-9}	30	585	11	BF CP_{40}
BF	36.46	3.5×10^{-7}	10	680	2.5	BF CP_{40}
BF	40.00	1.5×10^{-9}	30	746	23	BF CP_{40}
BF	40.10	4.3×10^{-8}	30	746	23	BF CP_{40}

silt size boundary is defined at 0.002 mm), except for one sandy clay core. The large proportion of silt relative to clay is an important characteristic of this formation, with clay mineralogy dominated by smectite (Timms and Acworth, 2005; Acworth and Timms, 2009).

Moisture content varied from 24.7 to 36.4 % by weight, and was consistent with site measured data on the core (Sect. S5), although not all the cores were fully saturated as received by the external laboratory. Bulk wet density varied from 1.71 to 1.88 g cm^{-3} and particle density from 2.47 to 2.58 g cm^{-3}. The K_v of cores tested in the CP module (Table 3) varied from 1.1×10^{-10} to 3.5×10^{-7} m s^{-1} ($n = 18$). Accelerations up to 100 g were applied during CP testing of semi-consolidated sediment cores and were more typically limited to 30–40 g. Figure 3 shows the measured influent and effluent rates and the calculated K_v values during a typical CP test as the g level was gradually increased. Steady-state flow (±10 % change over time with influent rate equal to effluent rate) was achieved at ~20 h (Fig. 3). However, a lower K_v value that was observed overnight (>12 h interval between samples) than was observed during the day (~1 h intervals between samples). The K_v values over shorter time periods, with minimal evaporative losses, were considered to be more reliable. Further experimentation and numerical modelling is required to adequately explain this anomalous data, which may be associated with evaporative losses over longer time periods of flow measurement or other transient processes within the system.

Table 4. Geometric mean, standard deviation (s_n) and confidence limits (CL %) analysis for K_v data using the CP method to test core from the clayey-silt formation at the CL, BF, and NR sites.

Site	n	K_v geometric mean ($m\,s^{-1}$)	$s_n \log K_v$	CL %	K_v confidence intervals ($m\,s^{-1}$) lower bound	upper bound
CL	5	1.3×10^{-9}	0.21	99	4.8×10^{-10}	2.4×10^{-9}
CL	9	1.6×10^{-9}	0.14	99	1.1×10^{-9}	2.0×10^{-9}
BF	6	1.3×10^{-8}	0.19	99	6.5×10^{-9}	2.1×10^{-8}
NR	3	1.2×10^{-8}	0.34	99	1.5×10^{-10}	8.5×10^{-8}
				90	3.4×10^{-9}	4.6×10^{-8}

Anomalous flow via preferential pathways could be readily identified by a flow rate of several orders of magnitude greater than otherwise observed. Anomalous flow was often observed along the interface of the cores and the liner during the early minutes of a test before sealing occurred and steady-state conditions were established. On one occasion a preferential flow path developed during the test, which caused very fast flow at accelerated gravity that was easily detected. A test failure like this could be readily identified and excluded from further evaluation.

A small uncertainty in K_v results for the CL site was calculated at a confidence limit of 99 % using the methods described in Sect. 4.3. By increasing the number of samples, the confidence bounds for K_v were narrowed from a range of 4.8×10^{-10} to $2.4 \times 10^{-9}\,m\,s^{-1}$ ($n = 5$) to a range of 1.1×10^{-9} to $2.1 \times 10^{-9}\,m\,s^{-1}$ ($n = 9$). Increasing the number of samples from five to nine decreased the standard deviation, although a similar geometric mean occurred with the increased sample number (Table 4). However, there was less certainty at two other core sites (BF and NR) due to limited and more variable K_v data. At the BF site the 99 % confidence interval had relatively wide K_v bounds for $n = 6$, while at NR site, a confidence interval of 90 % results in similarly wide K_v bounds for $n = 3$. However, such a small number of samples is not considered sufficient for statistical analysis. This evaluation of the results highlights the relative K_v variability and small sample set for the BF and NR sites, and the need for further testing, particularly at the NR site. This issue will be expanded in the following discussion.

5.2 Pore fluid pressure and stress conditions at accelerated gravity

How realistic the K_v measured by CP testing is of in situ conditions depends in part on the magnitude of stress and any structural changes that occur within the core matrix. Section S2 provides background on the definition and significance of hydrostatic pore pressure, centrifuge inertial (elevation) head, and gradients driving fluid flow. Section S2 also discusses the possibility that K values reported in this study could be biased on the high side, considering total stress at the base of the core under steady-state conditions.

During centrifuge testing effective stress is maximum at the base of the free draining core, where fluid pressure is zero, and thus effective stress is equal to total stress under hydrostatic conditions (no flow). In both testing methods in this study, the total stress was less than estimated in situ stress; however, the stress history of the core sample and effective stress dynamics were uncertain. It appears that the stresses during these tests were likely within an acceptable range to minimise structural changes including swelling and consolidation. There was no evidence of significant changes in core length due to consolidation of the samples during spot checks of core length with a digital calliper. However, further attention on these processes, including instrumentation to measure fluid pressures and core matrix changes during testing is required in future studies. A separate geotechnical study of these semi-consolidated sediments, including oedometer testing is in progress to better quantify the relationship between stress and permeability in these semi-consolidated materials. In future studies of semi-consolidated materials, measurement of consolidation state (over consolidation ratio) and pre-consolidation stress is recommended prior to centrifuge testing to ensure that an appropriate centrifuge stress is applied.

5.3 Comparison of in situ K_v and column testing methods at the CL site

A comparison of K_v from in situ and column testing methods are shown in Fig. 4 for the CL site. Results from the CP method (1.1×10^{-10} to $3.5 \times 10^{-9}\,m\,s^{-1}$, $n = 9$) were similar to K_v values from the independent and in situ method (1.6×10^{-9} to $4.0 \times 10^{-9}\,m\,s^{-1}$) confirming that the sequence is of low-permeability at the CL site with a reasonable level of confidence (Table 4). However, K_v from both in situ and CP methods were higher than the 1 g column tests of cores from 11.27 to 11.47 and 28.24 to 28.33 m BG from this site (1.4×10^{-9}, 1.1×10^{-10}, and $1.5 \times 10^{-10}\,m\,s^{-1}$, $n = 3$).

In situ K_v of the clayey-silt at the CL site was based on observed amplitude and phase changes of pore pressures (at hourly or 6-hourly intervals) due to five major rainfall events over 4 years (Timms and Acworth, 2005). The phase lag at the base of the clay varied between 49 and 72 days. The phase

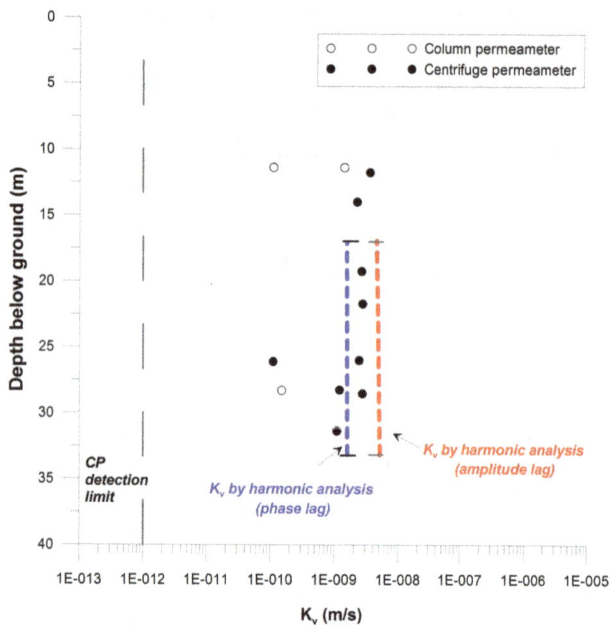

Figure 4. Vertical hydraulic conductivity (K_V) measurements by centrifuge permeameter and column permeameter compared with in situ K_V derived from pore pressure data at 6 h intervals over 4 years interpreted with harmonic analysis (after Timms and Acworth, 2005) for the Cattle Lane site with massive clayey-silt from the surface to 35 m depth.

lag pore pressure analysis resulted in a K_V value of $1.6 \times 10^{-9}\,\mathrm{m\,s^{-1}}$, while the change in amplitude over a vertical clay sequence of 18 m (from a 17 m depth piezometer to the inferred base of the aquitard at 35 m depth) resulted in a K_V value of $4.0 \times 10^{-9}\,\mathrm{m\,s^{-1}}$.

It is noted that the reliability of harmonic analysis related methods may be compromised by specific storage measurements. Jiang et al. (2013) relied on indirect specific storage values derived from downhole sonic and density log data from boreholes in the region, while Timms and Acworth (2005) calculated specific storage from barometric and loading responses that were recorded in the same groundwater level data set and boreholes used for harmonic analysis.

The reduced test times of CP testing may be attributed to the reduced time required to achieve steady-state flow with centrifugal forces driving flow. Alternatively, the relatively longer time required for $1g$ column testing may be attributed to deionised water interaction with clay that reduced infiltration rates into the cores (10–100 lower K_V result for 1 g column tests compared with CP tests). It is known that decreased ionic strength of influent (e.g. deionised water) causes a linear decrease in permeability, and that the relative concentrations of sodium and calcium can affect permeability due to swelling and inter-layer interactions (e.g. Shackelford et al., 2010; Ahn and Jo, 2009). Differences in K_V values from the two laboratory testing methods could be due to differences in test set-up (e.g. 45 vs. 100 mm diameter core) and stress changes that occur as discussed in Sects. 5.2 and S2.

CP testing was relatively rapid, typically with a few hours up to 24 h required for steady-state flow CP, compared with an average of 90 h (73, 96, and 100 h for the tests reported here) for 1 g column testing. In addition, an extended test of 830 h in the CP (unpublished data) verified that no significant changes occurred over extended testing periods. The CP technique can therefore reduce average testing time to $\sim 20\,\%$ of the time that would be required in 1 g laboratory testing systems, similar to the reduced time requirement of centrifuge methods for unsaturated hydraulic conductivity functions compared with 1 g column tests ASTM (2010). The relative time advantage of testing cores at accelerated gravity may be greater at lower K_V, due to the increased time required to establish steady-state flow conditions. The relative time advantage could be significant for contaminant transport testing, which requires several pore volumes of steady-state flow, compared to permeability testing where steady-state flow is established before one pore volume.

The similarity of K_V measurements with different scales at the CL site (Fig. 4) indicates that in this part of the alluvial deposit K is independent of vertical scale from centimetres to several metres. These K_V results from both in situ and laboratory methods provide an important constraint for evaluations of hydraulic connectivity, particularly as there is a general lack of K_V data for these sediments. Complimentary studies of hydraulic connectivity to quantify leakage rates include pore pressure monitoring and piezometer slug testing at various depth intervals along with hydrogeochemical and isotope tracer data. Recent geological studies of the alluvial sequence (Acworth et al., 2015) outlined in Sect. 2, and identification of dual porosity structures in the large diameter cores (Crane et al., 2015) indicate that it may be possible for vertical leakage to occur through clayey silts if a vertical hydraulic gradient were to be imposed. A diffusion dominated salt profile through the sequence suggest negligible vertical flow (Timms and Acworth, 2006); however, a proper assessment of flow connectivity requires vertical hydraulic gradients to also taken into account any salinity variations with depth and pore pressure variations that have occurred over at least the past decade.

5.4 Geological and regional context for permeability of a clayey-silt aquitard

The K_V measurements reported in this paper are important because there is a lack of aquitard data for alluvial groundwater systems globally. Even where many groundwater investigations have been completed (e.g. Murray–Darling Basin) there continues to be a lack of information on the thick clayey-silt sediments at various spatial scales.

The core samples for testing were randomly selected from the same lithostratigraphic formation, the upper 30 m of the alluvial sequence as described in Sect. 2. Although the alluvial sequence extends to over 100 m depth, we focussed this study on sediments defined by a low net-to-gross ratio (Larue and Hovadik, 2006) of < 0.4 that reflects that clay-rich part of the sequence (Timms et al., 2011). We assumed a log-normal distribution of K_v within this formation, which as noted by Fogg et al. (1998) might be justified within individual facies, but not over the full stratigraphic section. It was also assumed that the standard deviation of the samples tested is similar to the standard deviation of the total population of K_v results from the formation, which may only be known if a significantly large number of samples are tested.

K_v values for cores from the NR and BF sites were significantly larger than for the CL site, although additional data from the NR site are required to increased confidence intervals (Tables 3 and 4). Based on the data set currently available for each site there did not appear to be any significant K_v trend with depth, except at the CL site, with a possible decrease of K_v by a factor of 3 with depth increasing from 11 to 28 m BG. Further testing is in progress to better identify any spatially significant trends in K_v.

K_v results obtained from the CP for these clayey-silt aquitards were significantly larger than K_v for consolidated rock cores tested in this system (Bouzalakos et al., 2013). The relatively low g levels in this study (up to 80 g), compared to rock core testing (up to 520 g; Bouzalakos et al., 2013) were necessary for the shallow and semi-consolidated nature of the clayey-silt cores. In fact, steady-state flow was achieved at low g levels for K_v values that were at least 100 times higher than the current detection limit and uncertainty of the CP system (Sect. S4).

The vertical permeability of the clayey-silt aquitards in this region, and the relative importance of matrix flow and preferential flow through fractures and heterogeneities are critical to the sustainability of the groundwater resource. The K_v data reported in this study for these silty and semi-consolidated sediments are higher than reported for regional flow modelling in this area (McNeilage, 2006), indicating that the aquitards allow for significant recharge to underlying aquifers.

A regional groundwater flow model developed by McNeilage (2006), with a two-layer MODFLOW code, determined the dominant source of recharge to be diffuse leakage through the soil (and aquitards) in the Breeza groundwater management area. As in typical groundwater modelling practice (Barnett et al., 2012), the aquitard was not explicitly modelled, with water instead transferred from a shallow to a deeper aquifer using a vertical leakance value (units in day^{-1}).

The calibrated groundwater model indicated that approximately 70 % of the long-term average groundwater recharge (11 GL year^{-1}) was attributed to diffuse leakage in this area that included the CL and NR sites. This volume is equivalent to 20 mm year^{-1}, or a K_v of $\sim 6 \times 10^{-10}$ m s^{-1} assuming a unit vertical hydraulic gradient over an area of approximately 500 km^2. The actual K_v or leakance values were not reported. The calibrated leakance values were found to vary over 3 orders of magnitude across the Breeza area, with relatively high values in isolated areas in the south, centre, and north. In comparison, the K_v results on clayey-silt cores appear to be higher than the apparent K_v of the regional groundwater model, but with a similar degree of heterogeneity. The reasons for this discrepancy are not yet clear, but may be attributed to non-unique calibration of the regional flow model (e.g. underestimation of inter-aquifer leakance) or the lack of representative K_v values for this aquitard at a scale that accounts for heterogeneities and preferential flow paths.

The K_v results in this study are within the range of values reported elsewhere for semi-consolidated clay-silt sediments. For example, Neuzil (1994) reviewed aquitard K_v values for intact muds and lacustrine clays (10^{-8}–10^{-11} m s^{-1}) compared to consolidated materials such as shale with values as low as 10^{-16} m s^{-1} for argillite. A detailed study of a clayey-marl and limestone aquitard in France (Larroque et al., 2013) found a quasi-systematic bias of 1 order of magnitude between petrophysical K_v estimates (10^{-8}–10^{-10} m s^{-1}), compared with values (10^{-9}–10^{-11} m s^{-1}) from hydraulic diffusivity monitoring between 30 and 70 m BG. However, the empirical petrophysical relationships between porosity, pore size and intrinsic permeability do not adequately account for structural effects of clay materials. Field piezometer rising head tests ($n = 225$) indicated that K_v of a lacustrine clay aquitard around Mexico City was 10^{-8}–10^{-9} m s^{-1} in two areas, 100 times greater than matrix-scale permeability (Vargas and Ortega-Guerrero, 2004). However, in a third area of the Mexico City aquitard, the field tests were 10^{-10} m s^{-1} indicating the regional variability that can occur within clayey deposits.

Studies of glacial till aquitards in Canada, the US, and Denmark find that regional permeability is typically at least 2 orders magnitude greater than laboratory tests (van der Kamp, 2001; Fredericia 1990; Bradbury and Muldoon, 1990; Gerber and Howard, 2000), although one study (Husain et al., 1998) showed that for a thick glacial till aquitard in southern Ontario, Canada, the regional permeability is similar to the laboratory-obtained measurements, indicating the absence of significant permeable structures.

There is evidence of fracturing near the surface of the clayey aquitards that are the focus of this study. Fracture flow to a shallow pit and the freshening of porewater in the aquifers at 16 and 34 m depth during the irrigation season indicated rapid leakage had occurred at the BF site (Acworth and Timms, 2009). The dynamics of fracturing within ~ 2 m of the ground surface in these sediments was described by Greve et al. (2012). However, beyond the zone of fracturing near the ground surface, there appears to be insignificant groundwater flow. Solute profiles through the 30 m thick clayey deposit at the CL site indicate that downwards migration of saline water is limited to diffusion and that flow is insignificant (Timms and Acworth, 2006). On the basis of available evidence, the clayey sediments in this region may

Table 5. Linear flow velocity at natural gradient, unit gradient, and for various centrifuge permeameter set-ups (assuming a K_V of $10^{-8}\,m\,s^{-1}$).

	Natural gradient	Unit gradient	Centrifuge permeameter		
Core length × diameter (mm)	200 × 100	200 × 100	200 × 100	30 × 100	30 × 65
RPM	n/a	n/a	202	202	310
g level	1	1	30	30	70
Vertical fluid head gradient ($m\,m^{-1}$)	0.5	1	~0.2*	~0.5*	~0.5*
Flow ($mL\,h^{-1}$)	0.3	0.6	8.5	8.5	8.5
Time for 1 pore volume (h)	3333	1667	55.4	8.3	3.5
			Normalised		
Increased linear flow velocity			30	30	71
Reduced time for 1 PV			30	200	474

* Fluid head gradient depends on the depth of influent on the core and the length of the core.

lack preferential flow paths at some sites, and in other areas preferential flow may occur through features such as fractures and heterogeneity at a range of scales (Crane et al., 2015). Further work is required to determine permeability at a range of scales, and to better understand preferential flow paths. The current conceptual model on which the numerical models in this region are based (simple layered aquitard overlying an aquifer) do not allow for spatial variability in connectivity mechanisms that could be important across a large valley alluvial fill sequence. Multiple mechanisms for vertical connectivity, including matrix flow, fracture flow, and sedimentary heterogeneity, could be important in aquitards. The relative importance of each of these pathways for vertical flow would depend on the spatial scale and local setting in each aquitard.

5.5 Groundwater flow at natural gradient and accelerated conditions

To determine if accelerated flow conditions are realistic for hydrogeological environments, the linear flow velocity for various CP set-ups was compared with other flow scenarios. The rationale behind this comparison was that if the flow rate was consistent with scaling laws for physical modelling, with a unit gradient as a point of reference, then the results could be consider realistic for atypical in situ vertical hydraulic gradient. In Table 5, an in situ hydraulic gradient of 0.5 is compared with CP set-ups for 100 and 65 mm diameter cores of various lengths for an aquitard material with a K_V of $10^{-8}\,m\,s^{-1}$. The vertical flow rate varies from $0.3\,mL\,h^{-1}$ under in situ conditions to $8.5\,mL\,h^{-1}$ in the CP, such that linear flow velocities remain very low at 10^{-8}–$10^{-6}\,m\,s^{-1}$. The flow rate during centrifugation was N times greater than if a hydraulic gradient of 1 was applied to the core samples at $1\,g$. This increase in flow rate is consistent with scaling laws for physical modelling (Tan and Scott, 1987) providing further evidence that the K_V results are realistic.

The accelerated flow conditions, whilst realistic for hydrogeological environments, can also be an advantage for experimental studies of solute transport. K_V results of the order of $10^{-9}\,m\,s^{-1}$ were obtained in ~20 % of the time required for $1\,g$ column permeameter tests. Solute breakthrough experiments require longer testing periods of steady-state flow than for permeability testing. For example, Timms and Hendry (2008) and Timms et al. (2009) described continuous CP experiments over 90 days to quantify reactive solute transport during several pore volumes (PV) of flow. The comparisons of time required for 1 PV provided in Table 5 illustrate the possible advantages of CP for contaminant flow that may affect the structural integrity of the material.

6 Conclusions

An accurate and reasonable measurement of the vertical hydraulic conductivity (K_V) of aquitards is a critical concern for many applications. For example, following an empirical analysis of selected case studies, Bredehoeft (2005) reported that the collection of new field data may render the prevailing conceptual hydrogeological models invalid in 20–30 % of model analyses. Bredehoeft (2005) coined the term "conceptual model surprise" to explain this phenomenon. He then went on to explain that "often one does not have hydraulic conductivity values for confining layers because of the difficulties associated with acquiring such data".

The centrifuge technology described within this paper helps investigators to overcome some of the modelling limitations identified by Bredehoeft (2005). With centrifuge technology, realistic point-scale measures of hydraulic property data can be collected to develop more realistic numerical flow models to quantify the significance of transient drawdown, the associated release of water into adjacent aquifers over long time periods, and the possibility of preferential flow (Jiang et al., 2015). Realistic information on aquitard hydraulic properties could improve confidence in conceptual and numerical hydrogeological models of aquifer–aquitard systems.

In the absence of direct measurement of aquitard permeability there is a real risk that aquitard parameters may be ignored or misrepresented in analyses resulting in a corresponding underprediction of vertical connectivity via preferential flow paths and/or overprediction of aquifer storage and transmissivity. This is an especially important consideration in the analysis of aquifer tests that may not have been conducted for sufficient periods of time to identify distant boundary conditions or the characteristic effects of aquitard leakage and/or storage (Neuman and Witherspoon, 1968). In very low-permeability strata, however, there are practical limitations to pump tests and packer testing below about 10^{-8} m s^{-1}, depending on the equipment and the thickness of strata that is subject to testing. It is recognised that in many heterogeneous systems time lags for the propagation of drawdown responses through an aquitard can be significant (Kelly et al., 2013).

Core-scale measures of aquitard hydraulic conductivity are an integral component of hydrogeological studies concerning aquifer connectivity. The availability of core-scale facies measurements enables the up-scaling of bore log and geophysical data to determine upper and lower hydraulic conductivity bounds for regionally up-scaled aquitard units. Any differences between K values at various scales are important for indicating the possibility of preferential flow through heterogeneous strata or aquitard defects (e.g. faults and fractures). The availability of these bounded estimates helps to constrain the uncertainty analyses conducted on regional groundwater flow models to yield more confident predictions (Gerber and Howard, 2000).

Nevertheless, regional groundwater flow models generally use hydraulic resistance (leakance) values to transfer water vertically between aquifers (Barnett et al., 2012) rather than spatial discretisation of aquitards that control this transfer. While this simplification is justified in many models, such an approach is not capable of identifying rapid flow pathways through defects in the aquitards or the release of stored water from an aquitard to an aquifer and cannot resolve the vertical hydraulic head distribution across the aquitard to verify drawdown responses. An aquitard should be subdivided into at least three thinner layers to effectively model transient pressure responses (Barnett et al., 2012). Rather than assigning constant theoretical values for aquitard properties through these multiple layers a combination of realistic and rapid laboratory measurement and direct in situ measurements could improve confidence in conceptual understanding and model predictions.

Acknowledgements. Funding from the Australian Research Council and National Water Commission, through the National Centre for Groundwater Research and Training Program 1B is gratefully acknowledged. The contributions of N Baker and A Ainsworth of Broadbent and Sons, Huddersfield, UK, are acknowledged and J McCartney for a helpful discussion on the theory of fluid flow during centrifuge testing. We appreciated research support at the Breeza farm provided by M McLeod and S Goodworth of the NSW Department of Primary Industries. Clayey-silt cores were drilled by New South Wales Office of Water, with S McCulloch, H Studhome, and G Regmi. Experimental testing was assisted at UNSW by A Hartland, B Bambrook, M Aitkins, P King, S May, and T Meyers. Five reviewers and an editor are thanked for their constructive comments on earlier drafts of the paper.

Edited by: M. Riva

References

Acworth, R. I. and Timms, W.: Evidence for connected water processes through smectite dominated clays at Breeza, New South Wales, Aust. J. Earth Sci., 56, 81–96, 2009.

Acworth, R. I., Timms, W. A., Kelly, B. F. J., McGeeney, D., Ralph, T. J., Larkin, Z. T., and Rau, G. C.: Late Cenozoic Palaeovalley fill sequence from the Southern Liverpool Plains, New South Wales – implications for groundwater resource evaluation, Aust. J. Earth Sci., 62, 657–680, 2015.

Ahn, H. S. and Young Jo, H. Y.: Influence of exchangeable cations on hydraulic conductivity of compacted bentonite, Appl. Clay Sci., 44, 144–150, 2009.

API: Recommended Practices for Core Analysis. Recommended Practice 40, 2nd Edn., American Petroleum Institute Publishing Services, Washington DC, 1998.

APLNG: Groundwater Assessment, Australia Pacific LNG Upstream Project Phase 1, Q-LNG01-15-TR-1801, Australia Pacific LNG, Milton, Queensland, Australia, 266 pp., 2013.

AS: Methods of testing soil for engineering purposes, Standard methods 1289 2.1.1; 5.1.1; 5.3.2 and 6.7.3 Australian Standards, Sydney, 1991.

ASTM: Standard practice for thin-walled tube sampling of soils for geotechnical purposes, American Society for Testing and Materials International, ASTM D1587-08, West Conshohocken, PA, USA, 2008a.

ASTM: Standard test method for determining unsaturated and saturated hydraulic conductivity in porous media by steady state centrifugation, American Society for Testing and Materials International, ASTM D 6527-08, West Conshohocken, PA, USA, 2008b.

ASTM: Standard test methods for laboratory determination of density (unit weight) of soil specimens, American Society for Testing and Materials International, ASTM D7263-09, West Conshohocken, PA, USA, 2009.

ASTM: Standard test method for measurement of hydraulic conductivity of unsaturated soils, American Society for Testing and Materials International, ASTM D 7664-10, West Conshohocken, PA, USA, 2010.

Badenhop, A. M., and Timms, W. A.: Long-term salinity changes in an inland aquifer, NSW, Australia, in: Proceedings of the 34th Hydrology and Water Resources Symposium, Engineers Australia, Sydney, NSW, 43–51, 19–22 November 2012.

Barnett, B., Townley, L. R., Post, V., Evans, R. E., Hunt, R. J., Peeters, L., Richardson, S., Werner, A. D., Knapton, A., and Boronkay, A.: Australian Groundwater Modelling Guidelines, National Water Commission, Canberra, Australia, 2012.

Boldt-Leppin, B. E. J. and Hendry, J. M.: Application of harmonic analysis of water levels to determine vertical hydraulic conductivities in clay-rich aquitards, Ground Water, 41, 514–522, 2003.

Bouzalakos, S., Timms, W., Rahman, P., McGeeney, D., and Whelan, M.: Geotechnical centrifuge permeater for characterizing the hydraulic integrity of partially saturated confining strata for CSG operations, in: Reliable Mine Water Technology, Vol. I – 1–778, edited by: Brown, A., Figueroa, L., and Wolkersdorfer, C., Denver, Colorado, USA (Publication Printers), Proceedings of the International Mine Water Congress, Colorado School of Mines, 1193–1198, 5–9 August 2013.

Bouzalakos, S., Crane, R., Liu, H., and Timms, W. A.: Geotechnical and modelling studies of low permeability barriers to limit subsurface mine water seepage, in: 4th International Conference on Water Management in Mining, Vina del Mar, Chile, 161–171, 28–30 May 2014, 2014.

Bradbury, K. R. and Muldoon, M. A: Hydraulic conductivity determinations in unlithified glacial and fluvial materials, in: Ground Water and Vadose Zone Monitoring, edited by: Nielsen, D. M. and Johnson, A. I., ASTM STP 1053, American Society for Testing and Materials, Philadelphia, 138–151, 1990.

Bredehoeft, J.: The conceptualization model problem–surprise, Hydrogeol. J.,13, 37–46, 2005.

Broadbent: Operating Manual for Modular Geotechnical Centrifuge with GT2/0.65 Permeameter And GT6/0.75 Beam Environments, Broadbent and Sons. Ltd., Huddersfield, UK, 2011.

Cherry, J. A., Parker, B. L., Bradbury, K. R., Eaton, T. T., Gotkowitz, M. G., Hart, D. J., and Borchardt, M. A.: Role of Aquitards in the Protection of Aquifers from Contamination: A "State of the Science" Report, AWWA Research Foundation, Denver, CO, USA, 2004.

Conca, J. L. and Wright, J.: The UFA method for rapid, direct measurements of unsaturated transport properties in soil, sediment and rock, Aust. J. Soil. Res., 36, 1–25, 1998.

Crane, R. A., Cuthbert, M. O., and Timms, W.: Technical Note: The use of an interrupted-flow centrifugation method to characterise preferential flow in low permeability media, Hydrol. Earth Syst. Sci., 19, 3991–4000, doi:10.5194/hess-19-3991-2015, 2015.

Farley, C.: Aquitards and groundwater sustainability: three-dimensional mapping of aquitard architecture, unpublished honours thesis, School of Civil and Environmental Engineering, University of New South Wales, 2011.

Fogg, G. E., Noyes, C. D., and Carle, S. F.: Geologically based model of heterogeneous hydraulic conductivity in an alluvial setting, Hydrogeol. J., 6, 131–143, 1998.

Fredericia, J.: Saturated hydraulic conductivity of clayey tills and the role of fractures, Nordic Hydrol., 21, 119–132, 1990.

Garnier, J., Gaudin, C., Springman, S. M., Culligan, P. J., Goodings, D., Konig, D., Kutter, B., Phillips, R., Randolph, M. F., and Thorel, L.: Catalogue of scaling laws and similitude questions in geotechnical centrifuge modelling, Int. J. Phys. Model. Geotech., 3, 1–23, 2007.

Gerber, R. E. and Howard, K.: Recharge through a regional till aquitard: three-dimensional flow model water balance approach, Ground Water, 38, 410–422, 2000.

Gill, D. E., Corthesy, R., and Leite, M. H.: Determining the minimal number of specimens for laboratory testing of rock properties, Eng. Geol., 78, 29–51, 2005.

Greve, A. K., Andersen, M. S., and Acworth, R. I.: Monitoring the transition from preferential to matrix flow in cracking clay soil through changes in electrical anisotropy, Geoderma, 179–180, 46–52, 2012.

Grisak, G. E. and Cherry, J. A.: Hydrologic characteristics and response of fractured till and clay confining a shallow aquifer, Can. Geotech. J., 12, 23–43, 1975.

Head, K. H.: Manual of Soil Laboratory Testing, Pentech Press, London, UK, 1988.

Husain, M. M., Cherry, J. A., Fidler, S., and Frape, S. K.: On the long term hydraulic gradient in the thick clayey aquitard in the Sarnia region, Ontario, Can. Geotech. J., 35, 986–1003, 1998.

Jiang, Z., Mariethoz, G., Taulis, M., and Cox, M.: Determination of vertical hydraulic conductivity of aquitards in a multilayered leaky system using water-level signals in adjacent aquifers, J. Hydrol., 500, 170–182, 2013.

Jiang, Z., Mariethoz, G., Raiber, M., Timms, W. A., and Cox, M.: Application of 1D paleo-fluvial process modeling at a basin-scale to augment sparse borehole data: example of a Permian formation in the Galilee Basin, Australia, Hydrol. Proc., in press, 2015.

Josh, M., Esteban, L., Delle Piane, C., Sarout, J., Dewhurst, D. N., and Clenell, M. B.: Laboratory characterisation of shale properties, J. Petrol. Sci. Eng., 88, 107–124, doi:10.1016/j.petrol.2012.01.023, 2012.

Jougnot, D., Revil, A., Lu, N., and Wayllace, A.: Transport properties of the Callovo-Oxfodrian clay rock under partially saturated conditions, Water Resour. Res, 46, W08514, doi:10.1029/2009WR008552, 2010.

Kelly, B. F. J., Timms, W. A, Andersen, S. M., Ludowici, K., Blakers, R., Badenhop, A., McCallum, A. M., Rau, G. C., and Acworth, R. I.: Aquifer heterogeneity and response time: the challenge for groundwater management, Crop Pasture Sci., 64, 1141–1154, 2013.

Kelly, B. F. J., Timms, W., Ralph, T., Giambastini, B., Comunian, A., McCallum, A. M., Andersen, M. S., Acworth, R. I., and Baker, A.: A reassessment of the Lower Namoi Catchment aquifer architecture and hydraulic connectivity with reference to climate drivers, Aust. J. Earth Sci., 61, 501–511, 2014.

Larue, D. K. and Hovadik, J.: Connectivity of channelized reservoirs: A modelling approach, Petrol. Geosci., 12, 291–308, 2006.

Larroque, F., Cabaret, O., Atteia, O., Dupuy, A., and Franceschi, M.: Vertical heterogeneities of hydraulic aquitard parameters: preliminary results from laboratory and in situ monitoring, Hydrol. Sci. J., 58, 912–929, 2013.

McCartney, J. S. and Zornberg, J. G.: Centrifuge permeameter for unsaturated soils II: Measurement of the hydraulic characteristics of an unsaturated clay, J. Geotech. Geoenviron. Eng., 136, 1064–1076, 2010.

McNeilage, C.: Upper Namoi Groundwater Flow Model, NSW Department of Natural Resources, New South Wales, Parramatta, 2006.

Nakajima, H. and Stadler, A. T.: Centrifuge modeling of one-step outflow tests for unsaturated parameter estimations, Hydrol. Earth Syst. Sci., 10, 715–729, doi:10.5194/hess-10-715-2006, 2006.

Neuman, S. P. and Witherspoon, P. A.: Theory of flow in aquicludes adjacent to slightly leaky aquifers, Water Resour. Res., 4, 103–112, 1968.

Neuman, S. P. and Witherspoon, P. A.: Field determination of the hydraulic properties of leaky multiple aquifer systems, Water Resour. Res., 8, 1284–1298, 1972.

Neuzil, C. E.: Groundwater flow in low permeability environments, Water Resour. Res., 22, 8, 1163–1195, 1986.

Neuzil, C. E.: How permeable are clays and shales?, Water Resour. Res., 30, 145–150, 1994.

Nimmo, J. R. and Mello, K. A.: Centrifugal techniques for measuring saturated hydraulic conductivity, Water Resour. Res., 27, 1263–1269, 1991.

Parks, J., Stewart, M., and McCartney, J. S.: Validation of a centrifuge permeameter for investigation of transient infiltration and drainage flow processes in unsaturated soils, Geotech. Testing J., 35, GTJ103625, doi:10.1520/GTJ103625, 2012.

Potter, P. E., Maynard, J. B., and Pryor, W. A.: Sedimentology of Shale – Study Guide and Reference Source, Springer-Verlag, New York, 1980.

Rowe, R. K., Quigley, R. M., and Booker, J. R.: Clayey Barrier Systems for Waste Disposal Facilities, E&FN Spon, London, 1995.

Shackelford, C. D., Sevick, G. W., and Eykholt, G. R.: Hydraulic conductivity of geosynthetic clay liners to tailings impoundment solutions, Geotext. Geomembranes, 28, 149–162, 2010.

Smith, L. A., van der Kamp, G., and Hendry, M. J.: A new technique for obtaining high-resolution pore pressure records in thick claystone aquitards and its use to determine in situ compressibility, Water Resour. Res., 9, 732–743, 2013.

Sundaram, B., Feitz, A., de Caritat, P., Plazinska, A., Brodie, R., Coram, J., and Ransley, T.: Groundwater Sampling and Analysis – a Field Guide, Record 2009/27, Geoscience Australia, 95 pp., 2009.

Tan, T. S. and Scott, R. F.: Centrifuge scaling considerations for fluid-particle systems: discussion by R. N. Taylor and response, Geotechnique, 37, 131–133, 1987.

Timms, W. and Acworth, R. I.: Propagation of porewater pressure change through thick clay sequences: an example from the Yarramanbah site, Liverpool Plains, New South Wales, Hydrogeol. J., 13, 858–870, doi:10.1007/s10040-005-0436-7, 2005.

Timms, W. and Acworth, R. I.: Rethinking a conceptual model: advective vs. diffusive chloride flux in a low permeability clay sequence, International Association of Hydrogeologists Congress on Aquifer Systems Management, Dijon, France, 30 May–1 June 2006.

Timms, W. and Anderson, D.:Geotechnical centrifuge technology for characterising the interaction of tailings and pore water over decadal time periods, Australian Institute of Mining and Metallurgy, Conference Proceedings: Tailings and Mine Waste Management for the 21st Century, 247–262, 2015.

Timms, W. A. and Hendry, M. J.: Long term reactive solute transport in an aquitard using a centrifuge model, Ground Water, 46, 616–628, doi:10.1111/j.1745-6584.2008.00441.x, 2008.

Timms, W., Hendry, J., Muise, J., and Kerrich, R.: Coupling centrifuge modeling and laser ablation ICP-MS to determine contaminant retardation in clays, Environ. Sci. Technol., 43, 1153–1159, 2009.

Timms, W., Kelly, B. F. J., Blakers, R., Farley, C., Regmi, G., Larsen, J., and Bowling, A.: Implications of 3D geological architecture for surface-groundwater connectivity in the Mooki catchment, in: NSW International Association of Hydrogeologists Symposium 2011 – Uncertainty in Hydrogeology, edited by: McLean, W. and Milne-Holme, B., Sydney, Australia, 4–5 September 2011, available at: www.3pillarsnetwork.com.au/kb/iah-nsw-symposium-timms-.pdf (last access: 21 May 2015), 2011.

Timms, W., Acworth, I., Hartland, A., and Laurence, D.: Leading practices for assessing the integrity of confining strata: application to mining and coal seam gas extraction, in: International Water and Mining Association Symposium Proceedings, edited by: Clint, D., McCullough, C. D., Lund, M. A., and Wyse, L., Bunbury, Western Australia, 29 September to 4 October 2012, 139–148, 2012.

Timms, W., Whelan, M., Acworth, I., McGeeney, D., Bouzalakos, S., Crane, R., McCartney, J., and Hartland, A.: A novel centrifuge permeameter to characterize flow through low permeability strata, in: Proceedings of International Congress on Physical Modelling in Geotechnics (ICPMG), Perth, Balkema, 14–17 January 2014.

US EPA: Requirement for hazardous waste landfill design, construction and closure, EPA/625/4–89/022, Cincinnati, OH, USA, 1989.

van der Kamp, G.: Methods for determining the in situ hydraulic conductivity of shallow aquitards – an overview, J. Hydrol., 9, 5–6, 2001.

van der Kamp, G.: Determining the hydraulic properties of aquitards, 2nd Canadian Symposium on Aquitard Hydrogeology, University of Ottawa, Canada, 21–23 June 2011.

Vargas, C. and Ortega-Guerrero, A.: Fracture hydraulic conductivity in the Mexico City clayey aquitard: field piezometer rising-head tests, Hydrogeol. J., 12, 336–344, 2004.

Wiesner, T. and Acworth, R. I.: Groundwater contamination investigation using CCPTs, Water 99 Joint Congress, Brisbane, Australia, 6–8 July 1999.

Wright, M., Dillon, P., Pavelic, P., Peter, P., and Nefiodovas, A.: Measurement of 3-D hydraulic conductivity in aquifer cores at in situ effective stress, Ground Water, 40, 509–517, 2002.

Znidarčić, D., Miller, R., van Zyl, D., Fredlund, M., and Wells, S.: Consolidation Testing of Oil Sand Fine Tailings, Proceedings Tailings and Mine Waste, 2011, Vancouver, BC, 6 to 9 November 2011.

Zornberg, J. G. and McCartney, J. S.: Centrifuge permeameter for unsaturated soils I, theoretical basis and experimental developments, J. Geotech. Geoenv. Eng., 136, 1051–1063, 2010.

4

Streamflow forecasts from WRF precipitation for flood early warning in mountain tropical areas

María Carolina Rogelis[1] **and Micha Werner**[1,2]
[1]UNESCO-IHE, P.O. Box 3015, 2601 DA Delft, the Netherlands
[2]Deltares, P.O. Box 177, 2600 MH Delft, the Netherlands

Correspondence: Maria Carolina Rogelis (c.rogelisprada@unesco-ihe.org)

Abstract. Numerical weather prediction (NWP) models are fundamental to extend forecast lead times beyond the concentration time of a watershed. Particularly for flash flood forecasting in tropical mountainous watersheds, forecast precipitation is required to provide timely warnings. This paper aims to assess the potential of NWP for flood early warning purposes, and the possible improvement that bias correction can provide, in a tropical mountainous area. The paper focuses on the comparison of streamflows obtained from the post-processed precipitation forecasts, particularly the comparison of ensemble forecasts and their potential in providing skilful flood forecasts. The Weather Research and Forecasting (WRF) model is used to produce precipitation forecasts that are post-processed and used to drive a hydrologic model. Discharge forecasts obtained from the hydrological model are used to assess the skill of the WRF model. The results show that post-processed WRF precipitation adds value to the flood early warning system when compared to zero-precipitation forecasts, although the precipitation forecast used in this analysis showed little added value when compared to climatology. However, the reduction of biases obtained from the post-processed ensembles show the potential of this method and model to provide usable precipitation forecasts in tropical mountainous watersheds. The need for more detailed evaluation of the WRF model in the study area is highlighted, particularly the identification of the most suitable parameterisation, due to the inability of the model to adequately represent the convective precipitation found in the study area.

1 Introduction

Numerical weather prediction (NWP) models are fundamental to extend flood forecast lead times beyond the concentration time of a watershed. The significant advances in NWP and computer power during the last decades have led to the generation of high-resolution precipitation forecasts at the catchment scale, and quantitative precipitation forecasts (QPFs) from high-resolution NWPs are increasingly used in flood forecasting systems as a result (Xuan et al., 2009). Particularly for flashy watersheds with short times of concentration, such as those typically found in tropical mountainous watersheds, forecast precipitation is required to provide sufficient forecast lead times in support of timely warnings.

Despite the significant advances, NWP results contain noise, are contaminated by model biases, are too coarse to adequately resolve all features such as convection, and are influenced by uncertainty inherent in the initial conditions (Colman et al., 2013). Furthermore, weather forecasting in tropical mountains is highly challenging. In the tropics, local and mesoscale effects are more dominant than synoptic influences (except for tropical cyclones). In addition, there are limitations in the availability of surface and upper air monitoring networks (Laing and Evans, 2010), which influences modelling initialisation (Cuo et al., 2011). Additionally, the formation and movement of deep convection and mesoscale convective systems is affected by orography (Colman et al., 2013).

According to Habets et al. (2004) the potential of NWP precipitation forecast to be used by hydrological models for flood forecasting is mainly affected by (i) localisation of the events, since an error of a few kilometres can lead to the precipitation being forecast in the wrong watershed, (ii) timing of the events, since the response of the basin depends on previous events and on the timing of the present event, and (iii) precipitation intensity. This is especially true in flash-flood-prone watersheds, typical of tropical mountainous areas where location errors that are considered small at meteorological level can lead to a flood event in a watershed being completely missed (Vincendon et al., 2011).

In order to address the uncertain nature of NWP, ensemble prediction systems (EPSs) have been developed (Demeritt et al., 2007). In contrast to deterministic systems that produce one prediction, EPSs produce a suite or ensemble of predictions to reflect uncertainty, providing the capability to transform predictions into a probability distribution function (Leutbecher and Palmer, 2008; Demeritt et al., 2007). As EPS rainfall predictions often exhibit greater skill than deterministic predictions, the hope is then that EPS products will increase the skill and time horizon of flood forecasts (Demeritt et al., 2010).

The use of NWP ensembles to drive flood forecasting systems has increased and is a relevant research topic (Cloke and Pappenberger, 2009). Currently many flood forecasting centres use EPSs for representing uncertainty, but most of these are in Europe, Canada, the US and Australia (Demeritt et al., 2007). Experience of this kind of system in developing countries is very limited (Fan et al., 2014).

Rainfall forecasts provided by NWP require post-processing to correct bias and to reliably quantify uncertainty (Robertson et al., 2013). Several approaches have been used to produce ensemble rainfall forecasts by post-processing raw NWP. Robertson et al. (2013) present a method that uses a simplified version of the Bayesian joint probability modelling approach to produce forecast probability distributions for individual locations and forecast lead times; Theis et al. (2005) propose a methodology based on the hypothesis that some probabilistic information about a precipitation forecast at a certain time and location can be derived from its spatio-temporal neighbourhood in the model precipitation field. A set of forecasts is extracted from the spatio-temporal neighbourhood of a point and used to derive a probabilistic forecast at the central point of the neighbourhood; Bremnes (2004) proposes a method to produce probabilistic forecasts in terms of quantiles from NWP output using probit regression and quantile regression; Clark et al. (2004) use a two-stage approach that includes logistic regression and ordinary least squares regression to generate precipitation and temperature ensembles. Probabilistic forecasts of precipitation are especially challenging since precipitation has a mixed discrete–continuous probability distribution (Cortinas et al., 2002). In this study, a simple two-stage approach is used to produce precipitation ensembles from WRF forecasts, based on probit regression and quantile regression.

The assessment of the value of forecasts is a crucial issue. Verification is essential for the understanding of the abilities, weakness and value of forecasts, which leads to improving the forecast system. Several scales are proposed in the literature to assess the quality of forecasts as well as skill scores aimed at quantifying the relative accuracy of a forecast with respect to a reference forecast (Wilks, 2011), providing an estimation of its added value. Added value is often more important than a measure of skill (Jolliffe and Stephenson, 2012).

The objective of this paper is to assess the potential of NWP for flood early warning purposes in tropical mountainous watersheds, and the possible improvement that bias correction can provide. The main contribution of the research is focused on the assessment of the potential of WRF post-processed precipitation forecasts in providing skilful flood forecasts in watersheds characterised by rough topography and a tropical environment, where experience with precipitation forecasts, particularly ensemble approaches, has been very limited.

The study area is the upper part of a high tropical mountainous watershed located in Bogotá (Colombia). This corresponds to a páramo, which is a tropical high-mountain ecosystem, characterised by soils with a high water storage capacity and high conductivity. The hydrologic behaviour of páramo is complex, and there are still major gaps in knowledge (Sevink, 2011; Reyes, 2014; Buytaert et al., 2005, 2006). Additionally, hydrometeorological data are scarce.

In Colombia, NWP models forecasts are provided by the Instituto de Hidrología, Meteorología y Estudios Ambientales (IDEAM), the national hydro-meteorological institute. The raw output of these models is deterministic. In this study, the operational WRF model that is used by IDEAM for daily weather forecasts is used to provide precipitation forecasts that are post-processed and used to drive a hydrologic model of the study area. The discharges obtained from the hydrological model are used to assess the skill of the WRF model. This research assesses the current operational setup, in the meteorological agency of Colombia, which is based on the consideration that the orography has a significant effect in the mesoscale processes and that convection-allowing models such as the WRF may have the potential to provide skilful forecasts.

The paper is structured as follows: Sect. 2 describes the methods and data used for the assessment of the potential of the WRF model. In Sect. 3 the results are presented, including the bias correction of precipitation and the verification of precipitation and discharge forecasts. Section 4 discusses the results. Finally, the conclusions are presented in Sect. 5.

2 Methods and data

2.1 Study area

The Tunjuelo River basin is located in the south of the city of Bogotá (see Fig. 1). Its area is approximately 380 km^2 and the upper basin corresponds to a páramo area. The upper basin is composed of three watersheds – Chisacá, Mugroso and Curubital (see Fig. 1c). These discharge into two

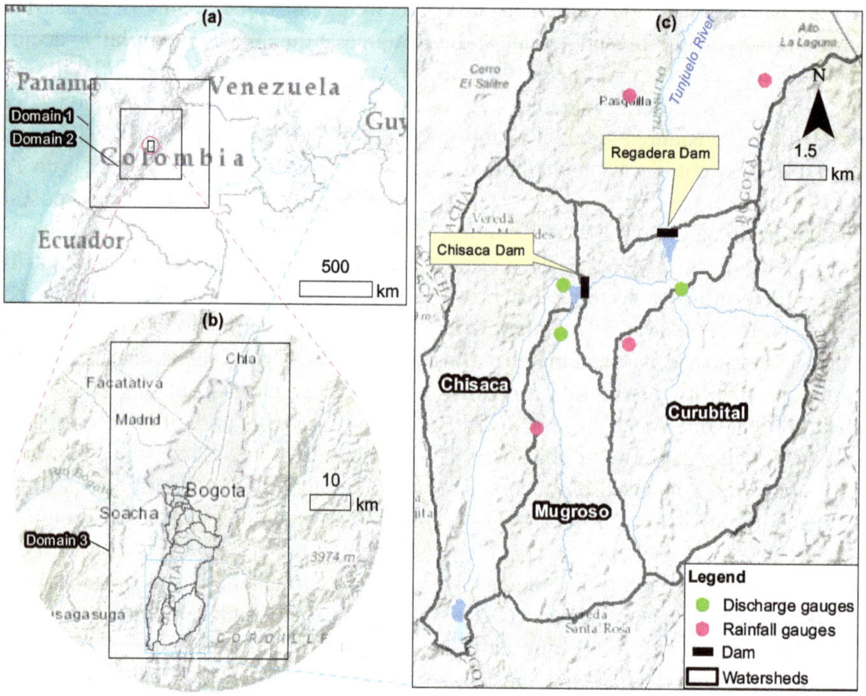

Figure 1. Study area. Service layer credits: Esri, DeLorme, NAVTEQ, TomTom, Intermap, increment P Corp., GEBCO, USGS, FAO, NPS, NRCAN, GeoBase, IGN, Kadaster NL, Ordnance Survey, Esri Japan, METI, Esri China (Hong Kong), swisstopo, and the GIS User Community.

reservoirs (Chisacá and Regadera) with volumes of 3.3 and 6.7 Mm^3 respectively. The reservoirs are operated to supply 1.2 $m^3 s^{-1}$ of water to the south of Bogotá. Flood waves in the urbanised lower basin are dominated by the discharge release of the two reservoirs.

The last major flood event in the basin occurred in 2002, causing the river to change its course to flow into two mining pits that since act as inline reservoirs. The peak release of the Regadera dam for that event reached 100 $m^3 s^{-1}$ (Rogelis, 2006), which caused flooding downstream. The current flood warning criteria include warning levels set on the water levels in the Regadera Reservoir. Forecasting of the input discharges to this reservoir, as well as to the Chisacá reservoir, which is located immediately upstream, is crucial.

The upper basin of the Tunjuelo river has a unimodal precipitation regime (rainy season April–November) (Bernal et al., 2007). The largest discharges in the upper Tunjuelo basin occur during the months May, June and July (Rogelis, 2006).

The monitoring network is shown in Fig. 1. Two tipping bucket telemetric rain gauges currently operate in the upper Tunjuelo River basin and three discharge gauges are available in the three watersheds that discharge into the two reservoirs of the upper basin with hourly records.

2.2 WRF model data and observed rainfall fields

The hydro-meteorological agency of Colombia, IDEAM, runs the Weather Research and Forecasting (WRF) model version WRFV3.1 for Bogotá, using the initial and boundary conditions provided by the GFS model (Ruiz, 2010). The WRF model has been used at IDEAM since 2007 to produce forecasts at national level, as well as in the Bogotá region at higher resolution (Arango and Ruiz, 2011).

The data set of WRF forecast used in this paper corresponds to 107 selected days when significant storms were recorded, during the period July 2009 to December 2002. For each of these days, deterministic forecasts with the WRF model were generated at 00:00, 06:00, 12:00 and 18:00 GMT. This choice of storms ignores days when the WRF model would have delivered false alarms, focusing exclusively on the days when high precipitation was effectively measured. This is due to operational reasons. The WRF forecasts have not been archived; therefore they had to be recreated, and to achieve a feasible setup of the experiment from a computational point of view, this choice had to be made.

The WRF model comprises three nested domains, centred in Bogotá. The coarsest domain covers most of the Colombian national territory with a spatial resolution of 15 km; the intermediate domain covers mainly the central and eastern Andean cordilleras with a spatial resolution of 5 km; and the finest domain covering Bogotá and the study area only has

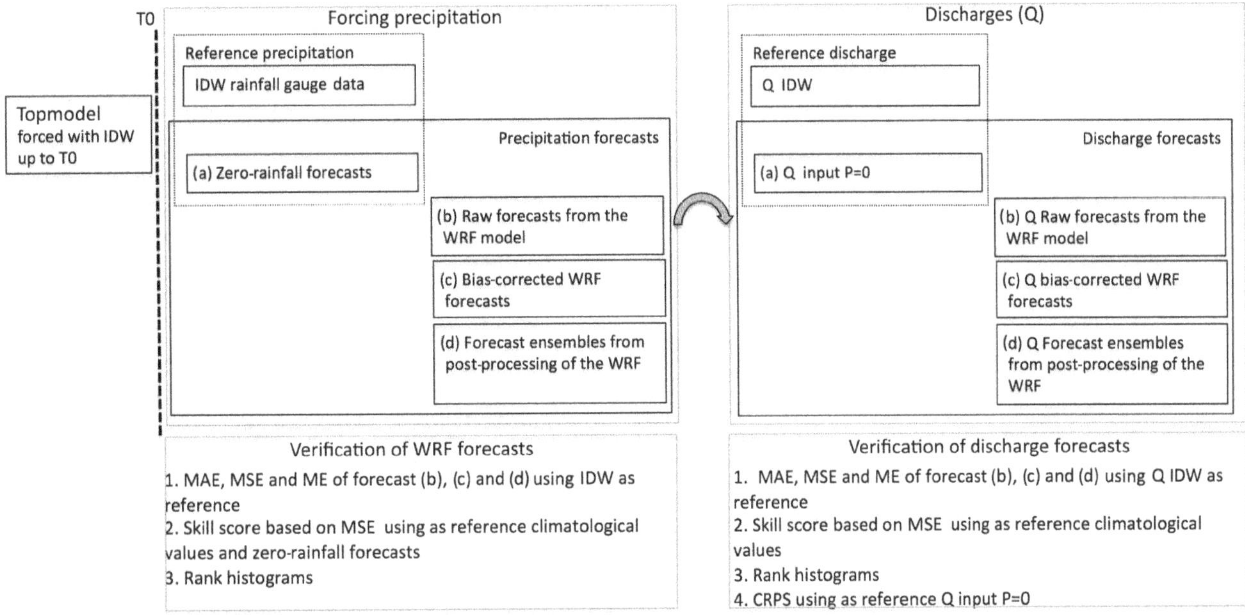

Figure 2. Overview of the methodology.

a spatial resolution of 1.67 km (Arango and Ruiz, 2011). The nested model domains have been set up using two-way communication (without smoothing). The parameterisation of the model corresponds to that used by IDEAM for its routine forecasts (Arango and Ruiz, 2011), using the Kain–Fritsch cumulus parameterisation scheme, except for the finest domain where convection is not parameterised but resolved. Microphysics is parameterised as the WRF Single-Moment 3-class scheme (WSM3). The lead time of the WRF model used operationally is 72 h, though in this paper only the first 48 h lead time was considered.

Observed hourly precipitation data were available from 18 tipping bucket rainfall gauges located in the city of Bogotá. These were used to produce rainfall fields through inverse distance weighing interpolation (IDW) for the entire area of the city that includes the study area.

Both WRF forecast rainfall fields and IDW rainfall fields were transformed to time series of mean average precipitation for the Chisacá, Mugroso and Curubital watersheds (see Fig. 1).

2.3 Methodology

The hydrology of the upper watersheds of the Tunjuelo river was modelled using the TOPMODEL (Beven and Kirkby, 1979). This was calibrated using as objective function the Kling and Gupta efficiency (Gupta et al., 2009) with the Shuffled Complex Evolution automatic search algorithm (Duan et al., 1992). The TOPMODEL was compared with two other hydrological models by Rogelis (2016), and found to provide a realistic representation of hydrological processes and good performance in the páramo watersheds for which residence times in the range from 1 to 2 h were found. Furthermore, the preference of TOPMODEL for páramo areas has been reported previously by other authors (Buytaert and Beven, 2011). The hydrological models were forced with precipitation input obtained from inverse distance interpolation (IDW) of hourly rainfall obtained from gauges up to the time of start of the forecast (T0). From T0 the model was forced using four different rainfall forecasts, (a) zero-rainfall forecasts, (b) raw forecasts from the WRF model, (c) bias-corrected WRF forecasts, and (d) precipitation forecast ensembles obtained from the post-processing of the WRF model, based on probit regression and quantile regression. The precipitation forecasts obtained from each of the three nested domains were used to force the hydrological model and assess their difference in forecast skill.

Both precipitation forecasts and the corresponding streamflow forecasts were verified through the use of standard skill scores and rank histograms (see Fig. 2). The precipitation forecasts were verified through the use of (see Fig. 2): (1) the mean absolute error (MAE), mean square error (MSE) and mean error (ME) using the IDW precipitation as reference, and (2) through the skill score based on the MSE using as reference climatological values and the zero rainfall forecasts. In the case of streamflow forecasts obtained from the WRF model, these were verified through (1) the MAE, MSE and ME using as reference the discharge obtained from forcing the model with IDW precipitation, (2) the skill score based on MSE using as reference climatological values, and (3) the continuous ranked probability skill score (CRPSS) using as reference a streamflow time series generated using the precipitation forecast equal to zero to assess the added value of the WRF forecasts. The procedures for the production of the different rainfall forecasts and the verification process are described in the following sections.

2.3.1 Generation of precipitation forecasts

The precipitation forecast to drive the models were generated under four strategies:

a. *Zero-rainfall forecasts.* After T0 values of zero precipitation are used to drive the TOPMODEL.

b. *Raw forecasts from the WRF.* The only post-processing of the WRF forecasts is the sampling of the grids to obtain the hourly mean areal precipitation for each watershed.

c. *Bias correction of WRF.* The time series of mean areal precipitation obtained from the WRF model are bias corrected through distribution-based scaling (DBS) (Yang et al., 2010). Mean areal precipitation was established using forecasts made for the 107 selected storm days as the biases during more extreme events is the prime interest. As reference values the observed time series of mean areal precipitation obtained from IDW are used, and the correction is carried out for each lead time of the WRF model. The DBS approach uses two steps: (1) correction of the percentage of wet time steps; a precipitation threshold of 0.1 mm (rainfall gauge accuracy) was used, below which a time step is considered to be dry, and (2) transformation of the remaining precipitation to match the observed frequency distribution. The gamma distribution is used to describe the probability distribution function of precipitation intensities given its ability to represent the asymmetrical and positively skewed distribution of precipitation. The distribution parameters are estimated using maximum likelihood estimation. To capture the main properties of normal precipitation, as well as of the extremes, the precipitation distribution is divided into two partitions separated by the 95th percentile. The resulting distribution corresponds to a double gamma distribution. The two sets of parameters are used to correct the WRF model outputs.

d. *Bias correction and generation of ensembles through post-processing of the rainfall forecast from the WRF model.* The reference and forecast time series were organised according to lead time and a two-stage post-processing model was applied to reflect the intermittent nature of rainfall (Rene et al., 2012; Clark et al., 2004). The first stage corresponds to the probit model (Scardovi, 2015; Kleiber et al., 2012; Bremnes, 2004) to simulate the probability of occurrence of precipitation. The second stage considers the probability distribution of the precipitation amount, given its occurrence, for which quantile regression was used. The mean precipitation time series (observed hourly mean areal precipitation for each watershed) is first disaggregated into a time series of occurrence (1 = wet time step and 0 = dry time step) and precipitation amounts, which is established for the wet time steps only. The time series of occurrence is used as the response variable for the probit regression model. The probit model allows the probability of occurrence of precipitation to be estimated using the WRF hourly mean areal precipitation over each watershed as a predictor or secondary variable. The time series of precipitation amounts is used as the response variable for the quantile regression model. This allows different quantiles of the response variable to be estimated using the WRF hourly mean precipitation for each watershed as predictor or secondary variable.

In the probit regression model, given binary observations of precipitation occurrence $y_1, \ldots, y_n$, and $x_{i1}, \ldots, x_{ik}$ covariates associated with the ith response, the probability that $y_i = 1$, p_i, is written as (Albert, 2009):

$$p_i = P(y_i = 1) = \Phi(x_{i1}\beta_1 + \cdots + x_{ik}\beta_k), \tag{1}$$

where $\beta = (\beta_1, \ldots, \beta_k)$ is a vector of unknown regression coefficients and $\Phi()$ is the cumulative density function of a standard normal distribution. If we assume a uniform prior for β, then the posterior density is given by

$$g(\beta \mid y) \propto \prod_{i=1}^{n} p_i^{y_i}(1 - p_i)^{1-y_i}. \tag{2}$$

The binary response corresponds to the occurrence $y_i = 1$ or non-occurrence of rain $y_i = 0$. A latent variable Z_i is introduced in such a way that if Z_i is positive for $y_i = 1$ and Z_i is negative for $y_i = 0$. This latent variable is related to the k covariates by the normal regression model:

$$Z_i = x_{i1}\beta_1 + \cdots + x_{ik}\beta_k + \epsilon_i, \tag{3}$$

where $\epsilon_1, \ldots, \epsilon_n$ are a random sample from a standard normal distribution. Then,

$$P(y_i = 1) = P(Z_i > 0) = \Phi(x_{i1}\beta_1 + \cdots + x_{ik}\beta_k). \tag{4}$$

This can be considered as a missing data problem where there is a normal regression model of latent data $Z_1, \ldots, Z_n$ and the observed responses are missing or incomplete and can only be observed if $Z_i > 0 (y_i = 1)$ or if $Z_i \leq 0 (y_i = 0)$.

In order to avoid distributional assumptions, quantile regression (QR) will be used to describe the probability of the amounts of precipitation given its occurrence (Bremnes, 2004). Let $r_1, \ldots, r_{n*}$ denote observed precipitation amounts of cases with observed precipitation above a given lower threshold and $z_1, \ldots, z_{n*}$ corresponding predictor values where $z = (z, \ldots, z)^T$. For linear quantile functions (Bremnes, 2004),

$$q_\theta(z_i; \beta) = \beta_0 + \sum_{k=1}^{K} \beta_k z_{ik}, \tag{5}$$

an estimate of the $q_\theta(z_i;\beta), 0<\beta<1$ quantile, is obtained by solving the following minimisation problem with respect to β:

$$\arg\min_{\beta} \sum_{i=1}^{n*} p_\theta \left(r_i - q_\theta(z_i;\beta)\right), \tag{6}$$

where the function $q_\beta(*)$, is defined in terms of the absolute deviation of residuals (u) by

$$p_\theta(u) = \begin{cases} u\theta & \text{if } u \geq 0 \\ u(\theta-1) & \text{otherwise.} \end{cases} \tag{7}$$

If several quantiles are of interest, the minimisation must be repeated for each quantile. A potential problem with using QR for the derivation of multiple conditional quantiles is that quantiles may cross, yielding predictive distributions that are not monotonically increasing, as a function of increasing quantiles (López López et al., 2014). In the present research study, the technique proposed by Muggeo et al. (2013) and implemented in the package quantregGrowth developed in the R environment (R Development Core Team, 2010) is used. This technique estimates nonparametric growth charts via quantile regression. Quantile curves are estimated via B-splines with a quadratic penalty on the spline coefficient differences, and non-crossing and monotonicity restrictions are set to obtain plausible estimates.

Two configurations were tested: (a) quantile regression applied to the raw precipitation data, and (b) quantile regression on the data transformed into normal domain through normal quantile transformation (NQT) (Bogner et al., 2012). For configuration (b) the time series of observed precipitation and forecast precipitation are transformed into the normal domain. After the derivation of the quantiles, the variables are back-transformed into original space. The rationale for using the transformation is that the joint distribution of transformed time series appears to be more linear, and can thus be better described by linear conditional quantiles (López López et al., 2014).

Back-transformation is, however, problematic if the quantiles of interest lie outside of the range of the empirical distribution of the untransformed variable in original space. To address this issue linear extrapolation in the tails of the distribution was used (López López et al., 2014).

To generate the probabilistic forecast of the occurrence of precipitation, the latent variable in the probit model is first sampled. The random value u is sampled from a uniform distribution. If $Z_i < u$ no precipitation occurs, if $Z_i \geq u$ precipitation is set to occur and the amount is computed with the quantile regression model. Given the occurrence of precipitation, the quantile regression model corresponding to the lead time under consideration is used to obtain the quantiles of the conditional distribution. The inverse cumulative function is approximated by a cubic spline and subsequently 50 uniform random numbers in the range 0–1 are generated and used to sample corresponding precipitation values from the inverse cumulative function.

2.3.2 Verification of forecasts

Deterministic forecasts of precipitation and discharge were verified using the mean absolute error (MAE), the mean squared error (MSE), the mean error (ME) and the skill score (SS) based on the MSE, using as reference forecasts provided by climatological values (subscript Clim in Eqs. 11 and 12) (Wilks, 2011) and zero forecasts (subscript 0 in Eqs. 13 and 14). The corresponding formulas are shown in Eqs. (8)–(14), where (y_k, o_k) is the kth of n pairs of forecast and observed values at a particular lead time. In the case of discharge, the discharge forecast obtained from the hydrological model driven with a precipitation input equal to zero, was used as reference. This reference was chosen given that this is the only choice when precipitation forecasts are not available, providing a method to assess the added value of WRF in comparison with current operation:

$$\mathrm{MAE} = \frac{1}{n}\sum_{k=1}^{n} |y_k - o_k|, \tag{8}$$

$$\mathrm{MSE} = \frac{1}{n}\sum_{k=1}^{n} (y_k - o_k)^2, \tag{9}$$

$$\mathrm{ME} = \frac{1}{n}\sum_{k=1}^{n} (y_k - o_k) = \bar{y} - \bar{o}, \tag{10}$$

$$\mathrm{MSE_{Clim}} = \frac{1}{n}\sum_{k=1}^{n} (\bar{o} - o_k)^2, \tag{11}$$

$$\mathrm{SS_{Clim}} = \frac{\mathrm{MSE} - \mathrm{MSE_{Clim}}}{0 - \mathrm{MSE_{Clim}}} = 1 - \frac{\mathrm{MSE}}{\mathrm{MSE_{Clim}}}, \tag{12}$$

$$\mathrm{MSE_0} = \frac{1}{n}\sum_{k=1}^{n} (0 - o_k)^2, \tag{13}$$

$$\mathrm{SS_0} = \frac{\mathrm{MSE} - \mathrm{MSE_0}}{0 - \mathrm{MSE_0}} = 1 - \frac{\mathrm{MSE}}{\mathrm{MSE_0}}. \tag{14}$$

In order to assess the performance of the probit model constructed to model the occurrence of precipitation from the WRF model, the ROC (relative operating characteristic) diagram was used. This is a discrimination-based graphical forecast verification display (Wilks, 2011). For perfect forecasts the ROC curve consists of two line segments coincident with the left boundary and the upper boundary of the ROC diagram. At the other extreme of forecast performance, random forecasts consistent with the sample climatological probabilities, the ROC curve will consist of the 45° diagonal connecting the points (0, 0) and (1, 1). ROC curves for real forecasts generally fall between these two extremes, lying above and to the left of the 45° diagonal. Forecasts with better discrimination exhibit ROC curves approaching the upper-left corner of the ROC diagram more closely, whereas forecasts

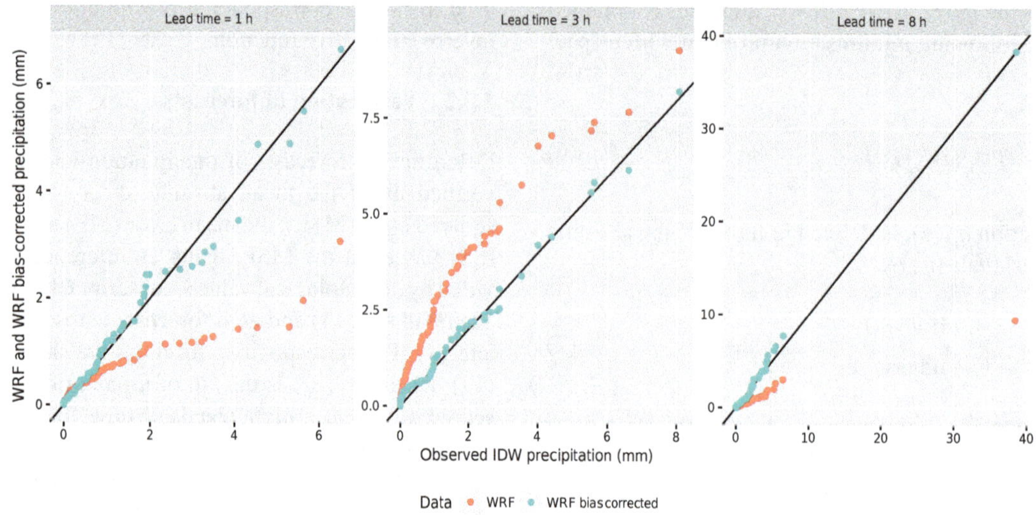

Figure 3. Q–Q plots (quantile–quantile plot, plot of the quantiles of the first data set against the quantiles of the second data set) for the Mugroso watershed comparing observed precipitation with WRF and WRF bias corrected precipitation.

with very little ability to discriminate the event exhibit ROC curves very close to the diagonal (Wilks, 2011). To summarise the ROC diagram, the area under the ROC curve was used (Wilks, 2011).

Regarding the quantile regression models, the pseudo-R-square measure (Koenker and Machado, 1999) was used to compare the models with normal quantile transformation and the ones based on raw data. The pseudo-R-square measure was proposed by Koenker and Machado (1999) as a goodness-of-fit indicator for quantile regression by comparing the sum of squared distances for the model of interest with the sum of squared distances between the observed and the fitted values that would be obtained if only the intercept term is included in the model (Hao and Naiman, 2007).

Ensemble forecasts of both precipitation and discharge were verified using rank histograms. Rank histograms make it possible to evaluate whether a collection of ensemble forecasts satisfy the consistency condition (Wilks, 2011). Rank histograms are constructed by accumulating the number of cases over space and time when the verifying analysis falls in any of $m+1$ intervals, where each of the $m+1$ intervals is defined by an ordered series of m ensemble members, including the two open-ended intervals (Jolliffe and Stephenson, 2012). Reliable or statistically consistent ensemble forecasts lead to a rank histogram that is close to flat (Jolliffe and Stephenson, 2012).

In order to assess the probabilistic forecasts of discharge obtained from the WRF precipitation ensembles, the CRPSS was used. The CRPSS is given by (Hersbach, 2000; Alfieri et al., 2014)

$$\mathrm{CRPSS} = \frac{\overline{\mathrm{CRPS}_{\mathrm{ref}}} - \overline{\mathrm{CRPS}_{\mathrm{forecast}}}}{\overline{\mathrm{CRPS}_{\mathrm{ref}}}}, \tag{15}$$

where

$$\mathrm{CRPS} = \int_{\infty}^{-\infty} \left[F(y) - F_0(y)\right]^2 \mathrm{d}y \tag{16}$$

and

$$F_0(y) = \begin{cases} 0, & y < \text{observed value} \\ 1, & y \geq \text{observed value.} \end{cases} \tag{17}$$

$F(y)$ is a stepwise cumulative distribution function (cdf) of the ensemble of each considered forecast. CRPSS ranges between 1 and $-\infty$. A CRPSS of 1 indicates a perfect forecast. Forecast ensembles are only valuable when CRPSS > 0, this is when the forecasts perform better than the reference. The discharge obtained from forecast precipitation equal to zero will be used as reference. Since this is a deterministic forecast, the $\mathrm{CRPS}_{\mathrm{ref}}$ corresponds to the mean absolute error of all forecast and observed pairs (Hersbach, 2000).

3 Results

3.1 Bias correction of precipitation forecasts through DBS

Figure 3 shows the quantile–quantile plots (Q–Q plots) comparing the WRF precipitation and the bias-corrected WRF precipitation with the observed precipitation obtained from IDW, for the 107 selected days of significant precipitation. The figure shows results for three lead times for the Mugroso watershed, as examples of the 12 that were analysed, results for the other two watersheds were similar and not shown here for brevity; 12 h was chosen as a first estimate of the usable lead time for the forecast, since NWP models describing

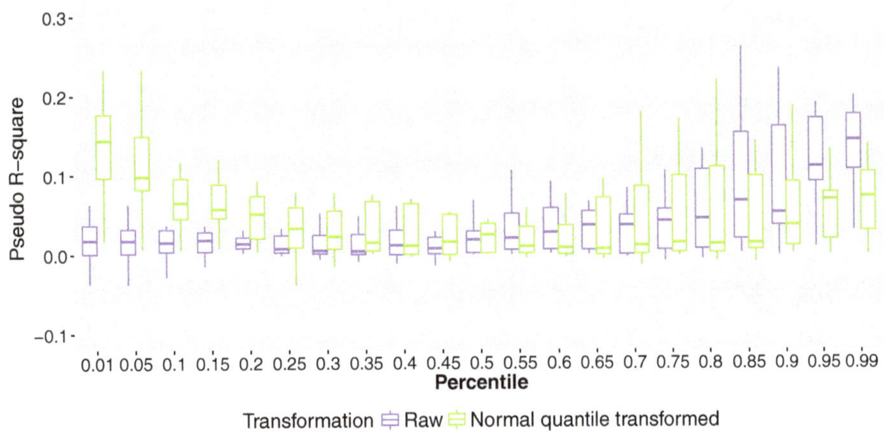

Figure 4. Pseudo-R square for the Mugroso watershed for the coarsest domain and lead times up to 12 h.

local–regional convective systems and orographic processes have been observed to significantly decrease their skill at lead times beyond 12–48 h (Cuo et al., 2011). The raw WRF and IDW time series were obtained from the corresponding rainfall fields as the mean areal precipitation calculated in the three watersheds of analysis by sampling the raster layers with the polygons of the watersheds.

The Q–Q plots show that both overestimation and underestimation of precipitations occurs. Underestimation is very noticeable for lead times of 1 h for values in the upper percentiles as shown in Fig. 3. In general for lead times between 2 and 5 h, overestimation of high precipitation values occurs, for larger lead times underestimation occurs and in the case of the lead time of 8 h a storm is entirely missed by the WRF model (see Fig. 3, Leadtime = 8 h).

3.2 Quantile regression model

Figure 4 shows the pseudo-R square for all quantile regression models for the Mugroso watershed and for the coarsest domain (resolution of 15 km), the results are similar for the other watersheds and domains. The R-square values are pooled for lead times up to 12 h. The figure shows that the values of the pseudo-R square up to the 50th percentile are higher when the normal quantile transformation is used, thus indicating a better fit of the quantile regression model. For higher percentiles the pseudo-R square of the models with normal quantile transformation is in approximately the same range of the raw data or lower. This means that for percentiles above the 50th percentile the fit of the quantile regressions is better for the raw data than for the normal transformed data.

3.3 Verification of deterministic precipitation forecasts and ensemble mean

Figure 5 shows the performance measures shown in Eqs. (8)–(14). The performance of all the WRF domains are shown for the three watersheds (Chisacá, Mugroso and Curubital). From top to bottom the figure shows the ME, the MAE and the MSE. The lowest two figures for each watershed show the skill score based on MSE using climatology as a reference, and the skill score using zero precipitation as a reference.

The ME of the raw WRF forecasts shows that in the three watersheds for lead times between 1 and 5 h the precipitation amounts are overpredicted (positive ME), while after 5 h the ME shifts to negative values, indicating underprediction. The bias-corrected WRF forecasts show values of ME very close to zero, which could be expected given the procedure applied (see Fig. 5). In the case of the mean of the WRF ensemble without transformation, the ME is close to zero for lead times up to 6 h. Beyond this lead time there is a tendency to underpredict precipitation. The behaviour of the ME for the mean of the ensemble with normal quantile transformation is very similar to that when the transformation is not applied.

The highest values of MAE (see Fig. 5) are obtained from the DBS bias-corrected forecast for lead times larger than 6 h, while the raw WRF forecasts exhibit the highest values at the shorter lead times. In the case of the MSE, the highest values are obtained from the bias-corrected WRF forecasts at lead times of 2 and 8 h. The increase in MSE when bias correction is applied is mainly attributed to the influence of a single (observed) high-precipitation event that was not forecast by the WRF model. These intense precipitation events being missed in the forecasts cause all WRF values higher than the 95th percentile to be significantly increased through the bias correction procedure, increasing their square difference with the observed value. These extremes are mainly present at lead times of 2 and 8 h due to extreme precipitation events in this area typically occurring in the mid-afternoon, some 2 to 8 h after the start of the 18:00 and 12:00 GTM WRF forecast runs (the time zone of the area is UTC − 5).

For both MSE and MAE, the best performance is obtained from the ensemble mean of the WRF forecasts. This is the case both with and without normal quantile transformation (see Fig. 5), which provide very similar performance. Re-

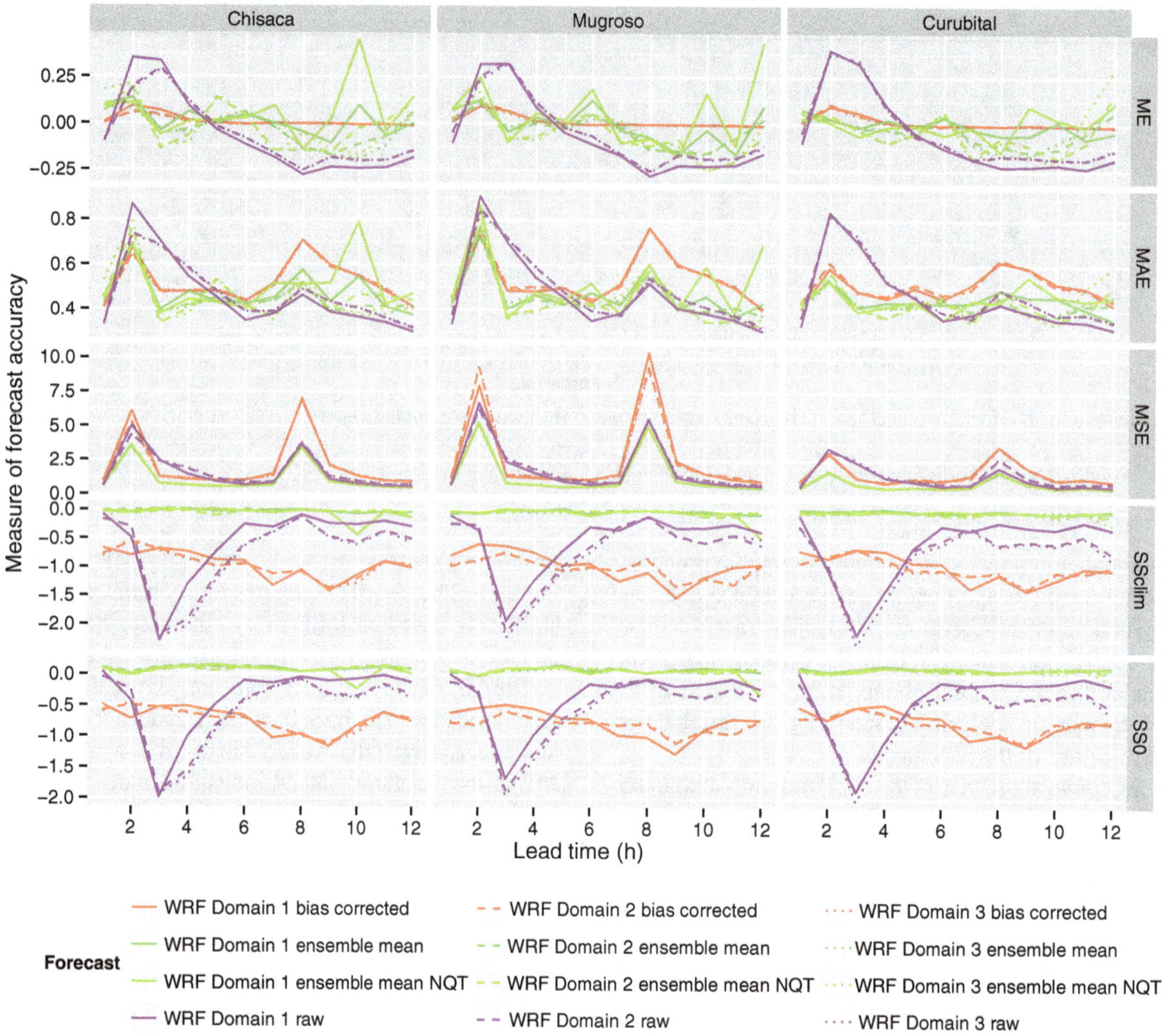

Figure 5. Accuracy measures for deterministic precipitation and ensemble mean obtained from the WRF model. Domain 1, domain 2 and domain 3 correspond respectively to the domains with resolutions 15, 5 and 1.67 km.

garding the skill score based on the MSE using the climatology as reference (see Eqs. 11 and 12) negative values are obtained for all forecasts, with the bias-corrected WRF and the raw WRF forecasts showing the worst skill. However, in the case of the ensemble mean, both for the normal quantile-transformed and raw ensembles, the values are very close to zero, albeit still negative. When zero precipitation is used as reference, the behaviour of the score is very similar to SS_{Clim}. However, most values for the ensemble mean forecast are positive, both with raw and with normal quantile-transformed data, with a maximum value of 0.16. This shows that the ensemble mean forecasts provide the best results in comparison to the other forecasts. In the case of the climatology being used as reference, they are approximately as good as the reference and compared to zero precipitation as reference; they provide an improvement up to 16 % over the reference.

Given that a single extreme storm which was entirely missed by the WRF forecast has a significant influence in the results shown in Fig. 5, (resulting in the poor performance at lead times of 2 and 8 h), the accuracy metrics were recalculated from the data set, but now excluding this storm. The results are shown in Fig. 6 only for the MSE, since the other metrics are less affected by the exclusion of the extreme storm. In Fig. 6 the reduction of the MSE is highly noticeable in all the forecasts (see Fig. 5 in comparison with Fig. 6), and the lowest values are exhibited by the ensemble mean of both raw and normal quantile-transformed data, which show very similar behaviour. This analysis clearly shows the sensitivity of performance measures based on squared residuals, such as MSE, to a single event with high precipitation being missed in the WRF forecasts.

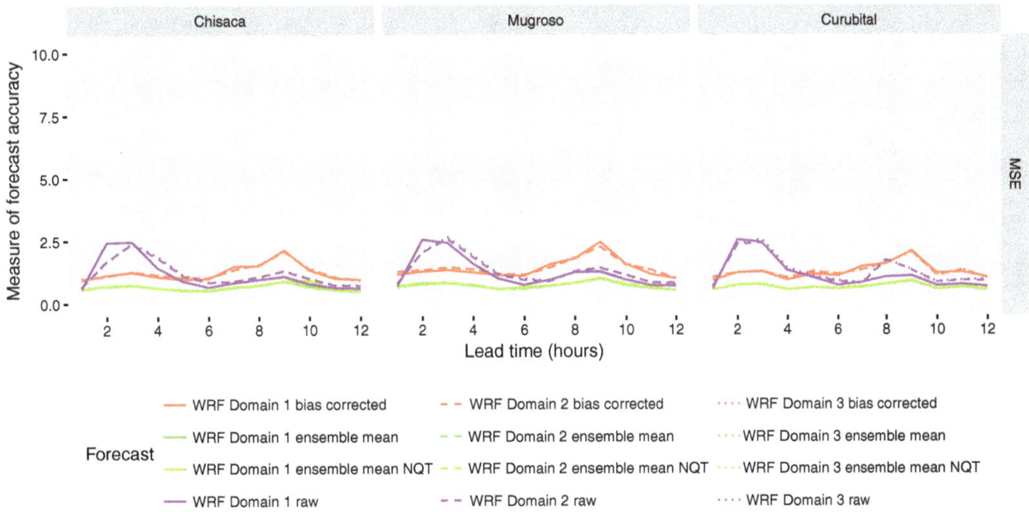

Figure 6. Accuracy measures for deterministic precipitation and ensemble mean obtained from the WRF model without the highest precipitation missed by the WRF model. Domain 1, domain 2 and domain 3 correspond respectively to the domains with resolutions 15, 5 and 1.67 km. The legend is the same as in Fig. 5.

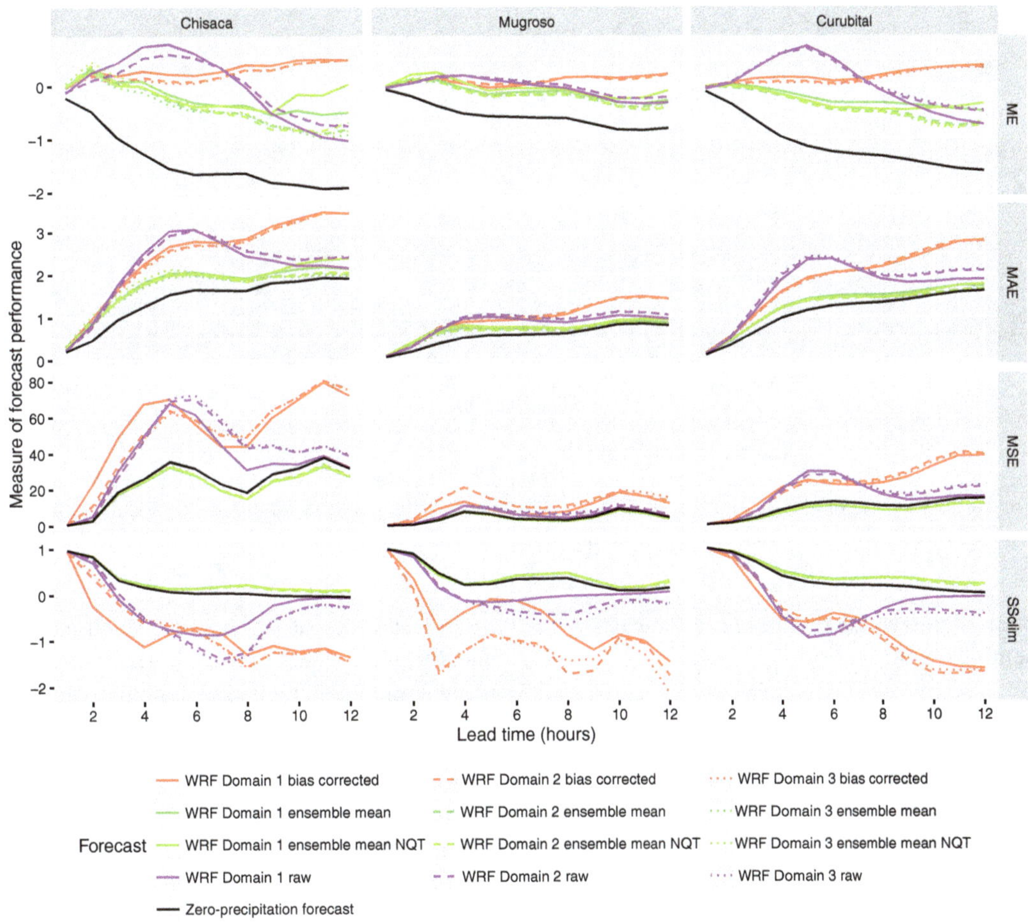

Figure 7. Performance measures for deterministic discharge and ensemble mean forecasts. Domain 1, domain 2 and domain 3 correspond respectively to the domains with resolutions 15, 5 and 1.67 km.

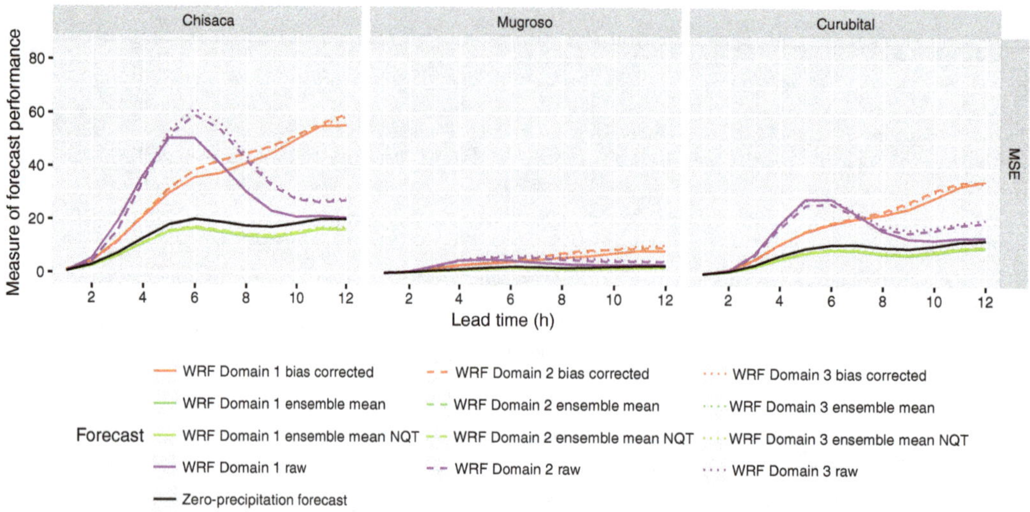

Figure 8. Performance measures for deterministic discharge and ensemble mean forecasts obtained excluding the storm with the highest precipitation missed by the WRF model. Domain 1, domain 2 and domain 3 correspond respectively to the domains with resolutions 15, 5 and 1.67 km. The legend is the same as in Fig. 7.

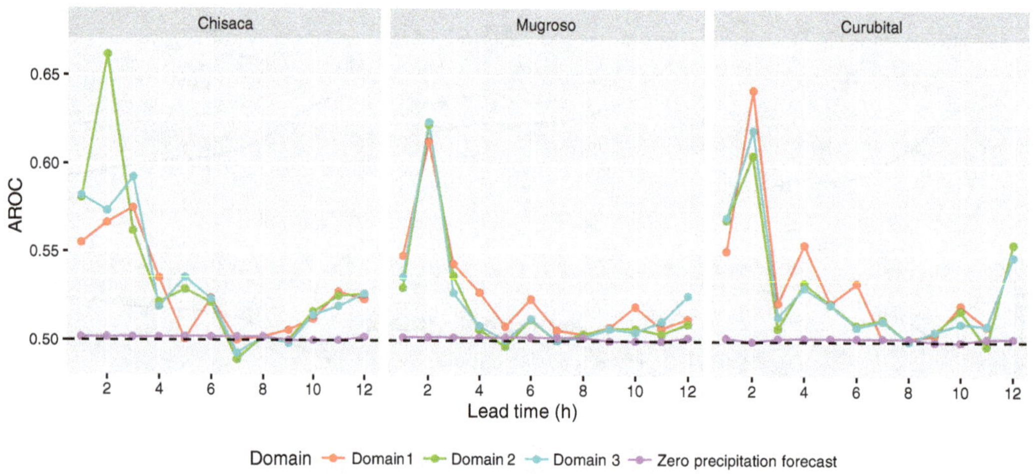

Figure 9. Area under the ROC curve (AROC) for the forecast of occurrence of precipitation with the probit model. Domain 1, domain 2 and domain 3 correspond respectively to the domains with resolutions 15, 5 and 1.67 km.

3.4 Verification of deterministic discharge forecasts and ensemble mean

The results for discharge using the precipitation forecasts are shown in Fig. 7. The structure of the figure is similar to that of Fig. 5, although the skill score is now determined using the climatology as reference, as the zero-precipitation forecast is considered as one of the forecasts. The ME shows that the raw WRF forecasts overestimate the discharges up to a lead time of some 8 h, while for larger lead times underestimation occurs. When the bias-corrected WRF forecasts are used, the ME is close to zero, slowly increasing with lead time. The use of the mean of the ensembles with and without normal quantile transformation produces ME close to zero up to lead times of 4 h, beyond which underestimation occurs (negative values of ME). In terms of the MSE the mean of the ensemble with and without normal quantile transformation produce the smallest values for all lead times, which is reflected in the SS_{Clim}, where they produce the highest values, although there is little improvement in the skill over that obtained from using a zero-precipitation series as forecast.

Figure 8 shows the MSE calculated using the stream flow simulations, but now excluding the storm with the highest precipitation that was missed by the WRF model. Only the MSE is shown, since the other metrics are less affected by the exclusion of the extreme storm. This shows a significant improvement in comparison with Fig. 7, particularly for the DBS bias-corrected data.

From the results shown in Figs. 5–8, the difference in performance for the three domains of the WRF model is not significant, both in the precipitation and in the discharge forecasts.

3.5 Verification of probabilistic forecasts

Figure 9 shows the area under the ROC curve (AROC), showing the skill of the forecasts to discriminate between the occurrence and non-occurrence of precipitation. Scores are shown for the three watersheds (Chisacá, Mugroso and Curubital in Fig. 1); for the three domains of the WRF model for lead times up to 12 h; and for a precipitation forecast of zero. The behaviour of the AROC is similar in the three watersheds and three domains, with the highest values found in the first 2 h, dropping to a value close to 0.5 after a lead time of about 5 h. The AROC values for a forecast of zero precipitation is very close to 0.5 for all lead times, and thus is almost equal to random forecasts consistent with the sample climatological probabilities (Wilks, 2011). This shows that the forecasts have some skill at short lead times in comparison to both zero-precipitation forecasts and random forecasts, although that skill is limited.

The rank histograms of the precipitation ensembles are shown in Fig. 10. The behaviour for all domains and all lead times is similar, therefore the figure only shows the lead time of 1 h. The rank histogram is approximately uniform from rank 1 to rank 5, and from rank 6 to rank 10 there is an increase in the frequency of the IDW precipitation values falling in these higher ranks. Similar behaviour in all rank histograms was observed when the storm with the highest precipitation missed by the WRF model was excluded. This suggests that the WRF ensemble is slightly underforecasting.

Figure 11 shows the rank histograms for the discharges obtained from the precipitation ensemble using the TOPMODEL. The rank histograms show approximate uniformity mainly for lead times up to 3 h. For longer lead times a slight underdispersion is observed. The figure shows lead times up to 6 h since for longer lead times the behaviour is similar to that of the lead time of 6 h.

Figure 12 shows the CRPS for the discharge ensembles, the MAE for the forecast produced by using zero as precipitation forecast and the corresponding CRPSS. The CRPSS exhibits values in the range 10.5–21 % (see Fig. 12) with most values in a range of 14–16 %. The same behaviour is observed in the three watersheds. The comparison between normal quantile-transformed quantile regressions and the quantile regressions with raw data again shows that the difference between the two is negligible.

4 Discussion

4.1 Evaluating precipitation forecasts from the WRF model

The Q–Q plots of the raw WRF forecasts and the precipitation obtained from IDW interpolation shows that the WRF model tends to underpredict precipitation for the first hour of lead time and to overpredict precipitation for larger lead times. The overpredicting behaviour has been observed in other tropical areas (Mourre et al., 2016). While the details of the implementation of the WRF model are out of the scope of this study, possible causes of the precipitation error found by other authors include; errors in the lateral boundary conditions (Ochoa et al., 2014); poor representation of the topography (Ochoa et al., 2014); and choice of convective treatment, microphysics and planetary boundary layer (Jankov et al., 2005). However, there are individual cases where the WRF model did not detect the occurrence of high precipitation. Missing or underpredicted forecasts of intensive precipitation have also been observed by other authors (Kryza et al., 2013; Liu et al., 2015), which may suggest the need for an improvement in model parameterisation and data assimilation.

No significant differences among three domains of the WRF model were found in the behaviour of the scores for the tests carried out in this study. The two-way communication between the nested domains implies that precipitation volumes are aligned between the different resolutions, with the small differences found likely being due to differing ratios between the WRF model grid cell sizes and the sizes of the watersheds. Other applications of NWP (e.g. Roberts et al., 2009) show that finer resolutions are capable of producing more accurate predictions, and that physics configuration, resolution and initial conditions highly influence the WRF model performance (Kryza et al., 2013). The similarity of results regardless of resolution found in this study may be also related to deficiencies in the parameterisation of the model, or to the inability to sufficiently resolve the topography. A more detailed review of the WRF model would be required to reveal possible deficiencies, and to suggest improvements. Furthermore, this analysis has only been focused on the forecast of high precipitation when this has been observed, which provides insight into the skill of the WRF model to detect events that can lead to floods. However, the behaviour of the WRF model under other conditions such as those that may lead to false alarms is planned for future research, which may reveal other potentialities and deficiencies to be taken into account in the improvement of the model.

The WRF model has been shown to be highly sensitive to the parameterisation of cumulus and microphysical processes (Remesan et al., 2014; Rama Rao et al., 2012). Parameter sensitivity is generally dependent on local conditions (Di et al., 2014), and the best configuration varies with time and rainfall threshold (Jankov et al., 2005). While it was our

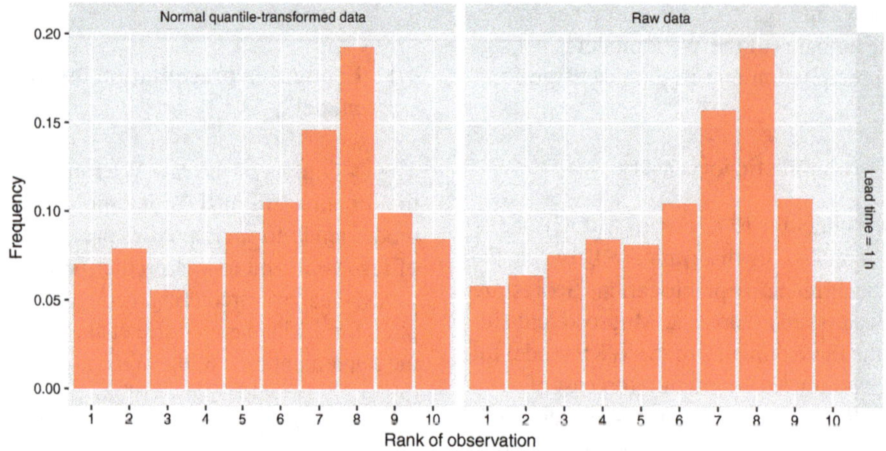

Figure 10. Rank histograms for the WRF ensemble for the Mugroso watershed for the finest domain (resolution of 1.67 km) and lead time of 1 h.

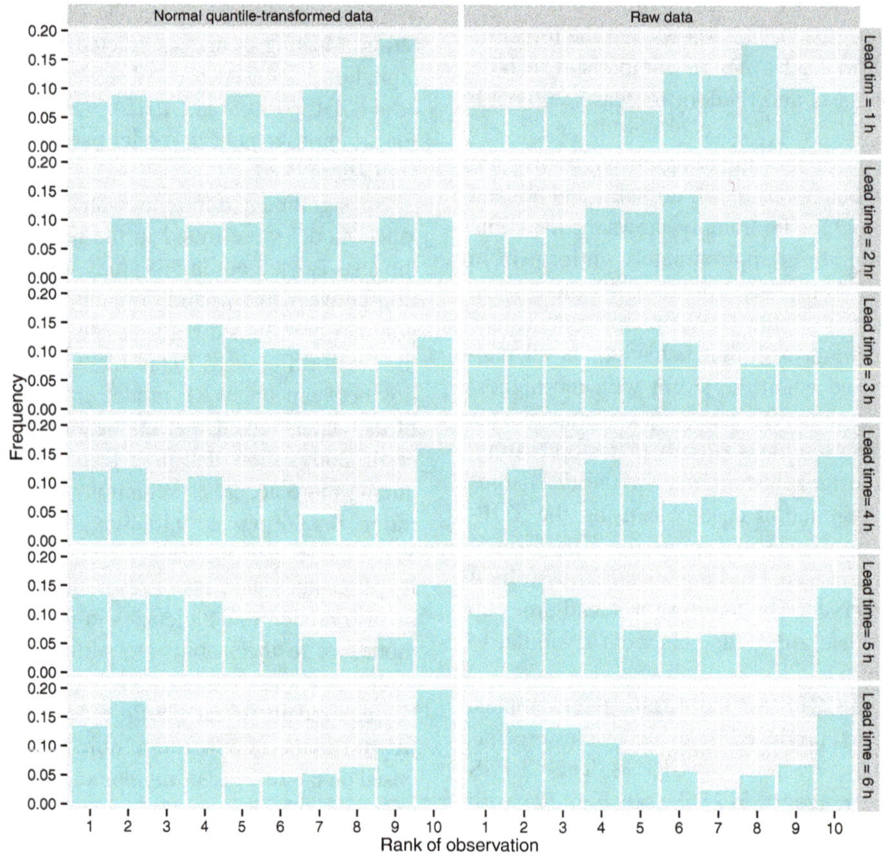

Figure 11. Rank histograms for the Q ensemble using as reference time series the discharge simulated with the TOPMODEL.

objective to test the configuration of the WRF model as it is used operationally by IDEAM and optimising the parameterisation of the WRF model is outside the scope of the present paper, further work could test various model parameterisations, assessing their sensitivity and searching for an optimal or a set of optimal configurations that help to improve parameterisation (in the case of the coarser model domains) and in particular resolving (in the case of the finest model domain) convective precipitation as frequently occurs in the case study area. Observing systems such as radars, gauges and satellites for data assimilation in NWP analysis could

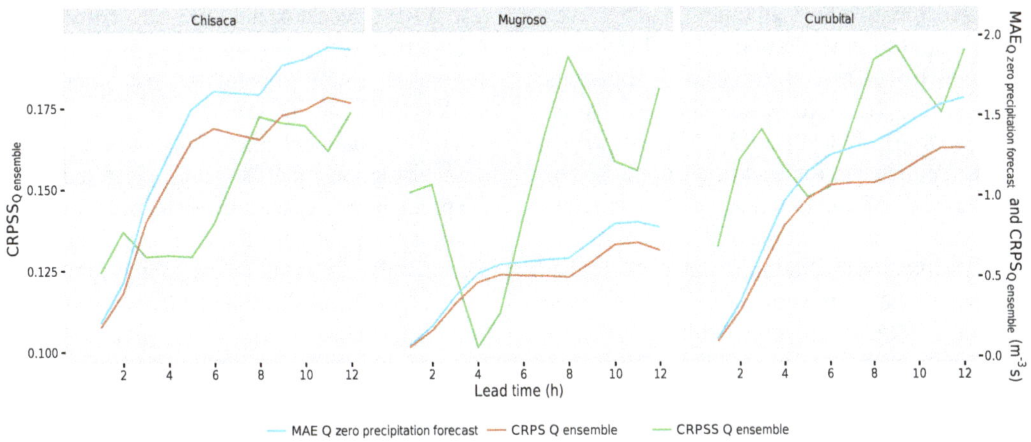

Figure 12. CRPSS for the discharge ensembles.

further improve the NWP forecasts (Rossa et al., 2011; Cuo et al., 2011; Yucel et al., 2015; Liu et al., 2015).

The application of the DBS approach has been reported previously to considerably reduce the differences in rainfall frequency between observations and forecasts (Yang et al., 2010). This reduction of differences was found in this study, as shown in Fig. 3. However, where high precipitation values were observed but not forecast in a single extreme event, the bias correction had the effect of reducing overall performance, particularly when measured by the MSE, since this is more sensitive to large residuals. This behaviour suggests that the DBS bias correction can introduce undesirable effects, which limit its effectiveness (Ehret et al., 2012). Similar behaviour was observed with the mean of the ensemble, although with less impact on the performance measures in comparison to the DBS approach. A significant difference between the ensemble means obtained from raw and normal quantile-transformed data were not found.

These biases are the result of the inability of the WRF model to adequately represent convective precipitation, failing to correctly predict when large events occur, or where it occurs. This has also been observed by other authors. For instance, Verkade et al. (2013) found that post-processing does not improve on all qualities at all lead times and at all levels of the verifying observations. The cause is rooted in the impossibility of the post-processing approaches to replace adequate model representation of physical processes (Haerter et al., 2011). Haerter et al. (2011) found that the conditions on climate model data to make the application of statistical bias correction schemes reasonable are that a realistic representation of the physical processes involved must be ensured; and that the quantitative discrepancies between the modelled and observed probability density function of the quantity at hand must be constant in time. Similarly, Chen et al. (2013) emphasise the impossibility any bias correction method being successful if there is no coherence between simulated and observed precipitation. The WRF data used in this study show limitations in fulfilling both these conditions, which leads to the post-processed precipitation being no more skilful than the sample climatology (as shown by negative values of SS_{Clim}) and providing only a modest improvement in comparison to the zero-precipitation forecast (as shown by most values of SS0 being positive, albeit low). However, an important result of this analysis is that despite the limitations, the ensemble mean outperforms the DBS bias correction and seems to be less sensitive to the presence of intense precipitation events that are not forecast by the WRF model. Post-processing the deterministic forecasts to generate the ensembles is clearly more sophisticated than just correcting the mean bias (Robertson et al., 2013). During the process, different distributions are used for each raw WRF forecast, which results in an improvement of skill that outperforms the bias correction through DBS. The magnitude of the bias of the ensemble mean is nearly always smaller than the raw forecasts, as well as of the DBS bias-corrected forecasts, particularly for lead times up to 6 h.

Regarding the verification of the precipitation ensembles, the rank histograms show that the IDW precipitation too often falls between ranks 6 and 10. This reflects overdispersion of the ensemble towards low values. This produces an underforecasting bias in the ensemble. The same behaviour of the rank histograms is observed in the ensembles obtained from quantile regression, both when the normal quantile transformation is used in the precipitation data and when raw data are used. This means that the transformation of the values does not improve the consistency of the ensemble.

The AROC of the occurrence of precipitation shows that values higher than 0.5 are obtained in all watersheds for lead times up to 5 h, indicating that the probit model has the ability to discriminate between the occurrence or non-occurrence of precipitation up to these lead times. In all domains and watersheds, the highest values of AROC are obtained for lead times between 2 and 3 h, in the range of 0.63–0.67. This means that there is some skill in the probit model to forecast occurrence of precipitation for lead times up to 5 h, even if the skill of the amount of precipitation is low. For longer times the probit model does not have any skill, consistent with other studies where the performance significantly reduces after a few hours (Liu et al., 2015). This behaviour of the probit model implies that for lead times up to 5 h the probability of occurrence of precipitation can be modelled from the WRF precipitation with some skill, meaning that the higher the WRF forecast precipitation value the higher the probability of occurrence of precipitation. However, even though the lead times at which the model still has skill are limited, this does imply that the current ability of NWP in tropical mountainous areas can provide a limited extension of lead time beyond the concentration time of small, flashy, watersheds.

4.2 Evaluating discharge forecast

The comparison of Figs. 7 and 8 shows that the biases in the precipitation forecasts are reflected in the biases of the discharge. When the single event with high precipitation that is not forecast by the WRF is excluded from the analysis, a reduction of bias in the forecast precipitation is observed, as well as a reduction in the bias of the forecast discharge. This is consistent with the findings of other studies where the errors in bias-corrected precipitation lead to amplified errors in modelled runoff (Teng et al., 2015), since in this analysis the increase in errors in the bias-corrected time series amplify the errors in the simulated discharges.

The forecasts generated with the post-processed ensemble precipitation forecasts exhibit better skill than the deterministic forecasts (based on raw, bias corrected and zero precipitation). Higher skill of the ensemble mean in comparison with deterministic forecasts has also been found in other flood forecasting systems driven by the WRF model (Calvetti et al., 2014) and other NWP models (Vincendon et al., 2011).

In terms of the lead time at which the coupled WRF-TOPMODEL forecasts provide value in these watersheds, a first limit is provided by the skill of the WRF model to forecast the occurrence of precipitation up to 5 h (see Fig. 9). A limit is also observed in the skill scores of the discharge forecast in Figs. 7 and 8. Beyond the lead time of 2 h, which is approximately the time of concentration of the watersheds, the skill (SS_{Clim} in Figs. 7 and 8) starts to drop significantly. For the discharge obtained from raw and DBS bias-corrected precipitation, 3 and 4 h are the respective lead times at which the added value is totally lost in comparison with the climatology (SS_{Clim} in Figs. 7 and 8 reaches zero). In the case of the ensemble means, the skill also decreases progressively after 2 h, with some value compared to climatology up to a lead time of some 6 h. Beyond this lead time, the values remain approximately constant in a range of 0.2 to 0.5 for the three watersheds. This behaviour indicates that the use of the ensemble mean provides some skill up to a lead time of 6 h. This is 3 h longer than the lead time provided by the raw WRF forecast or the DBS bias-corrected WRF forecast. A lead time of 6 h is consistent with other flood forecasting systems in small mountainous catchments driven by NWP models (Verkade and Werner, 2011).

The hydrological model produces ensemble results with rank histograms that do not reflect the overprediction of the precipitation ensemble, with approximately uniform rank histograms mainly for the first lead times (see Fig. 11). Approximately the same shape of the rank histograms is observed for ensembles obtained from quantile regression when no transformation is used in the precipitation data and when raw data are used. The minor influence of the transformation is consistent in all the performance assessments. This may be due to the relatively low improvement of the fit of the quantile regressions when the normal quantile transformation is used. According to Fig. 4, the most significant improvement is found in the low percentiles (up to the 25th percentile). For higher percentiles, the improvement is not significant and for some percentiles degradation of goodness of fit occurs. As floods occur for higher rainfall percentiles, the similarity in behaviour is logical.

The results of the CRPSS (see Fig. 12) show that forcing the hydrological model with WRF ensembles improves the forecasts in a range of 8.5–22 % in comparison with a forecast produced with the hydrological model forced with zero precipitation for lead times between 1 and 12 h. The positive values of the CRPSS imply added value to the forecasts, albeit modest. The CRPSS obtained from the ensembles is comparable to the CRPSS values found in other areas with other NWP models, albeit in the low range. For example, Robertson et al. (2013) found a CRPSS of 37 % on average for post-processed ensembles in Australia, where rainfall is predominantly produced by large-scale synoptic systems that are better predicted by NWP models. Therefore, given the high complexity of the meteorological conditions of the study area, and despite the relatively poor skill of the WRF model in predicting precipitation amounts, the WRF model does show promise at producing a benefit in its use for flood forecasting compared to not using precipitation forecasts. This is likely due to the skill found in the probit model in predicting the occurrence of rainfall.

Besides improvements in parameterisation and data assimilation, as described previously, bias correction of streamflow could provide a further skill improvement (Yuan and Wood, 2012).

5 Conclusions

This paper presents the assessment of WRF forecasts produced in a tropical high-mountain watershed in Bogotá, Colombia. The WRF forecasts were used to force a hydrological model of the study area. Simulated discharges were used to assess the value of the WRF forecasts for flood early warning. Although the skill of the forecasts developed is limited, results show that the streamflow forecasts obtained from a hydrological model forced by post-processed WRF precipitation do add value to the flood early warning system, particularly when compared to zero-precipitation forecasts.

The WRF model for the study area provides forecasts that are found to overpredict precipitation for the shorter lead times (up to 6 h), and that tend to fail to forecast high-intensity precipitation events. This behaviour may be due to parameterisation deficiencies, errors in boundary conditions and/or poor representation of the topography. A more detailed evaluation of the WRF model in this study area is recommended, including the assessing of other convective and microphysics schemes to identify the most suitable parameterisation. The use of satellite and soon to be available radar data may further help improve performance.

Bias correction of the WRF forecast precipitation, through distribution-based scaling (DBS) was shown to significantly reduce the differences in rainfall amounts between observations and forecasts. However, the DBS-corrected forecasts were found to be sensitive to failure of the WRF model to predict a small number of extreme events, resulting in a reduction of overall performance. This sensitivity was also observed with the ensemble mean forecasts, although for this post-processing technique the impact of these few missed events on the performance was lower when compared with DBS. This undesirable behaviour is the result of the inability of the WRF model to adequately represent convective precipitation, which cannot be corrected through simple post-processing.

Despite the limitations in the WRF forecasts, the ensemble mean was found to outperform the DBS bias correction, as well as being less sensitive to the presence of highly intense precipitation that are not correctly forecast by the WRF model. Although none of the precipitation forecasts that were assessed in this analysis showed added value when compared to climatology, the reduction of biases obtained from the ensembles show the potential of this method and model to provide usable precipitation forecasts. Although the skill of the quantitative precipitation models was found to be low, the probit model, which was applied to forecast the probability of occurrence of precipitation, based on the WRF forecast, showed that the WRF model does have some skill at short lead times (up to 5 h) in comparison to both zero-precipitation forecasts and random forecasts, although that skill is limited.

Increases in precipitation biases are reflected in the discharge forecasts. However, discharge forecasts generated with the post-processed ensembles did exhibit higher skill than deterministic forecasts (based either on raw, bias corrected or zero precipitation). Furthermore, the quality of these forecasts is better than that which could be obtained using zero precipitation as input to the hydrological model. The potentially usable lead times for discharge forecasts obtained from the WRF model found in this analysis (5–6 h) are in the range of lead times found in other studies with other NWP models. Despite the fact that the added value of the WRF model forecasts is modest, the results found do show there is some potential for increasing forecast skill in areas of high meteorological and topographic complexity, as well as how these can be improved.

Competing interests. The authors declare that they have no conflict of interest.

Acknowledgements. This work was funded by the UNESCO-IHE Partnership Research Fund – UPARF in the framework of the FORESEE project. We wish to express our gratitude to the IDEAM – Instituto de Hidrología, Meteorología y Estudios Ambientales for providing the WRF model for this analysis.

Edited by: Florian Pappenberger

References

Albert, J.: Bayesian Computation with R, 2nd Edn., Springer-Verlag, New York, 2009.

Alfieri, L., Pappenberger, F., Wetterhall, F., Haiden, T., Richardson, D., and Salamon, P.: Evaluation of ensemble streamflow predictions in Europe, J. Hydrol., 517, 913–922, 2014.

Arango, C. and Ruiz, J. F.: Implementación del modelo WRF para la sabana de Bogotá, Tech. rep., Instituto de Hidrología, Meteorología y Estudios Ambientales, Bogotá, available at: http://www.ideam.gov.co/documents/21021/21132/Modelo_WRF_Bogota.pdf/f1d34638-e9f8-4689-b5f4-31957c231c46 (last access: 20 November 2015), 2011.

Bernal, G., Rosero, M., Cadena, M., Montealegre, J., and Sanabria, F.: Estudio de la Caracterización Climática de Bogotá y cuenca alta del Río Tunjuelo, Tech. rep., Instituto de Hidrología, Meteorología y Estudios Ambientales IDEAM – Fondo de Prevención y Atención de Emergencias FOPAE, Bogotá, 2007.

Beven, K. J. and Kirkby, M. J.: A physically based, variable contributing area model of basin hydrology/Un modèle à base physique de zone d'appel variable de l'hydrologie du bassin versant, Hydrol. Sci. B., 24, 43–69, https://doi.org/10.1080/02626667909491834, 1979.

Bogner, K., Pappenberger, F., and Cloke, H. L.: Technical Note: The normal quantile transformation and its application in a flood forecasting system, Hydrol. Earth Syst. Sci., 16, 1085–1094, https://doi.org/10.5194/hess-16-1085-2012, 2012.

Bremnes, J. B.: Probabilistic forecasts of precipitation in terms of quantiles using NWP model output, Mon. Weather Rev., 132, 338–347, 2004.

Buytaert, W. and Beven, K.: Models as multiple working hypotheses: hydrological simulation of tropical alpine wetlands, Hydrol. Process., 25, 1784–1799, https://doi.org/10.1002/hyp.7936, 2011.

Buytaert, W., Wyseure, G., De Bièvre, B., and Deckers, J.: The effect of land-use changes on the hydrological behaviour of Histic Andosols in south Ecuador, Hydrol. Process., 19, 3985–3997, https://doi.org/10.1002/hyp.5867, 2005.

Buytaert, W., Célleri, R., De Bièvre, B., Cisneros, F., Wyseure, G., Deckers, J., and Hofstede, R.: Human impact on the hydrology of the Andean páramos, Earth-Sci. Rev., 79, 53–72, https://doi.org/10.1016/j.earscirev.2006.06.002, 2006.

Calvetti, L., José, A., and Filho, P.: Ensemble hydrometeorological forecasts using WRF hourly QPF and TopModel for a middle watershed, Adv. Meteorol., 2014, 484120, https://doi.org/10.1155/2014/484120, 2014.

Chen, J., Brissette, F. P., Chaumont, D., and Braun, M.: Finding appropriate bias correction methods in downscaling precipitation for hydrologic impact studies over North America, Water Resour. Res., 49, 4187–4205, https://doi.org/10.1002/wrcr.20331, 2013.

Clark, M., Gangopadhyay, S., Hay, L., Rajagopalan, B., and Wilby, R.: The Schaake Shuffle: a method for reconstructing space–time variability in forecasted precipitation and temperature fields, J. Hydrometeorol., 5, 243–262, 2004.

Cloke, H. L. and Pappenberger, F.: Ensemble flood forecasting: a review, J. Hydrol., 375, 613–626, https://doi.org/10.1016/j.jhydrol.2009.06.005, 2009.

Colman, B., Cook, K., and Snyder, B.: Numerical weather prediction and weather forecasting in complex terrain, in: Mountain Weather Research and Forecasting Recent Progress and Current Challenges, chap. 11, edited by: Chow, F. K., De Wekker, S. F., and Snyder, B. J., Springer, Dordrecht, 2013.

Cortinas, J. V. J., Brill, K. F., and Baldwin, M. E.: Probabilistic forecasts of precipitation type, in: 16th Conference on Probability and Statistics in the Atmospheric Sciences, 12–17 January 2002, Orlando, Florida, 2002.

Cuo, L., Pagano, T. C., and Wang, Q. J.: A review of quantitative precipitation forecasts and their use in short- to medium-range streamflow forecasting, J. Hydrometeorol., 12, 713–728, https://doi.org/10.1175/2011JHM1347.1, 2011.

Demeritt, D., Cloke, H., Pappenberger, F., Thielen, J., Bartholmes, J., and Ramos, M. H.: Ensemble predictions and perceptions of risk, uncertainty, and error in flood forecasting, Environ. Hazards, 7, 115–127, 2007.

Demeritt, D., Nobert, Ś., Cloke, H., and Pappenberg, F.: Challenges in communicating and using ensembles in operational flood forecasting, Meteorol. Appl., 17, 209–222, 2010.

Di, Z., Qingyun, D., Wei, G., Chen, W., Yanjun, G., Jiping, Q., Jianduo, L., Chiyuan, M., Aizhong, Y., and Charles, T.: Assessing WRF model parameter sensitivity: A case study with 5 day summer precipitation forecasting in the Greater Beijing Area, Geophys. Res. Lett., 42, 579–587, https://doi.org/10.1002/2014GL061623, 2014.

Duan, Q.-Y. Y., Sorooshian, S., and Gupta, V.: Effective and efficient global optimization for conceptual rainfall–runoff models, Water Resour. Res., 28, 1015–1031, https://doi.org/10.1029/91WR02985, 1992.

Ehret, U., Zehe, E., Wulfmeyer, V., Warrach-Sagi, K., and Liebert, J.: HESS Opinions "Should we apply bias correction to global and regional climate model data?", Hydrol. Earth Syst. Sci., 16, 3391–3404, https://doi.org/10.5194/hess-16-3391-2012, 2012.

Fan, F. M., Collischonn, W., Meller, A., and Botelho, L. C. M.: Ensemble streamflow forecasting experiments in a tropical basin: the São Francisco river case study, J. Hydrol., 519, 2906–2919, https://doi.org/10.1016/j.jhydrol.2014.04.038, 2014.

Gupta, H. V., Kling, H., Yilmaz, K. K., and Martinez, G. F.: Decomposition of the mean squared error and NSE performance criteria: Implications for improving hydrological modelling, J. Hydrol., 377, 80–91, https://doi.org/10.1016/j.jhydrol.2009.08.003, 2009.

Habets, F., LeMoigne, P., and Noilhan, J.: On the utility of operational precipitation forecasts to served as input for streamflow forecasting, J. Hydrol., 293, 270–288, 2004.

Haerter, J. O., Hagemann, S., Moseley, C., and Piani, C.: Climate model bias correction and the role of timescales, Hydrol. Earth Syst. Sci., 15, 1065–1079, https://doi.org/10.5194/hess-15-1065-2011, 2011.

Hao, L. and Naiman, D.: Quantile Regression, SAGE Publications, Thousand Oaks, California, 2007.

Hersbach, H.: Decomposition of the continuous ranked probability score for ensemble prediction systems, Weather Forecast., 15, 559–570, https://doi.org/10.1175/1520-0434(2000)015<0559:DOTCRP>2.0.CO;2, 2000.

Jankov, I., Gallus, W. A., Segal, M., Shaw, B., and Koch, S. E.: The impact of different WRF model physical parameterizations and their interactions on warm season MCS rainfall, Weather Forecast., 20, 1048–1060, https://doi.org/10.1175/WAF888.1, 2005.

Jolliffe, I. and Stephenson, D.: Forecast Verification: A Practitioner's Guide in Atmospheric Science, John Wiley & Sons, Ltd, West Sussex, England, 2012.

Kleiber, W., Katz, R. W., and Rajagopalan, B.: Daily spatiotemporal precipitation simulation using latent and transformed Gaussian processes, Water Resour. Res., 48, 1–17, 2012.

Koenker, R. and Machado, J. A. F.: Goodness of fit and related inference processes for quantile regression. (Statistical Data Included), J. Am. Stat. Assoc., 94, 1–22, 1999.

Kryza, M., Werner, M., Wałaszek, K., and Dore, A. J.: Application and evaluation of the WRF model for high-resolution forecasting of rainfall – a case study of SW Poland, Meteorol. Z., 22, 595–601, 2013.

Laing, A. and Evans, J.-L.: Introduction to Tropical Meteorology, The Comet Program, Boulder, USA, 2010.

Leutbecher, M. and Palmer, T. N.: Ensemble forecasting, J. Comput. Phys., 227, 3515–3539, 2008.

Liu, J., Wang, J., Pan, S., Tang, K., Li, C., and Han, D.: A real-time flood forecasting system with dual updating of the NWP rainfall and the river flow, Nat. Hazards, 77, 1161–1182, https://doi.org/10.1007/s11069-015-1643-8, 2015.

López López, P., Verkade, J. S., Weerts, A. H., and Solomatine, D. P.: Alternative configurations of quantile regression for estimating predictive uncertainty in water level forecasts for the upper Severn River: a comparison, Hydrol. Earth Syst. Sci., 18, 3411–3428, https://doi.org/10.5194/hess-18-3411-2014, 2014.

Mourre, L., Condom, T., Junquas, C., Lebel, T., Sicart, J., Figueroa, R., and Cochachin, A.: Spatio-temporal assessment of WRF, TRMM and in situ precipitation data in a tropical mountain environment (Cordillera Blanca, Peru), Hydrol. Earth Syst. Sci., 20, 125–141, https://doi.org/10.5194/hess-20-125-2016, 2016.

Muggeo, V. M. R., Sciandra, M., Tomasello, A., and Calvo, S.: Estimating growth charts via nonparametric quantile regression: a practical framework with application in ecology, Environ. Ecol. Stat., 20, 519–531, 2013.

Ochoa, A., Pineda, L., Crespo, P., and Willems, P.: Evaluation of TRMM 3B42 precipitation estimates and WRF retrospective precipitation simulation over the Pacific–Andean region of Ecuador and Peru, Hydrol. Earth Syst. Sci., 18, 3179–3193, https://doi.org/10.5194/hess-18-3179-2014, 2014.

Rama Rao, Y. V., Alves, L., Seulall, B., Mitchell, Z., Samaroo, K., and Cummings, G.: Evaluation of the weather research and forecasting (WRF) model over Guyana, Nat. Hazards, 61, 1243–1261, 2012.

R Development Core Team: A language and environment for statistical computing, Tech. rep., R Foundation for Statistical Computing, Vienna, Austria, available at: http://www.R-project.org (last access: 10 April 2015), 2010.

Remesan, R., Bellerby, T., Holman, I., and Frostick, L.: WRF model sensitivity to choice of parameterization: a study of the "York Flood 1999", Theor. Appl. Climatol., 122, 229–247, https://doi.org/10.1007/s00704-014-1282-0, 2014.

Rene, J.-R., Madsen, H., and Mark, O.: Probabilistic forecasting for urban water management: a case study, in: 9th International Conference on Urban Drainage Modelling, 4–6 September 2012, Belgrade, Serbia, 1–11, 2012.

Reyes, O.: Utilización de modelos hidrológicos para la determinación de cuencas en ecosistemas de páramo, Revista Ambiental Agua, Aire y Suelo, 56–65, available at: http://revistas.unipamplona.edu.co/ojs_viceinves/index.php/RA/article/view/432 (last access: 22 August 2015), 2014.

Roberts, N. M., Cole, S. J., Forbes, R. M., Moore, R. J., and Boswellc, D.: Use of high-resolution NWP rainfall and river flow forecasts for advance warning of the Carlisle flood, north-west England, Meteorol. Appl., 16, 23–34, 2009.

Robertson, D. E., Shrestha, D. L., and Wang, Q. J.: Post-processing rainfall forecasts from numerical weather prediction models for short-term streamflow forecasting, Hydrol. Earth Syst. Sci., 17, 3587–3603, https://doi.org/10.5194/hess-17-3587-2013, 2013.

Rogelis, M. C.: Sistema de alerta temprana del río Tunjuelo 2006, Tech. rep., Fondo de Prevención y Atención de Emergencias, Bogotá, 2006.

Rogelis, M. C.: Operational flood forecasting, warning and response for multi-scale flood risks in developing cities, PhD thesis, Delft University of Technology and of the Academic Board of the UNESCO-IHE Institute for Water Education, Leiden, the Netherlands, 2016.

Rossa, A., Liechti, K., Zappa, M., Bruen, M., Germann, U., Haase, G., Keil, C., and Krahe, P.: The COST 731 Action: a review on uncertainty propagation in advanced hydrometeorological forecast systems, Atmos. Res., 100, 150–167, https://doi.org/10.1016/j.atmosres.2010.11.016, 2011.

Ruiz, J. F.: ¿Como Interpretar Los Modelos De Pronóstico Del Estado Del Tiempo?, Tech. rep., Instituto de Hidrología, Meteorología y Estudios Ambientales, Bogotá, available at: http://bart.ideam.gov.co/wrfideam/GUIA_MODELOS.pdf (last access: 15 September 2014), 2010.

Scardovi, E.: Rainfall spatial predictions: a two-part model and its assessment, PhD thesis, University of Bologna, Bologna, available at: http://amsdottorato.unibo.it/6744/1/scardovi_elena_tesi.pdf (last access: 10 January 2016), 2015.

Sevink, J.: Páramo Andino Project Hydrology workshop in Merida, Venezuela, 2011.

Teng, J., Potter, N. J., Chiew, F. H. S., Zhang, L., Wang, B., Vaze, J., and Evans, J. P.: How does bias correction of regional climate model precipitation affect modelled runoff?, Hydrol. Earth Syst. Sci., 19, 711–728, https://doi.org/10.5194/hess-19-711-2015, 2015.

Theis, S. E., Hense, A., and Damrath, U.: Probabilistic precipitation forecasts from a deterministic model: a pragmatic approach, Meteorol. Appl., 12, 257–268, https://doi.org/10.1017/S1350482705001763, 2005.

Verkade, J. S. and Werner, M. G. F.: Estimating the benefits of single value and probability forecasting for flood warning, Hydrol. Earth Syst. Sci., 15, 3751–3765, https://doi.org/10.5194/hess-15-3751-2011, 2011.

Ruiz, J. F.: ¿Como Interpretar Los Modelos De Pronóstico Del Estado Del Tiempo?, Tech. rep., Instituto de Hidrología, Meteorología y Estudios Ambientales, Bogotá, available at: http://bart.ideam.gov.co/wrfideam/GUIA_MODELOS.pdf (last access: 15 September 2014), 2010.

Scardovi, E.: Rainfall spatial predictions: a two-part model and its assessment, PhD thesis, University of Bologna, Bologna, available at: http://amsdottorato.unibo.it/6744/1/scardovi_elena_tesi.pdf (last access: 10 January 2016), 2015.

Sevink, J.: Páramo Andino Project Hydrology workshop in Merida, Venezuela, 2011.

Teng, J., Potter, N. J., Chiew, F. H. S., Zhang, L., Wang, B., Vaze, J., and Evans, J. P.: How does bias correction of regional climate model precipitation affect modelled runoff?, Hydrol. Earth Syst. Sci., 19, 711–728, https://doi.org/10.5194/hess-19-711-2015, 2015.

Theis, S. E., Hense, A., and Damrath, U.: Probabilistic precipitation forecasts from a deterministic model: a pragmatic approach, Meteorol. Appl., 12, 257–268, https://doi.org/10.1017/S1350482705001763, 2005.

Verkade, J. S. and Werner, M. G. F.: Estimating the benefits of single value and probability forecasting for flood warning, Hydrol. Earth Syst. Sci., 15, 3751–3765, https://doi.org/10.5194/hess-15-3751-2011, 2011.

Verkade, J. S., Brown, J. D., Reggiani, P., and Weerts, A. H.: Post-processing ECMWF precipitation and temperature ensemble reforecasts for operational hydrologic forecasting at various spatial scales, J. Hydrol., 501, 73–91, https://doi.org/10.1016/j.jhydrol.2013.07.039, 2013.

Vincendon, B., Ducrocq, V., Nuissier, O., and Vié, B.: Perturbation of convection-permitting NWP forecasts for flash-flood ensemble forecasting, Nat. Hazards Earth Syst. Sci., 11, 1529–1544, https://doi.org/10.5194/nhess-11-1529-2011, 2011.

Wilks, D. S.: Forecast Verification, in: vol. 100, 2nd Edn., Elsevier, San Diego, California, USA, 2011.

Xuan, Y., Cluckie, I. D., and Wang, Y.: Uncertainty analysis of hydrological ensemble forecasts in a distributed model utilising short-range rainfall prediction, Hydrol. Earth Syst. Sci., 13, 293–303, https://doi.org/10.5194/hess-13-293-2009, 2009.

Yang, W., Andréasson, J., Phil Graham, L., Olsson, J., Rosberg, J., and Wetterhall, F.: Distribution-based scaling to improve usability of regional climate model projections for hydrological climate change impacts studies, Hydrol. Res., 41, 211–229, 2010.

Yuan, X. and Wood, E. F.: Downscaling precipitation or bias-correcting streamflow? Some implications for coupled general circulation model (CGCM)-based ensemble seasonal hydrologic forecast, Water Resour. Res., 48, 1–7, 2012.

Yucel, I., Onen, a., Yilmaz, K. K., and Gochis, D. J.: Calibration and evaluation of a flood forecasting system: Utility of numerical weather prediction model, data assimilation and satellite-based rainfall, J. Hydrol., 523, 49–66, https://doi.org/10.1016/j.jhydrol.2015.01.042, 2015.

5

Space–time analysis of rainfall extremes in Italy: clues from a reconciled dataset

Andrea Libertino, Daniele Ganora, and Pierluigi Claps

Department of Environment, Land and Infrastructure Engineering (DIATI), Politecnico di Torino, Torino, Italy

Correspondence: Andrea Libertino (andrea.libertino@polito.it)

Abstract. Like other Mediterranean areas, Italy is prone to the development of events with significant rainfall intensity, lasting for several hours. The main triggering mechanisms of these events are quite well known, but the aim of developing rainstorm hazard maps compatible with their actual probability of occurrence is still far from being reached. A systematic frequency analysis of these occasional highly intense events would require a complete countrywide dataset of sub-daily rainfall records, but this kind of information was still lacking for the Italian territory. In this work several sources of data are gathered, for assembling the first comprehensive and updated dataset of extreme rainfall of short duration in Italy. The resulting dataset, referred to as the Italian Rainfall Extreme Dataset (I-RED), includes the annual maximum rainfalls recorded in 1 to 24 consecutive hours from more than 4500 stations across the country, spanning the period between 1916 and 2014. A detailed description of the spatial and temporal coverage of the I-RED is presented, together with an exploratory statistical analysis aimed at providing preliminary information on the climatology of extreme rainfall at the national scale. Due to some legal restrictions, the database can be provided only under certain conditions. Taking into account the potentialities emerging from the analysis, a description of the ongoing and planned future work activities on the database is provided.

1 Introduction

Italy can boast of a role at the highest level in the development of meteorological observations (Brunetti et al., 2006), with 6 meteorological stations operating since the 18th century (Bologna, Milano, Roma, Padova, Palermo and Torino), and 15 stations with observation starting in the first half of the 19th century. The first attempts of performing a systematic collection of monthly rainfall data go back to as early as 1880 when the National Office for Meteorology and Climate was founded. The National Hydrographic Service (SIN) and the National Hydrographic and Mareographic Service (SIMN) collected annual maxima values for 1, 3, 6, 12 and 24 h durations in the Hydrological Yearbooks from 1917 to early 2000s (the final publication year depends on the local agencies of the SIMN). The D.Lgs 112/1998 dismantled the SIMN, transferring its tasks to the 19 administrative regions and the 2 autonomous provinces of Trento and Bolzano. These authorities were designated as local Operational Centres and Regional Environmental Agencies to deal with hydro-meteorological monitoring and civil protection issues.

In spite of the huge heritage of data, only a small fraction of the Italian rainfall data is available in a computer-readable format. Moreover, the dismantlement of the National Service led to a lack of updates for the national database of extreme rainfall that is still stuck, for some regions, at the beginning of the 1990s. This has led to a very fragmented framework: updated rainstorm hazard assessments are actually only available for some regions and only at the regional scale (see, e.g. Uboldi et al., 2014; Libertino et al., 2017). Various regional studies present different methodologies and are sometimes based on very different data densities and record lengths (e.g. Claps et al., 2016), but few updated analyses on short-duration rainfall in a regional framework are available (e.g. Rudari et al., 2005; Allamano et al., 2009).

In view of the assembling of the first comprehensive dataset of extreme rainfall of short duration in Italy several major sources of data have been analysed. The resulting dataset, referred to as the Italian Rainfall Extremes Database (I-RED), includes data from more than 4500 stations across the country, spanning the period between 1916 and 2014, and refers to annual maximum rainfall recorded in 1 to 24 consecutive hours (exact durations available are 1, 3, 6, 12 and 24 h).

The following sections describes the sources of the data, the work carried out for the merging of the database and the operations that are still required for making it suitable for nationwide robust rainfall frequency analyses. A preliminary analysis of the extreme rainfall regime at the national scale is also presented.

2 Merging the I-RED Dataset

2.1 Data sources

As a follow-up of the activities of the Italian National Group for the Prevention of the Hydrogeological Disasters (GNDCI) a comprehensive nationwide hydrological information system has been set up, within the "CUBIST project", funded by the Italian Ministry of Education and Research within the funding PRIN 2005 (Italian Research Projects of National Relevance). The database includes about 6000 pluviographs and pluviometers, 700 temperature stations and about 400 river basins (Claps et al., 2008) and is available at http://www.cubist.polito.it; (last access: 26 October 2017). In detail, the database includes rainfall data from 1900 to 2001 (depending on the region) and constitutes the first important attempt to make the large Italian hydrological heritage freely available in a computer-readable format. The annual maxima data for different durations included in the CUBIST dataset are extracted with a sliding-windows process from manual tipping-bucket rain gauge data, equipped with a recording system that writes on diagram paper. Further information on the characteristics of the stations can be found in Acquaotta et al. (2016). The number of data per year is not constant across the analysed period, as it has increased in time as more stations have been installed in recent years. Data availability decreases in the period of the Second World War, as many records were missed in that period. After 1980, with the progressive dismissal of the SIMN and the development of the local hydrographic authorities, data availability decreases rapidly until 2001, when the rain gauges still under the SIMN were taken over by the local operational centres.

After the late 1980s, indeed, the local environmental agencies started to support the SIMN in its work. Gradually, the 21 regional hydrological services took over the networks and the tasks of the national one. In this period most of the old manual tipping-bucket rain gauges have been substituted with automatic stations, similar to the one described in Acquaotta et al. (2016) for the Piedmont region. Each hydrological service adopted its own rules for the publication of the collected data and, even if the Italian law adopted an open-

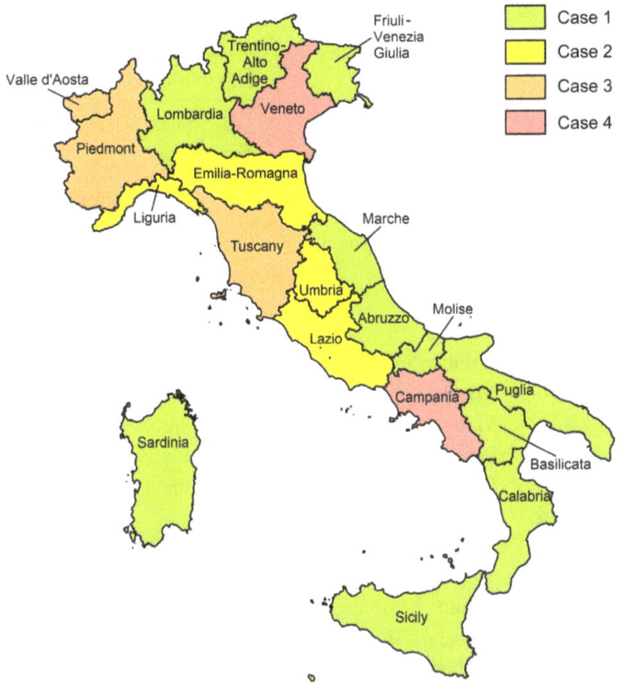

Figure 1. Names of the Italian regions and type of datasets provided by the regional authorities. The cases refer to the bullet list of Sect. 2.2.

source policy for the public data for non-commercial uses (under D.Lgs.82/2005, D.Lgs.36/2006, D.M.10/11/2011, L.221/2012, D.Lgs.179/2012 and L.114/2014), an updated database of the annual rainfall maxima for sub-daily duration at the national scale is still lacking. For the scope of this research, the different agencies have been contacted, and the regional annual maxima datasets for sub-daily durations were requested. The regions of Italy are shown in Fig. 1 together with the type of data provided, which will be described in the following section. Table 1 lists the names of the local authorities and the regional codes, aimed at identifying them in the database. The public availability of the original dataset is also reported.

2.2 Cleaning and merging operations

Merging and harmonizing the different datasets is a quite long and difficult operation, that is still ongoing. The different operational centres provided different types of datasets, with different temporal coverages and spatial reference systems. Duplicate stations are often present in the databases of neighbouring regions.

The first steps of this work have been carried out at the regional scale. For each region all the data falling inside the regional boundaries have been considered. These data, according to the setting of the databases of the local operational centres, could belong to one of these three categories:

Table 1. Regions of Italy with the assigned code and the related local Operational Center with references to the availability of digitized data.

CD	Region	Operational centre	Digitized data availability
1	Abruzzo	Ufficio Idrografico e Mareografico Regione Abruzzo	available upon request
2	Basilicata	Dipartimento Protezione Civile Regione Basilicata	available in[1]
3	Calabria	Centro Funzionale Multirischi - ARPACAL	available at[2]
4	Campania	Centro Funzionale Regione Campania	available upon request
5	Emilia-Romagna	ARPA Emilia-Romagna	available upon request
6	Friuli-Venezia Giulia	Ufficio Idrografico Regione Autonoma Friuli-Venezia Giulia	available upon request
7	Lazio	Centro Funzionale Regione Lazio	available upon request
8	Liguria	ARPAL-CFMI-PC	partially available at[3]
9	Lombardia	ARPA Lombardia	available at[4]
10	Marche	Dipartimento di Protezione Civile Regione Marche	available at[5]
11	Molise	Centro Funzionale Regione Molise	available upon request
12	Piedmont	ARPA Piemonte	partially available at[6]
13	Puglia	Dipartimento di Protezione Civile Regione Puglia	available at[7]
14	Sardinia	ARPAS	available upon request
15	Sicily	Osservatorio delle Acque Regione Siciliana	available upon request
16	Toscana	Servizio Irdrografico Regionale Toscana	available at[8]
17	Trento[10]	Centro Funzionale Provincia Autonoma di Trento	available at[9]
18	Bolzano–Alto Adige[10]	Ufficio Idrografico Provincia Autonoma di Bolzano–Alto Adige	available upon request
19	Umbria	Regione Umbria	available upon request
20	Valle d'Aosta	Centro Funzionale Regione Autonoma Valle d'Aosta	available upon request
21	Veneto	ARPAV	available upon request

[1] Manfreda, S., Sole, A. and De Costanzo, G.: Le precipitazioni estreme in Basilicata, Editrice Universo Sud, 2015.
[2] ARPACAL: Centro Funzionale Multirischi, http://www.cfd.calabria.it/, (last access: 1 August 2016).
[3] ARPAL: Consultazione Dati Meteoclimatici, http://www.cartografiarl.regione.liguria.it/SiraQualMeteo/Fruizione.asp, (last access: 1 August 2016).
[4] ARPA Lombardia: Progetto Strada, http://idro.arpalombardia.it/pmapper-4.0/map.phtml, (last access: 1 August 2016).
[5] Protezione Civile Regione Marche: Annali Idrologici Regione Marche, http://console.protezionecivile.marche.it, (last access: 1 August 2016).
[6] ARPA Piemonte: Banca dati meteorologica, http://www.regione.piemonte.it/ambiente/aria/rilev/ariaday/annali/meteorologici, (last access: 1 August 2016).
[7] Protezione Civile Puglia: Annali Idrologici – Parte I, http://www.protezionecivile.puglia.it/centro-funzionale/analisielaborazione-dati, (last access: 1 August 2016).
[8] SIR Toscana: Settore Idrologico Regionale, http://www.sir.toscana.it/, (last access: 1 August 2016).
[9] Centro Funzionale di Protezione Civile Provincia Autonoma di Trento: Meteotrentino, http://www.meteotrentino.it/, (last access: 1 August 2016).
[10] the autonomous provinces of Trento and Bolzano–Alto Adige, together, constitute the region Trentino-Alto Adige (CD: 22).

- data from the CUBIST database for the 1900–2001 period already available from the former national service
- data provided by the regional authority
- data provided by the regional authorities of the neighbouring regions that extend beyond their regional borders.

Observations dating from before 1916 have been discarded, as they are considered not significant and too unevenly distributed. Considering that most of the provided data have been validated from the related authorities, they are considered reliable and, at first, included directly in the I-RED. For information on the validation procedures, please refer to the Appendix A and to Barbero S. et al. (2017). In the presence of inconsistencies between the type (b) and type (c) data, preliminary manual merging was carried out. The sources of the inconsistencies could be various, according to the evolution of the monitoring systems of the different regions, and these inconsistencies are often due to the joint management of interregional basins. The different regional authorities often have adopted different codes and/or names for the same station, the first step has been thus to identify the presence of duplicate stations with the same or a similar name covering different time intervals. Sometimes, even for the same station, neighbouring regions can provide different data for the same years. This can be, for example, due to the fact that sometimes regions share rainfall data before their validation and official publication. If the same station was found in the database of more neighbouring regions, the first attempt of merging the series together was carried out by analysing the data recorded year by year. If the merging was not feasible, higher priority was given to the data provided by the authority of the considered region (that is usually also the owner of the network). This allowed the presence of duplicate series in the I-RED to be avoided.

Once the type (b) and type (c) datasets were merged for each region,, the resulting dataset has to be merged with the type (a) dataset. This operation has been quite complex, as the overlapping period between the different dataset was different for each region and because most of the authorities did not track the change in the name and code of the stations. The different procedures performed, according to the type of the dataset that the region has provided (as reported in Fig. 1) can be summarized as follows:

1. *Regions that digitized the whole SIMN database for their areal domain and provide a complete merged database.* The provided data were inserted in the I-RED without editing and without considering the CUBIST series. Only for the Abruzzo and Molise regions some preliminary refinement was needed, as the two regions were divided in 1963, and the databases of the two regions partially overlap. The stations were then divided according to the actual regional boundaries and the duplicate series removed.

2. *Regions that provided datasets including data from their actual regional network partially merged with subsets of digitized data from the SIMN Hydrological Yearbooks.* As not all the SIMN datasets were digitized from the local authorities, the dataset lacked part of the stations included in the CUBIST database. To maximize the available information, data from the regional databases and the CUBIST one were manually analysed and merged, in order to avoid duplicate values. Stations with the same name and similar coordinates were merged together in the presence of a 2-year consistent overlapping period. In the presence of inconsistencies between the values recorded by the two stations in the overlapping period, they were treated as different stations and renamed. If it was not possible to remove any doubt, the stations were considered as separate entities. For the Liguria region, the information in ARPAL (2013) was used to overcome the lack of information on the continuity of the series.

3. *Regions that provided two different datasets: one containing the whole digitized data from the SIMN stations and another containing the digitized data from their actual networks.* The data of the two databases were merged together, the overlapping period manually analysed to avoid overlapping, and the CUBIST database ignored. The operation was made possible by the collaboration of ARPA Piemonte, for Piedmont and Valle d'Aosta, and of the Università degli Studi di Firenze, for Tuscany.

4. *Regions that provided only the data recorded from the network they actually manage.* All the information concerning the SIMN stations was lacking. The provided dataset was therefore merged with the whole CUBIST database for the considered regions. Duplicate values were excluded when manually analysing the overlapping period, if present.

With the application of the above described rules, 20 complete regional datasets have been obtained. The regional datasets were finally merged together to generate the I-RED. After the merging phase some reliability check has been performed, in order to detect any problematic or incorrect information. These include the identification and removal of the duplicate data or stations, and reliability checks on the larger values of the dataset, comparing them to the absolute record-breaking events for all the durations (see Libertino, 2017), aimed at detecting inconsistencies in rainfall series. If any suspect value was found, its year of occurrence was compared, when referring to recent years, with the data from event reports or newspapers. If the data refers to a SIMN station, the Hydrological Yearbooks were consulted. If no evidence was found, the related authority was contacted. Most of the operations need human supervision, and thorough verification work. If it is not possible to remove any doubt the suspect value is discarded.

Due to the complexity of the check operations, further efforts and collaborations with the regional authorities are still ongoing to increase the consistency of the database. Nevertheless, to date (October 2017) the I-RED includes more than 4500 stations nationwide and constitutes the largest updated dataset of annual maxima for Italy.

Considering that most of the regional authorities supervise the use and widespread dissemination of their datasets in order to prevent improper use, a detailed description on how to access the I-RED is reported in the Data Availability section.

In the following, the spatio-temporal distribution of the assembled data will be described.

3 Main features of the I-RED database

The number of data available per year in the I-RED is reported in Fig. 2a, compared with that of the CUBIST database. As every station is related to a unique value of annual maxima for a given duration, the presence of a measurement implies the presence of a station. The number of available stations increases with time, and drastically grows after the dismissal of the SIMN and the development of the local agencies. The decrease after 2010 can be attributed to the fact that not all the regions have published the data for the most recent years.

The smaller size of the I-RED compared the CUBIST database in some years can be due to the following:

- The presence, before 1945, in the CUBIST database of data from territories lost by Italy after World War II (e.g. Istria) or from neighbouring countries, not included in the I-RED;

- The fact that some regional agencies could have decided for different reasons not to include data or stations from the SIMN in their datasets. Considering that these data are only contained in the CUBIST dataset, for the regions where the procedures (1) and (3) described in Section 2.2 are applied, they are not included in the I-RED.

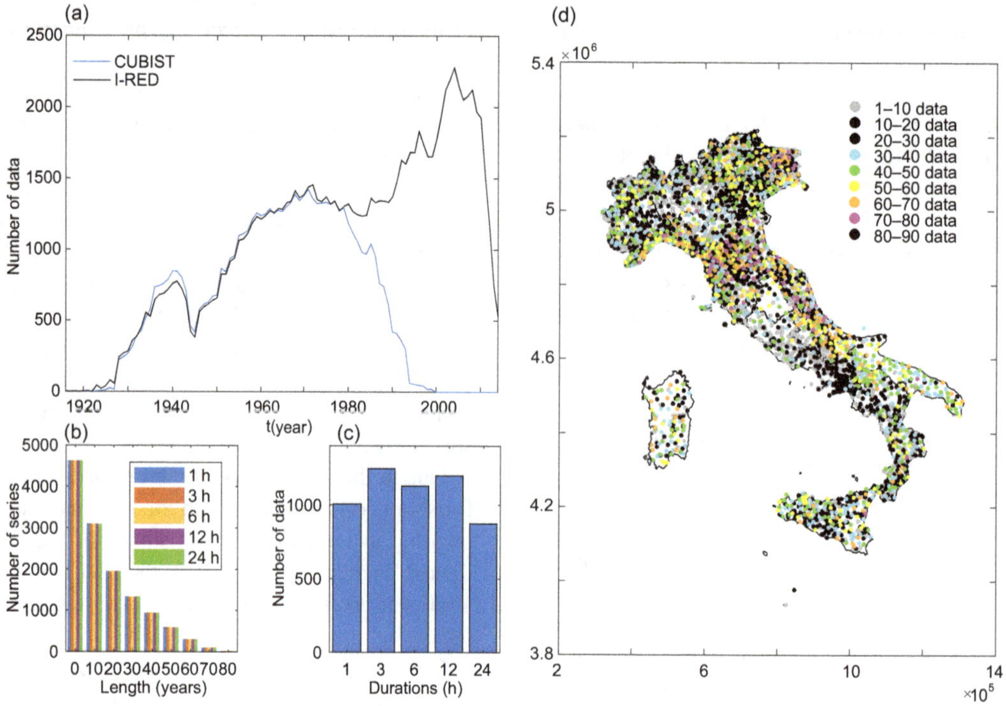

Figure 2. **(a)** Data availability per year in the I-RED and CUBIST databases (the smallest value across the five considered durations is reported per year). **(b)** Number of series longer than fixed threshold values in the I-RED databases per duration (null values are ignored). **(c)** Number of null values per duration. **(d)** Length of the series in the I-RED database represented in space: the colour refers to the minimum length among the five available durations. If more stations overlap due to the resolution of the picture, the one with the longer series appears on the top.

Considering the limited significance of the information loss, further efforts for including these data will be planned only in a future stage of the development of the database.

For a descriptive analysis of the rainfall data, all the assembled time series are classified according to their length. Results are shown in Fig. 2b. Considering the short life of the rain gauges installed by the regional operational centres, a large percentage of the series is shorter than 20 years but the contribution of the CUBIST database allows for a significant amount of longer series. The series with more than 80 years of data for the 1, 3, 6, 12, 24 h durations are 16, 14, 15, 14 and 17 respectively. In general, all the durations report a similar behaviour, despite some differences in the distribution of the null values as shown in Fig. 2c. The reasons that lead to missing data only for certain durations can be various and related to either the measuring, the recording or the storage of the data (e.g. missed reading of the record from the operator, data classified as not-valid in the validation phase).

The spatial distribution of the stations is shown in Fig. 2d. The colour scale refers to the number of the available data per series. The minimum number across the five durations is considered. One can clearly distinguish that, even if the whole national territory is represented, the density of the stations widely changes across the nation. To show the relevance of the non-uniformity, a gridded domain with a mesh size of 50 km is introduced. Figure 3 shows the total number of data per cell, i.e., in the sum across the whole period of the annual number of stations with available data falling in the cell. If data consistency changes for the different durations, the shortest one is considered. The non-uniformity of the network density clearly emerges, with some cells presenting almost 10 times the number of data of other cells. The most densely gauged cells can be found in the north-west of the country, in particular in the Liguria region, in northern Tuscany and in the north-east.

4 Descriptive statistical analysis of rainstorms in Italy

A preliminary descriptive analysis of the characteristics of extreme rainfalls at the national scale has been carried out on the newly developed I-RED database. Series with a minimum length of 20 years of data have been considered in this analysis. This length constraint leads to a subset of 1974 series available for the analysis, out of the original 4686. For each duration, the median of the series is depicted in Fig. 4. The median is used as a robust estimator of the central tendency of a series, less sensitive than the mean to the presence of outliers. As common methods of fitting distributions

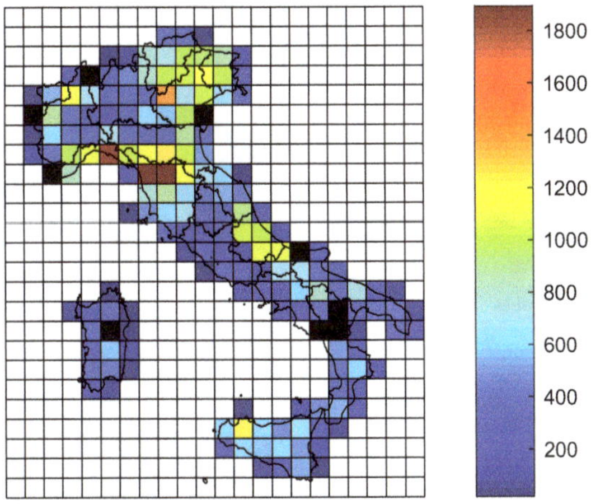

Figure 3. Total number of data per cell over a 50 km grid.

(e.g. product moments or L-moments) use mean values for representing the central tendency, maps of the mean for the different durations are attached in the Supplement.

Some geographical areas are characterized by clusters of large median values and these clusters appear consistent across the different durations. Furthermore, at the country-wide scale we observe that the coefficient of variation of the medians increases for increasing durations, suggesting a wider range of variability of the corresponding median values.

For each series, the sample L-moments (Hosking and Wallis, 1997) have then been computed to describe the shape of the empirical distribution of the records. The mean L-moment ratios among the different durations give information respectively on the dispersion (L-CV), skewness (L-skewness) and "peakedness" (L-kurtosis) of the empirical distributions. All the above statistics are mapped in Fig. 4. Considering that the L-moment ratios show similar behaviour for the considered durations, we decided for simplicity to include only the average ones in the paper. The maps for the different durations are reported in the Supplement. Figure 4f shows that the coastal areas and the islands are generally characterized by a higher variability in the annual maxima series, presenting larger L-CV values. The northern part of the peninsula, even if characterized by large median values, shows lower L-CV, which is typical of areas with large average rainfall values. It is harder to identify a precise spatial pattern in the distribution of the skewness and kurtosis values (Fig. 4g and h). Coastal and island areas seem to generally show larger skewness values, confirming the influence of the Mediterranean Sea on the climate of these areas. All the aforementioned maps have been also interpolated for visualization purposes with ordinary kriging; detailed results are reported in the Supplement.

The significance of the developed dataset also allows preliminary exploration of the rainfall events sometimes referred to as "black swans" (Blöschl et al., 2015), showing extraordinary intensities even when compared with the population of annual maxima. In Italy, many of these events have been studied as individual extraordinary events (e.g. Rebora et al., 2013; Fiori et al., 2014), due to the large recorded intensities and to their severe consequences, but the fragmented configurations of the national database have prevented a systematic treatment of this population of "extremes of the extremes" (Snorrason et al., 2002). A preliminary investigation on the occurrence of very extreme events at the national scale has been performed and summarized in Fig. 5a, which shows the spatial distribution of the record-breaking rainfall events for the considered durations from 1935 to 2015. A record-breaking event is defined as the annual value that exceeds all the previous ones. At this stage, only nationwide records are considered, pulling up all the data together year by year. Record-breaking rainfall amounts can provide a picture of the spatio-temporal distribution of the major weather anomalies in the country. Analysing record-breaking events has some advantages from both an operational and a statistical point of view. Due to the significant amounts recorded, these events can be easily verified by combining different sources of information, and, moreover, this kind of analysis does not require any assumption on the underlying probability distribution (Coumou et al., 2013). The spatial distribution of the events seems to suggest a clustering of these phenomena in some areas of the country: the eastern part of Liguria and northern part of Tuscany and the extreme south of Calabria. Localized events also occurred in Campania, Sicily and Sardinia. All of these areas are generally characterized by complex orography in the proximity of the coastline: a framework that can promote the development of particularly intense phenomena (Furcolo et al., 2015). On-site systematic analysis of the record-breaking events, which are expected to provide useful information for characterizing the extreme rainfall regime in the country (Lehmann et al., 2015), is now possible thanks to the consistency of the new I-RED database. Figure 5b shows the record-breaking evolution over time (for each duration); the occurrence of a new records is represented by an increasing step in the line. Figure 5c reports the same records, whose values are normalized by the 1935 values.

5 Conclusions

The first comprehensive dataset of extreme rainfall in Italy, called I-RED, has been presented here. It is a significant source of information, able to provide unprecedented knowledge on the characteristics of heavy precipitation in Italy and on the possible rainfall regime changes in the last century. Further efforts will be addressed to increase the spatial data homogeneity and coverage in time, by including the data of the most recent years and, eventually, by contacting the local authorities for requesting assistance in the merging of the se-

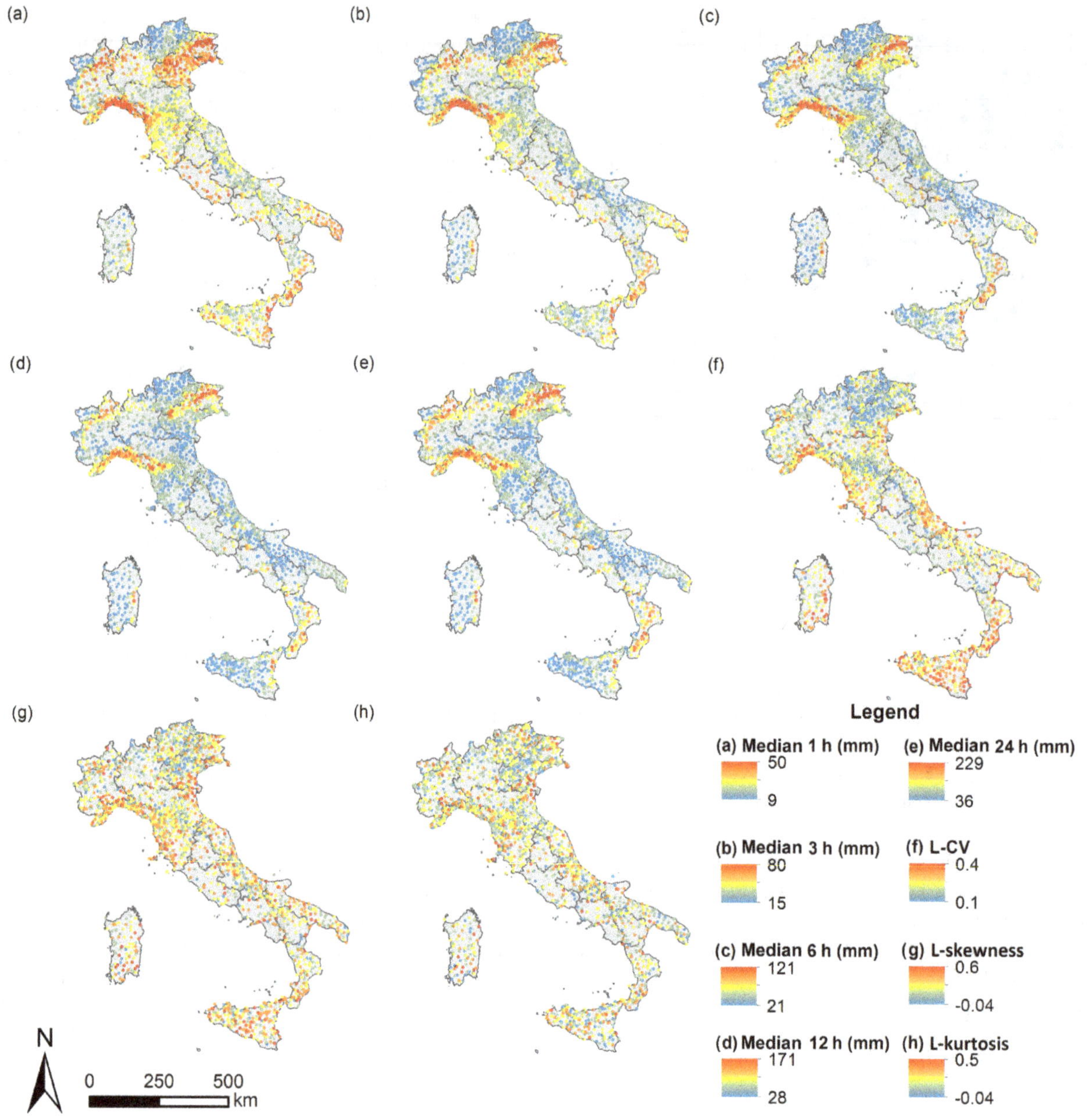

Figure 4. Median values of the I-RED series from 1 **(a)** to 24 h **(e)**. Average statistics for the five durations considered: **(f)** L-CV, **(g)** L-skewness and **(h)** L-kurtosis. Series with more than 20 data are considered.

ries. The final aim is to make the update of the database systematic and unsupervised. This can be done by strengthening the collaboration with the data providers, in the framework of joined projects, as did the one that led to the development of the ArCIS (Pavan et al., 2013) dataset, collecting updated rainfall and temperature data from a group of regional authorities in northern Italy. Collaborations with other projects, focused on different spatial or temporal scales, will be also explored in order to automatically and efficiently analyse the consistency of the I-RED dataset and to integrate it with the existing ones. A possible target is the SCIA dataset (Desiato et al., 2007) referring to the 24 h and daily scale. Joined projects with international institutions will be evaluated and endorsed in order to make available the I-RED database in larger frameworks for trans-boundary exchange of precipitation data. In the meantime the I-RED will be used for exploring the different outcomes provided by this preliminary analysis, e.g. assessing the influence of the spatial distribution of the stations on the observation of record-breaking extreme events, evaluating the presence of trends in the temporal distribution of the "black swans" and analysing the statistical predictability of these kind of events on such a wide and complex domain.

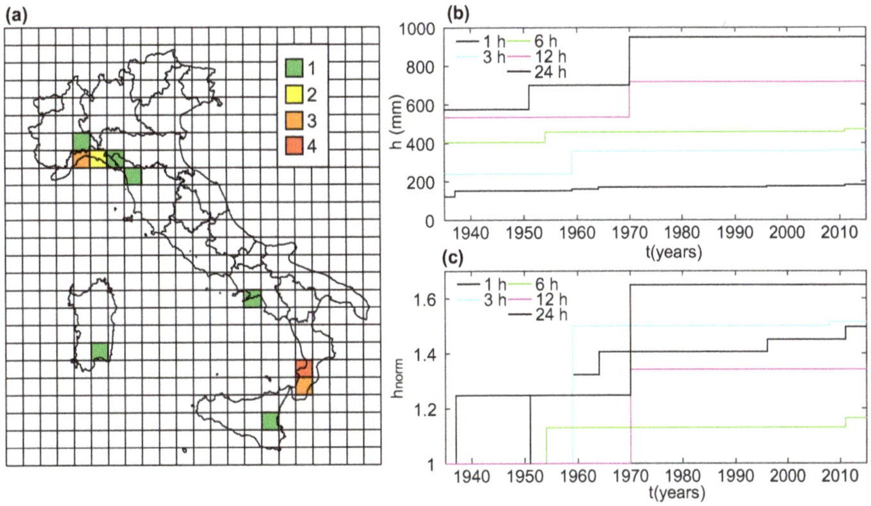

Figure 5. **(a)** Number of record-breaking events per cell over a 50 km grid. Record-breaking rainfall depths for the five considered durations from 1935 to 2015 **(b)** in absolute values and **(c)** normalized against the 1935 values.

Data availability. The original data can be requested to the authorities reported in Table 1. Some of the agreements signed with the data providers, aimed at monitoring the correct use of the data, restrict their use to the aims of the authors' project. Due to these legal restrictions, the full or partial access to the I-RED can be provided to the following:

- research individuals or groups in the framework of the authors' project;
- research individuals or groups not collaborating with the authors' project, upon evidence of permission received by the involved regional agencies, reported in Table 1.

For further details and queries, please contact the corresponding author.

Appendix A: Guidelines for the quality check of hydro-meteorological data

Extracted and translated in English from Barbero S. et al. (2017).

A1 Quality control

[...] The attribution of a certain level of quality to the measured data passes through the process of validation of the data themselves, which consists of analysing all the data collected in terms of completeness, reasonableness and eliminating erroneous values. Data validation (validity check) is only one of the quality control (QC) operational procedures consisting of a set of procedures and rules to ensure that a measurement system achieves and maintains the specific quality level initially established. The periodic calibration of the instruments, the periodic inspection of the sites and the preventive maintenance also belong to the QC process. The QC can be applied both in real time (real-time quality control) and in delayed time, according to the needs of sharing, using and storing data nationally and internationally (e.g. World Meteorological Organization, 1993, 2010a, b). Moreover, the QC must be integrated into effective and well-coordinated quality management (QM). The QM is, in fact, is expressed through the joint application of quality assurance (QA) and of the QC (World Meteorological Organization, 2014). The QA is the set of planned and systematics activities applied within a quality management system to provide the level of confidence with which the quality requirements are met. Basically, the QC is a system of activities to provide a quality product while the QA is a system of activity designed to verify that the quality control system is functioning properly. The main objectives of the QM are the identification, quantification and reduction of errors. Errors can be made for both technical reasons (i.e. due to the methods and technologies used) and procedural reasons (i.e. linked to unclear or ineffective management or to the lack of adequate preparation of the operators). Furthermore, errors can be made during detection (e.g. the sensor does not read correctly), when the observation is transcribed (transposition of digits, shifting of dates, etc.) and during data transmission and storage (computer errors, errors of digitization, etc.). Many of these errors can be prevented by an appropriate QA, others must be identified and corrected through the QC procedures.

A2 Levels of the validity check

The first level of data validation is performed on the raw data (or gross data), i.e. the data at the original temporal resolution with which they are transmitted or detected at the measuring station and consists of the application of basic procedures for verifying the validity of the data. These checks aim at indicating malfunctions, instability or interference. In the case of data coming from automatic measuring stations the validity checks are applied to the "meteorological message" coming from the station in the transcoding phase of the message that for the transmission must comply with certain rules. The checks carried out will therefore be related to the expected formats within a given message, to the date and time stamps, to the location of measuring station, to the codes of stations and sensors and to the presence of duplicate elements. This category of checks includes syntax controls (e.g. alphabetic characters appearing in a text that should be numeric) which, if incorrect, can mine the transcoding process; logical controls that refer to both the intrinsic characteristics of the magnitude (e.g. World Meteorological Organization, 1993) and to the limits imposed by technical characteristics of the instrument used, in terms of measuring range (e.g. for a rain gauge: 0–300 mm h^{-}1), resolution (e.g. for a rain gauge: 0.1 or 0.2 mm) and limits in the operating temperatures (e.g. 0–70 °C for an unheated or −30–70 °C for a heated rain gauge). The first level of controls can have three types of outcomes: control passed, suspected data or control not passed. In the first two cases the data is then subjected to subsequent checks; in the third the data is considered incorrect and discarded. The first-level controls are performed on the elementary data, i.e. the data to the temporal aggregation derived from the measurement station. At this stage it is appropriate also carry out internal consistency checks. These are checks that are based on the comparison between synchronous values of different variables somehow related (e.g. by physical laws), so as to highlight any inconsistencies between the data. The second level of data validation consists of a series of "consistency" checks:

- *Time consistency checks.* They are based on the verification of a maximum and minimum level of variability of data over time and they have the purpose of identifying any anomalies between temporally contiguous data or with respect to the values that have historically occurred at a given site. Concerning the allowed minimum variability, consistency verification procedures ascertaining the presence of persistence of measured values in the series, consisting in the lasting over time of the same or a similar value.

- *Cross-checks with other quantities recorded at the same station.* They are based on the control of the considered data with reference to other related quantities measured at the same site, e.g. temperature comparison with solar radiation.

- *Spatial consistency checks.* They are based on the hypothesis of gradual variability of the observed quantity in space and therefore on the existence of a sort of spatial correlation between the contemporaneous measures carried out in neighbouring stations. However, when dealing with rainfall, the hypothesis of gradual variability is less acceptable when smaller temporal aggregations are considered.

the validity checks are applied to the "meteorological message" coming from the station in the transcoding phase of the message that for the transmission must comply with certain rules. The checks carried out will therefore be related to

the expected formats within a given message, to the date and time stamps, to the location of measuring station, to the codes of stations and sensors and to the presence of duplicate elements. This category of checks includes syntax controls (e.g. alphabetic characters appearing in a text that should be numeric) which, if incorrect, can mine the transcoding process; logical controls that refer to both the intrinsic characteristics of the magnitude (e.g. World Meteorological Organization, 1993) and to the limits imposed by technical characteristics of the instrument used, in terms of measuring range (e.g. for a rain gauge: 0–300 mm h^{-}1), resolution (e.g. for a rain gauge: 0.1 or 0.2 mm) and limits in the operating temperatures (e.g. 0–70 °C for an unheated or −30–70 °C for a heated rain gauge). The first level of controls can have three types of outcomes: control passed, suspected data or control not passed. In the first two cases the data is then subjected to subsequent checks; in the third the data is considered incorrect and discarded. The first-level controls are performed on the elementary data, i.e. the data to the temporal aggregation derived from the measurement station. At this stage it is appropriate also carry out internal consistency checks. These are checks that are based on the comparison between synchronous values of different variables somehow related (e.g. by physical laws), so as to highlight any inconsistencies between the data. The second level of data validation consists of a series of "consistency" checks:

- *Time consistency checks.* They are based on the verification of a maximum and minimum level of variability of data over time and they have the purpose of identifying any anomalies between temporally contiguous data or with respect to the values that have historically occurred at a given site. Concerning the allowed minimum variability, consistency verification procedures ascertaining the presence of persistence of measured values in the series, consisting in the lasting over time of the same or a similar value.

- *Cross-checks with other quantities recorded at the same station.* They are based on the control of the considered data with reference to other related quantities measured at the same site, e.g. temperature comparison with solar radiation.

- *Spatial consistency checks.* They are based on the hypothesis of gradual variability of the observed quantity in space and therefore on the existence of a sort of spatial correlation between the contemporaneous measures carried out in neighbouring stations. However, when dealing with rainfall, the hypothesis of gradual variability is less acceptable when smaller temporal aggregations are considered.

- *Climatological checks.* They are based on comparing the quantity under examination with some parameters derived from the whole historical series (e.g. tests based on the comparison with percentiles calculated on specific time intervals). The data are validated, at first, using automatic procedures. However, for the evaluation of the so-called "suspicious" data, a manual revision by qualified personnel is required to decide for every case to validate the suspect data, reject it as not valid or fix it if possible.

Competing interests. The authors declare that they have no conflict of interest.

Acknowledgements. The authors thank Enrica Caporali and Valentina Chiarello for their assistance in preparing and screening the Tuscany regional dataset, Stefano Macchia for his contribution in collecting and cleaning the data and the insightful comments of Alberto Montanari, three anonymous reviewers and the handling editor that allowed the quality of the original manuscript to be significantly improved. Data providers reported in Table 1 are acknowledged.

Edited by: Matjaz Mikos

References

Acquaotta, F., Fratianni, S., and Venema, V.: Assessment of parallel precipitation measurements networks in Piedmont, Italy, Int. J. Climatol., 36, 3963–3974, 2016.

Allamano, P., Claps, P., Laio, F., and Thea, C.: A data-based assessment of the dependence of short-duration precipitation on elevation, Phys. Chem. Earth, 34, 635–641, 2009.

ARPAL: Atlante Climatico della Liguria, Final report, Regione Liguria, available at: https://www.arpal.gov.it/contenuti_statici/ /clima/atlante/Atlante_climatico_della_Liguria.pdf Genova, IT, 2013.

Barbero, S., Zaccagnino, M., Mariani, S., Lastoria, B., Braca, G., Bussettini, M., Casaioli, M., Marsico, L., Rotundo, R., Pavan, V., Ricciardi., G., Zenoni, E., Cicogna, A., Micheletti., S., Cazzuli., O., Di Priolo, S., Ranci, M., Rondanini, C., Bianco, G., Egiatti, G., Montanini, P., Saccardo, I., Campione, E., Pupillo, S., Iocca, F., Lazzeri, M., Tedeschini, M., Marzano, V., Schena, P., Licciardello, A., Manzella, F., Brunier, F., Ratto, S.: Linee guida per il controllo di validità dei dati idro-meteorologici (Guidelines for the quality check of hydrometeorological data), Resources in Italian, 156, ISPRA – Manuali e Linee Guida, Roma, 2017.

Blöschl, G., Gaál, L., Hall, J., Kiss, A., Komma, J., Nester, T., Parajka, J., Perdigão, R. A. P., Plavcová, L., Rogger, M., Salinas, J. L., and Viglione, A.: Increasing river floods: fiction or reality?, Wiley Interdiscip. Rev. Cogn. Sci, 2, 329–344, https://doi.org/10.1002/wat2.1079, 2015.

Brunetti, M., Maugeri, M., Monti, F., and Nanni, T.: Temperature and precipitation variability in Italy in the last two centuries from homogenised instrumental time series, Int. J. Climatol., 26, 345–381, 2006.

Claps, P., Barberis, C., Agostino, M. D., Gallo, E., Laguardia, G., Laio, F., Miotto, F., Plebani, F., Vezzù, G., Viglione, A., and Zanetta, M.: Development of an Information System of the Italian basins for the CUBIST project, in: EGU General Assembly 2008, 2008.

Claps, P., Caporali, E., Chiarello, V., R., D., De Luca, D., Giuzio, L., Libertino, A., Lo Conti, F., Manfreda, S., Noto, V., and Versace, P.: Stima operativa delle piogge estreme sul territorio nazionale: nuovi metodi e possibili sinergie, in: Atti del XXXV Convegno Nazionale di Idraulica e Costruzioni Idrauliche, 2016.

Coumou, D., Robinson, A., and Rahmstorf, S.: Global increase in record-breaking monthly-mean temperatures, Clim. Change, 118, 771–782, 2013.

Desiato, F., Lena, F., and Toreti, A.: SCIA: a system for a better knowledge of the Italian climate, B. Geofis. Teor. Appl., 48, 351–358, 2007.

Fiori, E., Comellas, A., Molini, L., Rebora, N., Siccardi, F., Gochis, D., Tanelli, S., and Parodi, A.: Analysis and hindcast simulations of an extreme rainfall event in the Mediterranean area: The Genoa 2011 case, Atmos. Res., 138, 13–29, https://doi.org/10.1016/j.atmosres.2013.10.007, 2014.

Furcolo, P., Pelosi, A., and Rossi, F.: Statistical identification of orographic effects in the regional analysis of extreme rainfall, Hydrol. Proc., 30, 1342–1353, 2015.

Hosking, J. R. M. and Wallis, J. R.: Regional frequency analysis: an approach based on L-moments, Cambridge University Press, 1997.

Lehmann, J., Coumou, D., and Frieler, K.: Increased record-breaking precipitation events under global warming, Clim. Change, 132, 501–515, 2015.

Libertino, A.: Advances in the space-time analysis of rainfall extremes, Ph.D. thesis, Politecnico di Torino, 2017.

Libertino, A., Allamano, P., Laio, F., and Claps, P.: Regional-scale analysis of extreme precipitation from short and fragmented records, Adv. Water Res., 112, 147–159, https://doi.org/10.1016/j.advwatres.2017.12.015, 2017.

Pavan, V., Antolini, G., Agrillo, G., Auteri, L., Barbiero, R., Bonati, V., Brunier, F., Cacciamani, C., Cazzuli, O., Cicogna, A., De Luigi, C., Maraldo, L., Marigo, G., Millini, R., Panettieri, E., Ratto, S., Ronchi, C., Saibanti, S., Sulis, A., Tomei, F., Tomozeiu, R., Torlai, I., and Villani, G.: TS5 The ARCIS project, Ital. J. Agrometeorol., 2, 51–55, 2013.

Rebora, N., Molini, L., Casella, E., Comellas, A., Fiori, E., Pignone, F., Siccardi, F., Silvestro, F., Tanelli, S., and Parodi, A.: Extreme rainfall in the mediterranean: what can we learn from observations?, J. Hydrometeorol., 14, 906–922, 2013.

Rudari, R., Entekhabi, D., and Roth, G.: Large-scale atmospheric patterns associated with mesoscale features leading to extreme precipitation events in Northwestern Italy, Adv. Water Res., 28, 601–614, 2005.

Snorrason, A., Finnsdóttir, H. P., and Moss, M. E.: The extremes of the extremes: extraordinary floods, ISBN=978-1-901502-66-4, IAHS publication, no. 271, Wallingford, International Association of Hydrological Sciences, 2002.

Uboldi, F., Sulis, A. N., Lussana, C., Cislaghi, M., and Russo, M.: A spatial bootstrap technique for parameter estimation of rainfall annual maxima distribution, Hydrol. Earth Syst. Sci., 18, 981–995, https://doi.org/10.5194/hess-18-981-2014, 2014.

World Meteorological Organization: Guide on the Global Data-processing System, 305, WMO, 1993.

World Meteorological Organization: Manual on the Global Data-processing System, 485,1, WMO, 2010a.

World Meteorological Organization: Manual on Global Observing System, 544,1, WMO, 2010b.

World Meteorological Organization: Guide to Meteorological Instruments and Methods of Observation, 8, WMO, 2014.

6

Assessing the impact of different sources of topographic data on 1-D hydraulic modelling of floods

A. Md Ali[1,2], D. P. Solomatine[1,3], and G. Di Baldassarre[4]

[1]Department of Integrated Water System and Knowledge Management, UNESCO-IHE Institute for Water Education, Delft, the Netherlands
[2]Department of Irrigation and Drainage, Kuala Lumpur, Malaysia
[3]Water Resource Section, Delft University of Technology, the Netherlands
[4]Department of Earth Sciences, Uppsala University, Sweden

Correspondence to: A. Md Ali (a.ali@unesco-ihe.org)

Abstract. Topographic data, such as digital elevation models (DEMs), are essential input in flood inundation modelling. DEMs can be derived from several sources either through remote sensing techniques (spaceborne or airborne imagery) or from traditional methods (ground survey). The Advanced Spaceborne Thermal Emission and Reflection Radiometer (ASTER), the Shuttle Radar Topography Mission (SRTM), the light detection and ranging (lidar), and topographic contour maps are some of the most commonly used sources of data for DEMs. These DEMs are characterized by different precision and accuracy. On the one hand, the spatial resolution of low-cost DEMs from satellite imagery, such as ASTER and SRTM, is rather coarse (around 30 to 90 m). On the other hand, the lidar technique is able to produce high-resolution DEMs (at around 1 m), but at a much higher cost. Lastly, contour mapping based on ground survey is time consuming, particularly for higher scales, and may not be possible for some remote areas. The use of these different sources of DEM obviously affects the results of flood inundation models. This paper shows and compares a number of 1-D hydraulic models developed using HEC-RAS as model code and the aforementioned sources of DEM as geometric input. To test model selection, the outcomes of the 1-D models were also compared, in terms of flood water levels, to the results of 2-D models (LISFLOOD-FP). The study was carried out on a reach of the Johor River, in Malaysia. The effect of the different sources of DEMs (and different resolutions) was investigated by considering the performance of the hydraulic models in simulating flood water levels as well as inundation maps. The outcomes of our study show that the use of different DEMs has serious implications to the results of hydraulic models. The outcomes also indicate that the loss of model accuracy due to re-sampling the highest resolution DEM (i.e. lidar 1 m) to lower resolution is much less than the loss of model accuracy due to the use of low-cost DEM that have not only a lower resolution, but also a lower quality. Lastly, to better explore the sensitivity of the 1-D hydraulic models to different DEMs, we performed an uncertainty analysis based on the GLUE methodology.

1 Introduction

In hydraulic modelling of floods, one of the most fundamental input data is the geometric description of the floodplains and river channels often provided in the form of digital elevation models (DEMs). During the past decades, there has been a significant change in data collection for topographic mapping techniques, from conventional ground survey to remote sensing techniques (i.e. radar wave and laser altimetry; e.g. Mark and Bates, 2000; Castellarin et al., 2009). This shift has a number of advantages in terms of processing efficiency, cost effectiveness and accuracy (Bates, 2012; Di Baldassarre and Uhlenbrook, 2012).

DEMs can be acquired from many sources of topographic information ranging from the high-resolution and accurate, but costly, lidar (Light Detection and Ranging) obtained from lower altitude, to low-cost, and coarse resolution, spaceborne data, such as ASTER (Advanced Spaceborne Thermal Emission and Reflection Radiometer) and SRTM (Shuttle Radar Topography Mission). DEMs can also be developed from traditional ground surveying (e.g. topographic contour maps) by interpolating a number of elevation points.

DEM horizontal resolution, vertical precision and accuracy differ considerably. This diversity is caused by the types of equipment and methods used in obtaining the topographic data. When used as an input to hydraulic modelling, the differences in the quality of each DEM subsequently result in differences in model output performance. In addition, resampling processes of raster data via a geographic information system (GIS) may also deteriorate the accuracy of the DEMs. The usefulness of diverse topographic data in supporting hydraulic modelling of floods is subject to the availability of DEMs, economic factors and geographical conditions of survey area (Cobby and Mason, 1999; Casas et al., 2006; Schumann et al., 2008).

To date, a number of studies have been carried out with the aim of evaluating the impact of accuracy and precision of the topographic data on the results of hydraulic models (e.g. Table 1).

Werner (2001) investigated the effect of varying grid element size on flood extent estimation from a 1-D model approach based on a lidar DEM. The study found that the flood extent estimation increased as the resolution of the DEM becomes coarser.

Horrit and Bates (2001) demonstrated the effects of spatial resolution on a raster-based flood model simulation. Simulation tests were performed at resolution sizes of 10, 20, 50, 100, 250, 500 and 1000 m and the predictions were compared with satellite observations of inundated area and ground measurements of flood-wave travel times. They found that the model reached a maximum performance at resolution of 100 m when calibrated against the observed inundated area. The resolution of 500 m proved to be adequate for the prediction of water levels. They also highlighted that the predicted flood-wave travel times are strongly dependent on the model resolution used.

Wilson and Atkinson (2005) set up a 2-D model, LISFLOOD-FP, using three different DEMs – contour data set, synthetic-aperture radar (SAR) data set, and differential global positioning system (DGPS) – used to predict flood inundation for 1998 flood event in the United Kingdom. The results showed that the contour data sets resulted in a substantial difference in the timing and the extent of flood inundation when compared to the DGPS data set. Although the SAR data set also showed differences in the timing and the extent, it was not as massive as for the contour data set. Nevertheless, the authors also highlighted a potential problem with the use of satellite remotely sensed topographic data in flood hazard assessment over small areas.

Casas et al. (2006) investigated the effects of the topographic data sources and resolution on 1-D hydraulic modelling of floods. They found that the contour-based digital terrain model (DTM) was the least accurate in the determination of the water level and inundated area of the floodplain; however, the GPS-based DTM led to a more realistic estimate of the water surface elevation and of the flooded area. The lidar-based model produced the most acceptable results in terms of water surface elevation and inundated flooded area compared to the reference data. The authors also pointed out that the different grid sizes used in lidar data have no significant effect on the determination of the water surface elevation. In addition, from an analysis of the time–cost ratio for each DEM used, they concluded that the most cost-effective technique for developing a DEM by means of an acceptable accuracy is from laser altimetry survey (lidar), especially for large areas.

Schumann et al. (2008) demonstrated the effects of DEMs on deriving the water stage and inundation area. Three DEMs at three different resolutions from three sources (lidar, contour and SRTM DEM) were used for a study area in Luxembourg. By using the HEC-RAS 1-D hydraulic model to simulate the flood propagation, the result shows that the lidar DEM derived water stages by displaying the lowest RMSE, followed by the contour DEM and lastly the SRTM. Considering the performance of the SRTM (it was relatively good with RMSE of 1.07 m), they suggested that the SRTM DEM is a valuable source for initial vital flood information extraction in large, homogeneous floodplains.

For the large flood-prone area, the availability of DEM from the public domain (e.g. ASTER, SRTM) makes it easier to conduct a study. Patro et al. (2009) selected a study area in India and demonstrated the usefulness of using SRTM DEM to derive river cross-section for the use in hydraulic modelling. They found that the calibration and validation results from the hydraulic model performed quite satisfactorily in simulating the river flow. Furthermore, the model performed quite well in simulating the peak flow, which is important in flood modelling. The study by Tarekegn et al. (2010) carried out on a study area in Ethiopia used a DEM which was generated from ASTER image. Integration between remote sensing and GIS technique were needed to construct the floodplain terrain and channel bathymetry. From the results obtained, they concluded that the ASTER DEM is able to simulate the observed flooding pattern and inundated area extents with reasonable accuracy. Nevertheless, they also highlighted the need for advanced GIS processing knowledge when developing a digital representation of the floodplain and channel terrain.

Schumann et al. (2010) demonstrate that near real-time coarse-resolution radar imagery of a particular flood event on the River Po (Italy) combined with SRTM terrain height data leads to a water slope remarkably similar to that derived by combining the radar image with highly accurate airborne laser altimetry. Moreover, they showed that this spaceborne flood wave approximation compares well to a hydraulic model thus allowing the performance of the latter,

Table 1. Summary of studies assessing the impact of topographic input data on the results of flood inundation models.

Author(s)	Numerical modelling (1-D*/1-D2-D**/2-D***)	Calibration^{+}/ validation^{++} data	Source of DEMs	Type of assessment	Study area
Horrit and Bates (2001)	LISFLOOD-FP**/NCFS**	SAR flood imagery^{+}	Lidar	Precision	River Severn, UK
Werner (2001)	HEC-RAS*	n/a	Laser altimetry data	Precision	River Saar, Germany
Wilson and Atkinson (2005)	LISFLOOD-FP**	SAR flood imagery^{++}	InSAR, topography & GPS	Accuracy	River Nene, UK
Casas et al. (2006)	HEC-RAS*	n/a	GPS, bathymetry, lidar & topography	Accuracy & precision	River Ter, Spain
Schumann et al. (2008)	REFIX*** & HEC-RAS*	Field data^{+}/1-D model output^{++}	Lidar, SRTM topography	Accuracy	River Alzette, Luxembourg
Schumann et al. (2010)	HEC-RAS*	Field data^{+}/lidar derived water levels^{++}	Lidar & SRTM	Accuracy	River Po, Italy
Yan et al. (2013)	HEC-RAS*	Field data^{+}/SAR flood imagery^{++}	Lidar & SRTM	Accuracy	River Po, Italy

calibrated on a previous event, to be assessed when applied to an event of different magnitude in near-real time.

Paiva et al. (2011) demonstrated the use of SRTM DEM in a large-scale hydrologic model with a full 1-D hydrodynamic module to calculate flow propagation on a complex river network. The study was conducted on one of the major tributaries of the Amazon, the Purus River basin. They found that a model validation using discharge and water level data is capable of reproducing the main hydrological features of the Purus River basin. Furthermore, realistic floodplain inundation maps were derived from the results of the model. The authors concluded that it is possible to employ full hydrodynamic models within large-scale hydrological models even when using limited data for river geometry and floodplain characterization.

Moya Quiroga et al. (2013) used Monte Carlo simulation sampling SRTM DEM elevation, and found a considerable influence of the SRTM uncertainty on the inundation area (the HEC-RAS hydraulic model of the Timis-Bega Basin in Romania was employed).

Most recently, Yan et al. (2013) made a comparison between a hydraulic model based on lidar and SRTM DEM. Besides the DEM inaccuracy, they also introduced uncertainty analysis by considering parameter and inflow uncertainty. The results of this study showed that the differences between the lidar-based model and the SRTM-based model are significant, but within the accuracy that is typically associated with large-scale flood studies.

Yet, the aforementioned studies explored the impact of topographic input data on the results flood inundation models by considering either the accuracy (or quality) or the precision (or resolution) of the DEMs (Table 1). When both accuracy and precision were considered (Casas, 2006), model results were not compared to observations via calibration and validation exercises.

This paper continues the presented line of research and deals with the assessment of the effects of using different DEM data source and resolution in a 1-D hydraulic modelling of floods. The novelty of our study is that both accuracy and precision of the DEM are explicitly considered and their impacts on hydraulic model results is evaluated in terms of both water surface elevation and inundation area. Furthermore, we compare model results via independent calibration and validation exercises and by explicitly considering parameter uncertainty and its potential compensation of inaccuracy of topographic data.

Hence, the goal of our paper is not to validate a specific approach for producing flood inundation maps, but rather to contribute to the existing literature with an original approach assessing the impact of topographic input data on hydraulic modelling of floods.

2 Study area and available data

2.1 Study area

The study area is located within the Johor River basin in the State of Johor, Malaysia. The river basin has a total area of 2690 km^2. The test site is a 30 km reach of the Johor River. The Johor River channel has a bankfull depth between 5 and 8 m and average slope around 0.03 %. The river reach under study is characterized by a stable main channel from 50 to 250 m wide. The study area consists of agricultural land, residential and commercial areas (see Fig. 1). As reported by Department of Irrigation and Drainage, Malaysia (DID, 2009), this test site has been experiencing some major historical flood events since 1948. The most recent ones happened in December 2006 and January 2007 when more than 3000 families were evacuated.

2.2 Hydraulic modelling

Flood inundation modelling was carried out by using the model code HEC-RAS, which was developed by the Hydrologic Engineering Center (HEC) of the United States Army Corps of Engineers (USACE, 2010). HEC-RAS is a 1-D model that can simulate both steady and unsteady flow conditions. In this study, all simulations were performed under unsteady flow conditions. To simulate open channel flows, HEC-RAS numerically solves the full 1-D Saint-Venant equations. The HEC-RAS model was set up using 32 cross-sections, whose topography is derived by different DEMs (see below). The observed flow hydrograph at an hourly time step was used as upstream boundary condition, while the friction slope was used as downstream boundary condition. The next section reports the different sources of topographic data used to define the geometric input. To develop flood inundation maps, the results were post-processed by using HEC-GeoRAS, an ArcGIS extension.

1-D hydraulic modelling does not properly simulate river hydraulics and floodplain flows. However, while 2-D models tend to schematize better flood inundation processes, they do not necessarily perform better when applied to real-world case studies because, besides model structure, many other sources of uncertainty affect model results (Werner, 2001; Bates et al., 2003; Pappenberger et al., 2005; Merwade et al., 2008; Di Baldassarre et al., 2009, 2010). A number of authors have carried out comparative studies and showed that the performance of 1-D models is often very close to that of 2-D models (e.g. Horrit and Bates, 2002; Castellarin et al., 2009; Cook and Merwade, 2009). Also, 1-D models are typically more efficient than 2-D models from a computation viewpoint, allowing for numerous simulations and uncertainty analysis to be carried out. In our case study, for a given flow, topography, river reach and a number of simulations, a HEC-RAS simulation (excluding post-processing GIS) took only 4 h to predict inundated area, whereas LISFLOOD-FP took around 26 h.

Anyhow, to properly test our model selection, we carried out a number of additional experiments (see Sect. 4.2) and compared the results of 1-D models to the results obtained with a 2-D model (LISFLOOD-FP; e.g. Hunter et al., 2006; Bates et al., 2010; Neal et al., 2012; Coulthard et al., 2013).

2.3 Digital elevation model

The required input data for the HEC-RAS include the geometry of the floodplain and the river, which is provided by a number of cross-sections. We identified several sources of DEM data for our study area (details are given below) with different spatial resolution and accuracy (Fig. 2):

Table 2. Information about the eight digital elevation models used as topographical input.

Model name	DEM type	Resolution (m)
Jhr L1	lidar	1 m
Jhr L2	(rescaled from lidar)	2
Jhr L20	(rescaled from lidar)	20
Jhr L30	(rescaled from lidar)	30
Jhr L90	(rescaled from lidar)	90
Jhr T20	Contour maps	20
Jhr A30	ASTER	30
Jhr S90	SRTM	90

1. DEMs derived from an original 1 m lidar data set (obtained from DID).

2. 20 m resolution DEM generated from the vectorial 1 : 25 000 cartography map obtained from DID with a permission of the Department of Survey and Mapping, Malaysia (DSMP).

3. 30 m resolution DEM derived from the globally and freely available ASTER data retrieved from the United States Geological Survey (USGS, http://earthexplorer.usgs.gov)

4. 90 m resolution DEM derived from the globally and freely available SRTM data retrieved from a Consortium for Spatial Information (CGIAR-CSI, www.cgiar-csi.org).

To analyse the influence of spatial resolution and separate it out from the impact of different accuracy, four additional DEMs were obtained by rescaling the original lidar DEM (1 m resolution) to the spatial resolutions of the DEMs derived from vectorial cartography (20 m), ASTER (30 m) and SRTM (90 m). Hence, a total of eight DEMs were used (see Table 2) to explore the impact of different topographic information on the hydraulic modelling of floods.

Given that the laser/radar waves used in the remote sensing techniques are not capable of penetrating the water surface and capture the river bed elevations, all the DEMs were integrated with river cross-section data derived from traditional ground survey. The ground survey of the river cross-sections within the study area was systematically carried out at about 1000 m intervals. Then, the flood simulation results across different data sets were compared to evaluate the effects of data spatial resolutions and data source differences.

3 Methodology

3.1 Evaluating DEM quality

First, the vertical error of each DEM was evaluated through comparison between the topographic data and 164 GPS

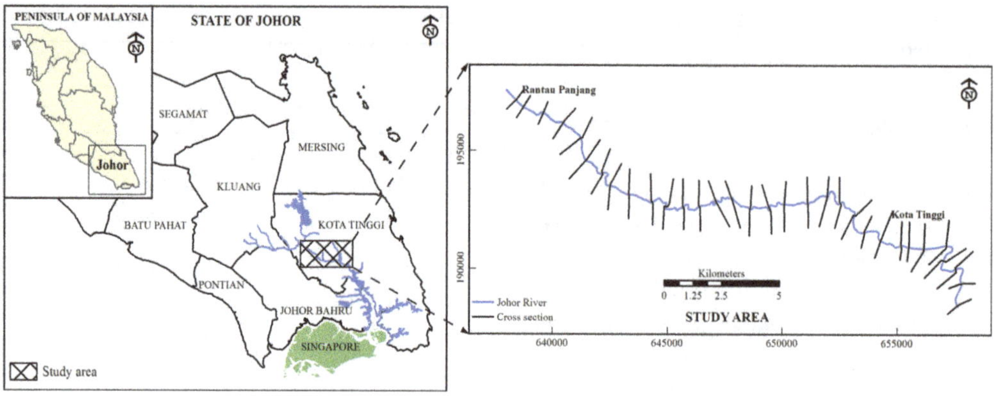

Figure 1. Layout map of study area: Johor River, Malaysia.

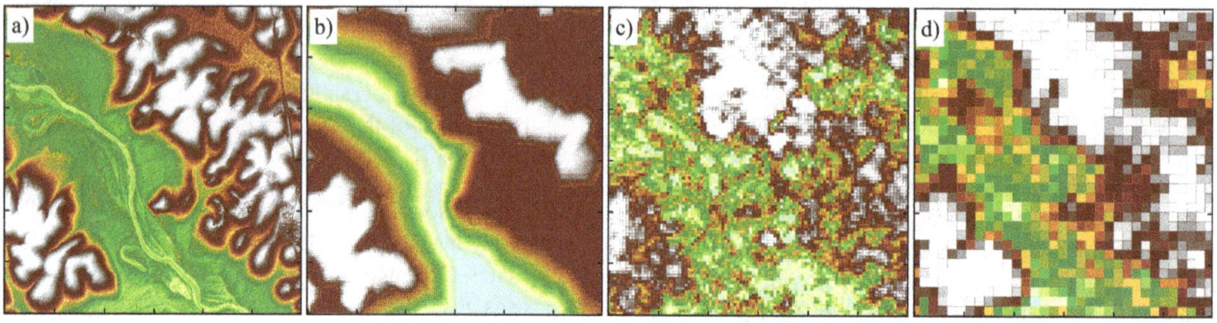

Figure 2. Original DEMs used in this study, based on **(a)** lidar data, **(b)** contour map, **(c)** ASTER data and **(d)** SRTM data.

ground points taken at random positions within the study area. The value of each reference elevation point was extracted from the study area using GPS survey equipment. The quality of each DEM is assessed by the root mean square error ($\mathrm{RMSE_{DEM}}$) and mean error ($\mathrm{ME_{DEM}}$). Thus

$$\mathrm{RMSE_{DEM}} = \sqrt{\frac{\sum_{i=1}^{n} (\mathrm{Elev_{GPS}} - \mathrm{Elev_{DEM}})^2}{n}}, \tag{1}$$

where $\mathrm{Elev_{GPS}}$ is the reference elevation (m) derived from GPS, $\mathrm{Elev_{DEM}}$ is the corresponding value derived from each DEM, and n corresponds to the total numbers of points.

3.2 Model calibration and validation

Then, data from two recent major flood events that occurred along the Johor River in 2006 and 2007 were used for independent calibration and validation of the models. The estimated peak flow of the 2006 event is approximately $375\,\mathrm{m^3\,s^{-1}}$, while the one of the 2007 event is around $595\,\mathrm{m^3\,s^{-1}}$. Both discharge data were measured and recorded at Rantau Panjang hydrological station. The 2006 flood data were used for the calibration exercise, while the 2007 flood data were used for model validation.

To assess the sensitivity of the different models to the model parameters, the Manning n roughness coefficients for all the models were sampled uniformly from 0.02 to $0.08\,\mathrm{m^{-1/3}s}$ for the river channel, and between 0.03 and $0.10\,\mathrm{m^{-1/3}s}$ for the floodplain, by steps of $0.0025\,\mathrm{m^{-1/3}s}$. The performance of the hydraulic models in producing the observed water levels was assessed by means of the mean absolute error (MAE):

$$\mathrm{MAE} = \frac{1}{T}\sum_{t=1}^{T} |O_t - S_t|, \tag{2}$$

where T is the number of steps in time series, O_t is the observed water level at time t, and S_t is the simulated water level at time t.

3.3 Quantifying the effect of the topographic data source on the water surface elevation and inundation area (sensitivity analysis)

The effects of DEM source and spatial resolution were further investigated by examining the sensitivity of model results in terms of maximum water surface elevation (WSE), inundation area and floodplain boundaries. For this additional analysis, the model results obtained with the most accurate and precise DEM source (lidar at 1 m resolution) were used as a reference. For WSE analysis, each model was compared to the reference model (Jhr L1, see Table 1) by means

Table 3. Statistics of errors (m) of each DEMs with respect to the GPS control points.

Model name	Min. error (m)	Max. error (m)	RMSE (m)
Jhr L1	−0.59	1.00	0.58
Jhr L2	−0.64	1.38	0.58
Jhr L20	−0.83	1.83	0.68
Jhr L30	−0.93	3.98	0.79
Jhr L90	−5.46	3.73	1.27
Jhr T20	−15.38	10.55	4.66
Jhr A30	−33.37	7.58	7.01
Jhr S90	−3.59	4.32	6.47

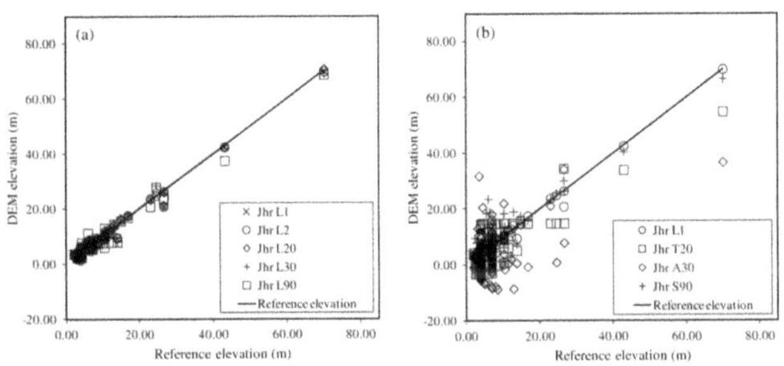

Figure 3. Comparison between GPS point elevations and elevations derived by the different DEMs: **(a)** lidar DEM at different resolution and **(b)** different sources of DEMs.

of the following measures:

$$\text{MAD}_{\text{WSE}} = \frac{1}{x}\sum_{x=1}^{x} |\text{WSE}_{\text{Ref}} - \text{WSE}_{\text{DEM}}|, \tag{3}$$

where WSE_{Ref} denotes the WSE simulated by the reference model (Jhr L1), WSE_{DEM} the WSE estimated by the models based on DEMs of lower resolution or different source (Table 1), and x corresponds to the total number of cross-sections where models results where compared.

To analyse the sensitivity to different topographic input in terms of simulated flood extent, we used the following measure of fit:

$$F(\%) = \frac{M_1 \cap M_2}{M_1 \cup M_2}.100, \tag{4}$$

where M_1 and M_2 are the simulated and observed (i.e. simulated by the reference model) inundation areas, and $\cup$ and $\cap$ are the union and intersection GIS operations respectively. F equal to 100 % indicates that the two areas are completely coincidental.

3.4 Uncertainty estimation – GLUE analysis

In hydraulic modelling, multiple sources of uncertainty can emerge from several factors, such as model structure, topography and friction coefficients (Aronica et al., 2002; Trigg

Table 4. Reported vertical accuracies of SRTM data.

Reference	Average height accuracy (m)	Continent
Rabus et al. (2003)	6.00	European
Sun et al. (2003)	11.20	European
SRTM mission specification (Rodríguez et al., 2005)	16.00	Global
Berry et al. (2007)	2.54 3.60	Eurasia Global
Farr et al. (2007)	6.20	Eurasia
Wang et al. (2011)	13.80	Eurasia

et al., 2009; Brandimarte and Di Baldassarre, 2012; Dottori et al., 2013). A methodological approach to estimate the uncertainty is the generalized likelihood uncertainty estimation (GLUE) methodology (Beven and Binley, 1992), a variant of Monte Carlo simulation. Although some aspects of this methodology are criticized in several papers (e.g. Hunter et al., 2005; Mantovan and Todini, 2006; Montanari, 2005; Stedinger et al., 2008), it is still widely used in hydrological modelling because of its ease of implementation and a common-sense approach to use only a set of the "best" models for uncertainty analysis (e.g. Hunter et al., 2005; Shrestha et al., 2009; Vázquez et al., 2009; Krueger et al., 2010; Jung and Merwade, 2012; Brandimarte and Woldeyes, 2013).

According to the GLUE framework (Beven and Binley, 1992), each simulation, i, is associated with the (generalized) likelihood weight, W_i, ranging from 0 to 1. The weight, W_i is expressed as a function of the measure fit, ε_i, of the behavioural models:

$$W_i = \frac{\varepsilon_{\text{max}} - \varepsilon_i}{\varepsilon_{\text{max}} - \varepsilon_{\text{min}}}, \tag{5}$$

where ε_{max} and ε_{min} are the maximum and minimum value of MAE of behavioural models. To identify the behaviour of the models, a threshold value (rejection criterion) has been set as follows:

1. simulations associated with MAE larger than 1.0 m and
2. Manning's n roughness coefficient of the floodplain smaller than the Manning's n roughness coefficient of the channel.

Then, the likelihood weights are the cumulative sum of 1 and the weighted 5th, 50th and 95th percentiles. The likelihood weights were calculated as follows:

$$L_i = \frac{W_i}{\sum_{i=1}^{n} W_i}. \tag{6}$$

For this study, the applications of uncertainty analysis considered only the parameter uncertainty and implemented for all DEMs.

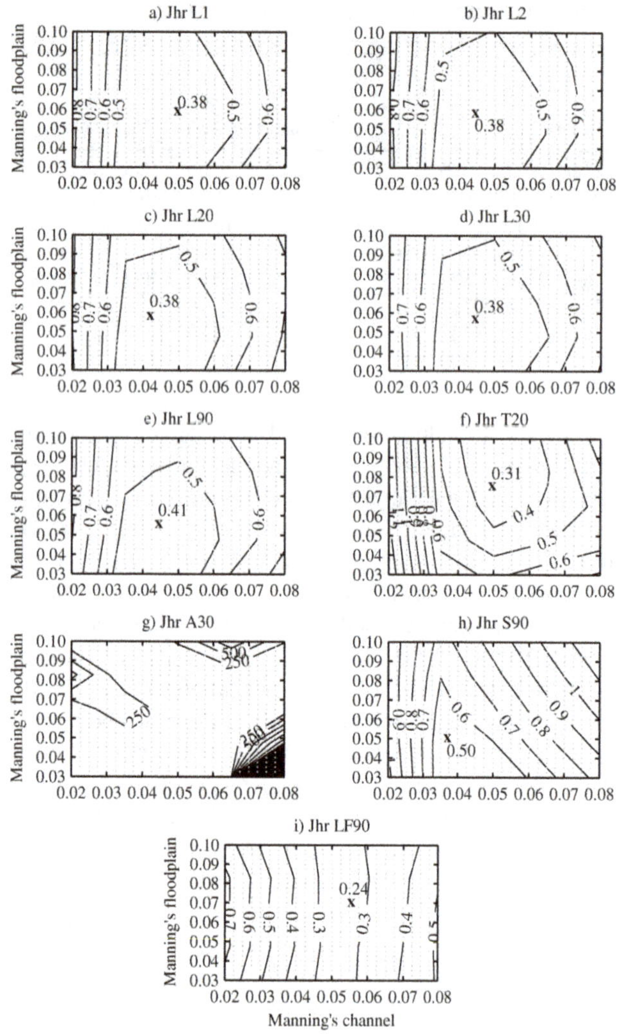

Figure 4. Model calibration: contour maps of MAE across the parameter space for **(a–h)** eight different 1-D models (HEC-RAS) and **(i)** for the 2-D model (LISFLOOD-FP).

Table 5. Model validation results.

Model name	Calibrated Manning's *n* roughness coefficient		MAE (m) (validation)
	Channel	Floodplain	
Jhr L1	0.0500	0.0575	0.40
Jhr L2	0.0450	0.0575	0.38
Jhr L20	0.0425	0.0575	0.37
Jhr L30	0.0450	0.0575	0.38
Jhr L90	0.0450	0.0550	0.39
Jhr T20	0.0500	0.0750	0.60
Jhr S90	0.0375	0.0500	0.60
Jhr LF90	0.0550	0.0700	0.52

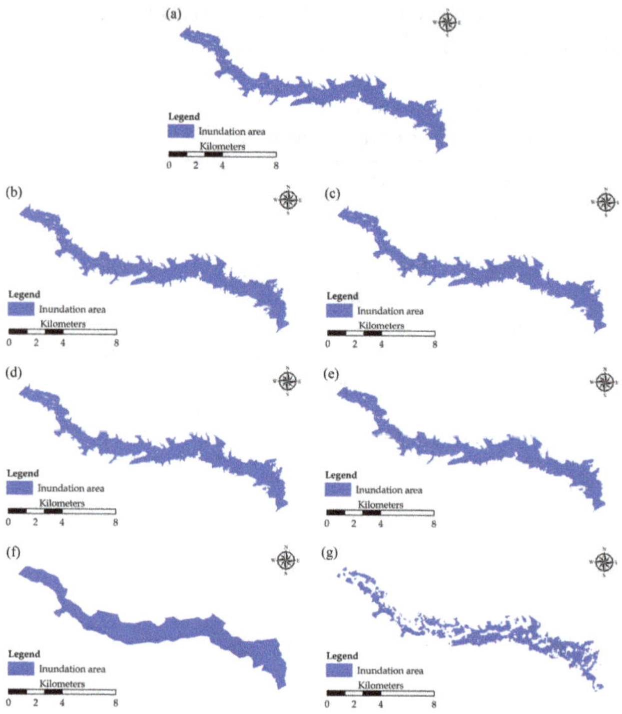

Figure 5. Effect of DEMs on Johor River. Inundation map resulting from **(a)** Jhr L1, **(b)** Jhr L2, **(c)** Jhr L20, **(d)** Jhr L30, **(e)** Jhr L90, **(f)** Jhr T20 and **(g)** Jhr S90.

4 Results and discussion

4.1 Quality of DEMs compared with the reference points

Table 3 shows the calculated statistical vertical errors for each different DEM for the same study area. As anticipated, lidar is not only the most precise DEM because of its highest resolution, but also the most accurate. The RMSE of each lidar DEM increased from 0.58 m (Jhr L1) to 1.27 m (Jhr L90) as the resolution of the DEMs reduced from 1 m (original resolution) to 90 m.

Overall, the terrain is considered well defined under the lidar DEMs even though the calculated errors are higher compared to the vertical accuracy reported in product specification (around 0.15 m). Figure 3 shows the distribution of each DEM compared to the GPS ground elevation.

Although lidar DEM gives the lowest error, it is useful to note that this type of DEM has a number of limitations as highlighted in the several papers (see Sun et al., 2003; Casas et al., 2006; Schumann et al., 2008):

1. it provides only discrete surface height samples and not continuous coverage,

2. its availability is very much limited by economic constraint,

Table 6. Effects of DEMs (source and resolution) on HEC-RAS simulations.

Model name	Inundation area (km^2)	Area difference (%)	F (%)	F (%)$^+$
Jhr L1	25.86	–	–	–
Jhr L2	25.78	−0.3	96.6	–
Jhr L20	25.96	0.4	92.9	–
Jhr L30	26.18	1.2	92.2	–
Jhr L90	25.84	−0.1	89.4	–
Jhr T20	29.23	13.0	73.7	74.2
Jhr S90	16.58	−35.9	48.9	49.6

$^+$ Overlap-fit percentage F (%) of the floodplain inundated area with those from lidar DEMs of the same resolutions (Jhr L20, Jhr L90).

3. it is unable to capture the river bed elevations due to the fact the laser does not penetrate the water surface, and
4. it is incapable of penetrating the ground surface in densely vegetated areas especially for the tropical region.

The RMSE value of the other DEMs is 4.66 m for contour maps, 7.01 m for ASTER and 6.47 m for SRTM. It is also noticeable that the RMSE of the SRTM DEM for this particular study area is within the average height accuracy found in other SRTM literature – either global or at particular continent (see Table 4). Whatever the case, it is proven that this type of DEM gives an acceptable result when used in large-scale flood modelling (e.g. Patro et al., 2009; Paiva et al., 2012; Yan et al., 2013).

Despite having the lowest vertical accuracies, the ASTER and contour DEMs are still widely used in the field of hydraulic flood research as they are globally available and free (e.g. Tarekegn et al., 2010; Wang et al., 2011; Gichamo et al., 2012). The differences in the vertical accuracies may be partly due to the lack of information in topographical flats areas such as floodplains. However, the further use of each DEM in this study is subject to its performance in the hydraulic flood modelling during the calibration and validation stages, which are described in the following section.

4.2 Model calibration and validation

Panels (a)–(h) of Fig. 4 show the model responses in terms of MAE provided by the eight HEC-RAS models in simulating the 2006 flood event. The models were built using the eight DEMs with different accuracy and precision (Table 2) as topographic input.

In general, all models (Fig. 4a–h) are seen to be more sensitive to the changing of Manning's n roughness coefficient of main channel than the Manning's n roughness coefficient of floodplain areas. The results of the calibration showed that the best-fit models based on lidar DEM with different resolutions (Jhr L2, Jhr L20, Jhr L30 and Jhr L90) generally gave good performances with only slight variations in the MAE value from 0.38 to 0.41 m. The optimum channel and floodplain Manning's n roughness coefficient are centred on similar values at $n_{\text{channel}} = 0.0425$ to 0.0500 and $n_{\text{floodplain}} = 0.0575$ for Jhr L1, Jhr L2, Jhr L20, Jhr L30 and Jhr L90. The best-fit models based on topographic map and SRTM also performed well with MAE of 0.31 and 0.50 m. On the other hand, ASTER-based model completely failed (the exceptionally high values of MAE in Fig. 4g are due to model instabilities) and was therefore eliminated from further analysis.

Table 7. Summary of mean absolute difference (MAD) in terms of water surface elevation simulated by the models.

Model name	MAD_{WSE}(m)
Jhr L1	–
Jhr L2	0.06
Jhr L20	0.05
Jhr L30	0.05
Jhr L90	0.08
Jhr T20	1.12
Jhr S90	0.76

Panel (i) of Fig. 4 shows the outcome of the additional experiment we carried out to test the appropriateness of selecting a 1-D model. In particular, a LISFLOOD-FP model was built using the lidar topography rescaled at 90 m, called here Jhr LF90. The specific topographic input was chosen as a trade-off between computational times and the need for an as-accurate-as-possible DEM for a proper comparison between 1-D and 2-D modelling. By comparing the calibration results of the LISFLOOD-FP model (Fig. 4i) to the corresponding (i.e. using the same topography) ones of the HEC-RAS model (Fig. 4e), one can observe that differences are not significant. Lastly, Fig. 4i shows that LISFLOOD-FP is also more sensitive to the main channel roughness coefficient than to the floodplain one.

The best-fit models, using the optimum Manning n roughness coefficients (Table 5), were then used to simulate the January 2007 flood event for model validation. This was carried out for all models except the ASTER-based model due to its poor performance (see Fig. 4g). Table 5 presents the MAE of each model obtained during model validation. It is noted that the MAE values for all lidar-based models (first five rows) with different resolutions remained almost the same with the difference within +0.03 m. The MAE values for the models based on topographic contour maps and SRTM DEM both provide MAE of 0.60 m.

The model validation exercise also supports the use of 1-D hydraulic models for this river reach. In particular, Table 5 shows that the LISFLOOD-FP model (Jhr LF90) provided a MAE of 0.52 m, while the corresponding HEC-RAS model (Jhr L90) provided a MAE of 0.39 m. Thus, the 1-D model performed even (slightly) better than the 2-D model.

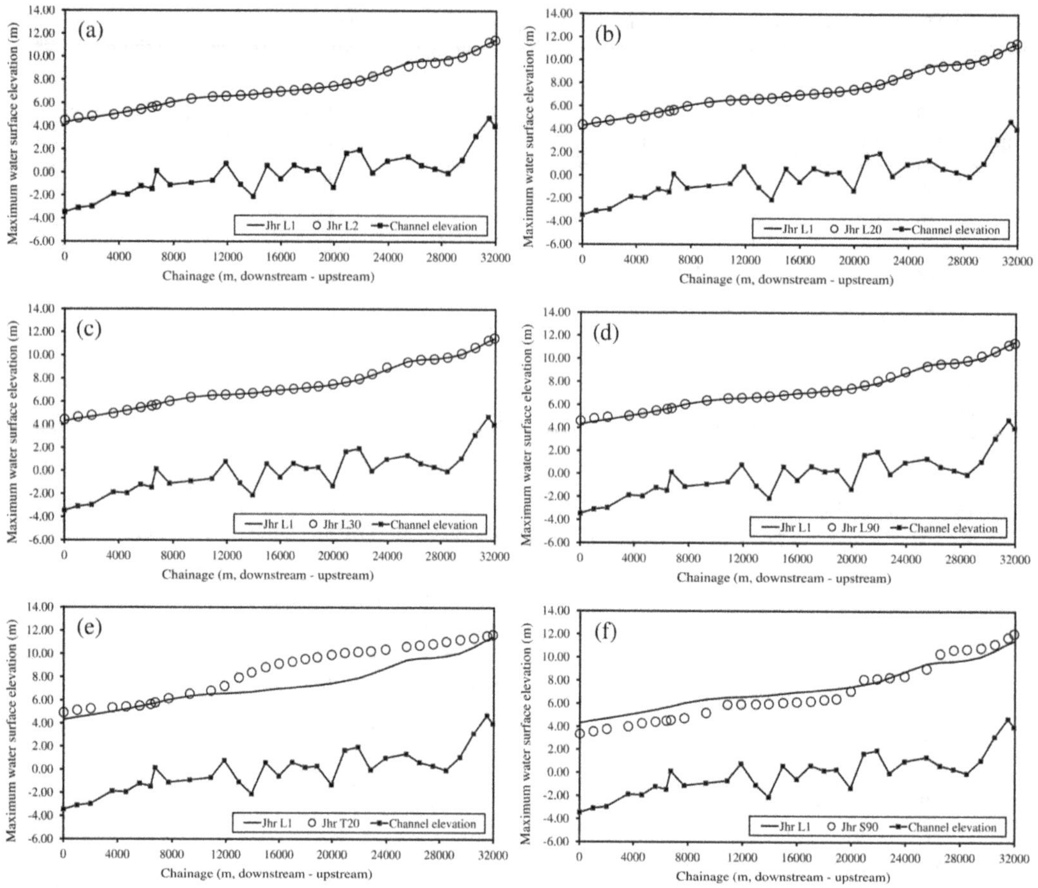

Figure 6. Maximum water surface elevation along the Johor River for the six hydraulic models compared to that simulated by the reference model.

The results of this first analysis suggest that the reduction in the resolution of lidar DEMs (from 1 to 90 m) does not significantly affect the model performance. However, the use of topographic contour maps (Jhr T20) and SRTM (Jhr S90) DEMs as geometric input to the hydraulic model produces a slight increase of model errors. For instance, Jhr L90 and Jhr S90 have the same resolution (90 m), but the different accuracy results in increased (tough not remarkably) errors in model validation (from 0.39 to 0.60 m). This limited degradation of model performance (Table 5), in spite of the much lower accuracy of topographic input (Table 2), can be attributed to the fact that models are compared to water levels observed in two cross-sections. A spatially distributed analysis (comparing the simulated flood extent and flood water profile along the river) might show more significant differences (see Sect. 4.3).

4.3 Quantifying the effect of the topographic data source on the water surface elevation and inundation area on 1-D model

4.3.1 Inundation area (sensitivity analysis)

This section reports an additional analysis aiming to better explore the sensitivity of model results to different topographic data (see Sect. 3.3). Figure 5 shows the simulated flood extent maps obtained from the seven different topographic input data. The floodplain areas simulated by the five lidar-based models (Jhr L1, Jhr L2, Jhr L20, Jhr L30 and Jhr L90) are very similar. In contrast, the floodplain areas simulated by the models based on topographic contour maps (Jhr T20) and SRTM DEM (Jhr S90) are substantially different (see Fig. 5 and Table 6).

Table 6 shows the comparison between the different models in terms of simulating flood extent. The aforementioned measure of fit F was found to decrease for both decreasing resolution and lowering accuracy. This sensitivity analysis also shows that the results of flood inundation models are more affected by the accuracy of the DEM used as topographic input than its resolution.

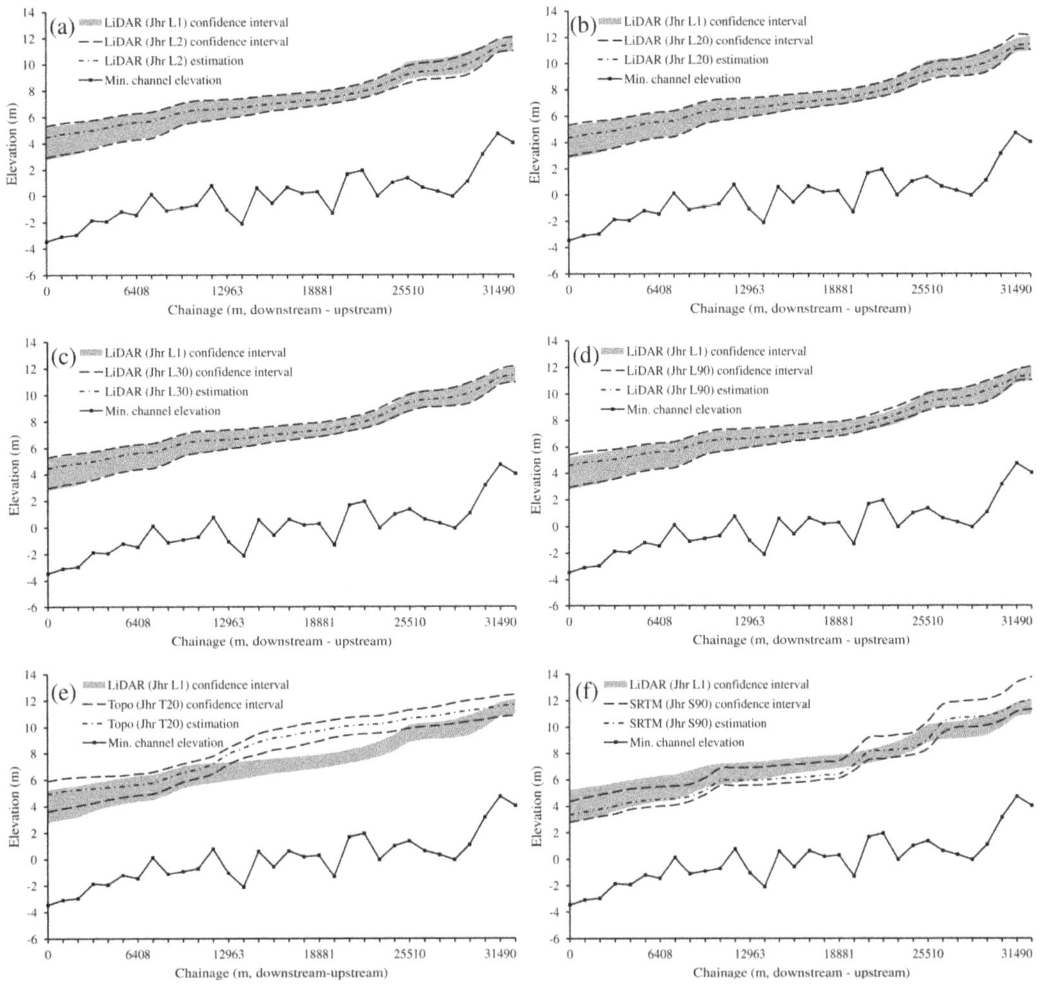

Figure 7. Comparison of uncertainty bounds (5th, 50th and 95th percentiles by considering parameter uncertainty only) between the reference model and the other models. The reference model uncertainty bounds are shown as grey areas, while the uncertainty bounds of the other six models are shown as dashed lines.

4.3.2 Water surface elevation

Figure 6 compares the flood water profiles simulated by the reference model (Jhr L1) with the flood water profiles (WSE) obtained from the other six models (Jhr L2, Jhr L20, Jhr L30, Jhr L90, Jhr T20 and Jhr S90). All these flood water profiles were obtained by simulating the 2007 flood event. Despite having different resolutions, the flood water profiles simulated from all lidar-based models portray similar flood water profiles to the reference model (see Fig. 6a to d). This is consistent with the findings about the inundation area (Fig. 5), whereas flood water profiles simulated by the models based on topographic contour maps and SRTM DEMs (see Fig. 6e and f) are rather different.

The discrepancies between the reference model (Jhr L1) and the other models as shown in Fig. 6 are quantified in terms of mean absolute difference (MAD). This shows that the re-sampled lidar data (Jhr L2, Jhr L20, Jhr L30 and Jhr L90) all have a low MAD: between 0.05 to 0.08 m. Higher discrepancies are found with the models based on SRTM DEM (0.76 m) and contour maps (1.12 m). The great differences obtained using the topographic contour maps may be partly due to the way that the DEM height is sampled. For instance, the contour DEMs in this study were based on topographic contours at 20 m intervals and required an interpolation technique to generate a DEM. Table 7 shows the MAD in terms of water surface elevation simulated by the models.

4.3.3 Uncertainty in flood profiles obtained from different DEMs by considering parameter uncertainty

To better interpret the differences that have emerged in comparing the results of models based on different topographic data, we carried out a set of numerical experiments to explore the uncertainty in model parameters. As mentioned, we varied Manning's n roughness coefficient between 0.02 and 0.08 $m^{-1/3}s$ for the river channel, and from 0.03 to

$0.10\,m^{-1/3}s$ for the floodplain, in steps of $0.0025\,m^{-1/3}s$. Then a number of simulations were rejected as described in Sect. 3.4. Figure 7 shows the uncertainty bounds for the different models. The width of these uncertainty bounds was found to be between 1.5 and 1.6 m for all models (only parameter uncertainty is considered here). Nevertheless, the model based on contour maps led to significant differences from the lidar-based model, even when the uncertainty induced by model parameters is explicitly accounted for (see Fig. 7e).

5 Conclusions

This study assessed how different DEMs (derived by various sources of topographic information or diverse resolutions) affect the output of hydraulic modelling. A reach of the Johor River, Malaysia, was used as the test site. The study was performed using a 1-D model (HEC-RAS), which was found to perform as well as a 2-D model (LISFLOOD-FP) in this case study. The sources of DEMs were lidar at 1 m resolution, topographic contour maps at 20 m resolution, ASTER data at 30 m resolution, and SRTM data at 90 m resolution. The lidar DEM was also re-sampled from its original resolution data set to 2, 20, 30 and 90 m cell size. Different models were built by using them as geometric input data.

The performance of the five lidar-based models (characterized by different resolutions ranging from 1 to 90 m; see Table 5) did not show significant differences – neither in the exercise of independent calibration and validation based on water level observations in an internal cross-section, nor in the sensitivity analysis of simulated flood profiles and inundation areas. Another striking result of our study is that the model based on ASTER data completely failed because of major inaccuracies of the DEM.

In contrast, the models based on SRTM data and topographic contour maps did relatively well in the validation exercise as they provided a mean absolute error of 0.6 m, which is only slightly higher than the ones obtained with lidar-based models (all around 0.4 m). However, this outcome could be attributed to the fact that validation could only be performed by using the water level observed in a two internal cross-sections. As a matter of fact, higher discrepancies emerged when lidar-based models were compared to the models based on SRTM data or topographic contour maps in terms of inundation areas or flood water profiles. These differences were found to be relevant even when parameter uncertainty was accounted for.

The study also showed that, to support flood inundation models, the quality and accuracy of the DEM is more relevant than the resolution and precision of the DEM. For instance, the model based on the 90 m DEM obtained by re-sampling the lidar data performed better than the model based on the 90 m DEM obtained from SRTM data. These outcomes are unavoidably associated with the specific test site, but the methodology proposed here can allow a comprehensive assessment of the impact of diverse topographic data on hydraulic modelling of floods for different rivers around the world.

Acknowledgements. The authors would like to thank to the Department of Irrigation and Drainage, Malaysia (DID) for providing useful input data used in this study. We also acknowledge the Public Service Department, Malaysia for providing a PhD Fellowship funding and study leave for the first author. We thank the Editor and the two anonymous reviewers as well as Fiona, Nagendra, Micah and Yan for their constructive comments that helped to improve the manuscript.

Edited by: H. Cloke

References

Aronica, G., Bates, P. D., and Horritt, M. S.: Assessing the uncertainty in distributed model predictions using observed binary pattern information within GLUE, Hydrol. Process., 16, 2001–2016, doi:10.1002/hyp.398, 2002.

Bates, P. D., Marks, K. J., and Horritt, M. S.: Optimal use of high-resolution topographic data in flood inundation models, Hydrol. Process., 17, 5237–5257, 2003.

Bates, P. D., Horritt, M. S., and Fewtrell, T. J.: A simple inertial formulation of the shallow water equations for efficient two-dimensional flood inundation modelling, J. Hydrol., 387, 33–45, doi:10.1016/j.jhydrol.2010.03.027, 2010.

Bates, P. D.: Integrating remote sensing data with flood inundation models: how far have we got?, Hydrol. Process., 26, 2515–2521, doi:10.1002/hyp.9374, 2012.

Berry, P. A. M., Garlick, J. D., and Smith, R. G.: Near-global validation of the SRTM DEM using satellite radar altimetry, Remote Sens. Environ., 106, 17–27, doi:10.1016/j.rse.2006.07.011, 2007.

Beven, K. and Binley, A.: The future of distributed models – model calibration and uncertainty prediction, Hydrol. Process., 6, 279–298, doi:10.1002/hyp.3360060305, 1992.

Brandimarte, L. and Di Baldassarre, G.: Uncertainty in design flood profiles derived by hydraulic modelling, Hydrol. Res., 43, 753–761, doi:10.2166/nh.2011.086, 2012.

Brandimarte, L. and Woldeyes, M. K.: Uncertainty in the estimation of backwater effects at bridge crossings, Hydrol. Process., 27, 1292–1300, doi:10.1002/hyp.9350, 2013.

Casas, A., Benito, G., Thorndycraft, V. R., and Rico, M.: The topographic data source of digital terrain models as a key element in the accuracy of hydraulic flood modelling, Earth Surf. Process. Landforms, 31, 444–456, doi:10.1002/esp.1278, 2006.

Castellarin, A., Di Baldassarre, G., Bates, P. D., and Brath, A.: Optimal cross-section spacing in Preissmann scheme 1-D hydrodynamic models, J. Hydraul. Eng.-ASCE, 135, 96–105, doi:10.1061/(ASCE)0733-9429(2009)135:2(96), 2009.

Cobby, D. M. and Mason, D. C.: Image processing of airborne scanning laser altimetry for improved river flood modelling, ISPRS J. Photogramm. Remote Sens., 56, 121–138, 1999.

Cook, A. and Merwade, V.: Effect of topographic data, geometric configuration and modeling approach on flood inundation mapping, J. Hydrol., 377, 131–142, 2009.

Coulthard, T. J., Neal, J. C., Bates, P. D., Ramirez, J., de Almeida, G. A. M., and Hancock, G. R.: Integrating the LISFLOOD-FP 2D hydrodynamic model with the CAESAR model: implications for modelling landscape evolution, Earth Surf. Process. Landforms, 38, 1897–1906, doi:10.1002/esp.3478, 2013.

Department of Irrigation and Drainage, Malaysia (DID): Master plan study on flood mitigation for Johor River basin, Malaysia, 2009.

Di Baldassarre, G., Schumann, G., and Bates, P. D.: A technique for the calibration of hydraulic models using uncertain satellite observations of flood extent, J. Hydrol., 367, 276–282, 2009.

Di Baldassarre, G., Schumann, G., Bates, P. D., Freer, J. E., and Beven, K. J.: Floodplain mapping: a critical discussion on deterministic and probabilistic approaches, Hydrolog. Sci. J., 55, 364–376, 2010.

Di Baldassarre, G. and Uhlenbrook, S.: Is the current flood of data enough? A treatise on research needs for the improvement of flood modelling, Hydrol. Process., 26, 153–158, doi:10.1002/hyp.8226, 2012.

Dottori, F., Di Baldassarre, G., and Todini, E.: Detailed data is welcome, but with a pinch of salt: Accuracy, precision, and uncertainty in flood inundation modelling, Water Resour. Res., 49, 6079–6085, doi:10.1002/wrcr.20406, 2013.

Farr, T. G., Rosen, P. A., Caro, E., Crippen, R., Duren, R., Hensley, S., Kobrick, M., Paller, M., Rodriguez, E., Roth, L., Seal, D., Shaffer, S., Shimada, J., Umland, J., Werner, M., Oskin, M., Burbank, D., and Alsdorf, D.: The shuttle radar topography mission, Rev. Geophys., 45, RG2004, doi:10.1029/2005RG000183, 2007.

Gichamo, T. Z., Popescu, I., Jonoski, A., and Solomatine, D.: River cross-section extraction from the ASTER global DEM for flood modelling, Environ. Modell. Softw., 31, 37–46, doi:10.1016/j.envsoft.2011.12.003, 2012.

Horrit, M. S. and Bates, P. D.: Effects of spatial resolution on a raster based model of flood flow, J. Hydrol., 253, 239–249, 2001.

Horrit, M. S. and Bates, P. D.: Evaluation of 1-D and 2-D models for predicting river flood inundation, J. Hydrol., 180, 87–99, 2002.

Hunter, N. M., Bates, P. D., Horritt, M. S., De Roo, A. P. J., and Werner, M. G. F.: Utility of different data types for calibrating flood inundation models within a GLUE framework, Hydrol. Earth Syst. Sci., 9, 412–430, doi:10.5194/hess-9-412-2005, 2005.

Hunter, N. M., Bates, P. D., Horritt, M. S., and Wilson, M. D.: Improved simulation of flood flows using storage cell models, P. I. Civil Eng.-Wat. M., 159, 9–18, 2006.

Jung, Y. and Merwade, V.: Uncertainty quantification in flood inundation mapping using generalized likelihood uncertainty estimate and sensitivity analysis, J. Hydrol. Eng., 17, 507–520, 2012.

Krueger, T., Freer, J., Quinton, J. N., Macleod, C. J. A., Bilotta, G. S., Brazier, R. E., Butler, P., and Haygarth, P. M.: Ensemble evaluation of hydrological model hypotheses, Water Resour. Res., 46, W07516, doi:10.1029/2009WR007845, 2010.

Mantovan, P. and Todini, E.: Hydrological forecasting uncertainty assessment: incoherence of the GLUE methodology, J. Hydrol., 330, 368–381, doi:10.1016/j.jhydrol.2006.04.046, 2006.

Marks, K. and Bates, P. D.: Integration of high resolution topographic data with floodplain flow models, Hydrol. Process., 14, 2109–2122, 2000.

Merwade, V., Olivera, F., Arabi, M., and Edleman, S.: Uncertainty in flood inundation mapping: current issues and future directions, J. Hydrol. Eng., 13, 608–620, 2008.

Montanari, A.: Large sample behaviors of the generalized likelihood uncertainty estimation (GLUE) in assessing the uncertainty of rainfall-runoff simulations, Water Resour. Res., 41, W08406, doi:10.1029/2004WR003826, 2005.

Moya Quiroga, V., Popescu, I., Solomatine, D. P., and Bociort, L.: Cloud and cluster computing in uncertainty analysis of integrated flood models, J. Hydroinf., 15, 55–69, doi:10.2166/hydro.2012.017, 2013.

Neal, J., Schumann, G., and Bates, P.: A subgrid channel model for simulating river hydraulics and floodplain inundation over large and data sparse areas, Water Resour. Res., 48, W11506, doi:10.1029/2012WR012514, 2012.

Paiva, R. C. D., Collischonn, E., and Tucci, C. E. M.: Large scale hydrologic and hydrodynamic modeling using limited data and a GIS based approach, J. Hydrol., 406, 170–181, doi:10.1016/j.jhydrol.2011.06.007, 2011.

Pappenberger, F., Beven, K. J., Horritt, M., and Blazkova, S.: Uncertainty in the calibration of effective roughness parameters in HEC-RAS using inundation and downstream level observations, J. Hydrol., 302, 46–69, 2005.

Patro, S., Chatterjee, C., Singh, R., and Raghuwanshi, N. S.: Hydrodynamic modelling of a large flood-prone system in India with limited data, Hydrol. Process., 23, 2774–2791, doi:10.1002/hyp.7375, 2009.

Rabus, B., Eineder, M., Roth, A., and Bamler, R.: The shuttle radar topography mission – a new class of digital elevation models acquired by spaceborne radar, ISPRS J. Photogramm. Remote Sens., 57, 241–262. doi:10.1016/S0924-2716(02)00124-7, 2003.

Rodríguez, E., Morris, C. S., Belz, J. E., Chapin, E. C., Martin, J. M., Daffer, W., and Hensley, S.: An assessment of the SRTM topographic products, Technical Report JPL D-31639, Jet Propulsion Laboratory, Pasadena, California, 143 pp., http://www2.jplnasa.gov/srtm/srtmBibliography.html (last access: 16 December 2013), 2005.

Schumann, G., Matgen, P., Cutler, M. E. J., Black, A., Hoffmann, L., and Pfister, L.: Comparison of remotely sensed water stages from LiDAR, topographic contours and SRTM, ISPRS J. Photogramm. Remote Sens., 63, 283–296, 2008.

Schumann, G., Di Baldassarre, G., Alsdorf, D., and Bates, P. D.: Near real-time flood wave approximation on large rivers from space: application to the River Po, Northern Italy, Water Resour. Res., 46, W05601, doi:10.1029/2008WR007672, 2010.

Shrestha, D. L., Kayastha, N., and Solomatine, D. P.: A novel approach to parameter uncertainty analysis of hydrological models using neural networks, Hydrol. Earth Syst. Sci., 13, 1235–1248, doi:10.5194/hess-13-1235-2009, 2009.

Stedinger, J. R., Vogel, R. M., Lee, S. U., and Batchelder, R.: Appraisal of the generalized likelihood uncertainty estimation (GLUE) method, Water Resour. Res., 44, W00B06, doi:10.1029/2008WR006822, 2008.

Sun, G., Ranson, K. J., Kharuk, V. I., and Kovacs, K.: Validation of surface height from shuttle radar topography mission using shuttle laser altimetry, Remote Sens. Environ., 88, 401–411. doi:10.1016/j.rse.2003.09.001, 2003.

Tarekegn, T. H., Haile, A. T., Rientjes, T., Reggiani, P., and Alkema, D.: Assessment of an ASTER generated DEM for 2D flood modelling, Int. J. Appl. Earth Obs. Geoinf., 12, 457–465. doi:10.1016/j.jag.2010.05.007, 2010.

Trigg, M. A., Wilson, M. D., Bates, P. D., Horritt, M. S., Alsdorf, D. E., Forsberg, B. R., and Vega, M. C.: Amazon flood wave hydraulics, J. Hydrol., 374, 92–105, 2009.

USACE: HEC-RAS River Analysis System User's Manual. Version 4.1, Hydrologic Engineering Center, Davis, California, 2010.

Vázquez, R. F., Beven, K., and Feyen, J.: GLUE based assessment on the overall predictions of a MIKE SHE application, Water Resour. Res., 23, 1325–1349, doi:10.1007/s11269-008-9329-6, 2009.

Wang, W., Yang, X., and Yao, T.: Evaluation of ASTER GDEM and SRTM and their suitability in hydraulic modelling of a glacial lake outburst flood in southeast Tibet, Hydrol. Process., 26, 213–225, doi:10.1002/hyp.8127, 2011.

Werner, M. G. F.: Impact of grid size in GIS based flood extent mapping using a 1-D flow model, Phys. Chem. Earth Pt. B, 26, 517–522, 2001.

Wilson, M. D. and Atkinson, P. M.: The use of elevation data in flood inundation modelling: a comparison of ERS interferometric SAR and combined contour and differential GPS data, Intl. J. River Basin Management, 3, 3–20, doi:10.1080/15715124.2005.9635241, 2005.

Yan, K., Di Baldassarre, G., and Solomatine D. P.: Exploring the potential of SRTM topographic data for flood inundation modelling under uncertainty, J. Hydroinf., 15, 849–861, doi:10.2166/hydro.2013.137, 2013.

7

Including effects of watershed heterogeneity in the curve number method using variable initial abstraction

Vijay P. Santikari and Lawrence C. Murdoch

Department of Environmental Engineering and Earth Sciences, Clemson University, Clemson, SC 29634, USA

Correspondence: Vijay P. Santikari (vsantik@g.clemson.edu)

Abstract. The curve number (CN) method was developed more than half a century ago and is still used in many watershed and water-quality models to estimate direct runoff from a rainfall event. Despite its popularity, the method is plagued by a conceptual problem where CN is assumed to be constant for a given set of watershed conditions, but many field observations show that CN decreases with event rainfall (P). Recent studies indicate that heterogeneity within the watershed is the cause of this behavior, but the governing mechanism remains poorly understood. This study shows that heterogeneity in initial abstraction, I_a, can be used to explain how CN varies with P. By conventional definition, I_a is equal to the cumulative rainfall before the onset of runoff and is assumed to be constant for a given set of watershed conditions. Our analysis shows that the total storage in I_a (I_{aT}) is constant, but the effective I_a varies with P, and is equal to the filled portion ofI_{aT}, which we call I_{aF}. CN calculated using I_{aF} varies with P similar to published field observations. This motivated modifications to the CN method, called variable I_a models (VIMs), which replace I_a with I_{aF}. VIMs were evaluated against conventional models CM0.2 ($\lambda = 0.2$) and CMλ (calibrated λ) in their ability to predict runoff data generated using a distributed parameter CN model. The performance of CM0.2 was the poorest, whereas those of the VIMs were the best in predicting overall runoff and watershed heterogeneity. VIMs also predicted the runoff from smaller events better than the CMs and eliminated the false prediction of zero-runoffs, which is a common shortcoming of the CMs. We conclude that including variable I_a accounts for heterogeneity and improves the performance of the CN method while retaining its simplicity.

1 Introduction

The estimation of runoff from a rainfall event is of primary importance in applied hydrology. It is necessary in the engineering design of small structures, post-event appraisals, environmental impact work, and other applications (Hawkins, 1993). One of the most popular techniques used for this purpose is the curve number (CN) method, which has been in use for more than half a century (D'Asaro and Grillone, 2012; Hawkins et al., 2008; Kent, 1968; Ponce and Hawkins, 1996; Rallison and Miller, 1982; Soil Conservation Service, 1956, 1972). The method uses a parameter called curve number, which is assumed to depend mainly on land cover, soil types, and antecedent conditions within a watershed.

Curve number varies spatially due to watershed heterogeneity and temporally due to changes in soil moisture, land cover, temperature, and other processes (Hawkins et al., 2008; Ponce and Hawkins, 1996; Rallison and Miller, 1982). CN also varies with the magnitude (D'Asaro and Grillone, 2012; Hawkins, 1993; Hjelmfelt Jr. et al., 2001) and spatiotemporal distribution of rainfall (Hawkins et al., 2008; Van Mullem, 1997). When heterogeneity is known at sufficient detail, CN variation can be accounted for by using a distributed parameter model, e.g., SWAT (Gassman et al., 2007). Otherwise this approach can introduce more parameters than can be reliably estimated from the available data (Soulis and Valiantzas, 2013) and can potentially cause large uncertainties in the predicted runoff. There are several ways to account for temporal variation of CN, each with its own advantages and shortcomings (Santikari, 2017). CN variation with the distribution of rainfall is usually ignored (Hawkins et al., 2008). The CN method is most commonly applied as an event-scale lumped-parameter model, which is simple but also limited in its ability to account for the variations of CN.

This diminishes the accuracy of its runoff predictions (e.g., Soulis and Valianzas, 2012).

The objective of this work is to improve the event-scale lumped-parameter application of the CN method by describing an approach for incorporating the spatiotemporal variations of CN. The investigation is described in two papers, which build on Santikari (2017). In this paper, effects of spatial variation of CN (heterogeneity) at the watershed scale are analyzed. Insights gained from this analysis are used to create modified models that account for heterogeneity. The modified models are evaluated using the runoff generated by a distributed parameter model applied to a hypothetical heterogeneous watershed. In a companion paper (Santikari and Murdoch, 2018) and in Santikari (2017), the modified models are refined by including an approach that accounts for the temporal variation of CN using antecedent moisture. The refined models, which account for spatial and temporal variability, are then evaluated using data from real watersheds.

1.1 Background

The CN method assumes that a rainfall event produces runoff (Q) when the event rainfall (P) exceeds the initial abstraction (I_a). I_a includes interception storage (by tree canopy, roof tops, and such), early infiltration, and surface depression storage. The effective rainfall, $P - I_a$, is partitioned between Q and further infiltration (F). This is given by mass balance as

$$P - I_a = F + Q \;\forall\; P \geq I_a. \tag{1}$$

Both F and Q are zero when $P \leq I_a$, and both increase with P when $P > I_a$. It is assumed that F has an upper limit, which is referred to as the potential maximum retention (S). In other words, S is the total storage available for infiltration after the runoff begins.

The conceptual basis that defines the curve number method comes from the following assumption (Hawkins et al., 2008; NRCS, 2003; Ponce and Hawkins, 1996; Rallison and Miller, 1982; Woodward et al., 2002):

$$\frac{Q}{P - I_a} = \frac{F}{S}, \tag{2}$$

i.e., the runoff coefficient (left-hand side) is equal to the fraction of storage filled in S (right-hand side). Equation (2) is developed using the reasoning that the equality holds at the end points ($P \leq I_a$ and $P \to \infty$) (Hawkins et al., 2008; Rallison and Miller, 1982; Woodward et al., 2002) and that the behavior of both ratios in the intermediate range is essentially the same (Fig. 1). When $P \leq I_a$, both Q and F are zero and therefore the ratios on either side of Eq. (2) are zero. When $P > I_a$, both ratios increase with P, whereas their rate of increase diminishes. At the limit of $P \to \infty$, both ratios approach unity.

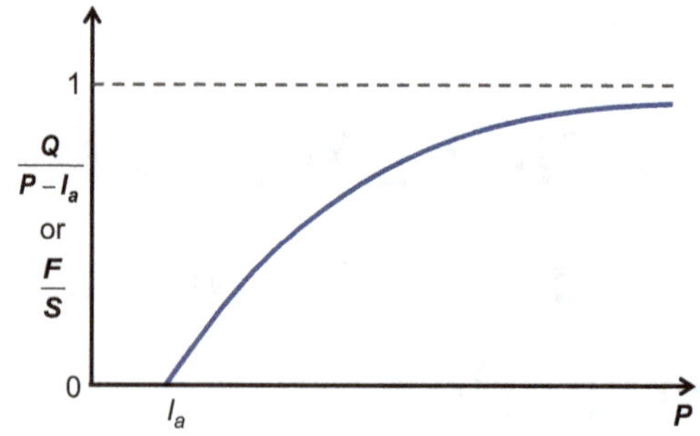

Figure 1. Presumed variation of the ratios in Eq. (2) with event rainfall (P). Q is event runoff, I_a is initial abstraction, F is cumulative infiltration after runoff begins, and S is potential maximum retention (modified from Rallison and Miller, 1982, Fig. 2).

To eliminate the need for an independent estimation of I_a (Ponce and Hawkins, 1996; Rallison and Miller, 1982), it is assumed that

$$I_a = \lambda S, \tag{3}$$

where λ is a dimensionless parameter called the initial abstraction ratio. Early field data suggested an optimum value of $\lambda = 0.2$ (Soil Conservation Service, 1956). However, more recent studies (Hawkins et al., 2008; Woodward et al., 2003) suggest that $\lambda = 0.05$ is more appropriate. Using Eqs. (1)–(3), I_a and F can be eliminated to give

$$Q = 0 \;\forall\; P \leq \lambda S,$$
$$Q = \frac{[P - \lambda S]^2}{P + (1 - \lambda)S} \;\forall\; P > \lambda S. \tag{4}$$

Since the value of λ is usually fixed (at 0.2 or 0.05), Eq. (4) requires only one parameter, S, which varies within the range $0 \leq S \leq \infty$.

For convenience (Hawkins et al., 2008; Ponce and Hawkins, 1996), S (units in mm) is mapped onto a dimensionless parameter called the curve number as

$$\mathrm{CN} = \frac{25\,400}{254 + S} \tag{5}$$

so that CN is 100 when S is zero, but approaches zero as S approaches infinity. In practice, when $\lambda = 0.2$, CN ranges from around 30 (for vegetated surfaces with highly permeable soils) to close to 100 (for impermeable surfaces or soils) (USDA, 1986). Tabulated CN values for various land uses, soil types, and management scenarios are available in handbooks and manuals (NRCS, 2003; USDA, 1986). CN can also be determined from field data by solving Eq. (4) for S as

$$S = \frac{1}{2\lambda^2}\left[2\lambda P + (1 - \lambda)Q - \sqrt{(1 - \lambda)^2 Q^2 + 4\lambda P Q}\right] \tag{6}$$

and then using Eq. (5). Conversely, when the CN of a watershed is known, Q can be estimated for a rainfall event using Eqs. (4) and (5).

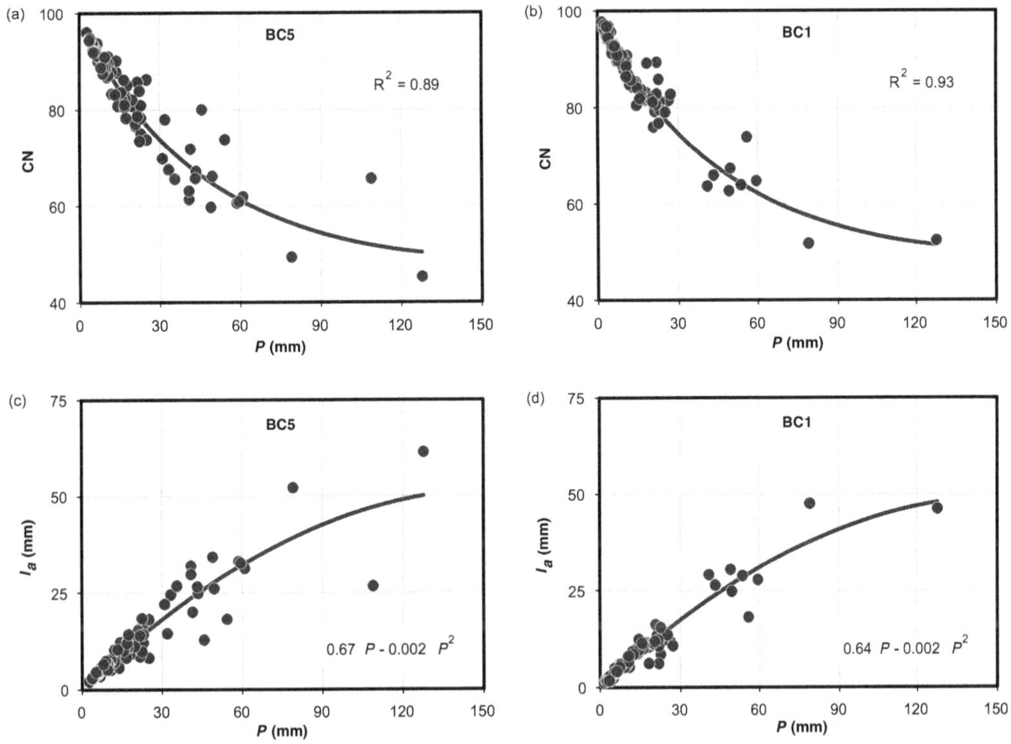

Figure 2. Variation of CN ($\lambda = 0.2$) with P in watersheds **(a)** BC5 and **(b)** BC1, near Greenville, SC. Variation of I_a with P in **(c)** BC5 and **(d)** BC1 (see Santikari, 2017, or Santikari and Murdoch, 2018, for study area description). Best-fit curves for I_a are quadratic functions of P with zero intercept. Corresponding best-fit curves for CN were derived from those of I_a using Eqs. (3) and (5).

The curve number method is appealing because it is based on an intuitive concept (Eq. 2), relies on only one parameter, has a large body of literature (Hawkins et al., 2008), and has a comprehensive database (NRCS, 2003; USDA, 1986). It has been included in many watershed and water-quality models such as SWAT (Soil and Water Assessment Tool) (Neitsch et al., 2005), CREAMS (Chemicals, Runoff and Erosion from Agricultural Management Systems), GLEAMS (Groundwater Loading Effects of Agricultural Management Systems) (Knisel and Douglas-Mankin, 2012), AnnAGNPS (Annualized Agricultural Non-point Source Pollution Model) (Bingner et al., 2011), EPIC (Environmental Policy Integrated Climate), APEX (Agricultural Policy/Environmental Extender) (Wang et al., 2012), and HydroCAD (HydroCAD, 2015). A physically based modeling framework, such as the diffusive-wave approximation for overland flow coupled with the Richard's equation for unsaturated subsurface flow, e.g., Panday and Huyakorn (2004), may improve accuracy and resolution of model predictions compared to the CN method, when the necessary input data, expertise, and computing resources are available. However, the CN method will likely remain popular for many applications in runoff modeling because of its ease of use, wide knowledge base, and less demand on computational resources than many physically based models.

1.2 CN variation with P

Curve number is assumed to be a watershed property that depends on the current conditions, but it also varies with P (e.g., Fig. 2a and b). This behavior was also observed in several previous studies (D'Asaro and Grillone, 2012; Hawkins, 1993; Hjelmfelt et al., 2001; Soulis et al., 2009), and it appears to be a common phenomenon. In 75 % of the watersheds, CN decreased with increasing P and asymptotically approached a constant value. Hawkins (1993) referred to this as *standard behavior*. In 20 % of the watersheds, CN decreased with P but an asymptote was not attained within the range of the observed P. This was referred to as *complacent behavior*. In about 5 % of the watersheds, the CN increased with P and asymptotically approached an apparent constant value. This behavior, referred to as *violent*, was often preceded by complacent behavior at smaller rainfalls. Hawkins (1993) hypothesized that the inverse relationship between CN and P may be due to some spurious correlation between them or due to a bias that inherently results from the selective omission of data from small storm events that failed to produce runoff. The reasoning is that large rainfalls always produce runoff but small rainfalls produce runoff only under wet conditions, when the CN is large. Therefore, small CN values for small rainfalls go unrecorded.

In watersheds showing a standard behavior, CN was treated as an asymptotic function of P as

$$\mathrm{CN} = \mathrm{CN}_\infty + (100 - \mathrm{CN}_\infty)\, e^{-kP}, \tag{7}$$

where $\mathrm{CN} = \mathrm{CN}_\infty$ is the asymptote and k is a calibration parameter (Hawkins, 1993). CN_∞ is the smallest possible value of CN for a watershed and is approached only at large values of P. To develop Eq. (7), measured values of Q, ideally for a large range of values of P, are needed. The usual procedure involves "frequency matching" the data (Hawkins, 1993), i.e., sorting the values of P and Q separately, and pairing them according to their rank. CN for each pair is then calculated using Eqs. (5) and (6). Frequency matching reduces the scatter of data points around the best-fit curve in a CN vs. P plot.

A standard behavior of CN was also observed in two watersheds (BC5 and BC1) near Greenville, South Carolina, USA (Fig. 2a and b). In these watersheds, CN (calculated using $\lambda = 0.2$) decreased from 97 to 50 as P increased from 2 to 128 mm. The data were characterized by a modest scatter ($R^2 = 0.9$) about the best-fit curve based on a quadratic function of P. Description of these watersheds is given by Santikari (2017) and Santikari and Murdoch (2018). The justification for using quadratic functions follows from the analysis of heterogeneity presented in Sect. 2.

The approach used in Fig. 2a and b avoids the commonly used frequency matching (e.g., Hawkins, 1993). Each CN value in the plot was calculated using the $P - Q$ pair from the same storm event. Frequency matching would significantly reduce the scatter in the plot, but it would also downplay the importance of CN variation due to antecedent conditions. Reducing the scatter by accounting for antecedent conditions, e.g., using antecedent moisture (Mishra et al., 2006), is a better approach.

The hypotheses given by Hawkins (1993) are valid, but insufficient to explain the standard and complacent behaviors. It may be true that small rainfalls produce runoff only under wet (large CN) conditions and therefore only the large CN values are recorded. However, if one has a large enough sample of storms, some of the larger storms also must have occurred during wet conditions. For the larger storms, therefore, one would expect to see the whole spectrum of CN values ranging from the largest to the smallest. However, this is not the case. As P increases, the values of CN decrease consistently (Fig. 2a and b).

1.3 Heterogeneity as a cause of CN variation with P

Soulis and Valiantzas (2012) hypothesized that the observed variation of CN with P in the standard and complacent cases is a consequence of watershed heterogeneity. They assumed a hypothetical heterogeneous watershed with two subareas characterized by different CNs. They then calculated the watershed runoff, for a range of values of P, as the area-weighted average of the runoffs from the subareas. Watershed CN calculated using this runoff varied with P akin to the standard behavior. The shape of the synthetically generated CN vs. P curve could be matched with the observations by adjusting the areas of the subareas and their respective CNs. This idea can also be extended to multiple subareas so that the heterogeneity within a watershed can be represented more accurately. However, this could lead to problems of over-parameterization, non-uniqueness, and non-convergence as pointed out by Soulis and Valiantzas (2012).

In a later paper, Soulis and Valiantzas (2013) suggested using spatial information on land cover and soils to delineate the areal extent of subareas and constrain their respective CNs. This approach would reduce the number of calibrated parameters by half because it only requires the calibration of the CNs for the subareas. In essence, the multiple-subarea approach is similar to a distributed modeling approach that calculates the watershed runoff as the area-weighted average of the runoffs from the subareas, e.g., SWAT (Gassman et al., 2007). The approach used by Soulis and Valiantzas (2013) attempts to match the observed and simulated values of CN, whereas that used by SWAT attempts to match the observed and simulated values of Q. Since CN and Q are uniquely related for given values of P and λ, these approaches are equivalent. A major implication of the work of Soulis and Valiantzas (2013) is that a distributed modeling approach can account for the standard and complacent behaviors of CN.

Using a single value of CN independent of P in a heterogeneous watershed can cause a systematic error in Q and lead to poor predictive ability of the method. This is because when CN is constant Q may be underestimated for small P and overestimated for large P (e.g., Soulis and Valianzas, 2012, 2013). This problem can be addressed either by treating CN as a function of P, e.g., asymptotic fitting (Hawkins, 1993), or by using a distributed modeling approach that accounts for heterogeneity in sufficient detail, e.g., SWAT (Gassman et al., 2007) or Soulis and Valianzas (2013). An understanding of the mechanism of how watershed heterogeneity leads to the variation of CN with P is also important. It could help in accounting for this variation without resorting to fine discretization or over-parameterization of the CN method. To accomplish this, an analysis of the effect of heterogeneity on I_a and S is performed, which can then be used to understand the effect on CN.

2 Reevaluation of initial abstraction

The quantities CN, I_a, and S are considered to be the properties of a watershed that depend on current conditions. In usual practice, CN estimated for a certain set of conditions is applicable to any rainfall event occurring in those conditions irrespective of the magnitude of P. However, in every watershed evaluated by the previous studies (D'Asaro and Grillone, 2012; Hawkins, 1993) the CN varied with P. If so, since I_a and S are inversely related to CN (Eqs. 3 and 5), one can expect that they too vary with P but inversely to that of CN. The calculated values of I_a (using Eqs. 3, 6, and $\lambda = 0.2$) for watersheds BC5 and BC1 near Greenville, SC, increase with P and appear to approach a constant at large values of P (Fig. 2c and d). A plot of S vs. P would be similar to the I_a vs. P plot, with the y coordinate scaled by $1/\lambda$.

To evaluate the link between heterogeneity in I_a and its variation with P, we looked at how the effective I_a of a heterogeneous watershed is determined and whether it is affected by the magnitude of P. Our analysis shows that there is an inconsistency between the theoretical definition of I_a and its calculated value at the watershed scale. It also shows how heterogeneity can cause I_a to vary with P and how this relates to variations of S and CN with P.

2.1 Problems with the current usage of I_a

By the theoretical definition of I_a, if runoff is detected in the hydrograph, it is assumed that I_a has been met for the watershed. Watersheds are heterogeneous combinations of various land-use–soil–slope complexes. These are referred to as hydrologic response units (HRUs) in SWAT (Gassman et al., 2007), and the same term is also used here. Each HRU is assumed to be homogeneous and is characterized by representative values of CN (CN_i) and I_a (I_{ai}). During a rainfall event, the HRU with the smallest of the I_{ai} values will be the first to generate runoff. Assuming that this runoff reaches the watershed outlet, by definition, the I_a of the watershed should be equal to the smallest of the I_{ai} values. This could even be zero if the watershed has surfaces such as open water bodies that cannot abstract the rainfall.

However, it is difficult to detect the exact moment of generation of runoff and determine the corresponding value of I_a, which is equal to the cumulative precipitation at that moment. There have been studies (Shi et al., 2009; Woodward et al., 2003) that tried to determine I_a from hydrographs. A problem with this approach is that there can be a time lag between runoff generation in headwaters and its detection at gauging station. Rainfall that occurs during this time lag is also included in I_a, leading to its overestimation. Another possible approach would be to collect observations from a large number of rainfall events and take I_a to be equal to the smallest P that produced runoff. This would eliminate the problem with the lag time, but Q needs to be insignificant to reduce the error in I_a. It should also be noted that I_a determined this way is only representative of the antecedent conditions of the smallest event that produced runoff.

It may be difficult to measure I_a directly, but it can be calculated for any event using Eqs. (6) and (3). However, in medium to large rainfall events, even the HRUs with larger values of I_{ai} will contribute to Q. Therefore, the calculated value of I_a in these events will also be influenced by the larger values of I_{ai}. So, the calculatedI_a tends to be greater than the smallest of the I_{ai} values. Moreover, it can be expected to increase with P as increasingly larger rainfalls generate runoff from HRUs with increasingly larger values of I_{ai}. Thus, there is an inconsistency between the definition of I_a and its calculated value at the watershed scale.

Spatial-scale effect on λ

Strictly adhering to the definition of I_a at the watershed scale may also cause a spatial-scale effect on λ. Let us refer to the CN of the watershed as CN_W and to I_a as I_{aW}. One of the common ways to determine CN_W is to calculate it as the area-weighted average of the CN_i values (NRCS, 2003) as

$$CN_W = \sum_{i=1}^{n} a_i CN_i, \tag{8}$$

where a_i is the fractional area of the ith HRU. Note that the fractional areas must add up to unity. By definition, I_{aW} is equal to the smallest of the I_{ai} values. Therefore, if $I_{a1} < I_{a2} < \ldots < I_{an}$, then

$$I_{aW} = I_{a1}. \tag{9}$$

From Eqs. (3) and (5) it can be shown that CN and I_a are related as

$$CN = \frac{25\,400}{254 + \left(\frac{I_a}{\lambda}\right)}. \tag{10}$$

If all the HRUs are assumed to have the same $\lambda = \lambda_i$, Eqs. (8)–(10) lead to

$$\begin{aligned} &CN_1 > CN_2 > \ldots > CN_n \\ &CN_W < CN_1 \\ &\frac{I_{aW}}{\lambda_W} > \frac{I_{a1}}{\lambda_i}, \\ &\lambda_W < \lambda_i, \end{aligned} \tag{11}$$

where λ_W is the effective initial abstraction ratio of the watershed. Therefore, if λ is assumed to be the same among the component HRUs, it will have a smaller value at the watershed scale. This implies that λ decreases with increasing spatial scale. Therefore, setting λ constant, equal to 0.2 or 0.05, for all the spatial scales contradicts the definition of I_a. In any case, it is probably more accurate to calculate runoff at the HRU scale (Q_i) and take the area-weighted average of Q_i values, rather than take the area-weighted average of the CN_i values and calculate Q at the watershed scale. It is also more appropriate because Q is runoff per unit area, whereas CN is a dimensionless index variable.

The inconsistencies in the usage of I_a are a direct result of heterogeneity in a watershed. Moreover, heterogeneity also appears to be responsible for the variation of I_{aW} with P (Fig. 2c and d). To verify this, a relationship between I_{aW} and the magnitude and areal distribution of I_{ai} values needs to be developed.

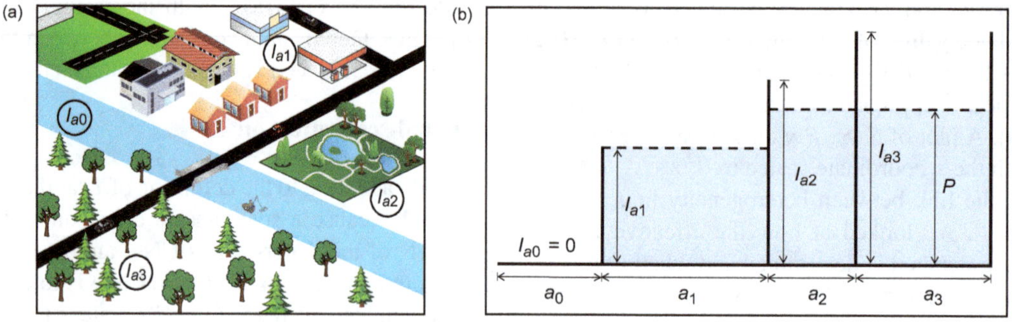

Figure 3. Spatial distribution of I_a in a heterogeneous watershed. **(a)** I_{ai} values of various HRUs mainly characterized by their land use types ($I_{a0} = 0 < I_{a1} < I_{a2} < I_{a3}$); **(b)** conceptual model in which each HRU is represented by a bin with height $= I_{ai}$, width $= a_i$, and unit thickness; shaded area indicates the filled portion during an event with rainfall $= P$.

2.2 I_a in a heterogeneous watershed

Consider a watershed with four HRUs mainly characterized by their land use types, viz. open waterbody (I_{a0}), urban area (I_{a1}), park (I_{a2}), and forest (I_{a3}) (Fig. 3a), such that $I_{a0} = 0 < I_{a1} < I_{a2} < I_{a3}$. An open waterbody generates runoff during every rainfall event. Other land use types generate runoff depending on the magnitude of the rainfall, with land uses of larger I_{ai} values requiring larger magnitudes. If each land use type is assumed to be directly connected to the drainage network, the number of land use types contributing to the runoff, in other words the runoff contributing area, increases with rainfall. This process can be conceptualized by representing the storage distribution of I_a as a series of bins where each bin corresponds to a HRU (Fig. 3b). The height and the width of a bin are given by I_{ai} and a_i, respectively, and all bins have unit thickness. In a rainfall event, only the bins with $I_{ai} \leq P$ are fully filled and contribute to runoff, whereas the others are partially filled and do not contribute to runoff. The total amount of filled storage in I_a (shaded area in Fig. 3b) increases with P until it reaches a constant value when all the bins are fully filled and the whole watershed is contributing to the runoff.

Consider a general case of a heterogeneous watershed with $n+1$ HRUs such that

$$I_{a0} = 0 < I_{a1} < I_{a2} < \ldots < I_{an}, \tag{12}$$

where I_{a0} represents open water bodies and other surfaces that cannot abstract rainfall. The areal average of the total initial abstraction (I_{aT}) is given by

$$I_{aT} = \sum_{i=0}^{n} a_i I_{ai}. \tag{13}$$

In a rainfall event, all the HRUs with $I_{ai} \leq P$ have their initial abstractions completely filled while the others are partially filled. Just by analyzing the runoff for that event, it is impossible to quantify the magnitudes of I_{ai} in HRUs that are partially filled. Because these HRUs have not contributed to the runoff, all that can be said is that their I_{ai} values are greater than P, but their magnitudes remain unknown. However, the information on the magnitudes of I_{ai} in HRUs that are completely filled should be present in the runoff data. In other words, it takes larger rainfalls to fill larger values of I_{ai} and gather information about their magnitude.

Then what is the effective initial abstraction of the watershed for a given rainfall event? Consider an event where the rainfall falls within the range: $I_{am} \leq P < I_{a(m+1)}$. HRUs with $I_{ai} \leq I_{am}$ have their initial abstractions completely filled and produce runoff, whereas HRUs with $I_{ai} \geq I_{a(m+1)}$ have their initial abstractions partially filled up to the level of P and do not produce runoff. The areal average of the filled portion (includes completely filled as well as partially filled HRUs) of the initial abstraction is given by

$$I_{aF} = \sum_{i=0}^{m} a_i I_{ai} + \left(1 - \sum_{i=0}^{m} a_i\right) P. \tag{14}$$

The first term on the right-hand side of Eq. (14) represents completely filled HRUs. The second term represents partially filled HRUs, all of which are filled to the level of P. Note that I_{aT} is the areal average of total initial abstraction, whereas I_{aF} is the areal average of the filled portion. Therefore,

$$\begin{aligned} I_{aF} &< I_{aT} \ \forall \ P < I_{an}, \\ I_{aF} &= I_{aT} \ \forall \ P \geq I_{an}. \end{aligned} \tag{15}$$

The conceptual model presented in Fig. 3 as well as in Eqs. (14) and (15) is intuitively appealing and also hints at the possibility that I_{aW} may be equal to I_{aF}. This is because I_{aF} increases with P and approaches a constant value (I_{aT}), similar to the observations in Fig. 2c and d. Equation (14) is also consistent with a distributed parameter model application of the CN method as described in Sect. 3.

2.3 Variation of I_{aF} with P

To investigate the variation of I_{aF} with P, Eqs. (14) and (15) are applied to the scenario presented in Fig. 3, where $n = 3$. A

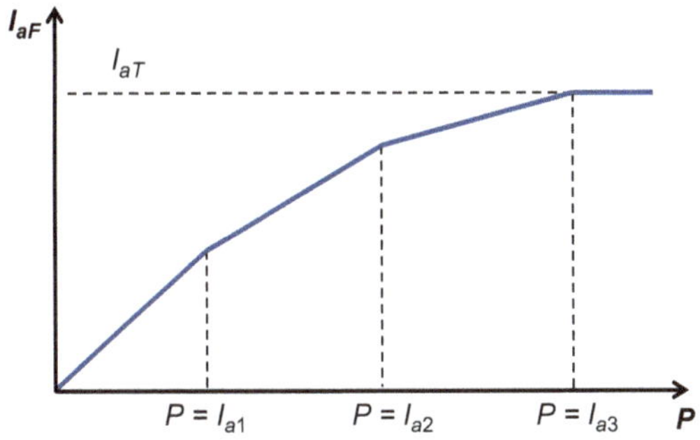

Figure 4. Variation of I_{aF} (Eqs. 14 and 15) with P for the scenario presented in Fig. 3.

plot of I_{aF} vs. P (Fig. 4) shows that I_{aF} increases with P and becomes constant ($I_{aF} = I_{aT}$) at large values of P ($P \geq I_{a3}$). The kink points joining the line segments occur when the initial abstraction of one of the HRUs becomes completely filled. At these points, P is equal to one of the I_{ai} values. In between these points ($I_{am} < P < I_{a(m+1)}$), the relationship between I_{aF} and P is linear with a slope of $\left(1 - \sum_{i=0}^{m} a_i\right)$. The slope abruptly changes across the kink points. It decreases with m and becomes zero when $m = n$. The maximum value the slope can take is unity. This occurs with the line segment passing through the origin, when HRUs with zero initial abstraction are absent (i.e., $a_0 = 0$). When these HRUs are present, however, the origin itself is a kink point where the slope abruptly jumps from unity to $1 - a_0$.

The analysis presented so far represents a discrete case where each HRU is homogeneous and has a finite area. The values of I_{ai} vary discontinuously across the HRUs. Their areal distribution can be represented by a plot of a_i vs. I_a (Fig. 5a). The smaller the area of HRUs, the more numerous they are and the more accurate the representation of the heterogeneity within the watershed is. The most ideal representation would occur when the HRUs shrink to points. Then the values of I_{ai} within the watershed vary continuously and therefore can be represented by a probabilistic distribution of areal occurrence (Fig. 5b). It is impractical to characterize the watershed at such fine scale, but it is worth understanding the properties of the initial abstraction at the finest resolution first and then making assumptions or simplifications later to suit the practical needs.

For the case of a continuous distribution of I_a, Eq. (14) takes the form

$$I_{aF} = \int_0^P I_a\, a(I_a)\, dI_a + \left(1 - \int_0^P a(I_a) dI_a\right) P, \tag{16}$$

where $a(I_a)$ is the probability density function of areal occurrence of I_a. The fractional area with initial abstraction $= I_a$ is given by $a(I_a)dI_a$. The upper limit of the integrals is set to P because the last initial abstraction to completely fill up would

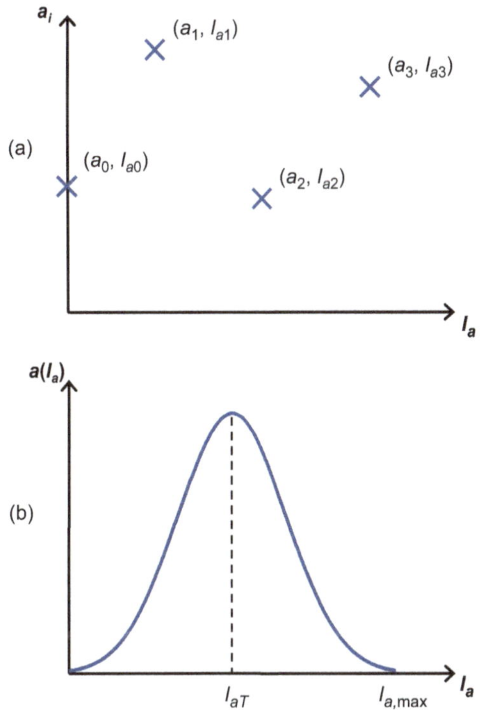

Figure 5. Representing areal distribution of I_a within a watershed **(a)** discrete case and **(b)** continuous case.

be equal to P. The areal average of total initial abstraction, I_{aT}, is given by

$$I_{aT} = \int_0^{I_{a,max}} I_a\, a(I_a)\, dI_a, \tag{17}$$

where $I_{a,max}$ is the maximum value of I_a within the watershed. Thus, I_{aT} is equal to the mean of the distribution (Fig. 5b). Equation (15) then becomes

$$\begin{aligned} &I_{aF} < I_{aT}\ \forall\ P < I_{a,max}, \\ &I_{aF} = I_{aT}\ \forall\ P \geq I_{a,max}. \end{aligned} \tag{18}$$

Unlike the discrete case, the slope of the I_{aF} curve for the continuous case decreases smoothly with increasing P (Fig. 6). This is because the line segments in the discrete case (Fig. 4) shrink to points in the continuous case. It follows from Eq. (16) that

$$\begin{aligned} &\left.\frac{dI_{aF}}{dP}\right|_{P=0} = 1, \\ &\left.\frac{dI_{aF}}{dP}\right|_{P \geq I_{a,max}} = 0. \end{aligned} \tag{19}$$

Thus, the I_{aF} curve is bounded by a line of slope $= 1$ passing through the origin and a line of slope $= 0$ with the intercept equal to I_{aT} (Fig. 6). The line of slope $= 1$ is referred to

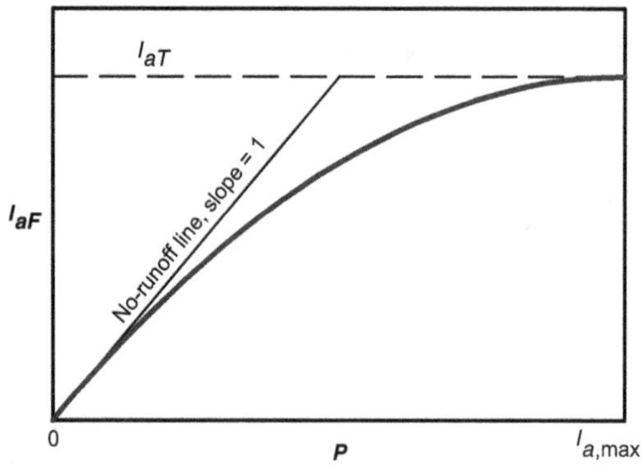

Figure 6. Variation of I_{aF} with P for a continuous distribution such as the one shown in Fig. 5b.

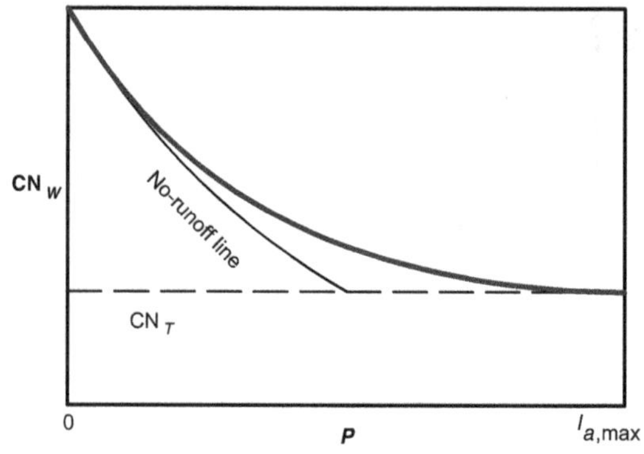

Figure 7. CN_W as a function of P when I_{aW} is assumed to be equal to I_{aF} (shown in Fig. 6).

as the no-runoff line because along this line $I_{aF} = P$. When the whole watershed is represented by a single HRU, the I_{aF} curve coincides with the no-runoff line until $I_{aF} = I_{aT}$. A comparison of Figs. 6 to 2c and d strengthens the case that I_{aW} is equal to I_{aF}.

2.4 Variation of CN_W with P

Let us hypothesize that $I_{aW} = I_{aF}$; i.e., the effective I_a of a watershed is equal to the area-weighted average of the filled portion of the initial abstraction. Then, if Eq. (10) is written for CN_W, I_a can be replaced by I_{aF}. Substituting Eq. (16) in Eq. (10) gives CN_W as a function of P. When plotted against P, CN_W starts at 100 when $P = 0$ and then decreases with increasing P (Fig. 7).

Differentiating Eq. (10) and using Eq. (19) gives

$$\left.\frac{\mathrm{d}(\mathrm{CN_W})}{\mathrm{d}P}\right|_{P=0} = -\frac{10}{\lambda},$$
$$\left.\frac{\mathrm{d}(\mathrm{CN_W})}{\mathrm{d}P}\right|_{P\geq I_{a,\max}} = 0, \tag{20}$$

where the constant 10 has units of 1/in. Thus, the CN_W vs. P curve is at its steepest at $P = 0$ and flattens with increasing P, and it becomes constant when $P \geq I_{a,\max}$. This constant, CN_T, is the smallest value CN_W can take and corresponds to the case $I_{aF} = I_{aT}$, when the initial abstractions of all the HRUs are fully filled. CN_W as a function of P is bounded by a curve corresponding to the condition $P = I_{aF}$, the no-runoff line, and a line of slope = 0 with the intercept equal to CN_T (Fig. 7).

The shape of the CN_W vs. P curve (Fig. 7) generated using Eqs. (10) and (16) is quite similar to the best-fit curves from field observations (Fig. 2a and b). Nearly 95 % of the watersheds evaluated in the previous studies (D'Asaro and Grillone, 2012; Hawkins, 1993) also had responses identical to Fig. 7, supporting the hypothesis that $I_{aW} = I_{aF}$. Thus, as also concluded by Soulis and Valiantzas (2012, 2013), the observed complacent and standard behaviors are caused by the inevitable presence of heterogeneity in a watershed. Moreover, complacent behavior appears to be a special case of standard behavior (Soulis and Valiantzas, 2012), where observations from larger rainfalls are unavailable. Therefore, it is probably more appropriate to refer to any "CN decreasing with P" trend as standard behavior. It also shows that assuming a partial source area whenever a complacent behavior is observed (D'Asaro and Grillone, 2012, 2015) can be misleading.

2.5 I_{aF} and CN_W curves for various distributions of I_a

The functional form of $a(I_a)$ defines the areal distribution of I_a within a watershed. We considered idealized functional forms of $a(I_a)$ that correspond to uniform, normal, triangular, and bimodal distributions (Table 1). In each $a(I_a)$, the maximum or other key value was constrained so that the total area under the distribution was unity. For example, the y coordinate of the apex in the triangular distribution must be equal to $2/I_{a,\max}$ (Table 1). In the case of normal distribution, however, the area under the curve is unity only when the limits are infinite. Therefore, a standard deviation (σ) much less than $I_{a,\max}$ was used so that the area under the curve within the range $0 \leq I_a \leq I_{a,\max}$ is approximately equal to unity.

For each distribution, the corresponding functional form of I_{aF} was determined using Eq. (16) and the results are presented in Table 1. For the general case of $a(I_a)$ as a polynomial, the corresponding I_{aF} is a polynomial two degrees higher than $a(I_a)$. For the normal distribution, I_{aF} is a combination of Gaussian and error functions (Table 1).

For the purpose of comparison, symmetrical versions of the distributions were considered such that all of them have the same minimum, mean, and maximum values of I_a (Fig. 8a). The minimum value of I_a was set to zero and the maximum value was $I_{a,\max}$. Therefore, the mean for all the distributions was $I_{a,\max}/2$.

The kurtosis (peakedness) of $a(I_a)$ has a major influence on the shapes of I_{aF} and CN_W plotted as functions of P (Fig. 8). The normal distribution has the greatest kurtosis,

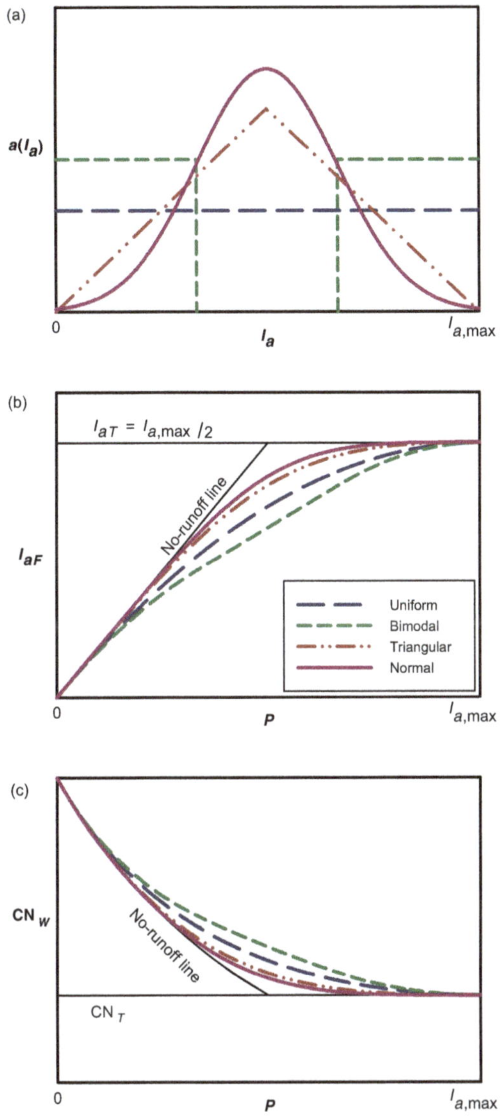

Figure 8. (a) Various symmetrical distributions of I_a with the same minimum (zero), mean ($I_{a,max}/2$), and maximum ($I_{a,max}$); **(b)** the corresponding I_{aF} curves calculated using Eq. (16); **(c)** the corresponding CN_W curves calculated using Eq. (10).

whereas the bimodal distribution has the least. As the kurtosis decreases, the I_{aF} and CN_W curves deviate further from the bounding lines (Fig. 8). When there is a gap in the distribution, as in the case of the bimodal distribution, the corresponding I_{aF} curve is linear for the range spanning the gap. This is consistent with the discrete case where I_{aF} was represented by line segments for the gaps in between the discrete values of I_{ai} (Fig. 4).

Skewness of $a(I_a)$ also affects I_{aF}, and this is illustrated by an idealized case where an initially uniform distribution is positively skewed (Fig. 9a and b). The mean of $a(I_a)$, which is equal to I_{aT} (Eq. 17), decreases with increasing positive skewness. This is important because a land use change such as conversion of forest to urban land is expected to increase the positive skewness (i.e., more low values of I_a). During the conversion, $I_{a,max}$ remains unchanged while some forested land remains. When the entire forest is converted, $I_{a,max}$ drops to a lower value.

The analysis also shows that a watershed cannot be characterized or compared with other watersheds using a single value of CN (such as CN_∞ used in asymptotic fitting, Eq. 7). Depending on the distribution of heterogeneity, the relative runoff potential of a watershed can be P dependent. This is illustrated by considering two uniform distributions, uni1 and uni2, where uni2 has a narrower range and a smaller mean than uni1 (Fig. 9c). For smaller values of P, $I_{aF,uni_1} < I_{aF,uni_2}$ (Fig. 9d) and therefore $CN_{W,uni_1} > CN_{W,uni_2}$. However, for larger values of P, the converse is true. Thus, the watershed with uni1 generates more runoff for smaller values of P, whereas the watershed with uni2 generates more runoff for larger values of P.

3 Effect of heterogeneity on S

Similar to the case of I_a, the presence of heterogeneity also causes the effective S of a watershed (S_W) to vary with P. The functional form of S_W depends not only on the potential maximum retentions of the HRUs (values of S_i) but also on the values of I_{ai}. S_W can be estimated using Eq. (2) if the quantities I_{aW}, Q_W, and F_W are known. A distributed modeling approach can be used to calculate these quantities for a heterogeneous watershed. Distributed CN models, e.g., SWAT (Gassman et al., 2007), commonly calculate Q_W as the area-weighted average of Q_is, and this assumption can also be extended to F_W. Thus,

$$Q_W = \sum_{i=0}^{n} a_i Q_i,$$

$$F_W = \sum_{i=0}^{n} a_i F_i. \tag{21}$$

Using Eq. (21) and applying mass balance (Eq. 1) at watershed and HRU scales gives Eq. (14) for I_{aW}. This shows that I_{aW} calculated using a distributed model is equal to I_{aF}.

Writing an expression for S_W in terms of I_{ai} and S_i values for a general case of a heterogeneous watershed is cumbersome. Therefore, it is only presented graphically for an example of a heterogeneous watershed. However, an expression for S_W can be presented in a compact form for a special case where all the values of I_{ai} are zero as

$$\text{if } I_{ai} = 0 \; \forall \; i \in \{0, 1, \ldots n\}$$

$$S_W = \frac{\sum_{i=0}^{n} \frac{a_i S_i}{P+S_i}}{\sum_{i=0}^{n} \frac{a_i}{P+S_i}}. \tag{22}$$

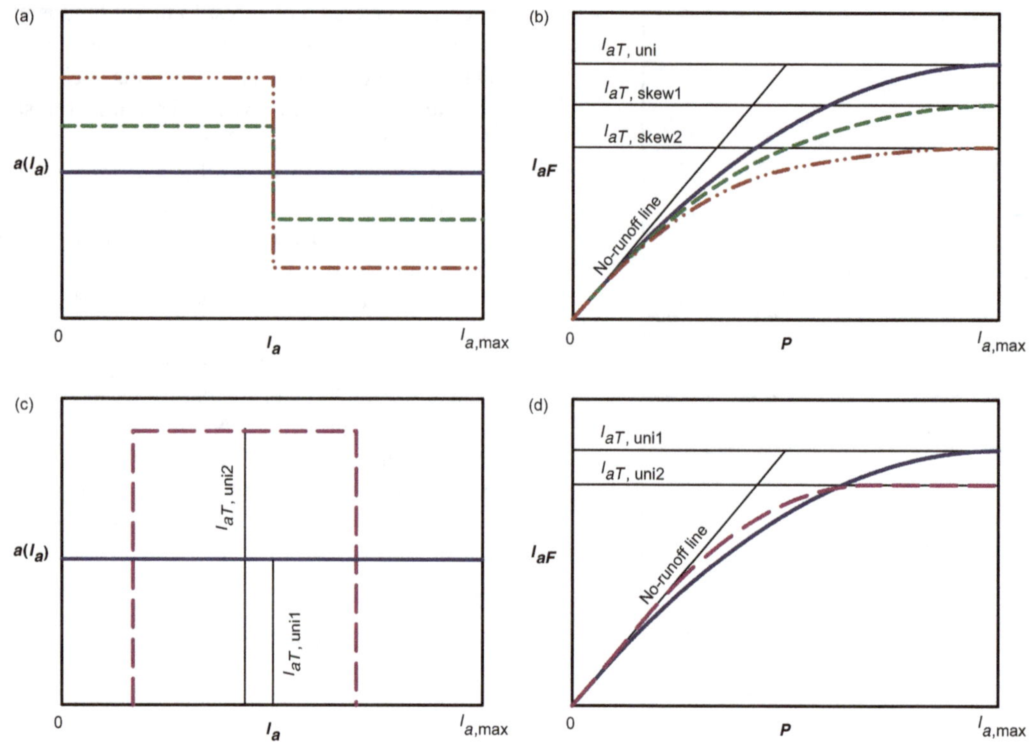

Figure 9. Effect of skewness, mean, and range of $a(I_a)$ on I_{aF} **(a)** uniform, uni (solid), and two positively skewed distributions, skew1 (dashed) and skew2 (dash, dot, dot); **(b)** I_{aF} as a function of P for the distributions shown in **(a)**; **(c)** uniform distributions uni1 (solid) and uni2 (dashed) where uni2 has a narrower range of values of I_a and a smaller mean than uni1; **(d)** I_{aF} as a function of P for the distributions shown in **(c)**.

Thus, S_W varies from the area-weighted harmonic mean $\left(\sum_{i=0}^{n} \frac{a_i}{S_i}\right)^{-1}$ when $P = 0$ to the area-weighted arithmetic mean $\left(\sum_{i=0}^{n} a_i S_i\right)$ when $P \gg S$.

To illustrate the effect of heterogeneity on S_W, an example watershed with the storage distribution shown in Table 2 was considered. The variation of S_W with P was analyzed for the cases of $\lambda_i = 0$ and $\lambda_i = 0.2$ (Fig. 10). In both cases, S_W increases with P and approaches the area-weighted arithmetic mean, S_∞, for large values of P. In the case of $\lambda_i = 0$, the slope of the curve is maximum at the origin and decreases monotonically with P. In the case of $\lambda_i = 0.2$, however, the slope is zero at the origin and generally increases with P until $P \approx I_{an} = 40$ mm ($P \approx I_{a,max}$ for the continuous case), where it reaches its maximum value. Thereafter the slope decreases monotonically with P, giving an S-shaped curve. In other words, the slope generally increases with P until the entire watershed area contributes to the runoff and decreases thereafter.

The similarities between I_{aW} and S_W are that they both increase with P and have an upper limit equal to the area-weighted arithmetic mean of their respective components. The difference is that I_{aW} reaches its upper limit of I_{aT} for a finite value of P ($P = I_{an}$ or $P = I_{a,max}$), whereas S_W requires large values of P ($P \gg S$) to reach its upper limit of S_∞. Moreover, S_W vs. P is an S-shaped curve when $\lambda_i > 0$. This shows that I_{aW} and S_W are not proportional, i.e., λ_W is not a constant even though λ_i values are assumed to be equal and constant.

4 Application

The analysis from previous sections shows that I_{aW} and S_W are functions of P and gives their functional forms. Incorporating these functions in the lumped-parameter application can potentially improve the performance of the CN method.

4.1 I_{aW} as a function of P

The distributed parameter modeling approach, Eq. (21) with the application of mass balance (Eq. 1) at watershed and HRU scales, shows that $I_{aW} = I_{aF}$. I_{aF} is given by Eq. (14) for the discrete case and Eq. (16) for the continuous case. All the distributions in Table 1, except the normal distribution, gave a zero-intercept polynomial for I_{aF}. Therefore, using a quadratic function of the form

Table 1. Functional forms of $a(I_a)$ and I_{aF} for various synthetic distributions.

Distribution	Graph	$a(I_a)$	I_{aF}
Uniform	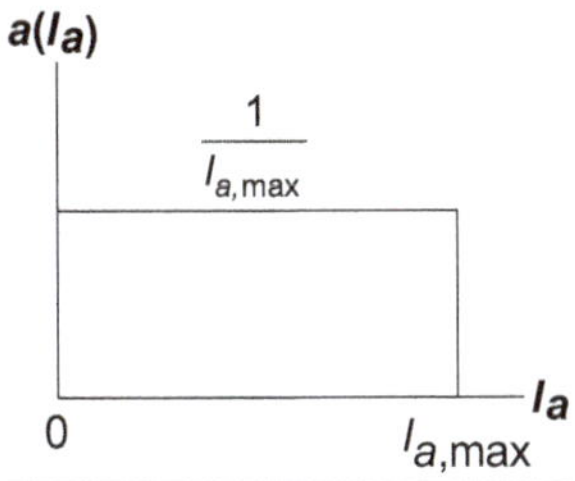	$\frac{1}{I_{a,max}}$	$P - \frac{P^2}{2I_{a,max}}$
Normal	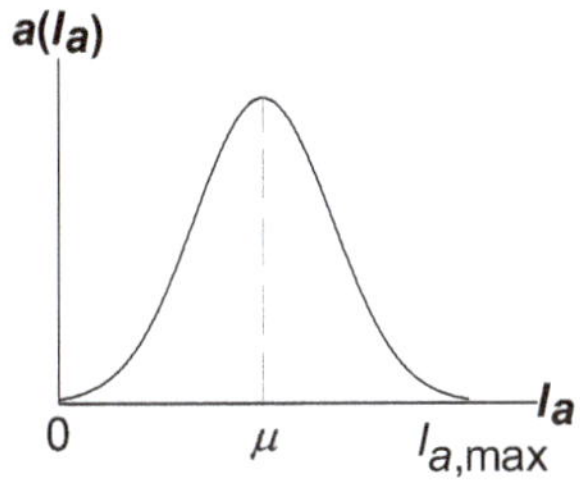	$\frac{1}{\sigma\sqrt{2\pi}}e^{-(I_a-\mu)^2/(2\sigma^2)}$	$P - \frac{\sigma}{\sqrt{2\pi}}\left[e^{-(P-\mu)^2/(2\sigma^2)} - e^{-\mu^2/(2\sigma^2)}\right]$ $-\frac{(P-\mu)}{2}\left[\mathrm{erf}\left(\frac{P-\mu}{\sqrt{2}\sigma}\right)+\mathrm{erf}\left(\frac{\mu}{\sqrt{2}\sigma}\right)\right]$
Triangular	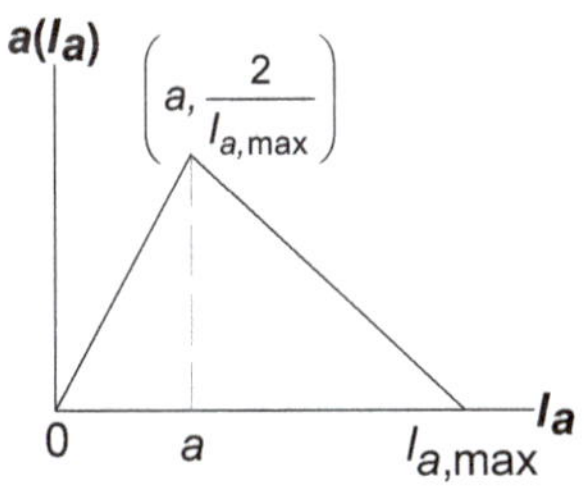	$\frac{2I_a}{aI_{a,max}}$ if $I_a \le a$ $\frac{2}{(I_{a,max}-a)}\left(1-\frac{I_a}{I_{a,max}}\right)$ if $a < I_a$	$P - \frac{P^3}{3aI_{a,max}}$ if $P \le a$ $\frac{1}{(I_{a,max}-a)}\left[-\frac{a^2}{3}+I_{a,max}P-P^2+\frac{P^3}{3I_{a,max}}\right]$ if $a < P$
Bimodal	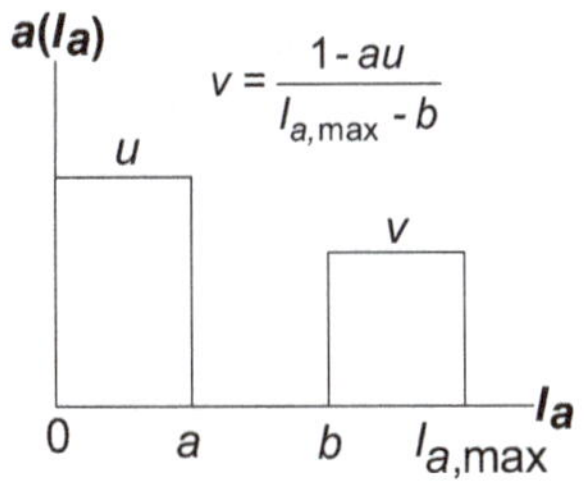	u if $I_a \le a$ 0 if $a < I_a < b$ v if $b \le I_a$	$P - \frac{uP^2}{2}$ if $P \le a$ $\frac{ua^2}{2} + (1-ua)P$ if $a < P < b$ $\frac{(ua^2-vb^2)}{2} + (1-ua+vb)P - \frac{vP^2}{2}$ if $b \le P$

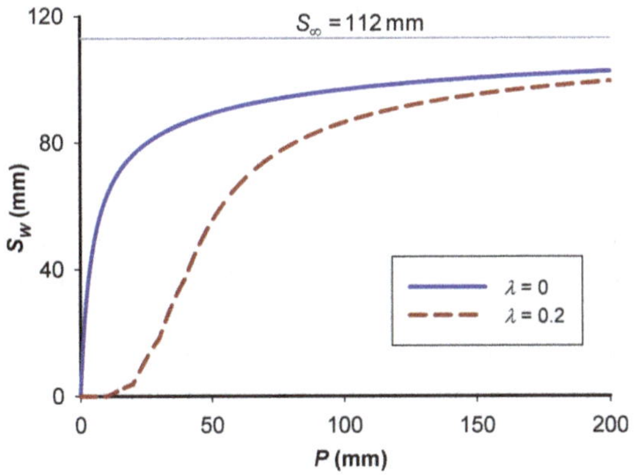

Figure 10. Variation of S_W with P in a heterogeneous watershed with the storage distribution shown in Table 2.

$$I_{aW} = c_1 P - c_2 P^2 \;\forall\; P \le I_{a,max}$$
$$I_{aW} = I_{aT} = c_1\left(I_{a,max}\right) - c_2\left(I_{a,max}\right)^2 \;\forall\; P > I_{a,max} \tag{23}$$

is an efficient way to describe I_{aW}. In Eq. (23), c_1 and c_2 are calibration parameters such that $0 \le c_1 \le 1$ and $c_2 \ge 0$. Since the slope of I_{aW} is zero at $P = I_{a,max}$ (Eq. 19), it follows from Eq. (23) that

$$I_{a,max} = \frac{c_1}{2c_2},$$
$$I_{aT} = \frac{c_1^2}{4c_2}. \tag{24}$$

Similarly, the slope of I_{aW} is unity at $P = 0$ so c_1 should be unity. However, it was kept as a free parameter in Eq. (23) to allow for the approximation of piecewise functions (e.g., I_{aF} for triangular and bimodal distributions in Table 1). Moreover, the analysis for the discrete case shows that when HRUs with zero initial abstraction are present, the origin is a kink point where the slope abruptly jumps from unity to $1 - a_0$. To avoid over-parameterization of the model, a polynomial of degree > 2 for I_{aW} was not considered.

4.2 S_W as a function of P

The sigmoid-shaped function of S_W, with the conditions that $S_W = 0$ when $P = 0$ and that the maximum slope occurs at $P = I_{a,max}$, requires at least two parameters to describe it. However, this along with Eq. (23) would also increase the number of calibrated parameters in the CN method, increasing its complexity and potentially causing non-uniqueness. A relatively simple approach is to assume that S_W is constant similar to the conventional CN method. Another approach is to assume that S_W is proportional to I_{aW}; i.e., Eq. (3) is applicable for a heterogeneous watershed.

Here the emphasis is placed on treating I_{aW} as a function of P while offering some flexibility on how S_W is treated. This is because the variation of I_{aW} with P had a significant impact on the model performance, whereas including the variation of S_W with P showed only marginal or no improvement. This may be because I_{aW} is a component of mass balance (Eq. 1) while S_W is not. F_W, which is the filled portion of S_W, is a component of mass balance and varies with P even if S_W is assumed to be a constant. Therefore, to maintain the simplicity of the CN method and avoid the problems of over-parameterization and non-uniqueness, modeling the sigmoid-shaped function of S_W is omitted.

4.3 Lumped-parameter models

Lumped-parameter application of the CN method was modified by treating I_{aW} as a function of P as described in the previous section. Modified lumped-parameter CN models were evaluated by comparing their performance with that of the conventional lumped-parameter CN models.

4.3.1 Conventional models (CMs)

Conventional CN models are defined by Eqs. (1) through (5) and by the assumption that I_{aW} and S_W are independent of P. In this study two types of conventional models, referred to as CM0.2 and CMλ, were used. In CM0.2, λ_W was fixed at 0.2, and, in CMλ, λ_W was determined by calibration. Thus, CM0.2 had one free parameter, S_W, whereas CMλ had two free parameters, λ_W and S_W.

4.3.2 Variable initial abstraction models (VIMs)

VIMs are defined by Eqs. (1), (2), (4), (5), and (23), and they have three free parameters. If S_W is assumed to be independent of P, then the model requires calibration of c_1, c_2, and S_W and is referred to as VIMS. If Eq. (3) is also included, then the model requires calibration of c_1, c_2, and λ_W and is referred to as VIMλ.

Table 2. Storage distribution in a hypothetical heterogeneous watershed used to illustrate the variation of S_W with P.

HRU	a_i	S_i (mm)
0	0.05	0
1	0.20	50
2	0.35	100
3	0.25	150
4	0.15	200

5 Evaluation

Lumped-parameter models described in the previous section were evaluated in their ability to predict runoff and account for watershed heterogeneity. Accounting for heterogeneity means that the model accurately predicts I_{aW} and S_W, as well as runoff from smaller events. This is because (i) I_{aW} and S_W as functions of P are directly related to heterogeneity, and (ii) the model's inability to account for their variation with P causes underestimation of runoff in smaller events.

Evaluation of lumped-parameter models requires the data for I_{aW}, Q_W, and S_W. This is generated using a distributed parameter model application of the CN method. The assumption is that a distributed parameter model accounts for heterogeneity, and therefore its estimates of I_{aW}, Q_W, and S_W are accurate.

5.1 Distributed parameter model

In a distributed parameter model, Eqs. (1) through (5) are applicable at the HRU scale, with the assumption that I_{ai} and S_i are independent of P. Once Q_i and F_i are calculated for each HRU, watershed-scale quantities I_{aW}, Q_W, F_W, and S_W are calculated using Eqs. (14), (21), and (2).

The distributed parameter model was applied to an idealized synthetic watershed with the storage distribution shown in Table 2, for the cases of $\lambda_i = 0$, 0.2, and 0.5. A range of values of P were synthetically generated such that they vary lognormally from 0.1 to 200 mm with a median of 8 mm. For each rainfall event, I_{aW}, Q_W, F_W, and S_W were calculated and used in the evaluation of the lumped-parameter models.

The reason for using a synthetic watershed here is that the heterogeneity can be precisely defined and used to evaluate the predictions of heterogeneity by the lumped-parameter models. In real watersheds, the heterogeneity has to be determined by calibration, and there can be non-uniqueness when multiple HRUs are present. Application of these modified models to data from real watersheds is discussed by Santikari (2017) and Santikari and Murdoch (2018).

5.2 Model evaluation criteria

Each lumped-parameter model was calibrated by minimizing the sum of the squared residuals between its predicted runoff (Q_W) and the baseline from the distributed parameter model. All the models were evaluated using the Nash–Sutcliffe efficiency parameter (NSE), the standard error of estimate (SEE), and the percent bias (PB) (McCuen, 2003; Moriasi et al., 2007). NSE can vary from $-\infty$ to 1. The calculations and observations are exactly equal when NSE $=1$. The calculations are only as good as the average observation when NSE $=0$. SEE is the root-mean-square residual adjusted to the degrees of freedom (Santikari, 2017). A smaller SEE indicates a better performance, and its ideal value is zero. PB indicates whether the model is over- (PB <0) or under-predicting (PB >0) on average. The optimal value for PB is zero.

NSE values were calculated for the model predictions of runoff (NSE_Q), initial abstraction (NSE_{Ia}), potential maximum retention (NSE_S), and runoff from events with P less than the median value (NSE_{Q50}). Relative NSE, rNSE (Krause et al., 2005), was used instead of NSE when the latter was strongly influenced by larger events. PB values were calculated for runoff from all the events (PB_Q) and runoff from events with P less than the median value (PB_{Q50}). SEE was calculated for runoff from all the events (SEE_Q).

NSE_{I_a} and NSE_S indicate how accurately a lumped-parameter model predicts the watershed heterogeneity. NSE_Q, SEE_Q, and PB_Q reflect the overall accuracy in a model prediction of runoff from all the events, whereas NSE_{Q50} and PB_{Q50} reflect the accuracy in predicting runoff from smaller events ($P<8$ mm). Conventional models tend to under-predict runoffs from smaller events because of the usage of constant I_a and S. They often falsely predict zero-runoffs because the runoff condition ($P>I_a$) cannot be overcome in smaller events. NSE_{Q50} and PB_{Q50} are used to expose this shortcoming.

6 Results and discussion

The results show that using variable initial abstraction improved the accuracy of model predictions of runoff and heterogeneity (Table 3). Based on their overall performance, the models can be arranged from the best to the worst as VIMλ > VIMS > CMλ > CM0.2. Results for the case of $\lambda_i=0$ are not presented in Table 3 because VIMλ, VIMS, and CMλ performed equally well while CM0.2 was the worst (i.e., VIMλ = VIMS = CMλ > CM0.2).

Variable I_a models predicted runoff better than the conventional models. It was not possible to determine relative model performance using NSE_Q because it was 1.0 for all the models. This was because NSE_Q was strongly influenced by a few larger events, and a good fit in these events was sufficient to render $\mathrm{NSE}_Q=1.0$. Therefore, rNSE_Q (Krause et al., 2005) was calculated instead and listed in Table 3. Larger events had greater influence on rNSE_Q as well, but the values varied slightly between the models (Table 3). rNSE_Q increased down the table whereas SEE_Q decreased, with both indicating an improvement in model performance. PB_Q was positive for all the models, indicating that they all under-predicted runoff. The extent of under-prediction, however, was smaller in variable I_a models than the conventional models.

Variable I_a models gave a better estimate of watershed heterogeneity than the conventional models as indicated by the higher values of NSE_{I_a} and NSE_S (Table 3). NSE_{I_a} was zero or negative in the conventional models, whereas it varied from 0.2 to 0.7 in the variable I_a models. NSE_S was negative in all the models, indicating that their estimates of S were poor. In the case of the conventional models this was due to using uniform I_a and S and thereby homogenizing the watershed. In the case of the variable I_a models, this was due to their inability to model the S-shaped function of S. Based on NSE_{I_a} and NSE_S, VIMλ was the best model in estimating watershed heterogeneity.

Variable I_a models also predicted runoff better than the conventional models in smaller rainfall events ($P<8$ mm) as indicated by NSE_{Q50} and PB_{Q50}. In both cases of $\lambda_i=0.2$ and 0.5, only HRU no. 0 (Table 2) produced runoff when $P<8$ mm. This was similar to the case of a partial source area. As CM0.2 and CMλ predicted an $I_a>8$ mm in both cases (Table 3), they falsely predicted zero-runoffs in all the events with $P<8$ mm because the runoff condition ($P>I_a$) could not be overcome. Therefore, their $\mathrm{PB}_{Q50}=100$ in both cases, indicating a 100 % under-prediction in small events. Their NSE_{Q50} was also poor with the same value in both cases. VIMS performed slightly better than the conventional models with 70–90 % under-predictions and with NSE_{Q50} varying from -0.8 to -1.8 (Table 3). VIMλ performed significantly better than all the other models with 30 % or less under-predictions and with NSE_{Q50} varying from 0.6 to 0.9. Even though there were under-predictions, there was no false prediction of zero-runoffs for any of the events in the variable I_a models.

In the models where λ_W was calibrated (CMλ and VIMλ), it was smaller than λ_i (Table 3). This shows that λ at the watershed scale tends to be smaller than that at the HRU scale in the lumped-parameter models. All the models under-predicted I_a or I_{aT} with CMλ being the most severe. There was also a corresponding over-prediction of S or S_∞ by all the models except for the case of $\lambda_i=0.2$ in CM0.2. Again, the most over-prediction of S occurred in CMλ. The under-prediction of I_a and the corresponding over-prediction of S is due to the transfer of storage from I_a to S, which generally improves the performance in the conventional models.

Table 3. The performance of lumped-parameter CN models that were calibrated to the runoff data generated using a distributed CN model for two cases of a synthetic watershed with the storage distribution shown in Table 2. SEE, I_a, and S are in mm. (SEE: standard error of estimate, PB: percent bias, NSE: Nash–Sutcliffe efficiency parameter, rNSE: relative NSE.)

Lumped model	rNSE_Q	SEE_Q	PB_Q	NSE_{I_a}	NSE_S	NSE_{Q50}	PB_{Q50}	λ_W	I_a or I_{aT}	$I_{a,max}$	S or S_∞
				Distributed model: $\lambda_i = 0.2$, $I_{aT} = 22$, $I_{a,max} = 40$, $S_\infty = 112$							
CM0.2	0.93	0.91	12.6	−1.8	−13	−2.9	100	0.20	19	–	97
CMλ	0.94	0.37	5.4	0.0	−26	−2.9	100	0.07	9	-	132
VIMS	0.97	0.13	2.1	0.2	−22	−0.8	71	–	12	64	121
VIMλ	1.00	0.06	0.2	0.4	−3	0.9	16	0.09	11	43	124
				Distributed model: $\lambda_i = 0.5$, $I_{aT} = 56$, $I_{a,max} = 100$, $S_\infty = 112$							
CM0.2	0.93	0.81	18.8	−1.4	−102	−2.9	100	0.20	31	–	155
CMλ	0.94	0.66	13.6	−0.3	−166	−2.9	100	0.11	21	–	197
VIMS	0.96	0.26	6.9	0.7	−83	−1.8	87	–	37	130	140
VIMλ	0.99	0.13	1.6	0.7	−9	0.6	33	0.21	33	96	153

6.1 Storage transfer from I_a to S

The storage in a watershed is distributed between I_a and S. I_a is the part of the storage that does not produce runoff while being filled, whereas S is the part that produces runoff while being filled. Using Eqs. (2) and (1) it can be shown that

$$S = \frac{(P - I_a)(P - I_a - Q)}{Q}. \tag{25}$$

For an observed storm event, P and Q are known and therefore are constants in Eq. (25), so decreasing I_a will increase S. However, the magnitude of increase in S will be greater than the magnitude of decrease in I_a. This is illustrated by differentiating Eq. (25) and using Eq. (4) to give

$$\frac{dS}{dI_a} = -\left(1 + \frac{2S}{P - I_a}\right) \;\forall\, P > I_a. \tag{26}$$

Thus, dS/dI_a is always negative and less than or equal to -1. If $(P - I_a) \gg S$ or $S \approx 0$, then $dS/dI_a \approx -1$, implying an equal transfer in storage between I_a and S. However, as P decreases, dS/dI_a becomes less than -1, implying that S changes more rapidly than I_a. In other words, the relative change of magnitude in S with respect to I_a is large for smaller P, decreases with increasing P, and approaches unity for large values of P.

Storage transfer is evident when the values of I_a and S for the models CM0.2 and CMλ are compared (Table 3). For the case of $\lambda_i = 0.2$, I_a decreased from 19 mm in CM0.2 to 9 mm in CMλ, whereas S increased from 97 to 132 mm, i.e., $dS/dI_a = -3.5$. Similarly, for the case of $\lambda_i = 0.5$, $dS/dI_a = -4.2$.

A transfer of storage from I_a to S improves the performance in the conventional models (i.e., CMλ > CM0.2) because (i) a smaller I_a reduces the percentage of events with falsely predicted zero-runoffs and (ii) it allows the model to mimic a variable I_a. Because of a larger I_a, CM0.2 falsely predicted zero-runoffs in 80 % of the events for $\lambda_i = 0.2$ and in 85 % of the events for $\lambda_i = 0.5$. In the case of CMλ, the percent of events with zero-runoffs dropped to 57 % and 81 %, respectively, because I_a for CMλ was smaller than CM0.2. Mimicking variable I_a can be explained by considering I_{aF} and F, which are the filled portions of I_a and S, respectively. I_{aF} and F have similar functional relationships with P (compare Fig. 6 to Fig. 1); i.e., they both increase with P and approach a constant for large values of P. In the conventional CN models, there is no provision to represent I_{aF} as a function of P. However, F is understood to be a function of P and is treated as such through Eq. (2) and Fig. 1. Therefore, by transferring the storage from I_a to S, CMλ uses F as a surrogate for I_{aF}, thereby partly mimicking the variable nature of I_{aF}.

Storage transfer from I_a to S also causes a decrease in λ_W (Table 3). Conversely, when λ_W decreases, storage is transferred from I_a to S. This is important because several studies (Baltas et al., 2007; D'Asaro and Grillone, 2012; Shi et al., 2009; Woodward et al., 2003) found that the optimal value of λ_W was much less than 0.2 and even close to zero in many watersheds. This shows that there is a positive correlation between a decrease in λ_W, storage transfer from I_a to S, and a general increase in model performance for the reasons mentioned above.

6.2 Application to real watersheds

The models were also evaluated using rainfall-runoff observations from 9 real watersheds located in different parts of the world (Santikari, 2017; Santikari and Murdoch, 2018). The models' ability to predict the observed runoff was assessed using NSE_Q. Results show that in all the watersheds VIMs performed better than CMs but the difference in performance, ΔNSE_Q, varied across the wa-

Table 4. Performance of the models for the cases of $\lambda_i = 0.2$ and 0.5, when the degree of heterogeneity in the synthetic watershed (Table 2) was increased by doubling the values of S_i for HRUs 3 and 4.

Lumped model	$\lambda_i = 0.2$		$\lambda_i = 0.5$	
	rNSE$_Q$	SEE$_Q$	rNSE$_Q$	SEE$_Q$
CM0.2	0.92	1.54	0.92	1.30
CMλ	0.95	0.19	0.94	0.38
VIMS	0.97	0.12	0.96	0.25
VIMλ	1.00	0.06	1.00	0.12

tersheds. Between VIMλ and CM0.2, $\Delta\text{NSE}_Q < 0.05$ in one watershed, $0.05 \leq \Delta\text{NSE}_Q < 0.7$ in six watersheds, and $\Delta\text{NSE}_Q \geq 0.7$ in two watersheds. Between VIMλ and CMλ, $\Delta\text{NSE}_Q < 0.05$ in three watersheds, $0.05 \leq \Delta\text{NSE}_Q < 0.1$ in four watersheds, and $\Delta\text{NSE}_Q \geq 0.1$ in two watersheds. Based on their performance, the models can be arranged from the best to the worst as VIMλ > VIMS > CMλ > CM0.2, which is consistent with results from their application to the synthetic watershed.

6.3 Effect of degree of heterogeneity

The degree of heterogeneity, defined as the sharpness of change in CN, I_a, or S between the HRUs, may affect the relative performance of the models. To verify this, the degree of heterogeneity of the synthetic watershed (Table 2) was increased by doubling the values of S_i for HRUs 3 and 4 while the others were left unchanged; i.e., the modified distribution was $S_0 = 0$ mm, $S_1 = 50$ mm, $S_2 = 100$ mm, $S_3 = 300$ mm, and $S_4 = 400$ mm. The models were applied to this modified synthetic watershed, for the cases of $\lambda_i = 0.2$ and 0.5, and their performances were assessed using rNSE$_Q$ and SEE$_Q$.

Comparing the results (Tables 3 and 4) shows that the performance of VIMs remained nearly the same, whereas the performance of CM0.2 decreased and that of CMλ increased. The relative order of performance remained unchanged, i.e., VIMλ > VIMS > CMλ > CM0.2.

The results from real watersheds (Santikari, 2017; Santikari and Murdoch, 2018) also show that the performance of CM0.2 was poor, $\text{NSE}_Q < 0.25$, in watersheds with a sharp change in CN. Therefore, CM0.2 is unsuitable when the degree of heterogeneity is large. CMλ performed moderately well on synthetic and real watersheds with a large degree of heterogeneity, possibly by transferring the storage (Sect. 6.1). So, CMλ is suitable for predicting overall runoff, but unreliable for predicting heterogeneity or runoff from small events. VIMs outperformed CMλ in synthetic (Table 4) as well as real watersheds (Santikari, 2017; Santikari and Murdoch, 2018) with a large degree of heterogeneity, and therefore they are more reliable.

6.4 Model suitability

One of the main objectives of this study was to improve the predictive ability of the CN method while maintaining its simplicity. Using the number of calibrated parameters as an indicator, the models can be arranged in the order of increasing complexity as CM0.2 (one) < CMλ (two) < VIMS = VIMλ (three). CM0.2 was the simplest but also had the poorest performance (Tables 3 and 4). Moreover, there is no justification in fixing λ_W at 0.2 or any other constant as its optimal value can vary from zero to one (Hawkins et al., 2008). Therefore, the usage of CM0.2 is not recommended.

CMλ predicted the overall runoff and the runoff from small events better than CM0.2. Often, the optimal λ_W is much smaller than 0.2 and this allows CMλ to partly mimic a variable I_{aF} by transferring storage from I_a to S. A smaller λ_W also reduces the false prediction of zero-runoffs, which are completely eliminated when $\lambda_W = 0$. Compared to the variable I_a models, CMλ is a poor predictor of runoff and watershed heterogeneity (Table 3). However, in watersheds with negligible I_{ai} values (or $\lambda_i \approx 0$) CMλ can perform as well as the variable I_a models and therefore may be preferable because of its simplicity.

Variable I_a models show that significant improvement in the model prediction of overall runoff and heterogeneity can be achieved by using one extra parameter (Table 3). This is because the functional form of I_{aF} (Eq. 23) is consistent with the observations (Fig. 2c and d) and the results from the theoretical analysis of heterogeneous watersheds (Eq. 16, Fig. 6, and Table 1). Using variable I_a also improved the runoff predictions in small events and eliminated the false prediction of zero-runoffs. Therefore, their application is recommended in heterogeneous watersheds with nonzero initial abstractions.

When the watershed heterogeneity is known in great detail such that the number of calibrated parameters ≤ 3, a distributed modeling approach (e.g., SWAT; Gassman et al., 2007 or Soulis and Valianzas, 2013) may be preferable over the variable I_a models. A distributed parameter model has advantages similar to the variable I_a models over the conventional models. It would inherently account for the variation of the CN method's parameters spatially and with P. It would also avoid the false prediction of zero-runoffs in small events because HRUs with larger CNs, which generate runoff even in small events, are explicitly considered. When the heterogeneity is unknown, however, the number of calibrated parameters (for values of CN$_i$ and a_i) in a distributed model with n HRUs is $2n - 1$. This number would increase further if values of λ_i are also calibrated. Therefore, when the number of calibrated parameters > 3, application of a variable I_a model should be considered.

6.5 Model limitation

A strength of the models proposed in this paper is that they provide a compact way to account for the spatial variation of CN, I_a, or S (watershed heterogeneity), but a limitation is that they do not account for the temporal variation. During dry periods, I_a and S increase whereas, CN decreases. The behavior is opposite during the wet periods. Changes in land cover introduce additional temporal variations. Therefore, the calibrated model parameters in this paper can be considered as temporal averages. The models may underpredict runoff during wet periods and overpredict during dry periods. A procedure to account for temporal variations using antecedent moisture is described by Santikari (2017) and Santikari and Murdoch (2018).

Another limitation of VIMs is that the CN values calculated using Eqs. (5) or (10) are incompatible with the standard CN values (NRCS, 2003; USDA, 1986) derived using CM0.2. However, this limitation is not unique to VIMs because any method, including CMλ, which involves an altered relationship between I_a and S (i.e., $\lambda \neq 0.2$) leads to CN values that are incompatible with those derived from CM0.2. Given that (i) CM0.2 is a poor predictor of runoff (Tables 3 and 4; Santikari, 2017; Santikari and Murdoch, 2018) and (ii) the evidence contradicts $\lambda = 0.2$ (Baltas et al., 2007; D'Asaro and Grillone, 2012; Shi et al., 2009; Woodward et al., 2003), the above-mentioned limitation is an acceptable compromise.

7 Conclusions

Watershed heterogeneity causes calculated values of I_a, S, and CN to vary with P. Therefore, using a single effective value of these quantities at the watershed scale can lead to systematic errors in the predictions of Q. This problem can be mitigated by treating I_a, S, or CN as functions of P. A theoretical analysis assuming spatial variation of I_a led to the following conclusions.

1. *Effective I_a of a watershed is equal to the filled portion of the total storage in I_a.* The total storage (called I_{aT}) is constant, whereas the filled portion (called I_{aF}) is a function of P (Eq. 16). Variation of I_{aF} with P (Fig. 6) is similar to the variation of calculated I_a (also called effective I_a or I_{aW}) with P (Fig. 2c and d). This shows that $I_{aW} = I_{aF}$, which is also supported by a distributed model using many HRUs (Eq. 21). The form of I_{aF} as a function of P depends on the spatial distribution of I_a within a watershed (Table 1, Figs. 8 and 9).

2. *λ decreases with increasing spatial scale.* Using CN_W, calculated as the area-weighted average of CN_i values (CNs of the HRUs), and the definition of I_a, it can be shown that $\lambda_W < \lambda_i$ (Eqs. 8 through 11). Even when λ_W was calibrated using CMλ, the result was $\lambda_W < \lambda_i$ (Table 3). This shows that in conventional models λ at the watershed scale tends to be smaller than that at the HRU scale; i.e., λ decreases with increasing spatial scale.

3. *Replacing I_a with I_{aF} can account for heterogeneity.* Heterogeneity causes the effective I_a of a watershed to vary with P, so to account for heterogeneity variable I_a models (VIMs) replace I_a with I_{aF}, which is a function of P (Fig. 6). For practical purposes, I_{aF} can be treated as a quadratic function of P (Eq. 23) with two free parameters c_1 and c_2 that need to be calibrated. In addition, the model also requires the calibration of either S (VIMS) or λ (VIMλ).

4. *Variable I_a models perform better than the conventional models.* Variable I_a models predict runoff and heterogeneity better than the conventional models CM0.2 ($\lambda = 0.2$) and CMλ (calibrated λ). They also eliminate the false prediction of zero-runoffs and improve runoff predictions in small events. Based on their overall performance, the models are arranged from the best to the worst as VIMλ > VIMS > CMλ > CM0.2.

5. *Storage transfer can improve model performance.* Storage transfer from I_a to S generally improves the model performance because the filled portions of I_a and S, I_{aF} and F, respectively, have similar functional relationships with P (compare Figs. 6 to 1). This enables a CN model to partly mimic a variable I_{aF} by using F as its surrogate. Storage transfer also lowers the threshold P for runoff generation, thereby reducing the false prediction of zero-runoffs. Storage transfer decreases λ_W (Eq. 3), and this can explain why the optimal value of λ_W from published studies is much less than 0.2 or even zero in many watersheds.

Author contributions. VS conceived the idea and performed the analysis. LM supervised the analysis and influenced the overall direction and content of the work. Both VS and LM wrote the paper.

Competing interests. The authors declare that they have no conflict of interest.

Acknowledgements. Primary funding for this study was provided by the USDA Natural Resources Conservation Service (NRCS-69-4639-1-0010) through the Changing Land Use and Environment (CLUE) project at Clemson University. Additional support was provided by the USDA Cooperative State Research, Education, and Extension Service under project number SC-1700278.

Edited by: Thomas Kjeldsen

Appendix A: List of symbols

a_i	fractional area of the ith HRU
$a(I_a)$	probability density function of areal occurrence of I_a
CM0.2	conventional curve number model with $\lambda = 0.2$
CMλ	conventional curve number model with calibrated λ
CN	curve number, applicable to any spatial scale
CN_i	curve number of the ith HRU
CN_T	curve number of a watershed when $I_{aF} = I_{aT}$
CN_W	curve number of a watershed
F	cumulative infiltration after runoff begins
HRU	hydrologic response unit
I_a	initial abstraction, applicable to any spatial scale
I_{aF}	areal average of the filled portion of I_{aT}
I_{ai}	initial abstraction of the ith HRU
I_{aT}	areal average of the total initial abstraction
I_{aW}	effective initial abstraction of a watershed
$I_{a,max}$	maximum value of I_a within a watershed
λ	initial abstraction ratio, applicable to any spatial scale
λ_i	initial abstraction ratio at HRU scale
λ_W	initial abstraction ratio at watershed scale
m	number of fully filled HRUs in which $I_{ai} \neq 0$
n	number of HRUs in which $I_{ai} \neq 0$
NSE	Nash–Sutcliffe efficiency parameter
P	event rainfall
PB	percent bias
Q	event runoff
R^2	coefficient of determination
S	potential maximum retention, applicable to any spatial scale
S_i	potential maximum retention of ith HRU
S_∞	maximum value of S_W, which occurs when P is infinitely large
S_W	effective potential maximum retention of a watershed
SEE	standard error of estimate
VIMλ	variable initial abstraction model in which λ is calibrated
VIMS	variable initial abstraction model in which S is calibrated

References

Baltas, E. A., Dervos, N. A., and Mimikou, M. A.: Technical note: Determination of the SCS initial abstraction ratio in an experimental watershed in Greece, Hydrol. Earth Syst. Sci., 11, 1825–1829, https://doi.org/10.5194/hess-11-1825-2007, 2007.

Bingner, R. L., Theurer, F. D., and Yuan, Y.: AnnAGNPS technical processes documentation 5.2, 2011.

D'Asaro, F. and Grillone, G.: Empirical investigation of Curve Number Method parameters in the Mediterranean area, J. Hydrol. Eng., 17, 1141–1152, https://doi.org/10.1061/(ASCE)HE.1943-5584.0000570, 2012.

D'Asaro, F. and Grillone, G.: Discussion: "Curve Number derivation for watersheds draining two headwater streams in lower coastal plain South Carolina, USA" by Epps T. H., Hitchcock D. R., Jayakaran A. D., Loflin D. R., Williams T. H., and Amatya D. M., J. Am. Water Resour. Assoc., 51, 573–578, https://doi.org/10.1111/jawr.12264, 2015.

Gassman, P. W., Reyes, M. R., Green, C. H., and Arnold, J. G.: The Soil and Water Assessment Tool – Historical development applications, and future research directions, T. ASABE, 50, 1211–1240, 2007.

Hawkins, R.: Asymptotic determination of runoff Curve Numbers from data, J. Irrig. Drain. Eng., 119, 334–345, https://doi.org/10.1061/(ASCE)0733-9437(1993)119:2(334), 1993.

Hawkins, R., Ward, T., Woodward, D., and Van Mullem, J.: Curve Number hydrology: State of practice, American Society of Civil Engineers, Reston, VA, 2008.

Hjelmfelt Jr., A. T., Woodward, D. A., Conaway, G., Plummer, A., Quan, Q. D., Van Mullen, J., Hawkins, R. H., and Rietz, D.: Curve numbers, recent developments, in: Proc. of the 29th Congress of the Int. As. for Hydraul. Res., 17–21 September 2001, Beijing, China, 2001.

HydroCAD: http://www.hydrocad.net, last access: August 2018.

Kent, K. M.: A method for estimating volume and rate of runoff in small watersheds, Technical Paper 149, USDA SCS, Washington, D.C., 1968.

Knisel, W. G. and Douglas-Mankin, K. R.: CREAMS/GLEAMS: Model use, calibration, and validation, T. ASABE, 55, 1291–1302, 2012.

Krause, P., Boyle, D. P., and Bäse, F.: Comparison of different efficiency criteria for hydrological model assessment, Adv. Geosci., 5, 89–97, https://doi.org/10.5194/adgeo-5-89 2005, 2005.

McCuen, R. H.: Modeling hydrologic change: Statistical methods, Lewis Publishers, Boca Raton, Florida, 2003.

Mishra, S. K., Sahu, R. K., Eldho, T. I., and Jain, M. K.: An improved I_a–S relation incorporating antecedent moisture in SCS-CN methodolog, Water Resour. Manage., 20, 643-660, https://doi.org/10.1007/s11269-005-9000-4, 2006.

Moriasi, D. N., Arnold, J. G., Van Liew, M., Bingner, R. L., Harmel, R. D., and Veith, T. L.: Model evaluation guidelines for systematic quantification of accuracy in watershed simulations, T. ASABE, 50, 885–900, 2007.

Neitsch, S. L., Arnold, J. G., Kiniry, J. R., and Williams, J. R.: Soil and Water Assessment Tool theoretical documentation (version 2005), Agricultural Research Service, Temple, TX, 2005.

NRCS: National engineering handbook: Part 630, Hydrology, USDA, Washington, D.C., 2003

Panday, S. and Huyakorn, P. S.: A fully coupled physically-based spatially-distributed model for evaluating surface/subsurface flow, Adv. Water Resour., 27, 361–382, https://doi.org/10.1016/j.advwatres.2004.02.016, 2004.

Ponce, V. and Hawkins, R.: Runoff Curve Number: Has it reached maturity?, J. Hydrol. Eng., 1, 11–19, https://doi.org/10.1061/(ASCE)1084-0699(1996)1:1(11), 1996.

Rallison, R. E. and Miller, N.: Past, present, and future SCS runoff procedure, in: Proceedings of the international symposium on rainfall runoff modeling: Rainfall–runoff relationship, 18–21 May 1981, Mississippi State University, Mississippi, edited by: Singh, V. P., Water Resources Publications, Littleton, Colorado, 353–364, 1982.

Santikari, V. P.: Accounting for Spatiotemporal Variations of Curve Number using Variable Initial Abstraction and Antecedent Moisture, chap. 5, in: Evaluating Effects of Construction-Related Land Use Change on Streamflow and Water Quality, Doctoral dissertation, https://tigerprints.clemson.edu/all_dissertations/1942 (last access: August 2018), 2017.

Santikari, V. P. and Murdoch, L. C.: Accounting for spatiotemporal variations of Curve Number using variable initial abstraction and antecedent moisture, in review, 2018.

Shi, Z., Chen, L., Fang, N., Qin, D., and Cai, C.: Research on the SCS-CN initial abstraction ratio using rainfall-runoff event analysis in the Three Gorges Area, China, Catena, 77, 1–7, https://doi.org/10.1016/j.catena.2008.11.006, 2009.

Soil Conservation Service: National engineering handbook: Section 4, Hydrology, USDA SCS, Washington, D.C., 1956.

Soil Conservation Service: National engineering handbook: Section 4, Hydrology, chap. 10, USDA SCS, Washington, D.C., 1972.

Soulis, K. X. and Valiantzas, J. D.: SCS-CN parameter determination using rainfall–runoff data in heterogeneous watersheds – the two-CN system approach, Hydrol. Earth Syst. Sci., 16, 1001–1015, https://doi.org/10.5194/hess-16-1001-2012, 2012.

Soulis, K. X., Valiantzas, J. D., Dercas, N. and Londra, P. A.: Investigation of the direct runoff generation mechanism for the analysis of the SCS-CN method applicability to a partial area experimental watershed, Hydrol. Earth Syst. Sci., 13, 605–615, https://doi.org/10.5194/hess-13-605-2009, 2009.

Soulis, K. X. and Valiantzas, J. D.: Identification of the SCS-CN parameter spatial distribution using rainfall-runoff data in heterogeneous watersheds, Water Resour. Manage., 27, 1737–1749, https://doi.org/10.1007/s11269-012-0082-5, 2013.

USDA: Urban hydrology for small watersheds, TR-55, USDA-NRCS, Washington, D.C., 1986.

Van Mullem, J. A.: Precipitation variability and curve numbers, in: Managing Water: Coping with Scarcity and Abundance, Proceedings of the 27th Congress of the International Association of Hydraulic Research, August 1997, San Francisco, CA, published by American Society of Civil Engineers, New York, 518–523, 1997.

Wang, X., Williams, J. R., Gassman, P. W., Baffaut, C., Izaurralde, R. C., Jeong J., and Kiniry, J. R.: EPIC and APEX: Model use, calibration, and validation, T. ASABE, 55, 1447–1462, 2012.

Woodward, D. E., Hawkins, R. H., and Quan, Q. D.: Curve Number method: Origins, applications and limitations, in: Hydrologic modeling for the 21st century: 2nd federal interagency hydrologic modeling conference, 28 July–1 August 2002, Las Vegas, Nevada, 2002.

Woodward, D. E., Hawkins, R., Jiang, R., Hjelmfelt Jr., A., Van Mullem, J., and Quan, Q.: Runoff curve number method: Examination of the initial abstraction ratio, in: World water & environmental resources congress, 23–26 June 2003, Philadelphia, Pennsylvania, USA, American Society of Civil Engineers, 1–10, 2003.

8

The Budyko functions under non-steady-state conditions

Roger Moussa[1] **and Jean-Paul Lhomme**[2]

[1]INRA, UMR LISAH, 2 place Viala, 34060 Montpellier, France
[2]IRD, UMR LISAH, 2 place Viala, 34060 Montpellier, France

Correspondence to: Roger Moussa (roger.moussa@inra.fr)

Abstract. The Budyko functions relate the evaporation ratio E / P (E is evaporation and P precipitation) to the aridity index $\Phi = E_\mathrm{p} / P$ (E_p is potential evaporation) and are valid on long timescales under steady-state conditions. A new physically based formulation (noted as Moussa–Lhomme, ML) is proposed to extend the Budyko framework under non-steady-state conditions taking into account the change in terrestrial water storage ΔS. The variation in storage amount ΔS is taken as negative when withdrawn from the area at stake and used for evaporation and positive otherwise, when removed from the precipitation and stored in the area. The ML formulation introduces a dimensionless parameter $H_\mathrm{E} = -\Delta S / E_\mathrm{p}$ and can be applied with any Budyko function. It represents a generic framework, easy to use at various time steps (year, season or month), with the only data required being E_p, P and ΔS. For the particular case where the Fu–Zhang equation is used, the ML formulation with $\Delta S \leq 0$ is similar to the analytical solution of Greve et al. (2016) in the standard Budyko space (E_p / P, E / P), a simple relationship existing between their respective parameters. The ML formulation is extended to the space $[E_\mathrm{p} / (P - \Delta S),\ E / (P - \Delta S)]$ and compared to the formulations of Chen et al. (2013) and Du et al. (2016). The ML (or Greve et al., 2016) feasible domain has a similar upper limit to that of Chen et al. (2013) and Du et al. (2016), but its lower boundary is different. Moreover, the domain of variation of $E_\mathrm{p} / (P - \Delta S)$ differs: for $\Delta S \leq 0$, it is bounded by an upper limit $1 / H_\mathrm{E}$ in the ML formulation, while it is only bounded by a lower limit in Chen et al.'s (2013) and Du et al.'s (2016) formulations. The ML formulation can also be conducted using the dimensionless parameter $H_\mathrm{P} = -\Delta S / P$ instead of H_E, which yields another form of the equations.

1 Introduction

The Budyko framework has become a simple tool that is widely used within the hydrological community to estimate the evaporation ratio E / P at catchment scale (E is evaporation and P precipitation) as a function of the aridity index $\Phi = E_\mathrm{p} / P$ (E_p is potential evaporation) through simple mathematical formulations $E / P = B_1(\Phi)$ and with long-term averages of the variables. Most of the formulations were empirically obtained (e.g., Oldekop, 1911; Turc, 1954; Tixeront, 1964; Budyko, 1974; Choudhury, 1999; Zhang et al., 2001; Zhou et al., 2015), but some of them were analytically derived from simple physical assumptions (Table 1): (i) the one derived by Mezentsev (1955) and then by Yang et al. (2008), which has the same form as the one initially proposed by Turc (1954) (noted hereafter as Turc–Mezentsev); (ii) the one derived by Fu (1981) and reworked by Zhang et al. (2004) (noted hereafter as Fu–Zhang). These two last formulations involve a shape parameter (respectively, λ and ω), which varies with catchment characteristics and vegetation dynamics (Donohue et al., 2007; Yang et al., 2009; Li et al., 2013; Carmona et al., 2014). When its value increases, actual evaporation gets closer to its maximum rate.

The Budyko framework was initially limited to steady-state conditions on long timescales, under the assumption of negligible change in soil water storage and groundwater. Hydrological processes leading to changes in water storage are not represented and the catchment is considered closed without any anthropogenic intervention: precipitation is the only input and evaporation and runoff Q the only outputs ($P = E + Q$). Recently, the Budyko framework has been downscaled to the year (Istanbulluoglu et al., 2012; Wang, 2012; Carmona et al., 2014; Du et al., 2016), the season (Gentine et al., 2012; Chen et al., 2013; Greve et al., 2016)

Table 1. Different expressions for the Budyko curves under steady-state conditions.

Reference	Equation $E / P = B_1(\Phi)$
Budyko (1974)	$\frac{E}{P} = \left\{\Phi \tanh\left(\frac{1}{\Phi}\right)\left[1 - \exp(-\Phi)\right]\right\}^{1/2}$
Turc (1954) with $\lambda = 2$, Mezentsev (1955), Yang et al. (2008)	$\frac{E}{P} = \Phi\left(1 + \Phi^{\lambda}\right)^{-\frac{1}{\lambda}}$
Fu (1981), Zhang et al. (2004)	$\frac{E}{P} = 1 + \Phi - \left(1 + \Phi^{\omega}\right)^{\frac{1}{\omega}}$
Zhang et al. (2001)	$\frac{E}{P} = \frac{1 + w\Phi}{1 + w\Phi + \Phi^{-1}}$
Zhou et al. (2015)	$\frac{E}{P} = \Phi\left(\frac{k}{1 + k\Phi^{n}}\right)^{1/n}$

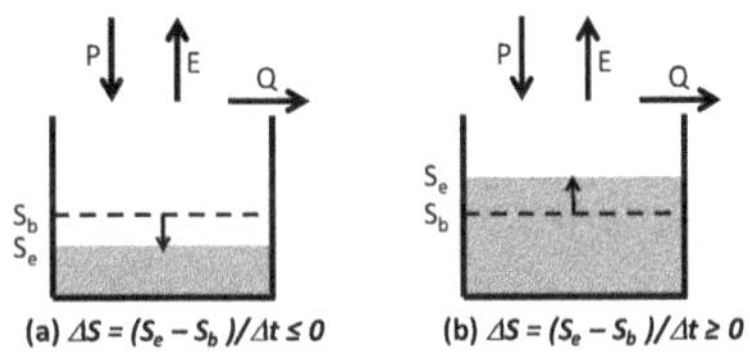

Figure 1. Representation of the change in soil water storage $\Delta S = (S_e - S_b) / \Delta t$ for the two cases considered in the paper: $\Delta S \leq 0$ and $\Delta S \geq 0$. S_b and S_e are, respectively, the storage at the beginning and the end of the time period Δt.

and the month (Zhang et al., 2008; Du et al., 2016). However, the water storage variation ΔS cannot be considered as negligible when dealing with these finer timescales or for unclosed basins (e.g., soil, groundwater, reservoir, snow, interbasin water transfer, irrigation; Jaramillo and Destouni, 2015). In these cases, the catchment is considered to be under non-steady-state conditions (Fig. 1) and the basin water balance should be written as $P = E + Q + \Delta S$. Table 2 shows some recent formulations of the Budyko framework extended to take into account the change in catchment water storage ΔS. Chen et al. (2013) (used in Fang et al., 2016) and Du et al. (2016) proposed empirical modifications of the Turc–Mezentsev and Fu–Zhang equations, respectively, precipitation P being replaced by the available water supply defined as $(P - \Delta S)$, with Du et al. (2016) including the interbasin water transfer into ΔS. Greve et al. (2016) analytically modified the Fu–Zhang equation in the standard Budyko space $(E_p / P, E / P)$ introducing an additional parameter, whereas Wang and Zhou (2016) proposed in the same Budyko space a formulation issued from the hydrological ABCD model (Alley, 1984), but with two additional parameters.

With the extension of the Budyko framework to non-steady-state conditions being a real challenge, this paper aims to propose a new formulation inferred from a clear physical rationale and compared to other non-steady formulations previously derived. The paper is organized as follows. First, we present the new formulation under non-steady-state conditions: its upper and lower limits, its generic equations under restricted evaporation in the Budyko space $(E_p / P, E / P)$ and in the space $[E_p / (P - \Delta S), E / (P - \Delta S)]$. Second, we compare the new formulation to the analytical solution of Greve et al. (2016) in the standard Budyko space and to the formulations of Chen et al. (2013) and Du et al. (2016) in the space $[E_p / (P - \Delta S), E / (P - \Delta S)]$.

2 New generic formulation under non-steady-state conditions

2.1 Upper and lower limits of the Budyko framework

In the Budyko framework, each catchment is characterized by the three hydrologic variables (P, E and E_p) which are represented in a 2-D space using dimensionless variables equal to the ratio between two of these variables and the third one. In the rest of the paper, following Andréassian et al. (2016), the space defined by $(\Phi = E_p / P, E / P)$ is called Budyko space and the one defined by $(\Phi^{-1} = P / E_p, E / E_p)$ is called Turc space. For steady-state conditions $(\Delta S = 0)$, it should be recalled that any Budyko function B_1 defined in the Budyko space $(E_p / P, E / P)$ generates an equivalent function B_2 in the Turc space expressed as

$$\frac{E}{E_p} = B_2\left(\Phi^{-1}\right) = \frac{B_1(\Phi)}{\Phi}, \tag{1}$$

and that any Budyko function verifies the following conditions under steady-state conditions: (i) $E = 0$ if $P = 0$; (ii) $E \leq P$ if $P \leq E_p$ (water limit); (iii) $E \leq E_p$ if $P \geq E_p$ (energy limit); (iv) $E \to E_p$ if $P \to \infty$.

First, we present the upper and lower limits in the Turc space under steady-state conditions, when all the water consumed by evaporation comes from the precipitation, the change in water storage ΔS being nil $(E = P - Q)$. Figure 2a represents the variation of maximum and minimum actual evapotranspiration, respectively, E_x and E_n, as a function of precipitation P with dimensionless variables. The upper solid line represents the dimensionless maximum evaporation rate E_x / E_p: it follows the precipitation up to $P / E_p = 1$ (the water limit is presented in bold blue on all graphs) and then is limited by potential evaporation $E_x / E_p = 1$ (the energy limit is in bold green). The lower solid line (in bold black) represents the dimensionless minimum evaporation rate E_n / E_p which follows the x axis: $E_n / E_p = 0$. The feasible domain between the upper and the lower limits is shown in grey. In the Budyko space, we have the following relationships: (i) when evaporation is maximum, for $E_p / P \leq 1$, $E_x / P = E_p / P$ and for $E_p / P \geq 1$, $E_x / P = 1$; (ii) when evaporation is minimum: $E_n / P = 0$. The corresponding Budyko non-dimensional graph is shown in Fig. 2b and represents the upper and lower limits of the feasible domain of $E / P = B_1 (E_p / P)$.

Under non-steady-state conditions, either a given amount of water ΔS stored in the area at stake participates in the

Table 2. Different expressions for the Budyko curves under non-steady-state conditions.

Reference	Steady-state conditions $B_1(\Phi)$	Non-steady-state conditions
Greve et al. (2016)	Fu–Zhang	$\frac{E}{P} = 1 + \frac{E_p}{P} - \left[1 + (1-y_0)^{\kappa-1}\left(\frac{E_p}{P}\right)^{\kappa}\right]^{1/\kappa}$ with κ and y_0 parameters
Chen et al. (2013)	Turc–Mezentsev	$\frac{E}{P-\Delta S} = \left[1 + \left(\frac{E_p}{P-\Delta S} - \Phi_t\right)^{-\lambda}\right]^{-\frac{1}{\lambda}}$ with λ and Φ_t parameters
Du et al. (2016)	Fu–Zhang	$\frac{E}{P-\Delta S} = 1 + \frac{E_p}{P-\Delta S} - \left[1 + \left(\frac{E_p}{P-\Delta S}\right)^{\omega} + \mu\right]^{\frac{1}{\omega}}$ with ω and μ parameters

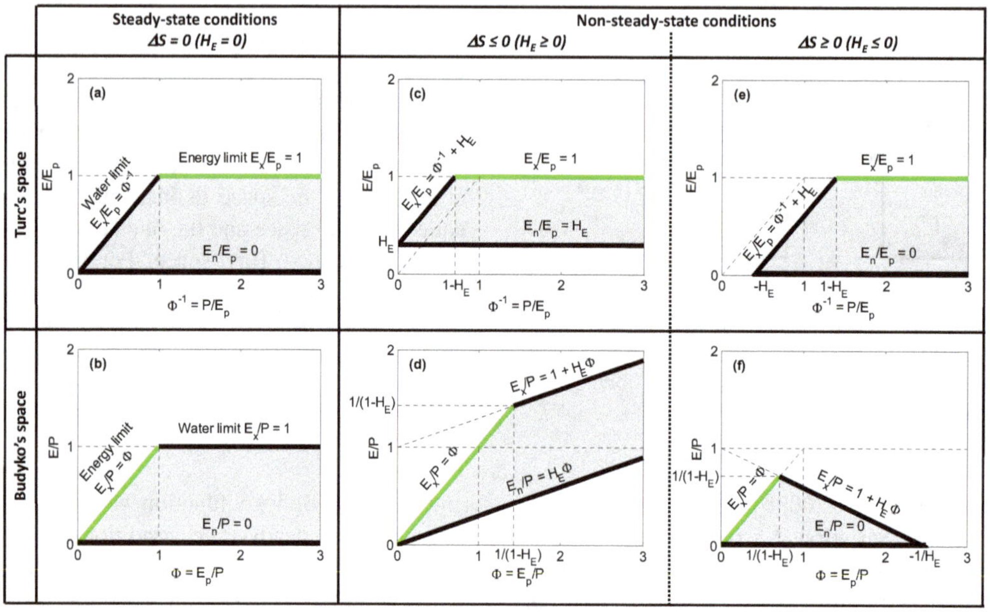

Figure 2. Upper and lower limits of the feasible domain (in grey) of evaporation in the Turc space ($P\,/\,E_\text{p}$, $E\,/\,E_\text{p}$) and in the Budyko space ($E_\text{p}\,/\,P$, $E\,/\,P$) (water limit in blue, energy limit in green and lower limit in black) when using the non-dimensional parameter H_E: (**a** and **b**) for steady-state conditions; (**c**, **d**, **e** and **f**) for non-steady-state conditions with a storage term ΔS (**c** and **d** for $\Delta S \leq 0$; and **(e)** and **(f)** for $\Delta S \geq 0$).

evaporation process (for instance, groundwater depletion for irrigation), or a given amount of the precipitation ΔS is stored in the area (soil water, ground water, reservoirs) following the water balance ($E = P - \Delta S - Q$). As shown in Fig. 1, the storage amount ΔS is taken as negative ($\Delta S \leq 0$) when withdrawn from the area and used for evaporation; it is taken as positive ($\Delta S \geq 0$) when removed from the precipitation and stored in the area. When ΔS is negative, $|\Delta S|$ should be lower than E_p because if $|\Delta S| \geq E_\text{p}$, evaporation would be systematically equal to E_p; when ΔS is positive, it should be necessarily lower than P. Consequently, $-E_\text{p} \leq \Delta S \leq P$. The variable ΔS is used in a dimensionless form, either as $H_\text{E} = -\,\Delta S\,/\,E_\text{p}$ or $H_\text{P} = -\,\Delta S\,/\,P$, which are positive when additional water is available for evapotranspiration and negative when water is withdrawn from precipitation. In the following, all the calculations are made with H_E ($-\Phi^{-1} \leq H_\text{E} \leq 1$), but a similar reasoning is conducted using H_P ($-1 \leq H_\text{P} \leq \Phi$) in Appendix A. Taking into account ΔS makes the upper and lower limits of the feasible domain different.

In the Turc space, the case where evaporation is at its maximum value is visualized as the upper limit in Fig. 2c and e (all the available water is used for evaporation). For both cases, $\Delta S \leq 0$ (Fig. 2c) or $\Delta S \geq 0$ (Fig. 2e), we have $E_x = P - \Delta S$ if $P - \Delta S \leq E_P$ and $E_x = E_P$ if $P - \Delta S \geq E_P$. Written with dimensionless variables, these equations transform into

$$\text{if } \frac{P}{E_\text{p}} \leq 1 + \frac{\Delta S}{E_\text{p}} \text{ then } \frac{E_x}{E_\text{p}} = \frac{P}{E_\text{p}} - \frac{\Delta S}{E_\text{p}} = \Phi^{-1} + H_\text{E}, \quad (2)$$

$$\text{if } \frac{P}{E_\text{p}} \geq 1 + \frac{\Delta S}{E_\text{p}} \text{ then } \frac{E_x}{E_\text{p}} = 1. \quad (3)$$

For the minimal value of evapotranspiration E_n, we have to distinguish two cases depending if $\Delta S \leq 0$ (Fig. 2c) or

$\Delta S \geq 0$ (Fig. 2e).

$$\text{if } \Delta S \leq 0 \text{ then } \frac{E_n}{E_\text{p}} = \frac{-\Delta S}{E_\text{p}} = H_\text{E}. \tag{4a}$$

$$\text{if } \Delta S \geq 0 \text{ then } \frac{E_n}{E_\text{p}} = 0 \tag{4b}$$

Translating the above equations into the Budyko space (Fig. 2d, f) yields the following for the upper limits:

$$\text{if } \frac{E_\text{p}}{P} \geq \frac{E_\text{p}}{E_\text{p}+\Delta S} \text{ then } \frac{E_x}{P} = 1 - \frac{\Delta S}{P} = 1 + H_\text{E}\frac{E_\text{p}}{P} = 1 + H_\text{E}\Phi, \tag{5}$$

$$\text{if } \frac{E_\text{p}}{P} \leq \frac{E_\text{p}}{E_\text{p}+\Delta S} \text{ then } \frac{E_x}{P} = \frac{E_\text{p}}{P} = \Phi. \tag{6}$$

Equation (5) has two limits: when $H_\text{E}=0$, $E_x / P = 1$, and when $\Delta S \to -E_\text{p}$, which corresponds to $H_\text{E}=1$, $E_x / P \to (1+\Phi)$. For the lower limits in the Budyko space we have

$$\text{if } \Delta S \leq 0 \text{ then } \frac{E_n}{P} = \frac{-\Delta S}{P} = H_\text{E}\frac{E_\text{p}}{P} = H_\text{E}\Phi, \tag{7a}$$

$$\text{if } \Delta S \geq 0 \text{ then } \frac{E_n}{P} = 0. \tag{7b}$$

Note that under steady-state conditions, the upper and lower limits are similar in both Turc and Budyko spaces, while this is not the case under non-steady-state conditions. It is also interesting to note that for the negative values of H_E the domain of variation of Φ is bounded $[0, -1 / H_\text{E}]$ and the possible space of the Budyko functions is limited to a triangle (Fig. 2f).

2.2 General equations with restricted evaporation

We now examine the case where the evaporation rate is lower than its maximum possible rate. In the Turc space, under non-steady-state conditions ($\Delta S \leq 0$ in Fig. 2c or $\Delta S \geq 0$ in Fig. 2e), Eq. (1) should be transformed to take into account the impact of water storage on the evaporation process. We search a mathematical formulation which transforms the upper and lower limits for the steady-state conditions (Fig. 2a) into the corresponding ones for the non-steady-state conditions (Fig. 2c if $\Delta S \leq 0$ and Fig. 2e if $\Delta S \geq 0$). The mathematical transformation is searched under the form $E / E_\text{p} = \alpha\ B_2(\Phi^{-1} + \gamma) + \beta$, which combines an x axis translation (γ), a y axis translation (β) and a homothetic transformation (α). This mathematical form is suggested by the way the physical domain of Turc's space is transformed when passing from steady-state conditions to non-steady-state conditions (Fig. 2a, c, e). Note that the reasoning can be conducted either in the Turc or the Budyko space, but the upper and lower limits and the transformation from steady- to non-steady-state conditions are easier to grasp in the Turc space than in the Budyko space. We distinguish the two cases corresponding to $\Delta S \leq 0$ and $\Delta S \geq 0$.

2.2.1 Case $\Delta S \leq 0$

In the Turc space, the lower limit $B_2(\Phi^{-1}) = 0$ in Fig. 2a transforms into $B_2(\Phi^{-1}) = H_\text{E}$ in Fig. 2c. Using the mathematical transformation described above, we obtain $(\alpha \times 0) + \beta = H_\text{E}$. Following a similar reasoning, the energy limit $B_2(\Phi^{-1}) = 1$ transforms into $B_2(\Phi^{-1}) = 1$, which yields $\alpha + \beta = 1$, and the water limit $B_2(\Phi^{-1}) = \Phi^{-1}$ transforms into $B_2(\Phi^{-1}) = H_\text{E} + \Phi^{-1}$, which yields $\alpha\ (\Phi^{-1} + \gamma) + \beta = H_\text{E} + \Phi^{-1}$. The resolution of these three equations gives $\alpha = 1 - H_\text{E}$, $\beta = H_\text{E}$ and $\gamma = \Phi^{-1} H_\text{E} / (1 - H_\text{E})$. Consequently, Eq. (1) should be transformed into

$$\frac{E}{E_\text{p}} = (1 - H_\text{E})\, B_2\left(\frac{\Phi^{-1}}{1 - H_\text{E}}\right) + H_\text{E}. \tag{8}$$

By introducing Eq. (1) into Eq. (8), we obtain the formulation in the Budyko space (Fig. 2d):

$$\frac{E}{P} = (1 - H_\text{E})\, \Phi B_2\left(\frac{\Phi^{-1}}{1 - H_\text{E}}\right) + H_\text{E}\Phi = B_1\left[(1 - H_\text{E})\,\Phi\right] + H_\text{E}\Phi. \tag{9}$$

The derivative of Eq. (9) is

$$\frac{d\left(\frac{E}{P}\right)}{d\Phi} = (1 - H_\text{E})\frac{dB_1\left[(1 - H_\text{E})\,\Phi\right]}{d\Phi} + H_\text{E}. \tag{10}$$

Given that $\frac{dB_1[(1-H_\text{E})\Phi]}{d\Phi} = 1$ for $\Phi = 0$ and $\frac{dB_1[(1-H_\text{E})\Phi]}{d\Phi} = 0$ when $\Phi \to \infty$, the derivative $\frac{d\left(\frac{E}{P}\right)}{d\Phi}$ (i.e., the slope of the curve) is equal to 1 for $\Phi = 0$, and when $\Phi \to \infty$, the derivative tends to H_E.

2.2.2 Case $\Delta S \geq 0$

Following the same reasoning as above, the lower limit, the energy limit and the water limit of $B_2(\Phi^{-1})$ in the Turc space in Fig. 2a (respectively, 0, 1 and Φ^{-1}) transform, respectively, into 0, 1, and $H_\text{E} + \Phi^{-1}$ in Fig. 2e. We obtain, respectively, the following three equations: $(\alpha \times 0) + \beta = 0$, $\alpha + \beta = 1$ and $\alpha\ (\Phi^{-1} + \gamma) + \beta = H_\text{E} + \Phi^{-1}$. The resolution gives $\alpha = 1$, $\beta = 0$ and $\gamma = H_\text{E}$. Consequently, Eq. (1) should be transformed into

$$\frac{E}{E_\text{p}} = B_2\left(\Phi^{-1} + H_\text{E}\right). \tag{11}$$

By introducing Eq. (1) into Eq. (11), we obtain the formulation in the Budyko space (Fig. 2f):

$$\frac{E}{P} = \Phi B_2\left(\Phi^{-1} + H_\text{E}\right) = (1 + H_\text{E}\Phi)\, B_1\left(\frac{\Phi}{1 + H_\text{E}\Phi}\right). \tag{12}$$

The derivative of Eq. (12) is

$$\frac{d\left(\frac{E}{P}\right)}{d\Phi} = H_\text{E} B_1\left(\frac{\Phi}{1 + H_\text{E}\Phi}\right) + (1 + H_\text{E}\Phi)\frac{d\left[B_1\left(\frac{\Phi}{1+H_\text{E}\Phi}\right)\right]}{d\Phi}. \tag{13}$$

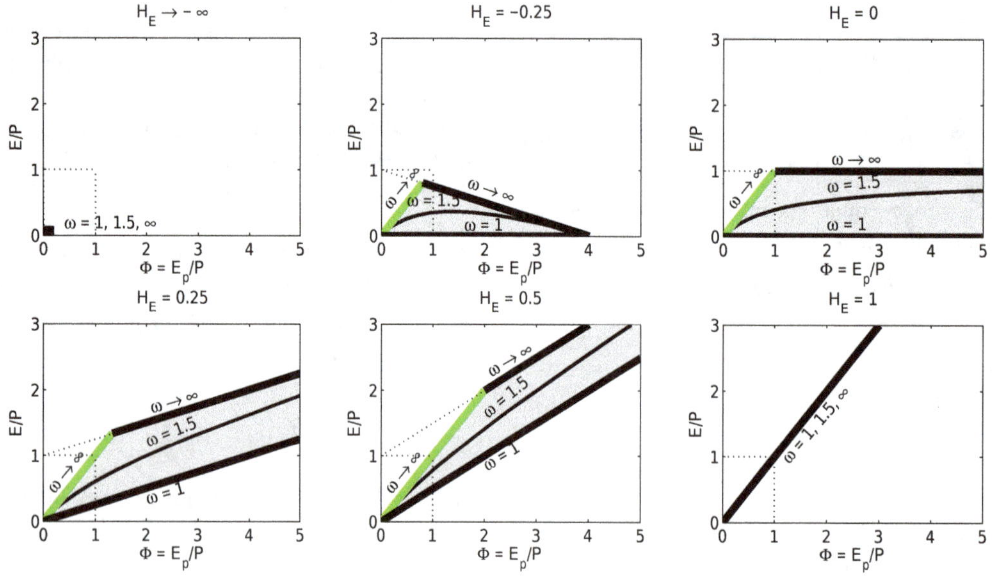

Figure 3. The ML formulation in the Budyko space with the Fu–Zhang relationship Eq. (14a, b) for $\omega = 1.5$ and for different values of H_E. The bold lines indicate the upper and lower limits of the feasible domain of evaporation (shown in grey).

Given that $B_1\left(\frac{\Phi}{1+H_E\Phi}\right) = 0$ and $\frac{d\left[B_1\left(\frac{\Phi}{1+H_E\Phi}\right)\right]}{d\Phi} = 1$ for $\Phi = 0$, the derivative $\frac{d\left(\frac{E}{P}\right)}{d\Phi}$ is equal to 1 for $\Phi = 0$. When $\Phi \to -1/H_E$, $B_1\left(\frac{\Phi}{1+H_E\Phi}\right) = 1$ and $\frac{d\left[B_1\left(\frac{\Phi}{1+H_E\Phi}\right)\right]}{d\Phi} = 0$, the derivative $\frac{d\left(\frac{E}{P}\right)}{d\Phi}$ tends to H_E.

In the following, these new generic formulae (Eqs. 8 and 9 for $\Delta S \leq 0$ and Eqs. 11 and 12 for $\Delta S \geq 0$) are called ML formulations (ML stands for Moussa–Lhomme).

2.2.3 Application

Any Budyko equation $B_1(\Phi)$ from Table 1 can be used in Eqs. (9) and (12) as detailed in Table S1 in the Supplement. It is worth noting that both the Turc–Mezentsev and Fu–Zhang functions, which are obtained from the resolution of a Pfaffian differential equation, have the following remarkable simple property: $F(1/x) = F(x)/x$. This means that the same mathematical expression is valid for B_1 and B_2: $B_1 = B_2$. Both Turc–Mezentsev and Fu–Zhang functions have similar shapes, and a simple linear relationship was established by Yang et al. (2008) between their parameters (see Table 1): $\omega = \lambda + 0.72$. The ML formulation is used hereafter with the Fu–Zhang function (Table 1) for comparison with the analytical solution of Greve et al. (2016) based upon the same function. Replacing B_1 by Fu–Zhang's equation, in Eq. (9) for $\Delta S \leq 0$ and in Eq. (12) for $\Delta S \geq 0$, in the Budyko space, gives

$$\text{if } \Delta S \leq 0 \text{ then } \frac{E}{P} = 1 + \Phi - \left[1 + (1 - H_E)^{\omega}\Phi^{\omega}\right]^{\frac{1}{\omega}}, \quad (14a)$$

$$\text{if } \Delta S \geq 0 \text{ then } \frac{E}{P} = 1 + (1 + H_E)\,\Phi - \left[(1 + H_E\Phi)^{\omega} + \Phi^{\omega}\right]^{\frac{1}{\omega}}. \quad (14b)$$

For $\Phi = 0$, and in both cases $\Delta S \leq 0$ and $\Delta S \geq 0$, we have $E/P = 0$. However, the upper limits of Φ differ: for $\Delta S \leq 0$, when $\Phi \to \infty$, $E/P \to \infty$, while for $\Delta S \geq 0$ the maximum value of Φ is $-1/H_E$ and corresponds to $E/P = 0$. Figure 3 shows some examples of the shape of the ML formulation in the Budyko space (Eq. 14a, b) for $\omega = 1.5$ and different values of H_E. Note that for the particular and unlikely case when $H_E \to -\infty$, upper and lower limits are reduced to the point ($E_p/P = 0$, $E/P = 0$). For $H_E = 0$, we obtain the curves corresponding to steady-state conditions, while for $H_E = 1$, upper and lower limits are superimposed, and the domain is restricted to the 1 : 1 line. We can easily verify that all functions in Table S1 of the Supplement give similar results.

2.3 The ML formulation in the space $[E_p/(P - \Delta S), E/(P - \Delta S)]$

As mentioned in the introduction, some authors (Chen et al., 2013; Du et al., 2016) have dealt with the non-steady conditions by modifying the Budyko reference space and replacing the precipitation P by $P - \Delta S$. Hereafter, we develop the ML formulations in this new space. The upper limits of the ML formulations can be obtained by transforming Eqs. (5) and (6) defined in the Budyko space. We get, respectively,

$$\text{if } \frac{E_p}{P - \Delta S} \geq 1 \text{ then } \frac{E_x}{P - \Delta S} = 1, \quad (15)$$

$$\text{if } \frac{E_p}{P - \Delta S} < 1 \text{ then } \frac{E_x}{P - \Delta S} = \frac{E_p}{P - \Delta S}. \quad (16)$$

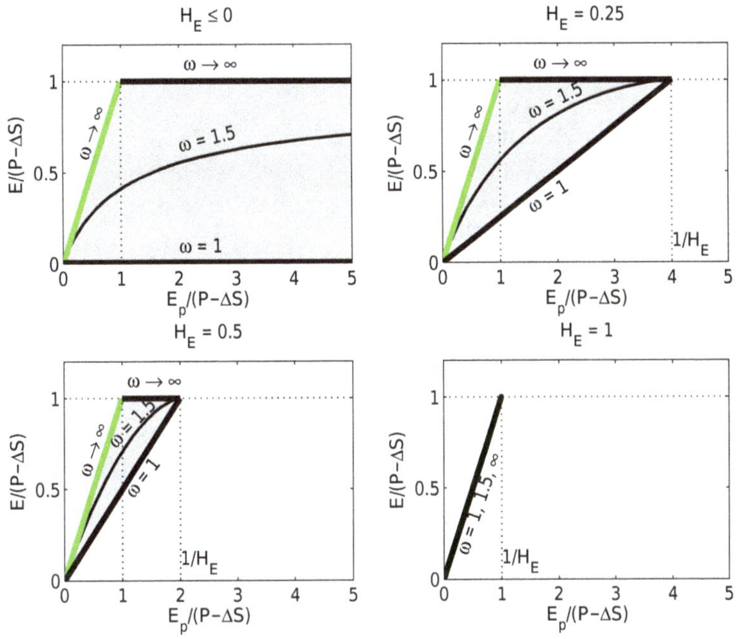

Figure 4. The ML formulation with the Fu–Zhang Eq. (21a, b) in the space $[E_p/(P-\Delta S),\ E/(P-\Delta S)]$ for $\omega = 1.5$ and four values of H_E. For $H_E \geq 0$, all curves have a common upper end at $\Phi' = 1/H_E$ corresponding to $E/(P-\Delta S) = 1$. The bold lines indicate the upper and lower limits of the feasible domain shown in grey. For $H_E \leq 0$, the curve is similar to the one under steady-state conditions.

The lower limits are obtained from Eq. (7a, b):

$$\text{if } \Delta S \leq 0 \text{ then } \frac{E_n}{P-\Delta S} = \frac{-\Delta S}{P-\Delta S} = H_E \frac{E_p}{P-\Delta S}, \quad (17a)$$

$$\text{if } \Delta S \geq 0 \text{ then } \frac{E_n}{P-\Delta S} = 0. \quad (17b)$$

In the new space, we put

$$\Phi' = \frac{E_p}{P-\Delta S} = \frac{\Phi}{1+H_E\Phi} \text{ or } \Phi = \frac{\Phi'}{1-H_E\Phi'}. \quad (18)$$

Consequently, the relationship between $E/(P-\Delta S)$, Φ' and E/P is given by

$$\frac{E}{P-\Delta S} = \frac{E}{P}\frac{P}{P-\Delta S} = \frac{1}{1+H_E\Phi}\frac{E}{P} = \left(1-H_E\Phi'\right)\frac{E}{P}. \quad (19)$$

Inserting Eqs. (9) and (12) into Eq. (19) and expressing Φ as a function of Φ' (Eq. 18) led to the ML formulation in the new space:

$$\text{if } \Delta S \leq 0 \text{ then } \frac{E}{P-\Delta S} = \frac{1}{1+H_E\Phi}\left\{B_1\left[(1-H_E)\Phi\right] + H_E\Phi\right\} = \left(1-H_E\Phi'\right)B_1\left[\frac{(1-H_E)\Phi'}{1-H_E\Phi'}\right] + H_E\Phi', \quad (20a)$$

$$\text{if } \Delta S \geq 0 \text{ then } \frac{E}{P-\Delta S} = \frac{1}{1+H_E\Phi}(1+H_E\Phi) \cdot B_1\left(\frac{\Phi}{1+H_E\Phi}\right) = B_1\left(\Phi'\right). \quad (20b)$$

Note that for $\Delta S \geq 0$, $E/(P-\Delta S) = B_1(\Phi')$ is independent of H_E and is identical to the steady-state condition $H_E = 0$. This is explained by the fact that the stored water ΔS being initially subtracted to the precipitation P, it does not participate in the evaporation process and consequently has no impact on the ratio $E/(P-\Delta S)$. For $\Delta S \leq 0$, and for $\Phi = 0$, i.e., $P \to \infty$, we have $\Phi' = 0$, $B_1 = 0$ and $E/(P-\Delta S) = 0$. When $\Phi \to \infty$, which corresponds to $P \to 0$, we have $\Phi' = 1/H_E$, $B_1 = 1$, and $E/(P-\Delta S) \to 1$.

Any Budyko formulation B_1 in Table 1 can be used with Eq. (20a, b), as shown in Table S2 of the Supplement. When the Fu–Zhang equation is used, Eq. (20a, b) become

$$\text{if } \Delta S \leq 0 \text{ then } \frac{E}{P-\Delta S} = 1 + (1-H_E)\Phi' - \left[\left(1-H_E\Phi'\right)^{\omega} + (1-H_E)^{\omega}\left(\Phi'\right)^{\omega}\right]^{1/\omega}, \quad (21a)$$

$$\text{if } \Delta S \geq 0 \text{ then } \frac{E}{P-\Delta S} = 1 + \Phi' - \left(1+\Phi'^{\omega}\right)^{\frac{1}{\omega}}. \quad (21b)$$

Figure 4 shows the ML formulation (Eq. 21a, b) in the space $[E_p/(P-\Delta S),\ E/(P-\Delta S)]$ for $\omega = 1.5$ and different values of H_E. For $H_E = 0$, we retrieve the original Fu–Zhang equation and when $\omega = 1$, we can easily verify that Eq. (21a, b) are equal to the lower limit of the domain $E/(P-\Delta S) = H_E\ E_p/(P-\Delta S)$ when $\Delta S \leq 0$, and $E/(P-\Delta S) = 0$ when $\Delta S \geq 0$.

2.4 The ML formulation using the dimensionless parameter H_P

A mathematical development, similar to the one of Sect. 2.1, 2.2 and 2.3, is conducted in Appendix A using the dimensionless parameter $H_P = -\Delta S / P = H_E \Phi$ (instead of $H_E = -\Delta S / E_P$) and yields another form of the ML formulation. Equivalent mathematical representations are obtained for $\Delta S \leq 0$ and $\Delta S \geq 0$ in the different spaces explored in Sect. 2.1, 2.2 and 2.3. Figures S1, S2 and S3 in the Supplement obtained with the parameter H_P correspond, respectively, to Figs. 2, 3 and 4 obtained with H_E. Similarly, Tables S3 and S4 in the Supplement (obtained with H_P) correspond to Tables S1 and S2 in the Supplement (obtained with H_E): they give the ML formulation applied to the different Budyko curves of Table 1 in the standard Budyko space $(E_p / P, E / P)$ and in the space $[E_p / (P - \Delta S), E / (P - \Delta S)]$. Significant differences appear concerning the mathematical equations and the shape of the feasible domain defined by its upper and lower limits. This is due to the fact that using H_E or H_P corresponds to different sets of data and different functional representations. Both approaches (H_E or H_P) can be used. When storage water contributes to enhancing evaporation ($\Delta S \leq 0$), ΔS is bounded by potential evaporation E_P and consequently represents a given percentage of E_P. Hence, it is more convenient to use $H_E = -\Delta S / E_p$ instead of $H_P = -\Delta S / P$, because H_E lies in the range [0, 1] which is not the case for H_P. Conversely, when precipitation water contributes to increase soil water storage ($\Delta S \geq 0$), ΔS is bounded by P and represents a percentage of precipitation P. Consequently, using H_P is more convenient because H_P lies in the range $[-1, 0]$. Moreover, in order to keep the parameter in the range [0, 1], $H'_P = -H_P$ could be preferred.

3 Comparing the new formulation with other formulae from the literature

3.1 In the standard Budyko space $(E_p / P, E / P)$

When evapotranspiration exceeds precipitation (corresponding herein to the case $\Delta S \leq 0$), Greve et al. (2016) analytically developed a Budyko-type equation where the water storage is taken into account through a parameter y_0 $(0 \leq y_0 \leq 1)$ introduced into the Fu–Zhang formulation (Table 2). In the Budyko space, this equation is written (Greve et al., 2016; Eq. 9) as

$$\frac{E}{P} = 1 + \Phi - \left[1 + (1 - y_0)^{\kappa - 1} \Phi^{\kappa}\right]^{1/\kappa}. \tag{22}$$

They used the shape parameter κ to avoid confusion with the traditional ω of Fu–Zhang equation. Despite different physical and mathematical backgrounds, Eqs. (14a) and (22) are exactly identical and a simple relationship between H_E and y_0 can be easily obtained. Equating Eqs. (14a) and (22) with

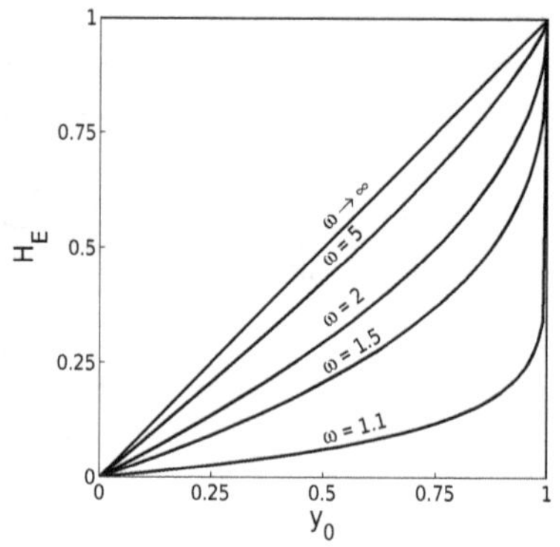

Figure 5. Relationship (Eq. 23) between the parameter H_E of the ML formulation (Eq. 14a) and the parameter y_0 of the Greve et al. (2016) (Eq. 22) for different values of ω with $\omega = \kappa$.

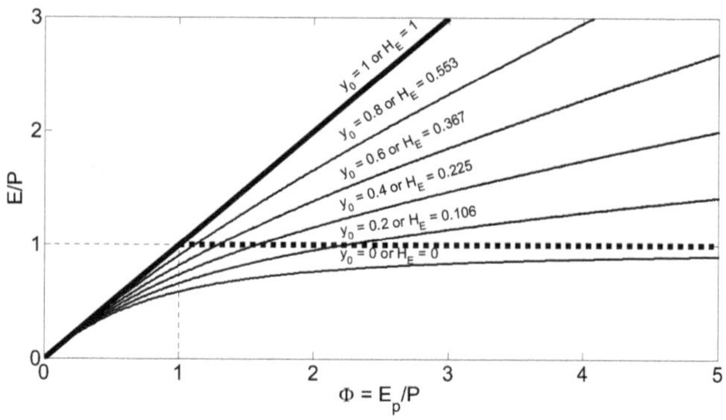

Figure 6. Example showing the similarity of the ML formulation Eq. (14a) and the equation of Greve et al. (2016) Eq. (22) (with $\omega = \kappa = 2$) for different values of y_0; the corresponding values of H_E are calculated using Eq. (23).

$\omega = \kappa$ yields

$$H_E = 1 - (1 - y_0)^{\frac{\omega - 1}{\omega}}. \tag{23}$$

The relationship between y_0 and H_E is independent from Φ. It is shown in Fig. 5 for different values of ω. For a given value of ω, we have $H_E < y_0$. For $\omega = 1$, we have $H_E = 0$, and when $\omega \to \infty$, we have $H_E = y_0$.

The derivative of Eq. (22) gives

$$\frac{d\left(\frac{E}{P}\right)}{d\Phi} = 1 - (1 - y_0)^{\kappa - 1} \Phi^{\kappa - 1} \left[1 + (1 - y_0)^{\kappa - 1} \Phi^{\kappa}\right]^{\frac{1 - \kappa}{\kappa}}. \tag{24}$$

For $\Phi = 0$, the derivative is equal to 1, and when $\Phi \to \infty$, the derivative tends to a value noted as m by Greve et al. (2016; Eq. 12):

$$m = 1 - (1 - y_0)^{\frac{\kappa - 1}{\kappa}}. \tag{25}$$

The value of the derivative (slope of the curve) is the same in both ML and Greve et al.'s (2016) formulations: for $\Phi = 0$ the derivative is equal to 1, and when $\Phi \to \infty$ we have $m = H_E$

(assuming $\omega=\kappa$). Greve et al. (2016; Sect. 4) show that y_0 is the maximum value of m reached when $\omega \to \infty$. Hence, substituting in Eq. (22) y_0 by its value inferred from Eq. (23) yields an equation identical to that obtained from the ML formulation (Eq. 14a).

Figure 6 compares the ML formulation Eq. (14a) with Greve et al.'s (2016) analytical solution Eq. (22) for $\omega=\kappa=2$ and different values of y_0 (0, 0.2, 0.4, 0.6, 0.8 and 1). The corresponding values of H_E (respectively, 0, 0.106, 0.225, 0.367, 0.553 and 1) are calculated using Eq. (23). The new ML formulation with $\omega=\kappa$, and only for $\Delta S \leq 0$, gives exactly the same curves as those obtained by Greve et al. (2016). Both formulations are identical and have the same upper and lower limits. Greve et al. (2016), however, did not mention the lower limit and limited the reasoning to positive values of y_0. Moreover, the case of $\Delta S \geq 0$ is not considered by Greve et al. (2016).

3.2 In the space $[E_p\,/\,(P-\Delta S),\,E\,/\,(P-\Delta S)]$

The formulations proposed by Chen et al. (2013) and Du et al. (2016) in the space $[E_p\,/\,(P-\Delta S),(E\,/\,(P-\Delta S)]$ are essentially empirical. The Chen et al. (2013) function (Table 2) is derived from the Turc–Mezentsev equation and written as

$$\frac{E}{P-\Delta S}=\left[1+\left(\frac{E_p}{P-\Delta S}-\Phi_t\right)^{-\lambda}\right]^{-\frac{1}{\lambda}}. \tag{26}$$

An additional parameter Φ_t is empirically introduced in order "to characterize the possible non-zero lower bound of the seasonal aridity index"; this parameter causes a shift of the curve $E\,/\,(P-\Delta S)$ along the horizontal axis such as for $E_p\,/\,(P-\Delta S)=\Phi_t$, we have $E\,/\,(P-\Delta S)=0$. The derivative of Eq. (26) when $E_p\,/\,(P-\Delta S)\to\infty$ is equal to 0. Similarly, the Du et al. (2016) function (Table 2) is an empirical modification of Fu–Zhang equation (Fu, 1981; Zhang et al., 2004) written as

$$\frac{E}{P-\Delta S}=1+\frac{E_p}{P-\Delta S}-\left[1+\left(\frac{E_p}{P-\Delta S}\right)^{\omega}+\mu\right]^{\frac{1}{\omega}}. \tag{27}$$

A supplementary parameter, noted here as μ (>-1), is added to modify the lower bound of the aridity index $E_P\,/\,(P-\Delta S)$. The parameter μ plays a similar role as Φ_t in Eq. (26). For $\mu=0$, Eq. (27) takes the original form of Fu–Zhang equation, $(P-\Delta S)$ replacing P. When μ becomes positive, the lower end of the curve $E\,/\,(P-\Delta S)$ shifts to the right. The function $E\,/\,(P-\Delta S)$ in Eq. (27) is equal to zero for the particular value of $E_p\,/\,(P-\Delta S)=\Phi_d$, such as

$$(1+\Phi_d)^{\omega}=1+(\Phi_d)^{\omega}+\mu. \tag{28}$$

Greve et al.'s (2016) formulation can be also written in the space $[E_p\,/\,(P-\Delta S),\,E\,/\,(P-\Delta S)]$. Inserting Eq. (22) into Eq. (20a) and expressing Φ as a function of Φ' (Eq. 18) leads

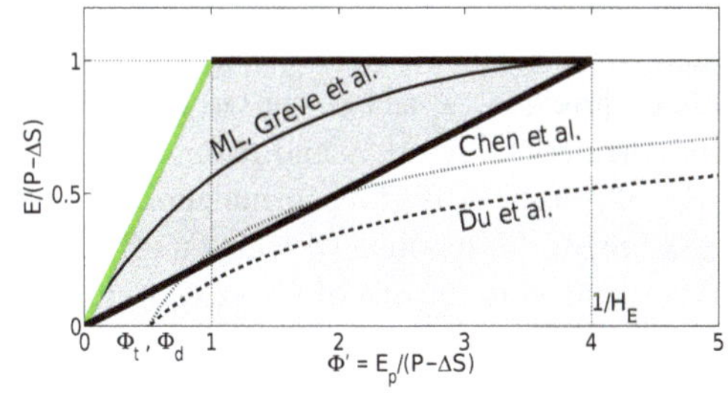

Figure 7. In the space $[E_p\,/\,(P-\Delta S),\,E\,/\,(P-\Delta S)]$, an example comparing the three formulations: Du et al. (2016) with $\omega=1.5$ and $\Phi_d=0.5$; Chen et al. (2013) with $\lambda=\omega-0.72=0.78$ and $\Phi_t=\Phi_d=0.5$; the ML formulation for $\Delta S\leq 0$ (Eq. 14a) with $\omega=1.5$ and $H_E=0.25$ (identical to Greve et al. 2016 formulation). The feasible domain of the ML formulation in dark grey is superimposed over the domains of both Chen et al. (2013) and Du et al. (2016) in light grey.

to

$$\frac{E}{P-\Delta S}=1+(1-H_E)\,\Phi'-\left[\left(1-H_E\Phi'\right)^{\kappa}+(1-y_0)^{\kappa-1}\left(\Phi'\right)^{\kappa}\right]^{1/\kappa}. \tag{29}$$

It can be mathematically shown that expressing $(1-y_0)$ in Eq. (29) as a function of H_E by inverting Eq. (23) (assuming $\omega=\kappa$) leads to the exact ML formulation of Eq. (21a). It is a direct consequence of the similarity of both formulations. Therefore, similar curves to those shown in Fig. 4 for the ML formulation with $H_E\geq 0$ are obtained with Greve et al.'s (2016) formulation.

For $\Delta S\geq 0$ (corresponding to $H_E\leq 0$; Fig. 4), the three formulations (Chen et al.'s, 2013, Du et al.'s 2016, and ML) have similar upper and lower limits. For $\Delta S\leq 0$, Fig. 7 shows an example of the curves obtained with Du et al.'s (2016) equation ($\omega=1.5$) and Chen et al.'s (2013) equation with $\lambda=0.78$ (such as $\lambda=\omega-0.72$ from Yang et al., 2008) and with $\Phi_t=\Phi_d=0.5$ (corresponding to $\mu=0.484$ from Eq. 28). Both Chen et al.'s (2013) and Du et al.'s (2016) formulations are compared to the ML formulation using the Fu–Zhang Eq. (14a) with $H_E=+0.25$. The ML and Greve et al. (2016) formulations are exactly identical if $\kappa=\omega=1.5$ and $y_0=0.578$ calculated from Eq. (23) for $H_E=+0.25$. The four formulations have similar upper limits but the lower limits are different. Both Chen et al.'s (2013) and Du et al.'s (2016) formulations have the x axis as the lower limit and $E\,/\,(P-\Delta S)$ tends to 1 when $\Phi'=E_p\,/\,(P-\Delta S)\to\infty$, while in the ML formulation with $\Delta S\leq 0$ (Fig. 2a) the feasible domain is a triangle, the domain of variation of Φ' being limited by 0 and $1\,/\,H_E$.

3.3 Discussion

All four formulations (ML; Greve et al., 2016; Chen et al., 2013; and Du et al., 2016) have two parameters each: one for the shape of the curve and another for its shift due to non-steady conditions: ω and H_E for the ML formulation (with the Fu–Zhang function), κ and y_0 for Greve et al. (2016), λ and Φ_t for Chen et al. (2013), ω and μ for Du et al. (2016). If $H_E = y_0 = \Phi_t = \mu = 0$, the four formulations are identical. For $\Delta S \leq 0$, the ML formulation with the Fu–Zhang equation (Eq. 14a) is identical to the one of Greve et al. (2016) in the Budyko space and also in the $[E_p / (P - \Delta S), E / (P - \Delta S)]$ space, provided the shape parameters are assumed to be identical ($\omega = \kappa$) (a simple relationship is established between H_E and the corresponding parameter y_0). Despite similar upper limits, the ML and Greve et al. (2016) formulations behave very differently from those of Chen et al. (2013) and Du et al. (2016) in the space $[E_p / (P - \Delta S), E / (P - \Delta S)]$. The ML formulation is different for $\Delta S \leq 0$ and $\Delta S \geq 0$, while those of Chen et al.'s (2013) and Du et al.'s (2016) do not distinguish the two cases $\Delta S \leq 0$ and $\Delta S \geq 0$. All the formulations have the same upper limits, but the domain of variation of Φ' differs: respectively, $[0, 1 / H_E]$ when $\Delta S \leq 0$ and $[0, \infty]$ when $\Delta S \geq 0$ for the ML formulation, $[\Phi_t, \infty]$ for Chen et al. (2013) and $[\Phi_d, \infty]$ for Du et al. (2016) The lower end of the curve $E / (P - \Delta S)$ corresponds, respectively, to (0, 0), (Φ_t, 0) and (Φ_d, 0) and the upper end to ($1 / H_E$, 1) when $\Delta S \leq 0$ and (∞, 1) when $\Delta S \geq 0$ for the ML formulation, (∞, 1) for the other two. Moreover, the ML formulation for $\Delta S \geq 0$ is reduced to a simple relationship $E / (P - \Delta S) = B_1 (\Phi')$ and is independent of H_E.

It is worth noting that for $\Delta S \leq 0$ the limits of Chen et al. (2013) and Du et al. (2016) functions are not completely sound from a strict physical standpoint: for very high precipitation, when $P \gg E_p$, Φ and Φ' should logically tend to zero and not to Φ_t and Φ_d; similarly, when $P \to 0$, i.e., $\Phi \to \infty$, it is physically logical that $\Phi' \to E_p / (-\Delta S) = 1 / H_E$, as predicted by our Eq. (20a). This tends to prove that the ML formulation, corroborated by the Greve et al. (2016) formulation, is physically more correct. Additionally, at simple glimpse, we note that the ML curves could be easily adjusted to the set of experimental points shown in Chen et al. (2013; Figs. 2 and 9) and in Du et al. (2016; Figs. 8 and 9).

4 Conclusion

The ML formulations constitute a general mathematical framework which allows any standard Budyko function developed at catchment scale under steady-state conditions (Table 1) to be extended to non-steady conditions (Table S1 in the Supplement). They take into account the change in catchment water storage ΔS through a dimensionless parameter $H_E = -\Delta S / E_p$ and the formulation differs according to the sign of ΔS (Eqs. 8 and 9 for $\Delta S \leq 0$ and Eqs. 11 and 12 for $\Delta S \geq 0$). Applications can be conducted at various time steps (yearly, seasonal or monthly) both in the Turc space (P / E_p, E / E_p) and in the standard Budyko space (E_p / P, E / P), with the only data required to obtain E being E_p, P and ΔS.

The new formulations are inferred from an evaluation of the feasible domain of evaporation in the Turc space, adjusted for the case where additional ($\Delta S \leq 0$) or restricted ($\Delta S \geq 0$) water is available for evaporation, and then transformed in the Budyko space. For $\Delta S = 0$, the ML formulations return the original equations under steady-state conditions, with similar upper and lower limits in both spaces. Under non-steady-state conditions, however, the upper and lower limits of the feasible domain differ when using either the Turc or the Budyko space. The ML formulations can be extended to the $[E_p / (P - \Delta S), E / (P - \Delta S)]$ space (Eq. 20a, b, Fig. 4). They can also be conducted using the dimensionless parameter $H_P = -\Delta S / P$ instead of H_E, which yields another form of the equations (Appendix A and the Supplement). It is shown that the ML formulation with $\Delta S \leq 0$ is identical to the analytical solution of Greve et al. (2016) in the standard Budyko space, a simple relationship existing between their respective parameters. On the other hand, the new formulation differs from those of Chen et al. (2013) and Du et al. (2016) in the space $[E_p / (P - \Delta S), E / (P - \Delta S)]$.

Appendix A: Scaling ΔS by P instead of E_p

Appendix A presents the set of equations when scaling the change in soil water storage ΔS by precipitation P instead of potential evaporation E_p, i.e., using $H_P = -\Delta S / P = H_E \Phi$ $(-1 \le H_P \le \Phi)$ instead of $H_E = -\Delta S / E_p$ $(-\Phi^{-1} \le H_E \le 1)$.

A1 Upper and lower limits of the Budyko framework

In the Turc space, the upper limits of evapotranspiration E_x / E_p are obtained from Eqs. (2) and (3):

$$\text{if } \frac{P}{E_p} \le 1 + \frac{\Delta S}{E_p} \text{ then } \frac{E_x}{E_p} = \frac{P}{E_p} - \frac{\Delta S}{E_p} = (1 + H_P)\,\Phi^{-1}, \tag{A1}$$

$$\text{if } \frac{P}{E_p} \ge 1 + \frac{\Delta S}{E_p} \text{ then } \frac{E_x}{E_p} = 1, \tag{A2}$$

and the lower limits of evapotranspiration E_n / E_p from Eq. (4a, b):

$$\text{if } \Delta S \le 0 \text{ then } \frac{E_n}{E_p} = -\frac{\Delta S}{E_p} = H_P \Phi^{-1}, \tag{A3a}$$

$$\text{if } \Delta S \ge 0 \text{ then } \frac{E_n}{E_p} = 0. \tag{A3b}$$

The translation in the Budyko space yields the following for the upper limits:

$$\text{if } \frac{E_p}{P} \ge \frac{E_p}{E_p + \Delta S} \text{ then } \frac{E_x}{P} = 1 - \frac{\Delta S}{P} = 1 + H_P, \tag{A4}$$

$$\text{if } \frac{E_p}{P} \le \frac{E_p}{E_p + \Delta S} \text{ then } \frac{E_x}{P} = \frac{E_p}{P} = \Phi, \tag{A5}$$

and the following for the lower limits:

$$\text{if } \Delta S \le 0 \text{ then } \frac{E_n}{P} = -\frac{\Delta S}{P} = H_E \frac{E_p}{P} = H_E \Phi = H_P, \tag{A6a}$$

$$\text{if } \Delta S \ge 0 \text{ then } \frac{E_n}{P} = 0. \tag{A6b}$$

In the Supplement, Fig. S1 shows the upper and lower limits of the feasible domain of evaporation in the Turc and Budyko spaces, drawn with the parameter $H_P = -\Delta S / P$. Figure S1 in the Supplement corresponds to Fig. 2 obtained with $H_E = -\Delta S / E_p$.

A2 General equations with restricted evaporation

We distinguish two cases: $\Delta S \le 0$ and $\Delta S \ge 0$. Substituting H_E by H_P / Φ in Eqs. (8), (9), (11) and (12) we obtain (if $\Delta S \le 0$ in the Turc space)

$$\frac{E}{E_p} = \left(1 - H_P \Phi^{-1}\right) B_2 \left(\frac{\Phi^{-1}}{1 - H_P \Phi^{-1}}\right) + H_P \Phi^{-1}, \tag{A7}$$

then in the Budyko space

$$\frac{E}{P} = B_1 (\Phi - H_P) + H_P. \tag{A8}$$

If $\Delta S > 0$ in the Turc space

$$\frac{E}{E_p} = B_2 \left[(1 + H_P)\,\Phi^{-1}\right], \tag{A9}$$

then in the Budyko space

$$\frac{E}{P} = (1 + H_P)\, B_1 \left(\frac{\Phi}{1 + H_P}\right). \tag{A10}$$

Replacing B_1 by Fu–Zhang's equation, in Eq. (A8) for $\Delta S \le 0$ and in Eq. (A10) for $\Delta S \ge 0$, gives the following in the Budyko space:

$$\text{if } \Delta S \le 0 \text{ then } \frac{E}{P} = 1 + \Phi - \left[1 + (\Phi - H_P)^{\omega}\right]^{\frac{1}{\omega}}, \tag{A11a}$$

$$\text{if } \Delta S \ge 0 \text{ then } \frac{E}{P} = 1 + \Phi + H_P - \left[(1 + H_P)^{\omega} + \Phi^{\omega}\right]^{\frac{1}{\omega}}. \tag{A11b}$$

In the Supplement, Fig. S2 shows an example of the ML formulation (Eq. A11a, b) in the Budyko space obtained with the parameter $H_P = -\Delta S / P$. It corresponds to Fig. 3 obtained with $H_E = -\Delta S / E_p$. Table S3 gives the ML formulation applied to the different Budyko curves of Table 1 with the parameter H_P (Eqs. A8 and A10). It corresponds to Table S1 in the Supplement obtained with H_E.

A3 The ML formulation in the space $[E_p / (P - \Delta S), E / (P - \Delta S)]$

Equations (15), (16), (17a) and (17b) yield the following for the upper limits:

$$\text{if } \frac{E_p}{P - \Delta S} \ge 1 \text{ then } \frac{E_x}{P - \Delta S} = 1, \tag{A12}$$

$$\text{if } \frac{E_p}{P - \Delta S} \le 1 \text{ then } \frac{E_x}{P - \Delta S} = \frac{E_p}{P - \Delta S}, \tag{A13}$$

and the following for the lower limits:

$$\text{if } \Delta S \le 0 \text{ then } \frac{E_n}{P - \Delta S} = \frac{-\Delta S}{P - \Delta S} = H_E \frac{E_p}{P - \Delta S} = \frac{H_P}{H_P + 1}, \tag{A14a}$$

$$\text{if } \Delta S \ge 0 \text{ then } \frac{E_n}{P - \Delta S} = 0. \tag{A14b}$$

In the new space, $[E_p / (P - \Delta S), E / (P - \Delta S)]$, we put

$$\Phi' = \frac{E_p}{P - \Delta S} = \frac{\Phi}{1 + H_P} \text{ or } \Phi = (1 + H_P)\,\Phi'. \tag{A15}$$

Consequently, the relationship between $E / (P - \Delta S)$ and E / P is given by

$$\frac{E}{P - \Delta S} = \frac{E}{P} \frac{P}{P - \Delta S} = \frac{1}{1 + H_p} \frac{E}{P}. \tag{A16}$$

Replacing H_E by H_P / Φ in Eq. (20a, b) we obtain

$$\text{if } \Delta S \leq 0 \text{ then } \frac{E}{P-\Delta S} = \frac{1}{1+H_P}\left[B_1(\Phi - H_P) + H_P\right] = \frac{1}{1+H_P} B_1\left[(1+H_P)\Phi' - H_P\right] + \frac{H_P}{1+H_P}, \quad \text{(A17a)}$$

$$\text{if } \Delta S \geq 0 \text{ then } \frac{E}{P-\Delta S} = \left(\frac{1}{1+H_P}\right) \cdot (1+H_P) B_1\left(\frac{\Phi}{1+H_P}\right) = B_1(\Phi'). \quad \text{(A17b)}$$

Using the Fu–Zhang equation for B_1 we get

$$\text{if } \Delta S \leq 0 \text{ then } \frac{E}{P-\Delta S} = 1 + \Phi' - \frac{H_P}{1+H_P} - \left[\left(\frac{1}{1+H_P}\right)^{\omega} + \left(\Phi' - \frac{H_P}{1+H_P}\right)^{\omega}\right]^{1/\omega}, \quad \text{(A18a)}$$

$$\text{if } \Delta S \geq 0 \text{ then } \frac{E}{P-\Delta S} = 1 + \Phi' - \left(1 + \Phi'^{\omega}\right)^{\frac{1}{\omega}}. \quad \text{(A18b)}$$

In the Supplement, Fig. S3 shows an example of the ML formulation (Eq. A18a, b) in the space $[E_p / (P - \Delta S), E / (P - \Delta S)]$ obtained with the parameter $H_P = -\Delta S / P$. It corresponds to Fig. 4 obtained with $H_E = -\Delta S / E_p$. Table S4 in the Supplement gives the ML formulation applied to the different Budyko curves of Table 1 in the space $[E_p / (P - \Delta S), E / (P - \Delta S)]$ with the parameter H_P (Eq. A17a and b). It corresponds to Table S2 in the Supplement obtained with H_E.

Appendix B: List of symbols

$B_1(\Phi)$	Relationship between E / P and Φ in the Budyko space $(E_p / P, E / P)$ such as $E / P = B_1(\Phi)$ [–]
$B_2(\Phi^{-1})$	Relationship between E / E_p and $\Phi^{-1} = P / E_p$ in the Turc space $(P / E_p, E / E_p)$ such as $E / E_p = B_2(\Phi^{-1})$ [–]
E	actual evaporation [LT^{-1}]
E_n	Lower limit of the feasible domain of evaporation [LT^{-1}]
E_p	Potential evaporation [LT^{-1}]
E_x	Upper limit of the feasible domain of evaporation [LT^{-1}]
H_E	$-\Delta S / E_p$ $(-P / E_p \leq H_E \leq 1)$ [–]
H_P	$-\Delta S / P$ $(-1 \leq H_P \leq E_p / P)$ [–]
m	Slope of the equation of Greve et al. (2016) when $\Phi \to \infty$ [–]
ML	New formulations: Eqs. (8) and (9) for $\Delta S \leq 0$ and Eqs. (11) and (12) for $\Delta S \geq 0$ (stands for Moussa–Lhomme)
P	Precipitation [LT^{-1}]
Q	Runoff [LT^{-1}]
y_0	Parameter in the Greve et al. (2016) equation accounting for non-steady-state conditions $(0 \leq y_0 \leq 1)$ [–]
κ	Shape parameter in the Greve et al. (2016) equation corresponding to ω in the Fu–Zhang equation [–]
ΔS	Water storage variation [LT^{-1}]
λ	Shape parameter in the Turc–Mezentsev equation $(\lambda > 0)$ [–]
μ	Parameter in the Du et al. (2013) equation [–]
Φ	Aridity index $(\Phi = E_p / P)$ [–]
Φ_d	Aridity index threshold in the Du et al. (2016) equation corresponding to $E / (P - \Delta S) = 0$ [–]
Φ_t	Aridity index threshold in the Chen et al. (2013) equation [–]
Φ'	$E_p / (P - \Delta S)$ [–]
ω	Shape parameter of the Fu–Zhang equation $(\omega > 1)$ [–]

Acknowledgements. The authors are very grateful to the reviewers, L. Gudmundsson and F. Jaramillo, and to the editor, M. Coenders-Gerrits, for their constructive comments of the manuscript. They also gratefully acknowledge the scientific and financial support of the UMR LISAH, as well as the always inspiring advice of A. Gaby.

Edited by: M. Coenders-Gerrits

References

Alley, W. M.: On the treatment of evapotranspiration, soil moisture accounting, and aquifer recharge in monthly water balance models, Water Resour. Res., 20, 1137–1149, 1984.

Andréassian, V., Mander, Ü., and Pae, T.: The Budyko hypothesis before Budyko: The hydrological legacy of Evald Oldekop, J. Hydrol., 535, 386–391, doi:10.1016/j.jhydrol.2016.02.002, 2016.

Budyko, M. I.: Climate and life, Academic Press, Orlando, FL, 508 pp., 1974.

Carmona, A., Sivapalan, M., Yaeger, M. A., and Poveda, G.: Regional patterns of interannual variability of catchment water balances across the continental U.S.: A Budyko framework, Water Resour. Res., 50, 9177–9193, doi:10.1002/2014WR016013, 2014.

Chen, X., Alimohammadi, N., and Wang, D.: Modeling interannual variability of seasonal evaporation and storage change based on the extended Budyko framework, Water Resour. Res., 49, 6067–6078, doi:10.1002/wrcr.20493, 2013.

Choudhury, B.: Evaluation of an empirical equation for annual evaporation using field observations and results from a biophysical model, J. Hydrol., 216, 99–110, 1999.

Donohue, R. J., Roderick, M. L., and McVicar, T. R.: On the importance of including vegetation dynamics in Budyko's hydrological model, Hydrol. Earth Syst. Sci., 11, 983–995, doi:10.5194/hess-11-983-2007, 2007.

Du, C., Sun, F., Yu, J., Liu, X., and Chen, Y.: New interpretation of the role of water balance in an extended Budyko hypothesis in arid regions. Hydrol. Earth Syst. Sci., 20, 393–409, doi:10.5194/hess-20-393-2016, 2016.

Fang, K., Shen, C., Fisher, J. B., and Niu, J.: Improving Budyko curve-based estimates of long-term water partitioning using hydrologic signatures from GRACE, Water Resour. Res., 52, 5537–5554, doi:10.1002/2016WR018748, 2016.

Fu, B. P.: On the calculation of evaporation from land surface (in Chinese), Sci. Atmos. Sin., 5, 23–31, 1981.

Gentine, P., D'Odorico, P., Lintner, B. R., Sivandran, G., and Salvucci, G.: Interdependence of climate, soil, and vegetation as constrained by the Budyko curve, Water Resour. Res., 39, L19404, doi:10.1029/2012GL053492, 2012.

Greve, P., Gudmundsson, L., Orlowsky, B., and Seneviratne, S. I.: A two-parameter Budyko function to represent conditions under which evapotranspiration exceeds precipitation, Hydrol. Earth Syst. Sci., 20, 2195–2205, doi:10.5194/hess-20-2195-2016, 2016.

Istanbulluoglu, E., Wang, T., Wright, O. M., and Lenters, J. D.: Interpretation of hydrologic trends from a water balance perspective: The role of groundwater storage in the Budyko hypothesis, Water Resour. Res., 48, W00H16, doi:10.1029/2010WR010100, 2012.

Jaramillo, F. and Destouni, G.: Local flow regulation and irrigation raise global human water consumption and footprint, Science, 350, 1248–1250, 2015.

Li, D., Pan, M., Cong, Z., Zhang, L., and Wood, E.: Vegetation control on water and energy balance within the Budyko framework, Water Resour. Res., 49, 969–976, doi:10.1002/wrcr.20107, 2013.

Mezentsev, V.: More on the computation of total evaporation, Meteorol. Gidrol., 5, 24–26, 1955.

Oldekop, E. M.: On evaporation from the surface of river basins, Collection of the Works of Students of the Meteorological Observatory, University of Tartu-Jurjew-Dorpat, Tartu, Estonia, p. 209, 1911.

Tixeront, J.: Prediction of streamflow, IAHS publication no. 63: General Assembly of Berkeley, IAHS, Gentbrugge, 118–126, 1964.

Turc, L.: Le bilan d'eau des sols: relations entre les précipitations, l'évaporation et l'écoulement, Ann. Agron., 5, 491–595, 1954.

Wang, D.: Evaluating interannual water storage changes at watersheds in Illinois based on long-term soil moisture and groundwater level data, Water Resour. Res., 48, W03502, doi:10.1029/2011WR010759, 2012.

Wang, X.-S. and Zhou, Y.: Shift of annual water balance in the Budyko space for catchments with groundwater-dependent evapotranspiration, Hydrol. Earth Syst. Sci., 20, 3673–3690, doi:10.5194/hess-20-3673-2016, 2016.

Yang, D., Shao, W., Yeh, P. J. F., Yang, H., Kanae, S., and Oki, T.: Impact of vegetation coverage on regional water balance in the nonhumid regions of China, Water Resour. Res., 45, W00A14, doi:10.1029/2008WR006948, 2009.

Yang, H., Yang, D., Lei, Z., and Sun, F.: New analytical derivation of the mean annual water-energy balance equation, Water Resour. Res., 44, W03410, doi:10.1029/2007WR006135, 2008.

Zhang, L., Dawes, W. R., and Walker, G. R.: Response of mean annual evapotranspiration to vegetation changes at catchment scale, Water Resour. Res., 37, 701–708, 2001.

Zhang, L., Hickel, K., Dawes, W. R., Chiew, F. H. S., Western, A. W., and Briggs, P. R.: A rational function approach for estimating mean annual evapotranspiration, Water Resour. Res., 40, W02502, doi:10.1029/2003WR002710, 2004.

Zhang, L., Potter, N., Hickel, K., Zhang, Y., and Shao, Q.: Water balance modeling over variable time scales based on the Budyko framework – Model development and testing, J. Hydrol., 360, 117–131, doi:10.1016/j.jhydrol.2008.07.021, 2008.

9

Design water demand of irrigation for a large region using a high-dimensional Gaussian copula

Xinjun Tu[1,2], Yiliang Du[1], Vijay P. Singh[3,4], Xiaohong Chen[1,2], Kairong Lin[1], and Haiou Wu[1]

[1]Center of Water Resources and Environment, Sun Yat-sen University, Guangzhou, 510275, China
[2]Center of Water Security Engineering and Technology in Southern China of Guangdong, Guangzhou, 510275, China
[3]Department of Biological and Agricultural Engineering, Texas A&M University, 2117 College Station, Texas 77843, USA
[4]Zachry Department of Civil Engineering, Texas A&M University, 2117 College Station, Texas 77843, USA

Correspondence: Xinjun Tu (eestxj@mail.sysu.edu.cn)

Abstract. Spatial and frequency distributions of precipitation should be considered in determining design water demand of irrigation for a large region. In Guangdong Province, South China, as a study case, an eight-dimensional joint distribution of precipitation for agricultural sub-regions was developed. A design procedure for water demand of irrigation for a given frequency of precipitation of the entire region was proposed. Water demands of irrigation in the entire region and its sub-regions using three design methods, i.e., equalized frequency (EF), typical year (TY) and most-likely weight function (MLW), were compared. Results demonstrated that the Gaussian copula efficiently fitted the high-dimensional joint distribution of eight sub-regional precipitation values. The Kendall frequency was better than the conventional joint frequency to analyze the linkage between the frequency of precipitation of the entire region and individual sub-regions. For given frequencies of precipitation of the entire region, design water demands of irrigation of the entire region among the MLW, EF and TY methods slightly differed, but those of individual sub-regions of the MLW and TY methods fluctuated around the demand lines of the EF method. The alterations of design water demand in sub-regions were more complicated than those in the entire region. The design procedure using the MLW method in association with a high-dimensional copula, which simulated individual univariate distributions, captured their dependences for multi-variables, and built a linkage between regional frequency and sub-regional frequency of precipitation, is recommended for design water demand of irrigation for a large region.

1 Introduction

Water demand of irrigation in a region is associated with exploitation and utilization of land and water resources, such as farm area, cultivation pattern, category of crops, canal technology, etc., and is also impacted by natural factors, for example precipitation volume and soil properties (Tarjuelo et al., 2005; Griffin, 2006; Leenhardt et al., 2004, 2011; Wriedt et al., 2009; Davidson and Hellegers, 2011). Among the many factors, precipitation, which is regarded as stochastic, is an uncertain factor influencing irrigation (Wisser et al., 2008; Thomas, 2008; Cai et al., 2011; Meza et al., 2012). In general, the more the precipitation, the less the water demand of irrigation, and vice versa. In practical regional water resources planning, water demands of irrigation are estimated in association with various frequencies of regional precipitation (Gohari et al., 2013).

For design water demand in a large region, it is important that the heterogeneity of those factors in geographical space is considered (Lankford, 2010; Leenhardt et al., 2011). Hence, water demand should generally be analyzed in separate sub-regions in which the factors are homogenous, and can eventually be summed up. However, for a given precipitation frequency of the entire region in the design water demand of irrigation, the difficulty is how to obtain a reasonable combination of precipitation frequencies of multiple sub-regions, even though other factors influencing irrigation have previously been demonstrated in water resources planning.

A method, named typical year (TY), has been used for design water demand of irrigation in China (Cai et al., 2001). A combination of observed sub-regional precipitation in a certain year in which the precipitation frequency for the entire region weighted by individual sub-regions was the nearest to the given frequency, was selected and was zoomed in or out. Due to the limited observed samples, the representation of the typical year has been questionable. In actuality, the frequency of regional precipitation in a large region corresponds to various combinations of sub-regional frequencies.

Moreover, design water demand of irrigation in a large region not only presents the stochastic characteristic of precipitation in the entire region and its sub-regions, but also addresses the relationship between sub-regional precipitation and heterogeneity. The multi-variable statistical simulation and joint design theory can be considered to improve the selection of an appropriate combination. Due to the multivariate dependence and the flexibility of unbounded marginal distributions, copula functions have been widely applied in the simulation and design of hydrological multi-variables, e.g., extreme properties of heavy precipitation (Zhang et al., 2011, 2013; Abdul Rauf and Zeephongsekul, 2014), floods (Zhang and Singh, 2006; Chowdhary et al., 2011; Zhang et al., 2015a) and droughts (Ganguli and Reddy, 2014; Zhang et al., 2015b; Tu et al., 2016). In recent years, they have been used to analyze floods or droughts encountered in multiple hydrological regions in China (Yan et al., 2010a; Xie et al., 2012; Tu et al., 2017b). These studies on multivariate hydrology mostly focused on bivariate and trivariate issues, but less on higher dimensional hydrological statistical analyses (Chen et al., 2015). A high-dimensional meta-Gaussian copula beyond three variables has been applied in other fields, e.g., economic analysis (Aussenegg and Cech, 2012; Creal and Tsay, 2015).

As a multidimensional copula and its marginal distributions are determined according to observed samples, the joint frequency and probability density for any combination of precipitation frequencies of sub-regions can be calculated. In a practical design combination, a large quantity of combinations can be randomly generated using the Monte Carlo method on the basis of the determined copula. It is clear that a given joint frequency corresponds to quite a number of combinations. In order to select one from various combinations, a simple method, based on the equalized frequency (EF) method, which refers to all marginal frequencies being identical, is used for design flood peak and volume (Liu et al., 2015).

Another improved method is that the combination can be exclusively determined by using the most-likely weight function (MLW) in association with the products of joint and marginal probability densities (Salvadori et al., 2011). The most-likely weight design method has been applied for the design combination of hydrological multivariables, such as flood and drought properties (Gräler et al., 2013; Zhang et al., 2015c), flood and tide or heavy precipitation and tide in coastal rivers (Corbella and Stretch, 2012; Lian et al., 2013; Zheng et al., 2013; Tu et al., 2017a), and the precipitation or streamflow of multiple regions (Yan et al., 2010b).

Furthermore, in practical water resources planning, the main concern is that water demand of irrigation of the entire region and that of sub-regions are designed for a given frequency of precipitation for the entire region, not for a given joint frequency of precipitation of sub-regions. The relationship between regional frequency and joint frequency of precipitation is also required to be investigated in determining design water demand of irrigation in a large region.

This paper considered water demand of irrigation of paddies in Guangdong Province, South China, as a case study of a large region. Annual precipitation data of eight agricultural sub-regions and their net quotas of irrigation in association with precipitation frequency were used. A high-dimensional meta-Gaussian copula and several conventional univariate distributions were applied to fit the joint and marginal distributions of sub-regional precipitation, respectively. Combinations of cumulative frequency for sub-regional precipitation were generated by using the Monte Carlo simulation method on the basis of the determined copula. The link between the joint frequency of sub-regional precipitation and the frequency of precipitation of the entire region was established. The methods, i.e., typical year, equalized frequency, and the most-likely weight function, were used for design combinations of sub-regional precipitation for given frequencies of precipitation of the entire region. Water demand of irrigation of the entire region and individual sub-regions among design methods was compared in order to improve design water demand of irrigation in a large region.

2 Methodology

2.1 Water demand of irrigation of paddy

In a paddy field, water demand of irrigation per unit area, i.e., the net quota of irrigation, is determined by various factors, such as crop and soil types, cultivation pattern and precipitation process, etc. In regional planning of water resources, as other factors are demonstrated in advance, the net quota of irrigation mainly changes with annual precipitation due to the stochastic property of precipitation. In practice, the net quota of irrigation per unit area $q(u)$ in association with the frequency of precipitation is determined via field experiments. Then, for a given frequency of annual precipitation, u, water demand of irrigation, $W(u)$, which refers to water withdrawn from river or other water sources engineering, can be calculated as

$$W(u) = \frac{Aq(u)}{\eta}, \tag{1}$$

where A and $q(u)$ refer to the paddy area and the net quota of irrigation, respectively, and η is the utilization coefficient

Table 1. Theoretical cumulative distribution and probability density function of three parameter univariate distribution.

Name	Cumulative distribution function	Probability density function	Parameter
GEV	$\exp\left\{-\left[1+\xi(x-\mu)/\sigma\right]^{-1/\xi}\right\}$	$\frac{1}{\sigma}\left[1+\xi(x-\mu)/\sigma\right]^{-(\xi+1)/\xi}\exp\left\{-\left[1+\xi(x-\mu)/\sigma\right]^{-1/\xi}\right\}$	$\mu\in R,\sigma>0,\xi\in R$
GLO	$1/\left\{1+\left[1-\xi(x-\mu)/\sigma\right]^{1/\xi}\right\}$	$\frac{1}{\sigma}\left[1+\xi(x-\mu)/\sigma\right]^{-(\xi+1)/\xi}/\left\{1+\left[1+\xi(x-\mu)/\sigma\right]^{1/\xi}\right\}^2$	$\mu\in R,\sigma>0,\xi\in R$
P-III	$\frac{\beta^\alpha}{\Gamma(\alpha)}\int_{a_0}^{x}(x-a_0)\exp\left[-\beta(x-a_0)\right]/\mathrm{d}x$	$\frac{\beta^\alpha}{\Gamma(\alpha)}(x-a_0)^{\alpha-1}\exp\left[-\beta(x-a_0)\right]$	$a_0\in R,\alpha>0,\beta>0$
GNO	$\Phi(y)\,y=-\ln\left[1-\xi(x-\mu)/\sigma\right]/\xi$	$\phi(y)/[\sigma-\xi(x-\mu)]\,y=-\ln\left[1-\xi(x-\mu)/\sigma\right]/\xi$	$\mu\in R,\sigma>0,\xi\in R$

of irrigation, which refers to the ratio of net water supply in the field to water withdrawn from river or other water sources engineering.

In a large region, for example Guangdong Province, China, the differences of annual precipitation and the net quota of irrigation among individual agricultural regions are required to be considered. Then, for a given regional frequency of annual precipitation, u_0, for a large region with sub-regions, water demand of irrigation, $W(u_0)$, can be deduced as follows:

$$W(u_0)=\sum_{i=1}^{d}\frac{A_i q_i(u_i)}{\eta}, \tag{2}$$

where A_i refers to the irrigation area of the ith sub-region and $q_i(u_i)$ is the net quota of irrigation per unit area at the frequency of precipitation of the ith sub-region, u_i.

Therefore, the key point to design water demand of irrigation for a given precipitation frequency of the entire region, u_0, is how to determine a combination of sub-regional precipitation frequencies, $\{u_1,\cdots,u_d\}$.

2.2 Annual precipitation distribution

If a large region consists of d sub-regions, let X_i be the series of annual precipitation in which $i=1,\cdots,d$ and $j=1,\cdots,n$ refer to sub-region d and n years, respectively. The theoretical cumulative distribution of annual precipitation in a sub-region, $F_{X_i}(x)$, can be fitted by several conventional three parameter univariate distributions, which have been widely used in hydrology, such as generalized extreme value (GEV), generalized logistic (GLO), Pearson III (P-III) and generalized normal (GNO). Their cumulative distribution and probability density functions are presented in Table 1.

The Kolmogorov–Smirnov (K–S) statistic, D, for the goodness-of-fit test of annual precipitation distribution of a sub-region was computed as (Dobric and Schmid, 2007; Massey, 2012; Tu et al., 2016)

$$D=\max_{1\le j\le n}\left\{F\left[x(j)\right]-\frac{j-1}{n},\frac{j-1}{n}-F\left[x(j)\right]\right\}, \tag{3}$$

The parameters of all recommended univariate distributions were estimated by the L-moment method. The critical values at the significance level of 0.05 for the goodness-of-fit test of all distributions were obtained by the Monte Carlo method with 5000 or more simulations. If the K–S statistic, D, which was computed from the samples, was less than the corresponding critical value, the tested distribution was accepted. The optimal distribution was selected from the accepted distributions by comparing their root-mean-square error (RMSE) and Akaike information criterion (AIC) values. In addition, empirical and theoretical distributions were compared to evaluate the goodness of fit of the observed samples of precipitation. In hydrological practice, the empirical distribution functions are defined and transformed by Gringorten (1963).

Subsequently, based on the areal weight method, the annual precipitation in the entire region, $X_0(j)$, was calculated as

$$X_0(j)=\sum_{i=1}^{d}\alpha_i X_i(j), \tag{4}$$

Where α_i refers to the areal ratio of the ith sub-region to the entire region. Then, the theoretical distribution of annual precipitation for the entire region, $F_{X_0}(x)$, can also be fitted by using the above recommended univariate distributions and the K-S test method.

2.3 Conventional design methods

2.3.1 Typical year method (TY)

In practical design, water demand of irrigation for a large region, a combination of observed sub-regional precipitation in a certain year, in which the precipitation of the entire region weighted by individual sub-regions is the nearest to that of a given frequency for the entire region, has been the only selection (Cai et al., 2001). The selected year was called the typical year corresponding to the given frequency of precipitation. Let $\widetilde{u}_0$ be a given cumulative distribution frequency (CDF) of precipitation of the entire region. Then, the corresponding precipitation can be calculated by using the inverse function of the frequency distribution as follows:

$$\widetilde{X}_0=F_{X_0}^{-1}(\widetilde{u}_0). \tag{5}$$

Further, the relative alteration, $R(j)\,j=1,\cdots,n$, of the observed precipitation in each year compared to the precipitation of the given frequency, was defined as

$$R(j) = \left| \frac{X_0(j) - \widetilde{X}_0}{\widetilde{X}_0} \right|. \tag{6}$$

Then,

$$J = \operatorname{argmin} R(j), \tag{7}$$

where J is the selected rank of a certain year, which corresponds to the typical year for the given frequency $\widetilde{u}_0$.

In addition, due to the limited length of annual precipitation observations, the relative alteration $R(J)$ of the typical selected year might be large. Namely, annual precipitation of the entire region weighted by sub-regional precipitation in a typical selected year, in terms of magnitude, may differ from that of the given frequency. Therefore, a scaling method was applied to zoom in or out for the sub-regional precipitation in the typical selected year as follows:

$$\beta = \frac{X_0(J)}{\widetilde{X}_0}, \tag{8}$$

$$\widetilde{X}_i(J) = \frac{X_i(J)}{\beta}, \tag{9}$$

where β is a scale coefficient and $X_i(J)$ and $\widetilde{X}_i(J)$ refer to the ith sub-regional precipitation before and after zooming in or out, respectively.

For a given precipitation frequency of the entire region, $\widetilde{u}_0$, individual sub-regional frequencies, $\widetilde{u}_i(J)\, i = 1, \cdots, d$, can eventually be deduced by their frequency distributions as follows:

$$\widetilde{u}_i(J) = F_{X_i}[X_i(J)]. \tag{10}$$

2.3.2 Equalized frequency method (EF)

In order to get a combination of sub-regional frequencies for a given precipitation frequency of the entire region, the equalized frequency method (Liu et al., 2015) is also used for downscaling precipitation for a large region. As the name implies, the EF assumes that the frequencies of sub-regional precipitation are identical. That is, for a given precipitation frequency $\widetilde{u}_0$, let $\widetilde{u}_1 =, \cdots, = \widetilde{u}_d$, and then $\widetilde{u}_i$ can be found as follows:

$$F_{X_0}^{-1}(\widetilde{u}_0) = \sum_{i=1}^{d} A_i F_{X_i}^{-1}(\widetilde{u}_i), \tag{11}$$

where $F_{X_0}^{-1}(\widetilde{u}_0)$ and $F_{X_i}^{-1}(\widetilde{u}_i)$ refer to the precipitation in the entire region and sub-regions calculated by using the inverse function of individual frequency distributions, respectively. In practical design, the sub-regional frequency $\widetilde{u}_i$ is determined by applying the method of successive search approximation within an available range.

2.4 Joint design based on copula function

2.4.1 Multidimensional joint distribution

For a large region, the difference of precipitation between any two sub-regions is associated with geographical location. Though annual precipitation in any sub-region is regarded as stochastic, there exists dependence between two sub-regions, in particular for adjacent sub-regions due to similar geographic and climate conditions. Herein, a multidimensional copula function was used for modeling the joint distribution of sub-regional precipitation. Assume that annual precipitation of each sub-region, $X_i, i = 1, \cdots, d$, is a continuous random variable with a d-dimensional joint distribution $H(X_i, \cdots, X_d)$ and individual marginal distribution functions $F_{X_i}(x), i = 1, \cdots, d$. Then, on the basis of the Sklar theorem (Nelsen, 2006), the joint distribution, $H(X_1, \cdots, X_d)$, can be defined as

$$\begin{aligned} H(X_1, \cdots, X_d) &= C\left[F_{X_1}(x), \cdots, F_{X_d}(x)\right] \\ &= C(u_1, \cdots, u_d), \end{aligned} \tag{12}$$

where $C(u_1, \cdots, u_d)$ is the d-dimensional copula function which is the joint distribution function of standard uniform random variables, and $u_i = F_{X_i}(x), i = 1, \cdots, d$, refer to individual CDFs of sub-regional precipitation.

The copula function, which accommodates different marginal distributions of individual variables and captures their dependence, has been widely applied in multivariate hydrology. More details on the theoretical properties of various copula families can be found in Nelsen (2006). Owing to its flexibility, accessibility and simple copula parameters in association with a correlation coefficient matrix, a d-dimensional meta-Gaussian copula was selected for modeling the joint distribution of multiple sub-regional precipitations. Its theoretical cumulative distribution function, $C(u_1, \cdots, u_d)$, and density function, $c(u_1, \cdots, u_d)$, were deduced as follows (Genest et al., 2007):

$$C(u_1, \cdots, u_d) = \int_{-\infty}^{b_1} \cdots \int_{-\infty}^{b_d} g(\omega_1, \cdots, \omega_d)\, d\omega_1, \cdots, d\omega_d, \tag{13}$$

$$c(u_1, \cdots, u_d) = |\Sigma|^{-1/2} \exp\left(-\frac{\zeta^T \Sigma^{-1} \zeta}{2} + \frac{\zeta^T \zeta}{2} \right), \tag{14}$$

where

$$g(\omega_1, \cdots, \omega_d) = (2\pi)^{-d/2} |\Sigma|^{-1/2} \exp\left(-\frac{\omega^T \Sigma^{-1} \omega}{2} \right), \tag{15}$$

where $b_1 = \Phi^{-1}(u_1), \cdots, b_d = \Phi^{-1}(u_d)$, in which $\Phi^{-1}(\cdot)$ refers to the inverse function of the standard normal distribution. $\omega = [\omega_1, \cdots, \omega_d]^T$ and $\zeta = [b_1, \cdots, b_d]^T$ are the matrices of variables in the integrand. The correlation coefficient matrix Σ was expressed as

$$\Sigma = \begin{bmatrix} 1 & \cdots & \rho_{1d} \\ \vdots & \ddots & \vdots \\ \rho_{d1} & \cdots & 1 \end{bmatrix} \rho_{ij} = \begin{cases} 1, i = j \\ \rho_{ji}, i \neq j \end{cases}, \tag{16}$$

where, $\rho_{ij} \in [-1, 1]$ refers to the correlation coefficient between any two sub-regional precipitations.

For the goodness-of-fit test of multi-dimensional meta-Gaussian copula, the Cramér–von Mises test statistic on the basis of the Rosenblatt transform was used (Rosenblatt, 1952; Genest et al., 2009). For the joint distribution of sub-regional precipitation with a d dimension, the goodness-of-fit test statistics, S_n^B, were formulated as

$$S_n^B = \frac{n}{3^d} - \frac{1}{2^{d-1}} \sum\nolimits_{j=1}^{n} \prod_{i=1}^{d} \left[1 - E_i^2(j)\right] + \frac{1}{n} \sum\nolimits_{k=1}^{n}$$
$$\sum\nolimits_{j=1}^{n} \prod_{i=1}^{d} \left\{1 - \max\left[E_i(j), E_i(k)\right]\right\}, \tag{17}$$

where E_i, i = 1, ⋯, d, refers to the pseudo-observations of individual sub-regional precipitation. Let $E_i = u_1$, and $E_i, i = 2, \cdots, d$ be assigned as (Rosenblatt, 1952; Dobric and Schmid, 2007)

$$E_i = C(u_i | u_1, \cdots, u_{i-1}) = \frac{\partial^{i-1} C(u_1, \cdots, u_i)}{\partial u_1 \cdots \partial u_{i-1}} / \frac{\partial^{i-1} C(u_1, \cdots, u_{i-1})}{\partial u_1 \cdots \partial u_{i-1}}, \tag{18}$$

A parametric bootstrap procedure for S_n^B, deduced from the literature, is addressed in Appendix D (Genest et al., 2009). In multivariate practice, the joint empirical distribution functions are defined (Dobric and Schmid, 2007; Genest et al., 2009; Tu et al., 2016) and transformed by Gringorten (1963).

In addition, the Kendall function, which is a univariate expression of multivariate information (Genest and Rivest, 1993; Barbe et al., 1996; Salvadori et al., 2011, 2013), has been shown to be an appropriate tool for calculating the copula-based joint frequency of multivariate events (Nappo and Spizzichino, 2009; Salvadori et al., 2011) and is widely applied in discussing the joint probability or return period of hydrological multivariables (Salvadori and Michele, 2004; Michele et al., 2013). The Kendall CDF, F_{K_c}, which was transformed from the joint CDF of eight sub-regional precipitation and was used in comparing it with the frequency of entire regional precipitation, was estimated as

$$F_{K_c}(q) = P\left[C\left[u_1(j), \cdots, u_d(j)\right] \le q\right] = \frac{1}{n} \sum\nolimits_{j=1}^{n} I(C \le q), \tag{19}$$

where $q \in (0, 1)$ is the probability level and n refers to the length of observed or simulated samples. The function $I(\cdot)$ is an indicator function, which is equal to 1 when the enclosed expression is true, and 0 otherwise.

2.4.2 Most-likely weight function (MLW)

In multivariate design practice, using sample data of annual precipitation of all sub-regions, the univariate distribution of the entire region, joint and marginal distributions of sub-regions, and parameters of all distributions can be determined by the recommended univariate models and Gaussian copula. The Monte Carlo method can be used to simulate new combinations of CDFs of precipitation using the determined distributions and parameters. Then, the CDFs of precipitation of the entire region corresponding to each combination of sub-regional CDFs can be achieved. However, there are a large number of combinations which lead to the CDFs of the entire region, which can be almost identical within a predefined small difference. That is, a given CDF of precipitation of the entire region can correspond to many combinations of sub-regional CDFs with enough further simulations. The design realization using the most-likely weight function was proposed by Salvadori et al. (2011) as follows:

$$[\widetilde{u}_1, \cdots, \widetilde{u}_d] = \operatorname{argmax} f(x_1, \cdots, x_d), \tag{20}$$

$$f(x_1, \cdots, x_d) = c(u_1, \cdots, u_d) f(x_1) \cdots f(x_d), \tag{21}$$

where $[\widetilde{u}_1, \cdots, \widetilde{u}_d]$ is eventually selected as the design combination of CDFs of sub-regional precipitation for a given CDF of the entire region, $\widetilde{u}_0$. $f(x_1, \cdots, x_d)$ refers to the product of joint probability densities, $c(u_1, \cdots, u_d)$, and their individual marginal probability densities, $f(x_i)\, i = 1, \cdots, d$. Therefore, the procedure for design combination of sub-regional precipitation for a given CDF of the entire region was as follows.

1. The joint and marginal distributions and parameters of sub-regional precipitation were determined by the goodness of fit of the recommended d-dimensional meta-Gaussian copula and univariate distributions for the observed samples.

2. Using the Monte Carlo method according to the determined d-dimensional joint distribution, large quantities of combinations of CDFs of sub-regional precipitation, $\left[u_1(j), \cdots, u_d(j)\right] j = 1, \cdots, m$, with the number of simulations, m, were generated, and the corresponding combinations of sub-regional precipitation, $\left[X_1(j), \cdots, X_d(j)\right] j = 1, \cdots, m$, and precipitation and CDFs of the entire region, $X_0(j)\, j = 1, \cdots, m$ and $u_0(j)\, j = 1, \cdots, m$, were calculated.

3. For a given CDF of precipitation of the entire region, $\widetilde{u}_0$, the allowable relative error, was defined as

$$\left| \frac{u_0(j) - \widetilde{u}_0}{\widetilde{u}_0} \right| \le R_e, \tag{22}$$

Table 2. Area, paddy field and precipitation of the sub-regions and entire region. The A0 region refers to the entire region of Guangdong Province.

Region	Area (10^3 km^2)	Paddy field (10^3 hm^2)	Annual precipitation (10^3 mm)			
			Maximum	Mean	Median	Minimum
A1	27.99	164.14	2.439	1.667	1.663	0.997
A2	33.37	187.63	2.379	1.704	1.672	1.102
A3	23.36	116.71	2.014	1.569	1.551	1.104
A4	9.36	106.59	2.457	1.676	1.659	0.965
A5	21.71	183.17	4.071	2.789	2.877	1.482
A6	14.29	126.43	2.573	1.725	1.665	0.724
A7	16.95	167.98	2.734	1.911	1.911	1.196
A8	11.53	108.08	2.275	1.514	1.487	0.747
A0	158.57	1160.73	2.421	1.835	1.845	1.152

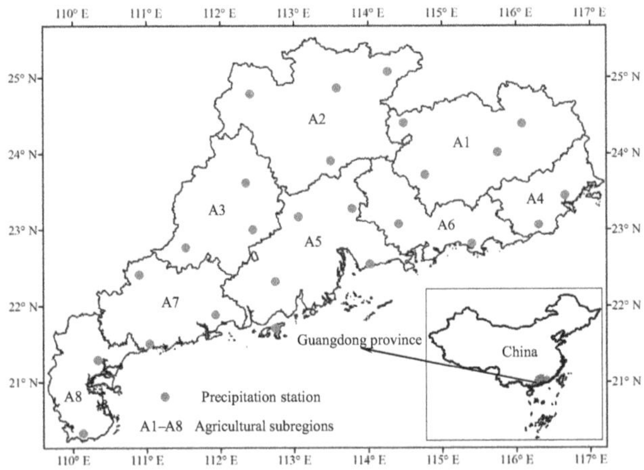

Figure 1. Location of the study region, agricultural sub-regions and precipitation stations.

where R_e refers to the threshold value of allowable relative error. Then, in the simulated $u_0(j)$, those which satisfied Eq. (22) were selected. That is, $u_0(k) k = 1, \cdots, l$ with the length of l were found to satisfy $u_0(k) \cong \widetilde{u}_0$ and the selected combinations, $[u_1(k), \cdots, u_d(k)] k = 1, \cdots, l$ and $[X_1(k), \cdots, X_d(k)] k = 1, \cdots, l$, corresponded to the given $\widetilde{u}_0$.

4. For all selected combinations, the products of the probability densities, $f[x_1(k), \cdots, x_d(k)] k = 1, \cdots, l$, on the basis of the joint probability densities, $c[u_1(k), \cdots, u_d(k)] k = 1, \cdots, l$ and marginal $f[x_1(k)], \cdots f[x_d(k)] k = 1, \cdots, l$, respectively, were calculated, and $[\widetilde{u}_1, \cdots, \widetilde{u}_d]$ with the maxima of the product from $[u_1(k), \cdots, u_d(k)] k = 1, \cdots, l$ was the design combination for the given CDF of precipitation of the entire region, $\widetilde{u}_0$.

In addition, in order to better present the relationship of the Kendall CDF of sub-regional precipitation and the CDF of precipitation of the entire region, a confidence interval (CI) was defined by the distance which deviated from the diagonal (Serinaldi, 2013; Volpi and Fiori, 2014) and transformed herein by the normal distribution. The CI with a probability of $1 - \alpha$ was involved.

3 Study region and data

The Guangdong Province as a study case is located in South China with a land area of 158.57×10^3 km^2 (illustrated in Fig. 1). The entire region mostly belongs to the monsoon climate zone, varying from the tropic to southern sub-tropic. Annual precipitation is abundant, but uneven in terms of spatial and temporal distribution. Since China's reform and opening up in the late 1970s, the water demand of the province has been increasing with the rapid development of the regional socio-economy. In 2015, the total water consumption of the province accounted for 44.31 billion m^3, in which approximately one-half was for irrigation.

According to the report of the *Irrigation Quota of Guangdong Province*, the entire region marked by A0, in terms of agriculture on the basis of climate, soil type, cropping system and other management measures, was zoned into eight sub-regions marked by A1–A8 in Fig. 1. Areal data of sub-regions and their paddy fields were used (see Table 2). Annual precipitation data of eight sub-regions for the period of 1953–2013, in terms of multi-site average, were transformed from 25 hydro-meteorological stations.

The net quotas of irrigation of individual sub-regions in association with precipitation frequency resulted from the previous field experiments of irrigation in the late twentieth century. According to the distribution and statistical properties on the basis of field experiments in a research report entitled Annual irrigation Quota in Guangdong Province (1999), the net quotas of irrigation per unit area in the paddy fields of individual sub-regions in association with the precipitation frequency are illustrated in Fig. 2. They ranged from 7221 to 8520 m^3 hm^{-2} in terms of annual average with coefficients

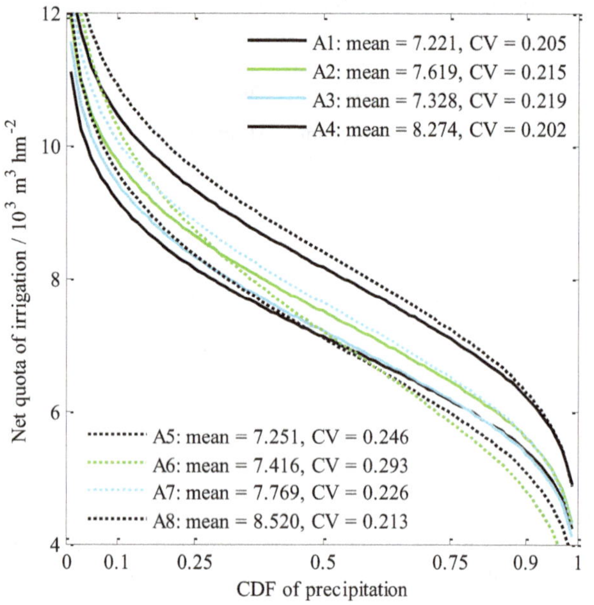

Figure 2. Net quota of irrigation for paddy fields in individual sub-regions.

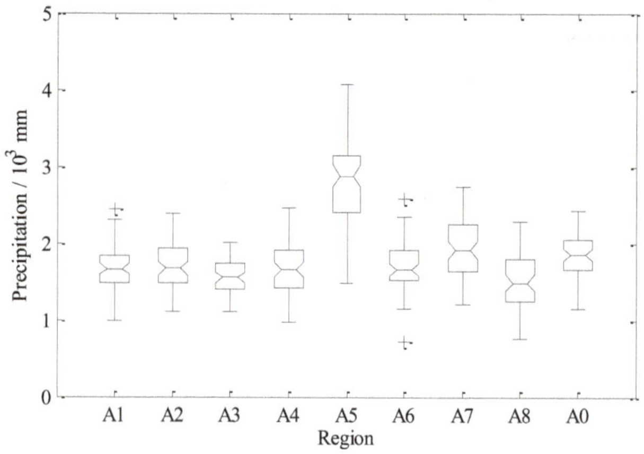

Figure 3. Box plot of precipitation of sub-regions and the entire region. The upper and lower boundaries of the box were set to the values of their percentiles as one-quarter and three-quarters, Q_1 and Q_3, respectively. The red solid line refers to the median. The upper and lower boundaries extended along the dashed line were further set to the values as $Q_1 + 15(Q_1 - Q_3)$ and $Q_3 - 15(Q_1 - Q_3)$, respectively. The values beyond extended boundaries were generally regarded as abnormal, marked by the red plus sign.

of variation of 0.205–0.293, and precipitation was regarded to follow the Pearson III distribution whose three parameters were transformed using the given mean values and coefficients of variation marked in Fig. 2.

The utilization coefficient of irrigation of Guangdong Province has been at a low level, which approximately accounted for 0.46 in terms of average value in 2000 according to the Water Resources Comprehensive Planning of Guangdong Province. In order to respond to tough national water management measures of China, the utilization coefficient is expected to be no less than 0.51 by 2020 in the Tough Water Management Assessing Performance of Guangdong Province. Therefore, the utilization coefficients of eight agricultural sub-regions were uniformly predefined as a fixed value of 0.51 in this paper.

4 Results and discussion

4.1 Univariate properties and distribution of precipitation

Annual precipitation in the entire region and individual sub-regions remarkably changed and was typically random. Precipitation in terms of average varied mostly from 1500 to 2000 mm, except for the A5 sub-region with a larger value of 2789 mm (see Table 2). The regional maximum of precipitation accounted for 4071 mm that occurred in 1973 in the A5 sub-region, and the regional minimum of precipitation less than 800 mm occurred in 1963 in the A6 sub-region. In general, the average precipitation of the entire region accounted for 1835 mm, with a maximum of 2421 mm and a minimum of 1152 mm. The general statistical characteristics of annual precipitation are illustrated in Fig. 3. All values in the range of extended boundaries were generally regarded as normal, and otherwise abnormal. The box plot shows that most samples of precipitation fell within the extended boundaries, except for several samples from the A1 and A6 sub-regions.

The statistics of the goodness-of-fit test of four alternative univariate distributions were smaller than those of the significance level of 0.05 (see Table 3), which implied that the GEV, GLO, P-III and GNO distributions fitted annual precipitation of the sub-regions and the entire region. The RMSE and AIC values among the distributions slightly differed and those of the GNO distribution were the smallest for most regions. Then, as illustrated in Fig. 4, all lines of the theoretical CDF almost overlapped the points of the empirical CDF. They demonstrated that the GNO satisfactorily fitted the frequency distributions of annual precipitation for the entire region and sub-regions. As shown in Table 4, the shape parameters of the GNO in the A1 and A2 sub-regions were clear minus, that in the A5 sub-region was clear positive and others were close to zero, which implied significantly left-skewed, right-skewed and normal distributions, respectively.

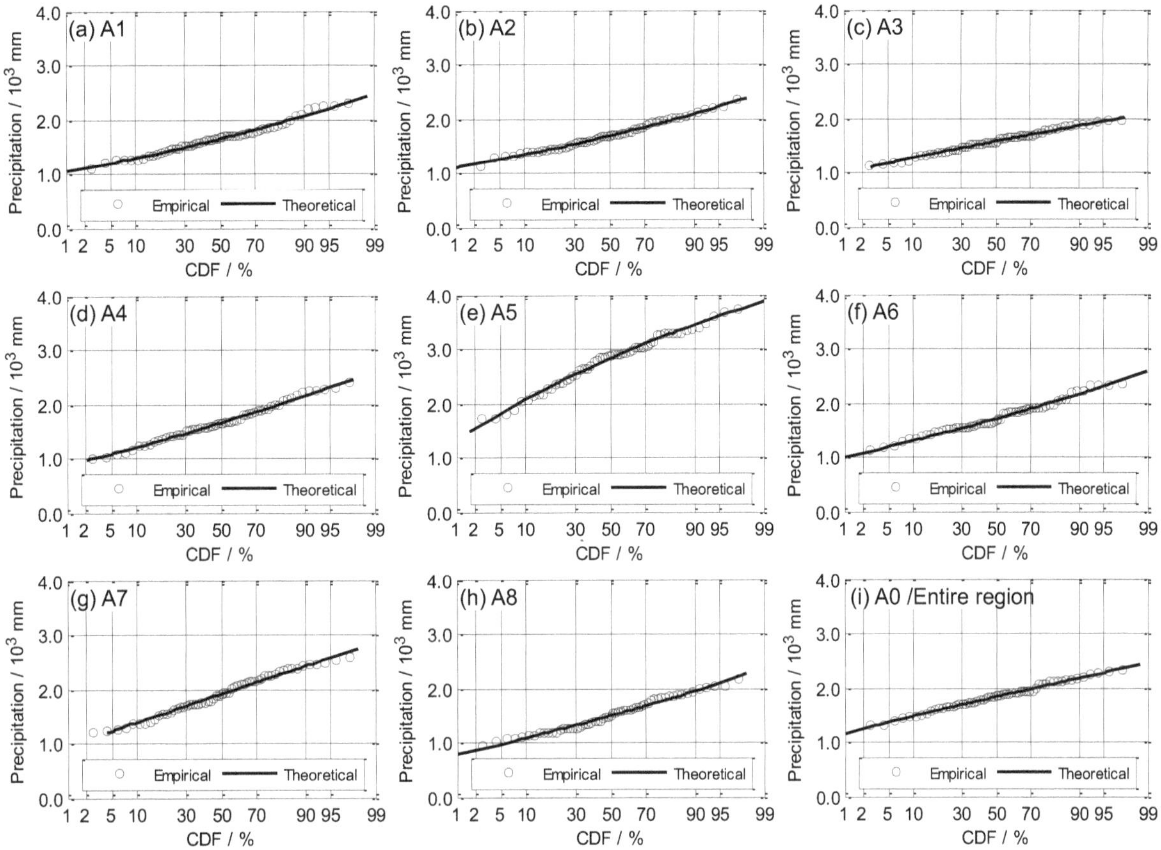

Figure 4. Empirical and theoretical CDFs of precipitation of sub-regions and the entire region using the GNO distribution. On the x axis, the Hessian scale transferred by using the standard normal distribution was used.

Table 3. Goodness-of-fit test of univariate distributions of precipitation of sub-regions and the entire region. The standard (value) refers to the statistic of the significance level of 0.05.

Region	GEV				GLO				P-III				GNO			
	Statistic	Standard	RMSE	AIC	Statistic	Standard	RMSE	AIC	Statistic	Standard	RMSE	AIC	Statistic	Standard	RMSE	AIC
A1	0.065	0.095	0.028	−426	0.053	0.099	0.022	−457	0.065	0.096	0.027	−430	0.064	0.096	0.027	−431
A2	0.055	0.095	0.019	−472	0.075	0.099	0.026	−434	0.058	0.095	0.020	−468	0.058	0.095	0.020	−467
A3	0.051	0.095	0.019	−473	0.064	0.098	0.026	−438	0.049	0.096	0.020	−468	0.049	0.094	0.020	−468
A4	0.060	0.094	0.021	−464	0.057	0.099	0.022	−455	0.059	0.096	0.021	−465	0.059	0.094	0.021	−465
A5	0.087	0.095	0.030	−419	0.091	0.098	0.030	−420	0.087	0.095	0.028	−426	0.087	0.095	0.028	−426
A6	0.076	0.094	0.031	−413	0.090	0.098	0.032	−410	0.079	0.095	0.031	−414	0.079	0.094	0.031	−415
A7	0.063	0.095	0.025	−440	0.081	0.097	0.039	−388	0.068	0.095	0.028	−426	0.068	0.095	0.028	−426
A8	0.066	0.095	0.028	−429	0.088	0.098	0.040	−384	0.069	0.095	0.029	−421	0.070	0.096	0.030	−420
A0	0.076	0.094	0.022	−454	0.067	0.098	0.023	−452	0.072	0.095	0.022	−458	0.072	0.095	0.022	−458

4.2 Eight-dimensional joint distribution of sub-regional precipitation

A matrix of correlation coefficients between sub-regional precipitations is illustrated in Fig. 5. Due to the geographical distance and direction, the coefficients varied from different pairs of sub-regions, but most of them were quite large except between A8 and other sub-regions, such as both A8–A6 and A8–A4, which were less than 0.1. In actuality, the correlation coefficient implied dependence between sub-regional precipitation values. The larger the coefficient was, the larger the dependence was.

The Q-Q plot of empirical and theoretical joint CDFs showed that the sample points fell near the diagonal of 1 : 1, though more fell in the lower tail (see Fig. 6). For the goodness-of-fit test of the eight-dimensional meta-Gaussian copula, the p-value accounted for 0.262 beyond the significance level of 0.05. This demonstrated that the high-dimensional Gaussian copula better fitted the joint distribution of precipitation of eight sub-regions. However, eight-

Table 4. Parameters of the GNO distribution of precipitation of sub-regions and the entire region.

Parameter	A1	A2	A3	A4	A5	A6	A7	A8	A0
Location	1.645	1.684	1.575	1.663	2.831	1.711	1.915	1.497	1.841
Scale	0.302	0.289	0.234	0.369	0.534	0.337	0.406	0.339	0.275
Shape	−0.141	−0.136	0.052	−0.073	0.156	−0.082	0.018	−0.100	0.045

	A1	A2	A3	A4	A5	A6	A7	A8
A1	1	0.839	0.651	0.745	0.604	0.765	0.435	0.148
A2	0.839	1	0.722	0.611	0.677	0.599	0.599	0.238
A3	0.651	0.722	1	0.572	0.753	0.610	0.709	0.309
A4	0.745	0.611	0.572	1	0.466	0.646	0.407	0.032
A5	0.604	0.677	0.753	0.466	1	0.708	0.603	0.264
A6	0.765	0.599	0.610	0.646	0.708	1	0.338	0.079
A7	0.435	0.599	0.709	0.407	0.603	0.338	1	0.489
A8	0.148	0.238	0.309	0.032	0.264	0.079	0.489	1

Figure 5. Matrix of correlative coefficients of precipitation between two sub-regions.

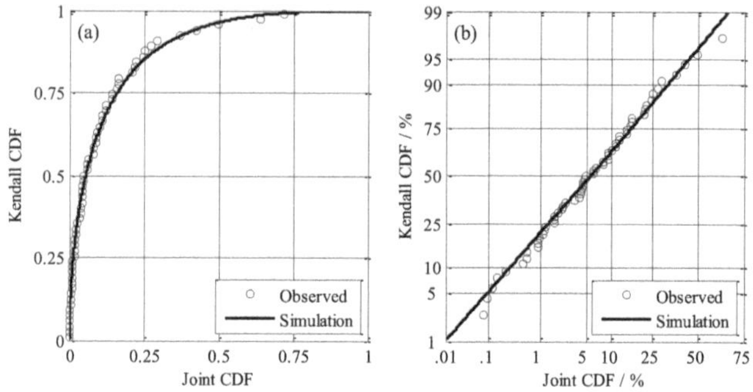

Figure 7. Comparison of conventional joint CDF and Kendall CDF in subfigures **(a)** with the general axes and **(b)** with the Hessian axes.

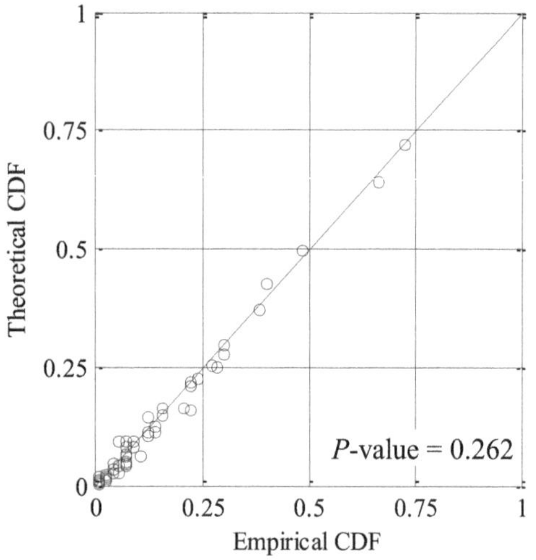

Figure 6. *Q*–*Q* plot of empirical and theoretical CDFs of sub-regional precipitation using the eight-dimensional Gaussian copula.

dimensional joint CDFs of sub-regional precipitation on the basis of observed data were limited and the maximum was less than 0.75.

When the conventional joint CDF of sub-regional precipitation was transformed into the Kendall CDF, the CDF was indeed enlarged (see Fig. 7a). For example, using observed data, the minimum of 0.00005 and the maximum of 0.71957 for the conventional joint CDF corresponded to 0.00916 and 0.99084 for the Kendall CDF, respectively. Using the Hessian axes in which the scales of dual axes were transferred following the standard normal distribution (see Fig. 7b), the conventional joint CDF and the Kendall CDF showed a linear relationship, which demonstrated that it was appropriate to use the Kendall instead of the conventional joint CDF.

4.3 Relationship between sub-regional joint and entire region CDFs of precipitation

According to the determined eight-dimensional Gaussian copula, a million combinations of CDFs of sub-regional precipitation were generated by using Monte Carlo simulation. Correspondingly, the conventional joint and Kendall CDFs of sub-regions and the CDFs of the entire region were achieved (illustrated in Fig. 8). Comparing the conventional joint and the entire region CDFs (see Fig. 8a), the combination points tended to happen in the northwest and had an up-convex lower boundary. When the CDF of the entire region was given a certain value, the corresponding joint CDFs varied within the limited upper bound on which the joint CDFs were less than the given value of the entire region.

Using the Kendall CDF instead of the conventional joint CDF (see Fig. 8b), the combination points scattered near the diagonal of 1 : 1 with a concave upper lower boundary. It also showed that most observed samples fell within the envelope of CI with a probability of 0.95. In addition, between the Kendall CDF and the CDF of the entire region, there were large correlation coefficients of 0.9221 and 0.9153 for the observed and simulated samples, respectively. These showed that the Kendall CDF instead of the conventional CDF was convenient to analyze the relationship of the CDF between the entire region and the sub-regional joint CDF.

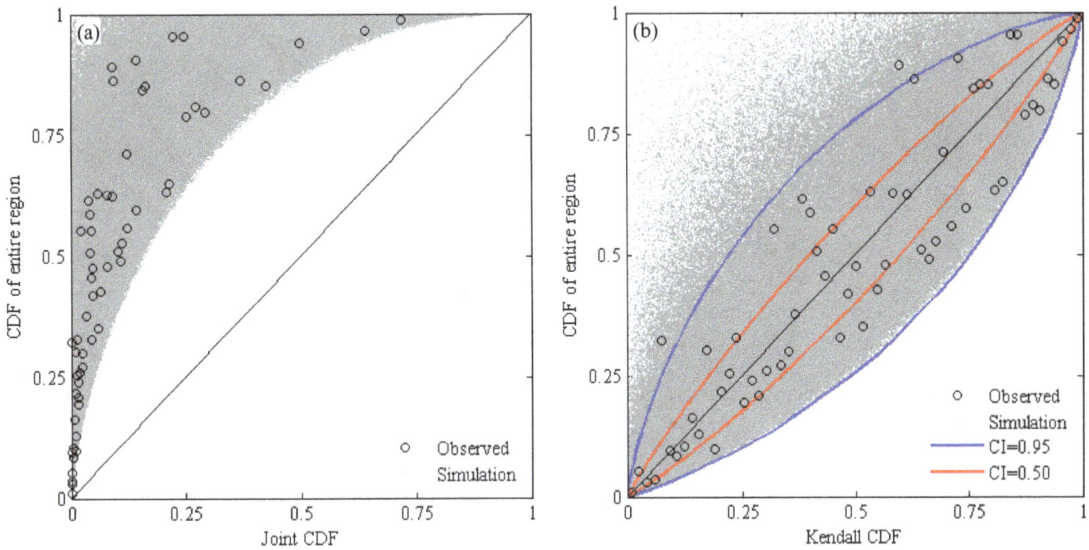

Figure 8. CDF of precipitation of the entire region responding to **(a)** the conventional joint CDF and **(b)** the Kendall CDF of eight sub-regions.

4.4 Design combinations of the CDF of sub-regional precipitation

The given CDFs of precipitation of the entire region can be predefined to change in the range from 0.05 to 0.95 with a step of 0.05, which refers to the alteration of regional precipitation from extreme dry to extreme wet. Considering the uncertainty of Monte Carlo simulation, those simulated combinations (see grey points in Fig. 9), in which the allowable relative error of their calculated CDFs of the entire region compared to a given CDF was less than 0.05 %, were selected to be an alternate for further design combination.

Using the EF method (see the blue dashed line in Fig. 9), design points consisted of a better smooth curve which mostly fell within 0.5 of the CI. The Kendall CDF was very close to the CDF of the entire region, but that of the former was larger than that of the latter, with lower CDFs being less than 0.55 in the study case, and vice versa in larger CDFs. Using the TY method (see the red circles in Fig. 9), design points were irregularly scattered on the two sides of the diagonal of 1 : 1, and even several points were beyond 0.95 of the CI, for example for the given CDFs of 0.8 and 0.9. Using the MLW method (see the blue triangles in Fig. 9), design points fell within the range between 0.5 and 0.95 of the CI, and design Kendall CDFs were larger than the given CDFs.

In general, if the CDF of sub-regional precipitation was equalized, the differences between the design Kendall CDF and the given CDF of the entire region were almost no more than a CI of 0.5. On the basis of maximum joint probability density, design Kendall CDFs were larger than the given CDFs and preferred to the upper limited bound. By zooming in or out according to the typical year, design points almost

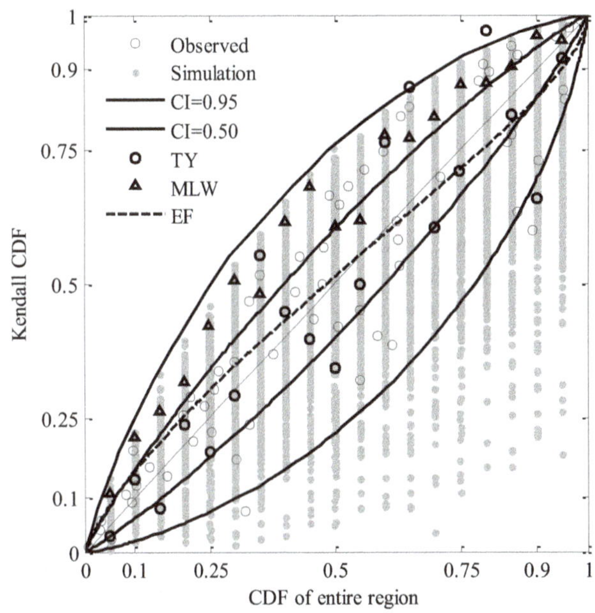

Figure 9. Design Kendall CDFs for given CDFs of precipitation of the entire region which varied from 0.05 to 0.95 with a step of 0.05.

fell within 0.95 of the CI, but they were relatively scattered in comparison to other methods.

In addition, between individual CDFs of eight sub-regional precipitations and the CDF of the entire region, the design points maintained a smooth curve and were undifferentiated for all sub-regions when using the EF method (see the blue dashed line in Fig. 10). The design individual CDFs of sub-regions were very close to the CDF of the entire region, but the former were larger than the latter in the lower

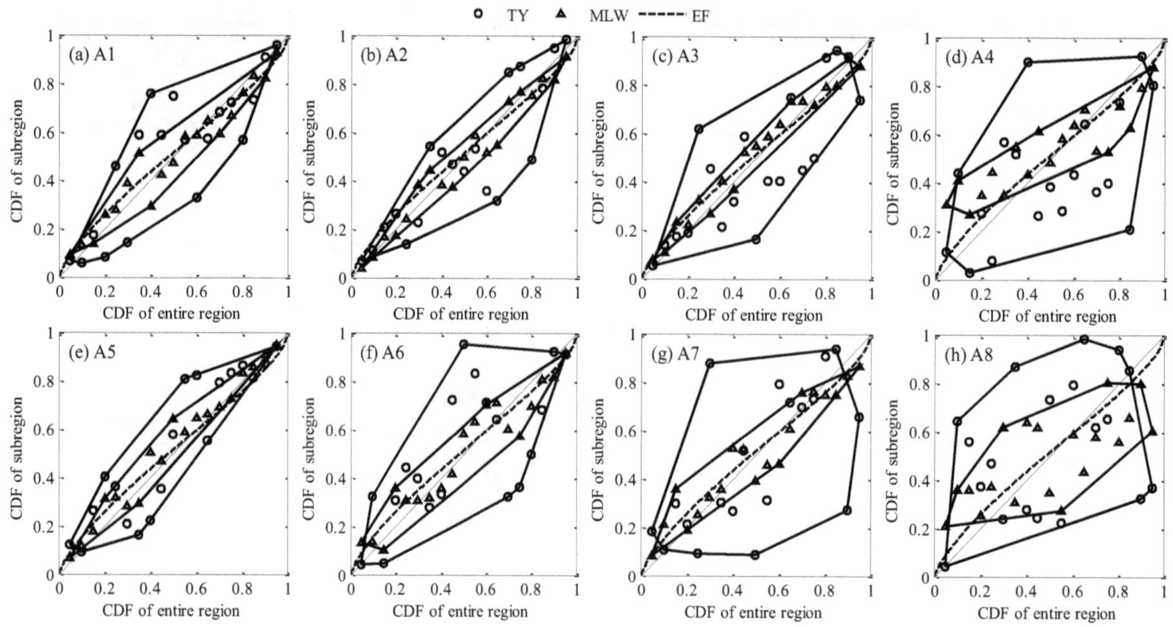

Figure 10. Design CDFs of sub-regions for given CDFs of precipitation of the entire region.

CDFs and vice versa in the larger CDFs. Using TY and MLW methods, design points were scattered on the two sides of the diagonal of 1 : 1. Design CDFs differed from sub-regions, but their differences were undetermined. However, as seen from the envelope of design points of individual sub-regions, the ranges of the MLW design were comparatively narrow and concentrated around the diagonal of 1 : 1 except for the A8 sub-region, but those of the TY design were much wider, in particular for the A4, A7 and A8 sub-regions.

4.5 Design water demand of irrigation

For given CDFs of precipitation of the entire region, the water demand of irrigation of the entire region for all selected simulations and design points are illustrated in Fig. 11a. As the given CDF changed from 0.05 to 0.95, the average value of water demand decreased from 22.79 to 12.78 billion m^3, correspondingly representing the range from extremely dry to extremely wet (see the black dashed line in Fig. 11a). The difference between the maximum and minimum demands for a given CDF (see the blue and red dashed lines in Fig. 11a) varied in the range of 1.15–1.72 billion m^3. Design demands of the MLW and EF methods (see the blue and black solid lines in Fig. 11a) were slightly smaller than the average values, but those of the TY method (see the red line in Fig. 11a) fluctuated around the line of average value. Compared to the average values (see Fig. 11b), the maximum and minimum demands increased and decreased by 3.0 %–7.5 % and 2.5 %–3.8 %, respectively. Design demands of the MLW and EF methods decreased within 1.4 % and 2.1 %, respectively, and those of the TY method increased or decreased within 2.8 % or 2.1 %, respectively. These values demonstrated that the differences of water demand among three design methods for the entire region were quite small.

Design water demands of irrigation in individual sub-regions are illustrated in Fig. 13. As the given CDF changed from 0.05 to 0.95, sub-regional demands of the EF method smoothly decreased from 2.16–3.73 to 1.16–2.29 billion m^3, correspondingly representing the range from extremely dry to extremely wet (see the black lines in Fig. 12). Then, the demands of the TY and MLW methods fluctuated around the lines of the demand of the EF method, and the fluctuations of the former were remarkably larger than those of the latter (see the red and blue lines in Fig. 12). Compared to the EF method (see Fig. 13), the increase or decrease in the water demand of the MLW design accounted for less than 13 % in most sub-regions, except for the A8 sub-region with a maximum of 26.1 %, but that of the TY method accounted for 15.4 %–24.3 % for the A1, A2, A3 and A5 sub-regions and 39.8 %–45.7 % for the A4, A6, A7 and A8 sub-regions. These values demonstrated that the alterations of design water demand in sub-regions were more complicated in comparison with those in the entire region.

5 Conclusions

Using Guangdong Province of South China as a case study of a large region, a high-dimensional meta-Gaussian copula was applied for fitting the joint distribution of multiple regional precipitation. A large number of combinations of CDFs of precipitation of eight sub-regions were generated

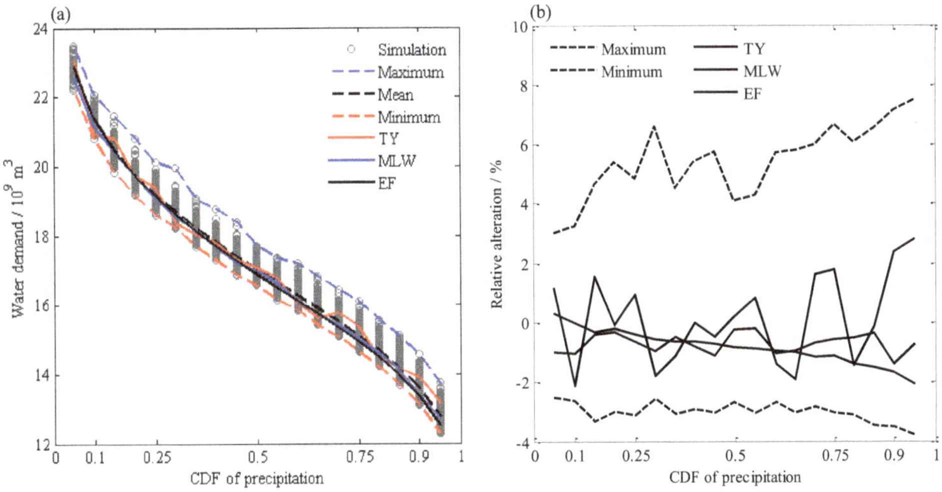

Figure 11. (a) Design water demand of irrigation of the entire region and **(b)** their relative alteration compared to the average demand.

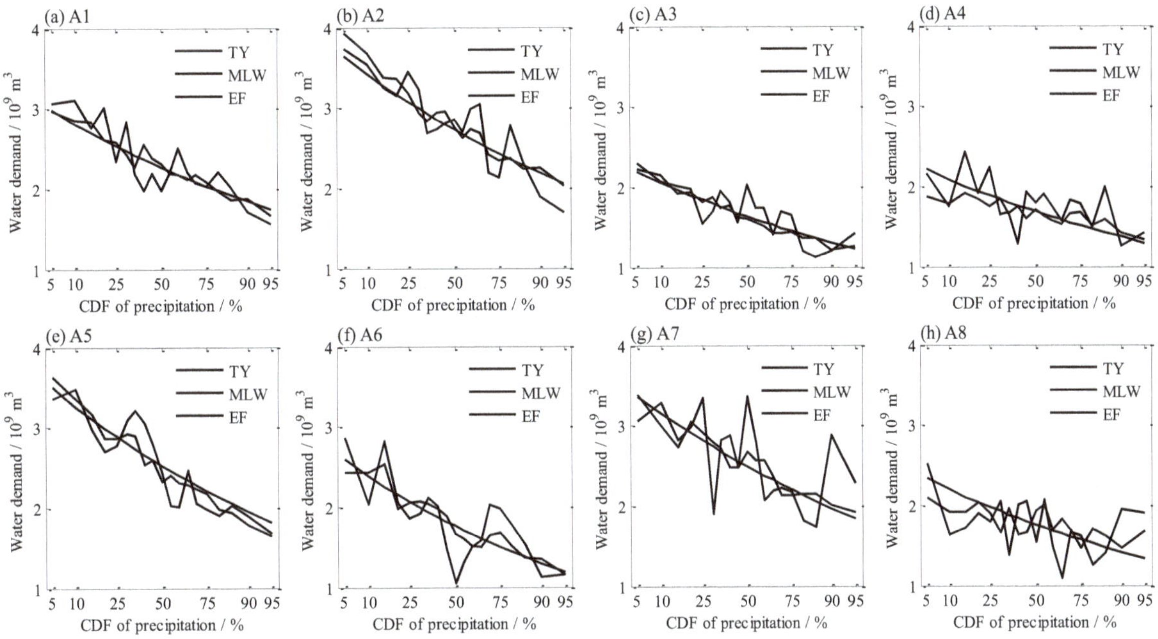

Figure 12. Design water demand of irrigation in sub-regions for given CDFs of precipitation of the entire region.

by using the Monte Carlo method. The relationship among the CDFs of the entire region, the conventional joint CDF, and the Kendall CDF of sub-regions was determined. Three design methods, including the EF and TY design methods, and a new design procedure of the MLW method in association with the joint probability density, were used for design combinations of sub-regional CDFs for given CDFs of precipitation of the entire region. Then, design water demands of irrigation of the entire region and individual sub-regions were compared. The main conclusions of this study are as follows.

1. The frequency distributions of annual precipitation of the entire region and of sub-regions were fitted well by the GNO distribution. The shape parameters in the A1 and A2 sub-regions were a clear minus, those in the A5 sub-region were a clear positive, and others were close to zero, which implied significantly left-skewed, right-skewed and normal distributions, respectively. The eight-dimensional Gaussian copula satisfactorily fitted the joint distribution of sub-regional precipitation.

2. There was a clear linear dependence between the conventional joint and Kendall CDFs of sub-regional pre-

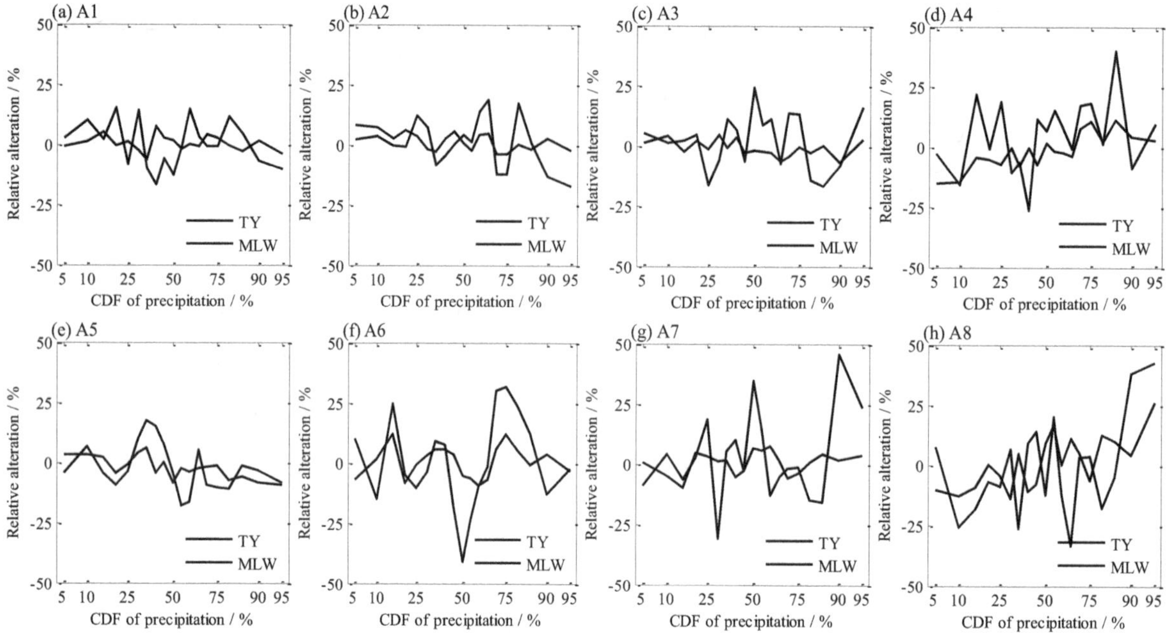

Figure 13. Relative alteration of design water demand compared to the EF method for given CDFs of precipitation of the entire region.

cipitation when both of them were transferred by the standard normal distribution. Comparing the Kendall CDFs of sub-regions and the CDF of the entire region, most observed samples fell within the envelope of CI of a probability of 0.95 around the diagonal of 1 : 1, and there were greater dependences between them with correlation coefficients of 0.9221 and 0.9153 for the observed and simulated samples, respectively. The use of the Kendall CDF instead of the conventional joint CDF can better link the joint frequency of sub-regions and the univariate frequency of the entire region. However, any one given CDF of the entire region corresponded to a large number of joint CDFs varying from very small to limited large CDFs. That is, there was an upper bound in larger values of the joint CDFs of sub-regions corresponding to given CDFs of the entire region.

3. For given CDFs of precipitation of the entire region, design Kendall CDFs and individual CDFs of eight sub-regions of the EF method maintained a smooth curve and were very close to their diagonal of 1 : 1. The design Kendall CDFs of the MLW method which were larger than the given CDFs of the entire region fell between 0.5 and 0.95 probabilities for the CI far from the diagonal, and those of the TY method were irregularly scattered on the two sides of the diagonal. Then, design CDFs of individual sub-regions of the MLW and TY methods were also scattered on the two sides of the diagonal, but they differed for individual sub-regions. The change ranges of the MLW design were comparatively narrow and were concentrated around the diagonal, but those of the TY design were much wider.

4. For given CDFs varying from 0.05 to 0.95 representing the range from extremely dry to extremely wet, the simulated water demand of irrigation of the entire region in terms of the average value accounted for from 22.79 to 12.78 billion m^3. Design demands of the MLW and EF methods were slightly smaller than the average values, and those of the TY method fluctuated around the average values. Compared to the average demand, design demands of the MLW, EF and TY methods decreased or increased, respectively, within 1.4 %, 2.1 %, and 2.8 %, which demonstrated that the differences of design demand of the entire region among the three methods were quite small.

5. For given CDFs varying from 0.05 to 0.95 representing the range from extremely dry to extremely wet, design water demands of individual sub-regions of the EF method decreased smoothly from 2.16–3.73 to 1.16–2.29 billion m^3, and those of the MLW and TY methods fluctuated around the demand lines of the EF method, but the fluctuations of the TY method were remarkably larger than those of the MLW method. Compared to the EF method, the increase or decrease in the water demand of the MLW design accounted for less than 13 % in most sub-regions, except for the A8 sub-region with a maximum of 26.1 %, but those of the TY method accounted for 39.8 %–45.7 % for the A4, A6, A7 and A8 sub-regions. These values demonstrated that the alterations of design water demand in sub-regions were more

complicated in comparison with those in the entire region.

All in all, in practical planning of regional water resources, using the EF method can realize water demand of irrigation for the entire region and its sub-regions for a given frequency of precipitation, but it is arbitrary that the series of sub-regional precipitation are regarded to be undifferentiated for a large region, for example the case region. The TY method was constrained by the limited observed data of precipitation and it cannot be chosen when several different combinations of sub-regional precipitation made the frequency of precipitation of the entire region approximately identical. Therefore, a design procedure using the MLW method in association with a high-dimensional copula, which simulated individual univariate distributions, captured their dependences for multi-variables and built a linkage between regional frequency and sub-regional frequency of precipitation, is recommended for design water demand of irrigation for a large region.

Author contributions. XJT conceived this study and developed the method and wrote the paper. YLD conducted modelling experiments and performed the analysis. VPS revised the manuscript. XHC and KRL contributed comments on the method and writing. HOW conducted data analysis.

Competing interests. The authors declare that they have no conflict of interest.

Acknowledgements. Support by the National Key R&D Program of China (2017YFC0405900), the National Natural Science Foundation of China (51479217; 51879288), and the Outstanding Youth Science Foundation of NSFC (51822908) is gratefully acknowledged.

Edited by: Luis Samaniego

References

Abdul Rauf, U. F. and Zeephongsekul, P.: Copula based analysis of rainfall severity and duration: a case study, Theor. Appl. Climatol., 115, 153–166, 2014.

Aussenegg, W. and Cech, C.: A new copula approach for high-dimensional real world portfolios, Working Paper Series by the University of Applied Sciences bfi Vienna, 68, 1–33, https://www.researchgate.net/publication/267391266, 2012.

Barbe, P., Genest, C., Ghoudi, K., and Remillard, B.: On Kendall's process, J. Multivar. Anal., 58, 197–229, 1996.

Cai, J. B., Liu, Y., Cai, L. Z., and Wu, Y. Y.: A method of determining suitable probability for irrigation project design, Irrig. Drain., 20, 30–31, 2001 (in Chinese).

Cai, X., Hejazi, M. I., and Wang, D.: Value of probabilistic weather forecasts: assessment by real-time optimization of irrigation scheduling, J. Water Resour. Plann. Manage., 137, 391–403, 2011.

Chen, L., Singh, V. P., Guo, S., and Zhou, J.: Copula-based method for multisite monthly and daily streamflow simulation, J. Hydrol., 528, 369–384, 2015.

Chowdhary, H., Escobar, L. A., and Singh, V. P.: Identification of suitable copulas for bivariate frequency analysis of flood peak and flood volume data, Hydrol. Res., 42, 193–215, 2011.

Corbella, S. and Stretch, D. D.: Multivariate return periods of sea storms for coastal erosion risk assessment, Nat. Hazards Earth Syst. Sci., 12, 2699–2708, https://doi.org/10.5194/nhess-12-2699-2012, 2012.

Creal, D. D. and Tsay, R. S.: High dimensional dynamic stochastic copula models, J. Econometrics, 189, 335–345, 2015.

Davidson, B. and Hellegers, P.: Estimating the own-price elasticity of demand for irrigation water in the Musi catchment of India, J. Hydrol., 408, 226–234, 2011.

Dobric, J. and Schmid, F.: A goodness of fit test for copulas based on Rosenblatt's transformation, Comput. Stat. Data Anal., 51, 4633–4642, 2007.

Ganguli, P. and Reddy, M. J.: Evaluation of trends and multivariate frequency analysis of droughts in three meteorological subdivisions of western India, Int. J. Climatol., 34, 911–928, 2014.

Genest, C. and Rivest, L. P.: Statistical inference procedures for bivariate Archimedean copulas, J. Am. Stat. Assoc., 88, 1034–1043, 1993.

Genest, C., Favre, A. C., Béliveau, J., and Jacques, C.: Metaelliptical copulas and their use in frequency analysis of multivariate hydrological data, Water Resour. Res., 43, 223–236, 2007.

Genest, C., Remillard, B., and Beaudoin, D.: Goodness–of–fit tests for copulas: a review and a power study, Insur. Math. Econ., 44, 199–213, 2009.

Gohari, A., Eslamian, S., Abedi-Koupaei, J., Bavani, A. M., Wang, D. B., and Madani, K.: Climate change impacts on crop production in Iran's Zayandeh-Rud River Basin, Sci. Total Environ., 442, 405–419, 2013.

Gräler, B., van den Berg, M. J., Vandenberghe, S., Petroselli, A., Grimaldi, S., De Baets, B., and Verhoest, N. E. C.: Multivariate return periods in hydrology: a critical and practical review focusing on synthetic design hydrograph estimation, Hydrol. Earth Syst. Sci., 17, 1281–1296, https://doi.org/10.5194/hess-17-1281-2013, 2013.

Griffin, R. C.: Achieving water use efficiency in irrigation districts, J. Water Resour. Plann. Manage., 132, 434–442, 2006.

Gringorten, I. I.: A plotting rule for extreme probability paper, J. Geophys. Res., 68, 813–814, 1963.

Lankford, B.: Localising irrigation efficiency, Irrig. Drain., 55, 345–362, 2010.

Leenhardt, D., Trouvat, J. L., Gonzalès, G., Peramaud, V., Prats, S., and Bergez, J. E.: Estimating irrigation demand for water management on a regional scale: II. Validation of ADEAUMIS, Agr. Water Manage., 68, 207–232, 2004.

Leenhardt, D., Angevin, F., Biarnès, A., Colbach, N., and Mignolet, C.: Describing and Locating Cropping Systems on a Regional Scale, in: Sustainable Agriculture, Vol. 2, Springer, Netherlands, 85–95, 2011.

Lian, J., Xu, K., and Ma, C.: Joint impact of rainfall and tidal level on flood risk in a coastal city with a complex river network: a case study of Fuzhou City, China, Hydrol. Earth Syst. Sci., 17, 679-689, doi: 10.5194/hess-17-679-2013, 2013. Liu, Z., Guo, S., Li, T., Hu, Y., and Li, L.: Interval estimation method for design flood region composition, J. Hydraul. Eng., 46, 543-550, doi: 10.13243/j.cnki.slxb.20140928, 2015. (in Chinese)

Liu, Z., Guo, S., Li, T., Hu, Y., and Li, L.: Interval estimation method for design flood region composition, J. Hydraul. Eng., 46, 543–550, 2015 (in Chinese).

Massey, F. J.: The Kolmogorov–Smirnov test for goodness of fit, J. Am. Stat. Assoc., 46, 68–78, 2012.

Meza, F. J., Wilks, D. S., Gurovich, L., and Bambach, N.: Impacts of climate change on irrigated agriculture in the Maipo Basin, Chile: reliability of water rights and changes in the demand for irrigation, J. Water Resour. Plann. Manage., 138, 421–430, 2012.

Michele, C., Salvadori, G., Vezzoli, R., and Pecora, S.: Multivariate assessment of droughts: frequency analysis and dynamic return period, Water Resour. Res., 49, 6985–6994, 2013.

Nappo, G. and Spizzichino, F.: Kendall distributions and level sets in bivariate exchangeable survival models, Inform. Sci., 179, 2878–2890, 2009.

Nelsen, R. B.: An Introduction to Copulas, Springer, New York, 2006.

Rosenblatt, M.: Remarks on a multivariate transformation, Ann. Math. Stat., 23, 470–472, 1952.

Salvadori, G. and Michele, C. D.: Frequency analysis via copulas: Theoretical aspects and applications to hydrological events, Water Resour. Res., 40, 229–244, 2004.

Salvadori, G., De Michele, C., and Durante, F.: On the return period and design in a multivariate framework, Hydrol. Earth Syst. Sci., 15, 3293–3305, https://doi.org/10.5194/hess-15-3293-2011, 2011.

Salvadori, G., Durante, F., and Michele, C. D.: Multivariate return period calculation via survival functions, Water Resour. Res., 49, 2308–2311, 2013.

Serinaldi, F.: An uncertain journey around the tails of multivariate hydrological distributions, Water Resour. Res., 49, 6527–6547, 2013.

Tarjuelo, J. M., Olalla, F. D. S., and Pereira, L. S.: Land and water use: environmental management tools and practices – Preface, Agr. Water Manage., 77, 1–3, 2005.

Thomas, A.: Agricultural irrigation demand under present and future climate scenarios in China, Global Planet. Change, 60, 306–326, 2008.

Tu, X., Singh, V. P., Chen, X., Ma, M., Zhang, Q., and Zhao, Y.: Uncertainty and variability in bivariate modeling of hydrological droughts, Stoch. Environ. Res. Risk A., 30, 1317–1334, 2016.

Tu, X., Du, Y., Chen, X., Chai, Y., and Qing, Y.: Modeling and design on joint distribution of precipitation and tide in the coastal city, Adv. Water Sci., 28, 49–58, 2017a (in Chinese).

Tu, X., Du, X., Singh, V. P., Chen, X., Du, Y., and Li, K.: Joint Risk of Interbasin Water Transfer and Impact of the Window size of Sampling low flows under Environmental Change, J. Hydrol., 554, 1–11, 2017b.

Volpi, E. and Fiori, A.: Hydraulic structures subject to bivariate hydrological loads: Return period, design, and risk assessment, Water Resour. Res., 50, 885–897, 2014.

Wisser, D., Frolking, S., Douglas, E. M., Fekete, B. M., Vorosmarty, C. J., and Schumann, A. H.: Global irrigation water demand: Variability and uncertainties arising from agricultural and climate data sets, Geophys. Res. Lett., 35, L24408, https://doi.org/10.1029/2008GL035296, 2008.

Wriedt, G., Velde, M. V. D., Aloe, A., and Bouraoui, F.: Estimating irrigation water requirements in Europe, J. Hydrol., 373, 527–544, 2009.

Xie, H., Luo, Q., and Huang, J.: Synchronous asynchronous encounter analysis of multiple hydrologic regions based on 3D copula function, Adv. Water Sci., 23, 186–193, 2012 (in Chinese).

Yan, B., Guo, S., Chen, L., and Liu, P.: Flood encountering risk analysis for the Yangtze River and Qingjiang River, J. Hydraul. Eng., 41, 553–559, 2010a (in Chinese).

Yan, B., Guo, S., Guo, J., Chen, L., Liu, P., and Chen, H.: Regional design flood composition based on copula function, J. Hydraul. Eng., 29, 60–65, 2010b (in Chinese).

Zhang, L. and Singh, V. P.: Bivariate flood frequency analysis using the copula method, J. Hydrol. Eng., 11, 150–164, 2006.

Zhang, Q., Li, J., Chen, X., and Bai, Y.: Spatial Variability of Probability Distribution of Extreme Precipitation in Xinjiang, Acta Geogr. Sin., 66, 3–12, 2011 (in Chinese).

Zhang, Q., Li, J., Singh, V. P., and Xu, C.: Copula-based spatio-temporal patterns of precipitation extremes in China, Int. J. Climatol., 33, 1140–1152, 2013.

Zhang, Q., Gu, X., Singh, V. P., Kong, D., and Chen, X.: Spatiotemporal behavior of floods and droughts and their impacts on agriculture in China, Global Planet. Change, 131, 63–72, 2015a.

Zhang, Q., Sun, P., Li, J., Singh, V. P., and Liu, J.: Spatiotemporal properties of droughts and related impacts on agriculture in Xinjiang, China, Int. J. Climatol., 35, 1254–1266, 2015b.

Zhang, Q., Xiao, M. Z., and Singh, V. P.: Uncertainty evaluation of copula analysis of hydrological droughts in the East River basin, China, Global Planet. Change, 129, 1–9, 2015c.

10

The rainfall erosivity factor in the Czech Republic and its uncertainty

Martin Hanel[1,2], **Petr Máca**[1], **Petr Bašta**[1], **Radek Vlnas**[1], **and Pavel Pech**[1]

[1]Faculty of Environmental Sciences, Czech University of Life Sciences, Kamýcká 1176, Prague 6, Czech Republic
[2]T. G. Masaryk Water Research Institute, Podbabská 30, Prague 6, Czech Republic

Correspondence to: Martin Hanel (hanel@fzp.czu.cz)

Abstract. In the present paper, the rainfall erosivity factor (R factor) for the area of the Czech Republic is assessed. Based on 10 min data for 96 stations and corresponding R factor estimates, a number of spatial interpolation methods are applied and cross-validated. These methods include inverse distance weighting, standard, ordinary, and regression kriging with parameters estimated by the method of moments and restricted maximum likelihood, and a generalized least-squares (GLS) model. For the regression-based methods, various statistics of monthly precipitation as well as geographical indices are considered as covariates. In addition to the uncertainty originating from spatial interpolation, the uncertainty due to estimation of the rainfall kinetic energy (needed for calculation of the R factor) as well as the effect of record length and spatial coverage are also addressed. Finally, the contribution of each source of uncertainty is quantified. The average R factor for the area of the Czech Republic is 640 MJ ha^{-1} mm h^{-1}, with values for the individual stations ranging between 320 and 1520 MJ ha^{-1} mm h^{-1}. Among various spatial interpolation methods, the GLS model relating the R factor to the altitude, longitude, mean precipitation, and mean fraction of precipitation above the 95th percentile of monthly precipitation performed best. Application of the GLS model also reduced the uncertainty due to the record length, which is substantial when the R factor is estimated for individual sites. Our results revealed that reasonable estimates of the R factor can be obtained even from relatively short records (15–20 years), provided sufficient spatial coverage and covariates are available.

1 Introduction

Erosion is a natural geological phenomenon resulting from the removal and transportation of soil particles by water, wind, ice, and gravity. Soil erosion by water is a widespread problem throughout Europe (Van der Knijff et al., 2000; Panagos et al., 2015). Erosion is usually driven by a combination of factors like climate (e.g. long dry periods followed by heavy rainfall), topography (steep slopes), inappropriate land use, land cover patterns (e.g. sparse vegetation), ecological disasters (e.g. forest fires), and soil characteristics (e.g. a thin layer of topsoil, a silty texture, or low organic matter content).

Although measurements of soil erosion exist, they are often used as a basis for development, modification, or verification of soil erosion models (applicable at larger scales), which relate soil loss to indicators of relevant factors. A classic example is the Universal Soil Loss Equation (USLE, Wischmeier and Smith, 1978) or the Revised USLE (Renard et al., 1997). Both methods express the long-term average annual soil loss as a product of rainfall erosivity factor (R factor), soil erodibility factor, slope length factor, slope steepness factor, cover-management factor, and support practice factor. The experimental data indicate that when factors other than rainfall are held constant, soil losses from cultivated fields are directly proportional to an erosivity index (EI30) calculated as the total rainfall kinetic energy times the maximum 30 min intensity (Renard et al., 1997). The R factor is then obtained as a long-term average annual rainfall erosivity index. In addition to a point estimate of the R factor, a spatial information (map) is often required for practical purposes. A number of such maps have been released recently at national (Lu and Yu, 2002; Yin et al., 2007; Bonilla and Vidal, 2011; Oliveira et al., 2013; Borrelli et al., 2016; Panagos et al., 2016a; Meddi et al., 2016) or larger (Panagos et al., 2015) scales.

Because of large spatiotemporal variability of rainfall, long records from a dense network of stations are in general required in order to provide reliable estimates of the R factor and/or to develop rainfall erosivity maps. For instance, Wischmeier and Smith (1978) or Verstraeten et al. (2006) recommend at least 22 years of data to be used for estimation of the R factor. On the other hand, lack of high-resolution rainfall data in combination with a need for soil erosion risk assessment often leads to situations where the R factor has to be estimated from shorter records. For instance, a number of stations used recently for the derivation of rainfall erosivity maps for Europe (Panagos et al., 2015) were shorter than 20 years and several records shorter than 10 years were considered. Similarly, the comparison of the spatial interpolation methods in the Ebro Basin (Angulo-Martínez et al., 2009) was based on 10 years of data (for a large number of stations) and Catari et al. (2011) assessed the uncertainty in the estimated R factor considering a 13-year record for eight stations. In general, the uncertainty in the estimated R factor increases as the length of the available record decreases, but can be reduced by combination of data from different sites (Catari et al., 2011) or by considering covariates that are better sampled and/or whose variation over space and time is smaller (Goovaerts, 1999). For the spatial interpolation of the R factor, variables like longitude, latitude, and elevation (Goovaerts, 1999; Angulo-Martínez et al., 2009) or long-term precipitation (Lee and Lin, 2014) are often considered.

In addition to spatial and temporal variability, the expressions for the rainfall kinetic energy (needed for estimation of the erosivity index) and spatial interpolation are also relevant sources of uncertainty for the development of an R factor map. The rainfall kinetic energy can be estimated by a number of expressions (see e.g. van Dijk et al., 2002). Therefore, several authors assessed the effect on the estimated R factor. For instance, Catari et al. (2011) mentioned a variation due to kinetic energy calculation of about 10 %. Similarly, many spatial models can be used to predict the R factor values over the area. The differences between several spatial interpolation methods have been reported e.g. by Angulo-Martínez et al. (2009). Catari et al. (2011) compared the contribution of different sources of variability to the overall uncertainty in the estimated basin average R factor in north-eastern Spain, concluding that while the uncertainty in the annual erosivity index is dominated by the temporal variability (explaining more than 40 % of variation) for the long-term R factor, the kinetic energy calculation becomes more important.

Several maps of the R factor for the whole of Europe have been released in recent decades. For instance, Van der Knijff et al. (2000) applied simple relationships between seasonal and annual rainfall and the R factor and Panagos et al. (2015) used many high-resolution station datasets (with various record lengths and observation periods) to derive an R factor map based on a Gaussian process model. The values derived for the Czech Republic from these maps range between 350 and 700 MJ ha^{-1} mm h^{-1}.

A number of estimates of the R factor have been also published specifically for the Czech Republic. See e.g. the overview by Krása et al. (2014), who also mention results from a number of national projects. Official guidelines for soil erosion risk assessment recommend the value 200 MJ ha^{-1} mm h^{-1} to be used for all agricultural land in the Czech Republic. Only recently was this value increased to 400 MJ ha^{-1} mm h^{-1} (Janeček et al., 2007, 2012a). These values are relatively small with respect to the neighbouring countries and some of the research published for the area. For instance, Janeček et al. (2006) published values of the R factor in the range of 430–1060 MJ ha^{-1} mm h^{-1} and concluded that considerably larger values of the R factor (450–600 MJ ha^{-1} mm h^{-1}) should be used for practical application of USLE in the Czech Republic (instead of 200 MJ ha^{-1} mm h^{-1}). Even larger values of the R factor are reported by Krása et al. (2015). On the other hand, Janeček et al. (2012b), using a regression between the daily erosion index (EI30) and daily precipitation in order to predict annual EI30, report values of the R factor between 150 and 1200 MJ ha^{-1} mm h^{-1}, with an average for arable land of 300–400 MJ ha^{-1} mm h^{-1}. Similarly, Janeček et al. (2013) derived an R factor for the Czech Republic from daily data considering the fraction of erosive events in each year for each station and the areal-average annual sum of the erosivity index. In addition, they excluded years with the largest and years with the smallest erosivity index from the analysis. This resulted in a recommendation to use 400 MJ ha^{-1} mm h^{-1} for all agricultural land in the Czech Republic. The trends in annual rainfall erosivity index were studied by Hanel et al. (2016), who found a significant increasing trend ($\approx$4 % per decade) in 51-year records for 11 stations (more than half of the considered stations). However, recent values of the erosivity index were not exceptional when compared to a longer (91-year) record available at a single station. Considerable projected increases in the rainfall erosivity index in an ensemble of regional climate model simulations for the Czech Republic were reported by Svoboda et al. (2016b).

The present paper compares several methods of spatial interpolation of the R factor over the Czech Republic and evaluates the bias and uncertainty due to expression for the kinetic energy, spatial model, record length, and spatial coverage. The paper is structured as follows. The study area and data considered for calculation and spatial interpolation of the R factor are given in Sect. 2. Section 3 describes the methods used for estimation of the R factor, spatial interpolation, and uncertainty assessment. The results are presented and discussed in Sect. 4. The paper is closed with concluding remarks (Sect. 5).

2 Study area and data

2.1 Study area

Because of relatively complex orography and a combination of Atlantic, Mediterranean, and continental effects, the precipitation patterns over the area of the Czech Republic are rather variable (see Fig. 1). The precipitation is mostly due to enhanced westerly flows in winter (Atlantic influence), except for the eastern part of the Czech Republic, which is typically influenced by the Mediterranean Sea. The Mediterranean influences are also dominant over the whole area in summer (e.g. Brádka, 1972; Brázdil, 1980).

The mean annual total precipitation varies from about 400 mm in the western part of the Czech Republic up to more than 1400 mm in the mountains to the north (Tolasz et al., 2007). Almost two-thirds of the annual total falls in the warm half of the year (April–September). The maximum precipitation amounts can be quite large at short timescales, thus contributing considerably to the annual total precipitation. Štekl et al. (2001) point to historical records of 237 mm in 1 h (25 May 1872), 345 mm in 1 day (29 July 1897), 537 mm in 3 days (6–8 July 1997), and 617 mm in 5 days (4–8 July 1997).

2.2 Pluviograph records

A database of 10 min digitized pluviograph records was used in our study. Since a considerable amount of precipitation falls as snow during winter, the data are in general available only for mid-May to mid-September (further referred to as the year). This is, however, the period when heavy precipitation usually occurs. The data were digitized by the Czech Hydrometeorological Institute (CHMI). Detailed identification and reconstruction of unreadable, damaged, or missing records comprised part of the digitization process (see Květoň et al., 2004, for details). Consistency of the aggregated 10 min data with daily records from control ombrometers was further checked by Hanel and Máca (2014). Identified inconsistent days were marked as unreliable and the years with more than 10 % of unreliable data were excluded from the analysis. A set of 96 stations covering the study area for the period 1989–2003 was selected (based mainly on data availability) in order to provide reasonable spatial coverage and record length for the spatial interpolation of rainfall erosivity. The distance between neighbouring stations is 5–42 km (17 km on average) and the stations are located at altitudes from 150 to 1118 m (450 m on average). Note that although a significant positive trend in the annual erosivity index has been reported by Hanel et al. (2016) for several stations in the Czech Republic, no clear trend is observed (not shown) in the considered period (1989–2003). Detailed information on the considered stations is included in the Supplement to this paper.

To analyse the temporal variability and the effect of the number of stations used for the spatial interpolation, the longest available record (C2TREB01 – Třeboň), with 80 years of data and an additional 40 stations (usually with a slightly different time period than 1989–2003), were also considered (see Sect. 3.3). The location of the stations is indicated in Fig. 1.

The database of 10 min digitized pluviograph records was also used for the estimation of average at-site rainfall event characteristics (event depth, event duration, maximum 10 min intensity, mean event rate, time to peak, depth to peak, and number of events), which were later linked with selected spatial interpolation models (see Hanel and Máca, 2014, for details of the considered rainfall event characteristics).

2.3 Gridded precipitation data

Gridded (1 km resolution) statistics of monthly precipitation for the period 1989–2003 (the same as for the station data) provided by the CHMI were considered as covariates for the spatial interpolation models. Specifically, we used the average (May–September) precipitation (r_{mea}), the coefficient of variation of monthly (May–September) precipitation (r_{cv}), and the mean fraction of precipitation above the 95 % quantile of monthly (May–September) precipitation (r_{p95}). Please note that these gridded statistics are based on much larger numbers of stations (usually more than 750 stations were available for each month and year) than were used in our study for spatial interpolation of the R factor.

3 Methods

3.1 Rainfall erosivity factor

A standard methodology for the assessment of the at-site rainfall erosivity factor (Wischmeier and Smith, 1978) was applied. A continuous 6 h interval without precipitation is used to separate individual rainfall events. To be considered erosive, the cumulative rainfall of an event should be greater than 12.7 mm, or the event should have at least one peak greater than 6.35 mm in 15 min. The latter criterion was modified to 8.5 mm in 20 min in order to match the temporal resolution of our dataset. A similar modification was reported by Meusburger et al. (2012). The proportion of erosive events that are included on the basis of the intensity criterion is, however, small.

The rainfall erosivity factor is defined as a long-term average annual sum of the event erosivity index (EI30 ($\mathrm{MJ\,mm\,ha^{-1}\,h^{-1}\,yr^{-1}}$), Wischmeier and Smith, 1978),

$$\mathrm{EI30} = I_{30} \sum_i e_i v_i, \quad (1)$$

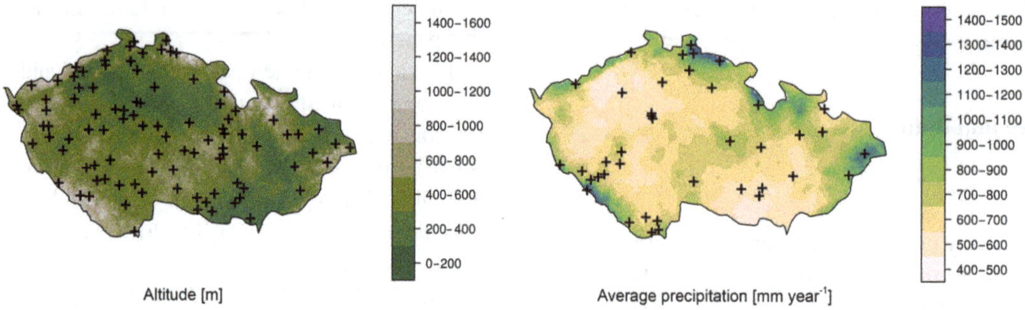

Figure 1. The study area with altitudes and locations of 96 stations used for derivation of the R factor map (left panel) and mean annual precipitation together with locations of additional stations used for uncertainty assessment (right panel).

where

$$e_i = 28.3\left[1 - 0.52\exp(-0.042r_i)\right] \quad (2)$$

is the unit rainfall energy ($\mathrm{MJ\,ha^{-1}\,mm^{-1}}$) (van Dijk et al., 2002), v_i and r_i are the rainfall volume (mm) and intensity ($\mathrm{mm\,h^{-1}}$) during a time interval i, respectively, and I_{30} is the maximum rainfall intensity during a period of 30 min in the event ($\mathrm{mm\,h^{-1}}$).

The unit rainfall energy in Eq. (2) is from van Dijk et al. (2002), who assessed many expressions for its calculation. To assess the related uncertainty, the R factor was also estimated using additional 14 expressions for e_i (see Appendix A for their definition).

3.2 Spatial interpolation models

An at-site dataset of rainfall erosivity factors was further explored using spatial interpolation models. Following the classification of spatial interpolation models for rainfall erosivity provided by Angulo-Martínez et al. (2009), selected local, geostatistical, mixed, and global models were tested.

The local models, which described the spatial distribution of rainfall erosivity using local at-site information of the R factor, were represented by the model based on inverse distance weighting (IDW; see e.g. Angulo-Martínez et al., 2009; Meusburger et al., 2012). The IDW model predicts the R factor as a weighted average of R factor estimates from a selected number of stations. Weights are inversely proportional to the distance (between the interpolated location and the corresponding station) raised to the power r. The number of considered stations and the value of parameter r were estimated during the model cross-validation (see Sects. 3.2.1 and 4).

The geostatistical models were represented by simple kriging (SK), ordinary kriging (OK), simple cookriging (SC), and ordinary cokriging (OC) models (Goovaerts, 1997, 1999). The arithmetic mean of at-site R factor values was used as the estimator of mean values for simple kriging and cokriging models. The cokriging models were tested using a set of seven rainfall event characteristics (see Sect. 2.2) as a cokriging variate. Their spatial covariance structures were described by the Matérn model (Minasny and McBratney, 2005; Haskard, 2007; Pardo-Iguzquiza and Chica-Olmo, 2008).

The mixed models were represented using a set of different regression kriging models, further denoted as RK (Hengl et al., 2004, 2007). Their spatial covariance structures of residuals were described using the Matérn model. The spatially varied means of R factors were estimated using three types of inputs:

1. location information represented by longitude (x), latitude (y), and altitude (z);
2. spatial rainfall information expressed by the combinations of r_{mea}, r_{cv}, and r_{p95}; and
3. information consisting of combinations of both previous types of spatial covariates.

The global spatial interpolation models were represented by the set of generalized linear models (GLS). The deterministic components were estimated using the same types of inputs as in the case of mixed models. Their random error terms were represented using the Matérn and exponential models; the heteroscedasticity of the GLS model residuals was studied by means of the exponential variance function (see p. 206 in Pinheiro and Bates, 2000). The parameters of all tested GLS models were estimated with the restricted maximum likelihood (REML) method (Kitanidis, 1993; Pinheiro and Bates, 2000; Minasny and McBratney, 2007).

3.2.1 Model selection

A standard leave-one-out procedure (Minasny and McBratney, 2007; Angulo-Martínez et al., 2009) was used for the validation of spatial interpolation models. The following indices were considered:

– Willmott's agreement index (WI)

$$\mathrm{WI} = 1 - \frac{\sum_{i=1}^{n} [R_{\mathrm{S}}(\boldsymbol{x}_i) - R_{\mathrm{I}}(\boldsymbol{x}_i)]^2}{\sum_{i=1}^{n} \left[|R_{\mathrm{S}}(\boldsymbol{x}_i) - \overline{R}_{\mathrm{S}}(\boldsymbol{x}_i)| + |R_{\mathrm{I}}(\boldsymbol{x}_i) - \overline{R}_{\mathrm{I}}(\boldsymbol{x}_i)|\right]^2}, \quad (3)$$

– Mean absolute error (MAE)

$$\mathrm{MAE} = \frac{1}{n}\sum_{i=1}^{n} |R_{\mathrm{S}}(\boldsymbol{x}_i) - R_{\mathrm{I}}(\boldsymbol{x}_i)|, \quad (4)$$

– Relative mean absolute error (rMAE)

$$\mathrm{rMAE} = \frac{1}{n}\sum_{i=1}^{n} \frac{|R_{\mathrm{S}}(\boldsymbol{x}_i) - R_{\mathrm{I}}(\boldsymbol{x}_i)|}{R_{\mathrm{S}}(\boldsymbol{x}_i)}, \quad (5)$$

– Root mean square error (RMSE)

$$\mathrm{RMSE} = \sqrt{\frac{1}{n}\sum_{i=1}^{n} [R_{\mathrm{S}}(\boldsymbol{x}_i) - R_{\mathrm{I}}(\boldsymbol{x}_i)]^2}, \quad (6)$$

where $R_{\mathrm{S}}(\boldsymbol{x}_i)$ is the rainfall erosivity calculated for station i, $R_{\mathrm{I}}(\boldsymbol{x}_i)$ is the spatially interpolated value of the rainfall erosivity for the same station, and $\overline{R}_{\mathrm{S}}(\boldsymbol{x}_i)$ and $\overline{R}_{\mathrm{I}}(\boldsymbol{x}_i)$ are the mean rainfall erosivities calculated for the station data and obtained from spatial interpolation, respectively.

The choice of the validation criteria follows the discussion about RMSE and MAE presented by Willmott and Matsuura (2005), who recommended the use of MAE for the estimation of average model error over the RMSE, since the RMSE can be influenced by outlying observations. The WI was also considered for consistency with other relevant studies (for example, see Angulo-Martínez et al., 2009).

The model selection aimed at identification of robust spatial interpolation methods from the considered groups and also at identification of optimal structures and parameters of the particular models, including various combinations of covariates, covariance models, and the number of nearest stations considered for local, geostatistical, and mixed models.

3.3 Uncertainty assessment

The uncertainty in the derived R factor map originates from the formulation of the kinetic energy term, the spatial interpolation model, and spatial and temporal variability. While the effect of the former two can be assessed by direct comparison of results for different expressions of rainfall kinetic energy and different spatial interpolation models, the effect of spatial and temporal variability is evaluated here by simple bootstrap resampling procedures, which are briefly summarized in the rest of this section.

3.3.1 Temporal variability

Record length influences the width of the confidence interval around the R factor estimate. Due to the temporal variability, sufficient record length is required in order to provide an estimate of the R factor such that the long-term average R factor (further denoted the "true R factor") would be covered by the estimated confidence interval. Therefore, we derived the confidence intervals for various record lengths together with the probability that the "true R factor" will lie within the corresponding confidence interval (further denoted the "coverage probability") using our longest available record, i.e. station C2TREB01 (Třeboň) with 80 years of data. This can be done using a nested bootstrap procedure in which a sample of required length (e.g. 10, 20, or 80 years) is drawn (with replacement) from the original record and further resampled. This allows for determination of the confidence interval and examination whether the "true R factor" is included. To obtain the coverage probability, the whole procedure has to be repeated many times. Here we evaluated the coverage probabilities for the record lengths of 10–80 years. For details, see Appendix B.

3.3.2 Record length and spatial coverage

Long records from a dense network of stations should be ideally available to derive an R factor map. In reality, however, long records are often available only for a relatively small number of stations and a balance between record length and spatial coverage has to be found. It is then not clear whether longer records or better spatial coverage should be preferred.

Specifically, we considered (a) what the relationship of the error in the estimated R factor with the spatial and temporal coverage is and how spatial and temporal coverage influences it, (b) the width of the confidence interval around the estimated R factor, and (c) the coverage probability (i.e. the probability that the estimated confidence interval will include the "true R factor").

To be able to assess these questions, a simulation study was conducted. The procedure is fully described in Appendix C, and here we provide only a general overview. As a reference, a synthetic dataset of the monthly (May–September) erosivity index (EI30) was created by permutation of 10 years of data available for 120 stations as follows: first, a 100-year long sequence of the months May–September was created and a random year (from the available period) was assigned to each month in each year. Data for each of the 120 stations were then rearranged according to this year–month sequence and the data were aggregated by years. This resulted in a dataset of 100 years for 120 stations. This procedure preserves the annual cycle of the erosivity index and its spatial variability, while it assumes independence of the erosivity index between individual months. This dataset is further denoted the "full dataset". Please note that although many different replications of this "full dataset" can be obtained, only one replication is used in this study. We do not expect the results (presented further) to vary significantly when different replication is considered.

The R factor in this "full dataset" was estimated and a simple GLS model of the form $R \sim \text{NAVY} + r_{\text{mea}} + Y$ was fitted. This model was used to predict the R factor for an additional 62 locations (coincident with real stations, but independent of the "full dataset"). This is further denoted the "validation dataset". The assessment of the effects of spatial and temporal coverage was based on a repetitive resampling of the subsets of the "full dataset", fitting a GLS model and predicting the R factor for the "validation dataset" (for details, refer to Appendix C).

Within the simulation study, we evaluated the RMSE, the width of the 90 % confidence interval, and the coverage probability for record lengths of 5, 10, 15, 20, 30, ..., 100 years for 10, 20, ..., 120 stations. Since it has been frequently noted that areal averaging is in general preferred to spatial interpolation in the Czech Republic (Janeček et al., 2006, 2012b, 2013), at least for agricultural areas (in general at low altitudes), we also assessed this "areal-average model" (i.e. a constant R factor for all locations). The "areal-average model" was applied for the whole "validation dataset" and also considering only the stations with altitudes below 600 m.

3.3.3 Comparison of uncertainties

To compare the contribution of different sources of uncertainty, the R factor was predicted for the "validation data" using

- the best GLS model of the R factor considering 15 different expressions for kinetic energy (see Appendix A);
- selected classes of spatial interpolation models;
- GLS models based on different numbers of stations and years (see Sect. 3.3.2).

The coefficient of variation (CV) for each set of the estimated R factors was calculated to summarize the variability due to different sources, similarly as done by Catari et al. (2011) or Panagos et al. (2015). For comparison, we also evaluated the CV for the R factor estimates, considering various record lengths for the C2TREB01 (Třeboň) station (see Sect. 3.3.1). Finally, the CV for the R factor estimate for each of the 96 stations (representing natural variability) was evaluated using a simple bootstrap resampling of the annual erosivity values.

4 Results and discussion

4.1 R factor

The estimated R factor (considering the kinetic energy relationship proposed by van Dijk et al., 2002, Eq. 2) for the 96 stations used for spatial interpolation ranges between 320 (U1KOPI01 – Kopisty, north-western Czech Republic) and 1520 (O1RASK01 – Raškovice, north-eastern Czech Republic) MJ ha^{-1} mm h^{-1}, and averages 640 MJ ha^{-1} mm h^{-1}. Note that the second-largest R factor (station H2DEST01 – Deštné v Orlických horách) equals 1108 MJ ha^{-1} mm h^{-1}. R factor values for individual stations can be found in the Supplement to this paper. Figure 2 shows average R factor values for subsets of stations based on the maximum station elevation included in the subset. For instance, the average R factor for the elevations up to 300 m is slightly less than 550 MJ ha^{-1} mm h^{-1} and for elevations up to 600 m slightly more than 600 MJ ha^{-1} mm h^{-1}. The average contribution of individual months to the annual total R factor is 17 % for May, 19 % for June, 28 % for July, 26 % for August, and 10 % for September.

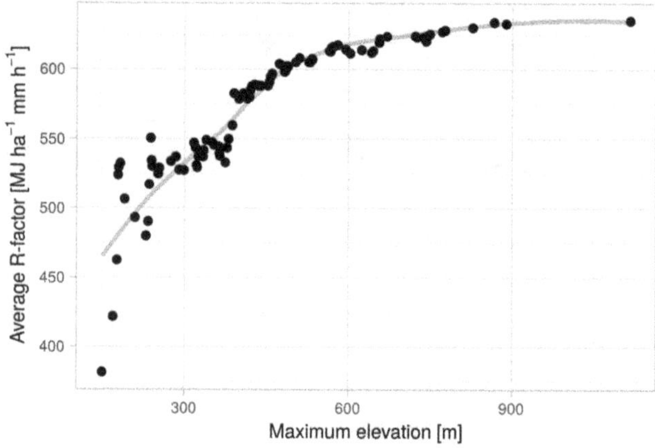

Figure 2. Average R factor for subsets of stations with smaller elevations than that plotted on the horizontal axis. The leftmost point corresponds to a station with the lowest altitude, the rightmost point to the overall average R factor.

The at-site R factor estimates correspond well to those published by Janeček et al. (2006) and Krása et al. (2015), but are considerably larger than values recommended by the official guidelines for the Czech Republic (Janeček et al., 2007, 2012a) and those published by Janeček et al. (2012b, 2013). In part, differences might be due to the modifications to the standard USLE methodology considered by Janeček et al. (2012b, 2013), e.g. calculation of the R factor as a trimmed mean (excluding years with the two smallest and two largest annual erosivity values) of the annual erosivity index. Perhaps in small measure, the differences can also be attributed to the different time period used for R factor assessment (here 1989–2003; in other studies, the series from 1960 and earlier were often considered). The estimated average R factor, after reduction for temporal resolution (from 10 to 30 min; Panagos et al., 2016b), becomes 525 MJ ha^{-1} mm h^{-1}, which almost equals the average R factor from the rainfall erosivity map of Europe – 524 MJ ha^{-1} mm h^{-1} for the Czech Republic (Panagos et al.,

Table 1. Cross-validation indices for the best variants of each spatial interpolation model. AVst – at-site average; SDst – at-site standard deviation; AVsp – spatial average; SDsp – standard deviation of the spatial R factor. The best cross-validation indices are marked in bold font.

	WI	RMSE	MAE	rMAE	AVst	SDst	AVsp	SDsp
IDW	0.78	164.33	123.68	0.20	625.69	162.15	649.46	165.37
SK	0.79	163.01	124.78	0.21	626.98	163.73	647.33	174.49
OK	0.79	165.96	126.48	0.21	630.72	174.62	656.59	135.70
SC	0.73	169.72	131.15	0.22	631.26	139.77	644.53	151.57
OC	0.79	158.78	121.86	0.21	632.86	158.90	656.32	138.13
RK_{XYZ}	0.79	171.44	131.34	0.22	636.86	191.49	653.69	187.78
$RK_{r_{mea},r_{cv}}$	0.89	126.89	93.59	0.15	636.87	185.67	626.22	168.13
$RK_{xyz,r_{mea},r_{cv}}$	0.90	121.91	93.13	0.15	636.18	183.74	632.77	158.78
GLS_M	0.90	115.38	90.57	0.15	635.34	175.12	628.27	149.29
GLS_E	**0.91**	**114.37**	**89.95**	**0.14**	635.64	177.47	627.80	153.35

2015). In addition to different temporal resolutions, Panagos et al. (2015) also considered different stations and kinetic energy formulation (i.e. that of Brown and Foster, 1987; see Fig. 5 for a comparison with van Dijk et al., 2002, used here). Although the spatial distribution of the R factor over the Czech Republic appears rather homogeneous in Panagos et al. (2015), the range of the at-site R factors (after correction for temporal resolution) corresponds well to that from Panagos et al. (2015) when the station with the maximum rainfall erosivity (O1RASK01) is left out. The effect of this station on the resulting R factor map is further discussed in the following section.

4.2 Model selection

The estimated at-site R factor is positively correlated with average precipitation (r_{mea}), coefficient of variation of monthly precipitation (r_{cv}), and the mean fraction of precipitation above the 95 % quantile of monthly precipitation (r_{p95}) derived from the gridded data (see Sect. 2.3), with correlation coefficients 0.75, 0.44, and 0.54, respectively. Further, a positive correlation was also found with altitude (0.32) and longitude (0.47), and a weak negative correlation (−0.25) with latitude. These variables have been primarily used also as covariates in the relevant spatial interpolation methods. Note that correlation may be substantially weaker or that different variables may be relevant in other regions, especially regions with complex topography (see e.g. Capra et al., 2015 or Porto, 2016).

To explore these relationships further, we analysed a set of GLS models with various combinations of fixed component covariates and spatial stochastic covariance structures. The number of fixed-term covariates ranged from one to five. Two theoretical models of spatial covariance and a heteroscedastic error model were considered (see Sect. 3.2). The best GLS model according to the cross-validation results of WI, RMSE, and rMAE had four fixed-term covariates: r_{mea}, r_{p95}, altitude, and longitude, exponential spatial correlation structure, and exponential heteroscedastic error. The coefficient of determination for regression between at-site R factor values and R factor values predicted by this GLS model equals 0.7, which is comparable to results of Meusburger et al. (2012). This was followed by a GLS model with fixed-term covariates longitude, latitude, altitude, r_{mea}, and r_{cv} and a Mathérn spatial covariance structure, which had the lowest value of MAE. These two GLS models are further referred to as GLS_E and GLS_M, respectively. The GLS_E model with estimated coefficients reads as

$$R = 10.246 r_{mea} + 2902.462 r_{p95} - y - 0.130 z + 3649.792, \quad (7)$$

with y the latitude (km) (in the Czech S-42–Pulkovo 1942/Gauss–Kruger zone 3 projection) and z the altitude (m). The parameters are transformed such that the equation can be directly applied to observed covariates. Please note that r_{mea} and r_{p95} are based on (1 km) gridded data (see Sect. 2.3). Calculating these indices using station data would likely lead to a positive bias in the resulting R factor. This is due to area reduction effects influencing areal-average rainfall event characteristics (see e.g. Svoboda et al., 2016a, for a discussion).

Figure 3 further demonstrates the effect of excluding station(s) with a large R factor on the interpolated R factor in the case of the GLS_E model. Although the mean and maximum R factors decrease and local R factor patterns change slightly when stations with a large R factor are excluded, the differences are small (especially with respect to other sources of uncertainty; see further), suggesting our model is rather robust.

In addition to GLS models, other spatial interpolation methods were considered. Table 1 presents the cross-validation indices, spatial average, and standard deviation of the interpolated R factor. The estimated average R factor ranges from 626 to 657 MJ ha^{-1} mm h^{-1}. Comparing the R factor estimated by different spatial interpolation models, the largest similarities were found between the GLS_E, GLS_M, and $RK_{xyz,r_{mea},r_{cv}}$. The GLS_M and $RK_{xyz,r_{mea},r_{cv}}$ models had

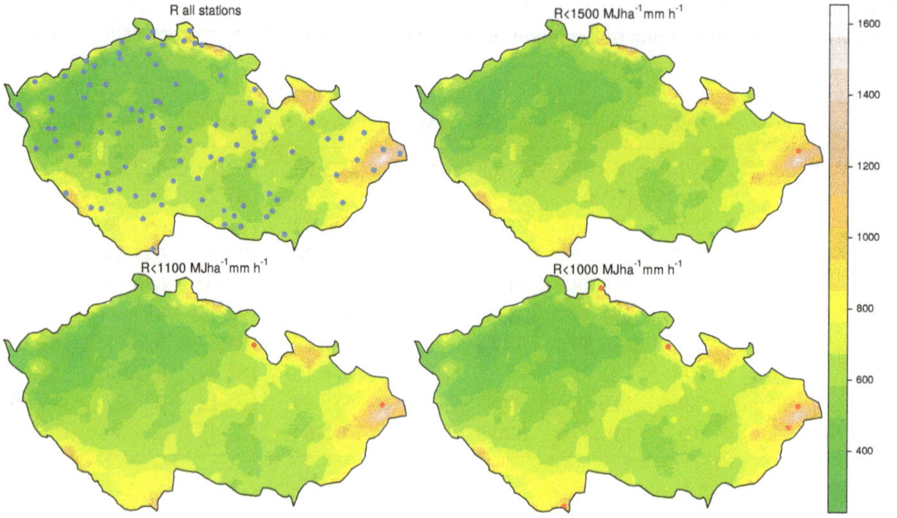

Figure 3. *R* factor estimated with the GLS_E model using all available stations (top left panel) and the GLS_E model fitted considering only stations with an *R* factor below 1500 MJ ha^{-1} mm h^{-1} (top right panel), 1100 MJ ha^{-1} mm h^{-1} (bottom left panel), and 1000 MJ ha^{-1} mm h^{-1} (bottom right panel). Blue points in the top left panel indicate the full set of 96 stations; red dots in the other panels correspond to stations excluded from the fitting procedure.

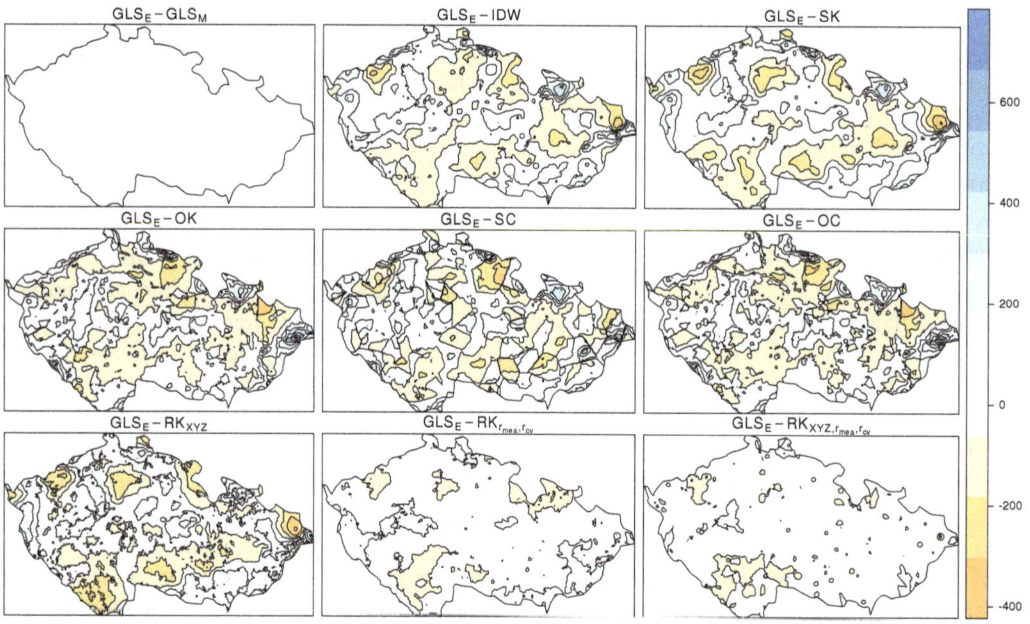

Figure 4. Differences (MJ ha^{-1} mm h^{-1}) between the GLS_E and other spatial interpolation models. GLS_E – generalized linear model with exponential covariance structure; GLS_M – generalized linear model with Matérn covariance structure; IDW – inverse distance weighting; SK – simple kriging; OK – ordinary kriging; SC – simple cokriging; OC – ordinary cokriging; RK – regression kriging.

fixed-term inputs formed from longitude, latitude, altitude, r_{mea}, and r_{cv}. The correlation coefficient between estimates of GLS_E and GLS_M equals 0.99, and between the GLS_E model and $RK_{xyz,r_{mea},r_{cv}}$ models equals 0.97. These similarities were confirmed on the rasters of differences between values of the GLS_E model and the remaining spatial interpolation models (see Fig. 4). The median absolute difference between the GLS_E and GLS_M models was 20.0 MJ ha^{-1} mm h^{-1}, with the standard deviation of differences 19.2 MJ ha^{-1} mm h^{-1}. For the differences between the GLS_E and the best RK model, the median absolute difference and standard deviation were roughly double.

The largest differences were found between the GLS_E model and the spatial interpolation models, which did not take into account the long-term rainfall characteristics. The largest difference in R factor estimates ($741\,\mathrm{MJ\,ha^{-1}\,mm\,h^{-1}}$) was found between the $\mathrm{GLS_E}$ and OK models. Large values of MAE, rMAE, and RMSE and small values of WI for the IDW, SK, OK, SC, and OC models show that the spatial distribution of the R factor could not be sufficiently described by the models, which emphasized the stochastic component of the R factor, or explain the R factor using local information. The spatial interpolation models with fixed-term covariates based on r_{mea} or r_{cv} or r_{p95} were superior to the models without a fixed component linked to the long-term rainfall characteristics (cf. the cross-validation results of RK_{XYZ}).

Including the stochastic information obtained from rainfall event characteristics in simple cokriging and ordinary cokriging models also did not improve the spatial interpolation of the R factor. The presented SC and OC models were selected from seven different types of cokriging models. They differed according to the rainfall event characteristic, which was used for the second cokriging variate. The best SC and OC models were those which linked their stochastic component with the maximum 10 min intensity and on-site R factor. Including these rainfall event characteristics did not, however, improve the spatial interpolation of the R factor over the spatial models based on the long-term rainfall event characteristics (see Table 1).

The success of models including long-term precipitation characteristics might be surprising in part because daily or subdaily rainfall data are in general preferred for calculation of the R factor over monthly or annual data (Angulo-Martínez et al., 2009). The potential of long-term rainfall characteristics for R factor estimation is also stressed by Lee and Lin (2014), who explored relationships between rainfall and erosivity indices at daily, monthly, and annual timescales and concluded that the relationship between the annual erosivity index and annual rainfall is closer than that for the other timescales.

The average R factor varies considerably when different formulas for kinetic energy are considered (see Fig. 5). The smallest average R factor is obtained by the RWa relationship ($500\,\mathrm{MJ\,ha^{-1}\,mm\,h^{-1}}$) and the largest value by the standard USLE ($790\,\mathrm{MJ\,ha^{-1}\,mm\,h^{-1}}$). The average ($640\,\mathrm{MJ\,ha^{-1}\,mm\,h^{-1}}$) corresponds well to the value estimated using the van Dijk formula (the difference is $6\,\mathrm{MJ\,ha^{-1}\,mm\,h^{-1}}$). The range between estimates for individual stations is proportional to the estimated R factor (corresponding roughly to 40 %). This also has an effect on the estimated spatial distribution of the R factor values.

4.3 Temporal variability

Wischmeier and Smith (1978) used a 22-year record to derive the rainfall erosivity factor because of "apparent cyclical patterns in rainfall data". The same is repeated by Renard et al. (1997) with a remark that longer records are advisable especially in the case that the coefficient of variation of annual precipitation is large. This recommendation is often mentioned in rainfall erosivity studies. However, due to data availability, shorter records are often considered, at least in addition to longer records (e.g. Angulo-Martínez et al., 2009; Meusburger et al., 2012; Oliveira et al., 2013; Lee and Lin, 2014; Panagos et al., 2015). For a 105-year record from Belgium, Verstraeten et al. (2006) tested whether rainfall erosivity derived from running 10- and 22-year averages is significantly different than that from using the overall (105-year) mean. They concluded that while a 22-year period is sufficient, reliable estimates of the R factor cannot be based on 10 years of data. Apart from their study, the actual effect of the sample size on the estimate of the R factor was seldom investigated.

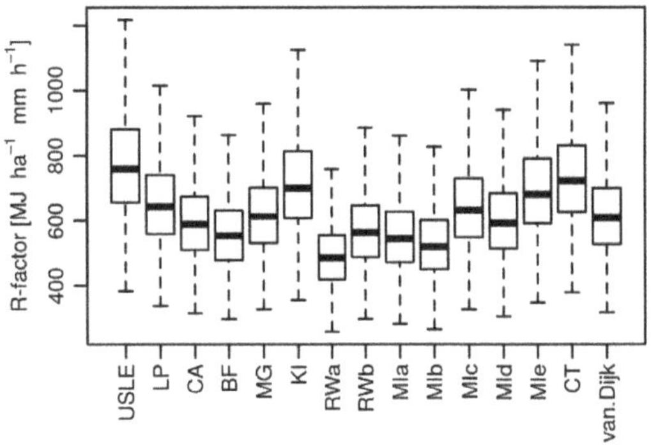

Figure 5. Estimated R factor ($\mathrm{MJ\,ha^{-1}\,mm\,h^{-1}}$) considering different kinetic energy formulations (see Appendix A).

Figure 6 (left panel) gives the estimated confidence intervals (grey area) together with the coverage probability, i.e. the probability that the long-term mean R factor ($669\,\mathrm{MJ\,ha^{-1}\,mm\,h^{-1}}$) will lie within the confidence intervals for record lengths between 10 and 80 years for the Třeboň station (C2TREB01). The 90 % confidence interval for record length 10 years ranges from 400 to $950\,\mathrm{MJ\,ha^{-1}\,mm\,h^{-1}}$ (i.e. ±40 %), narrows to 500–$870\,\mathrm{MJ\,ha^{-1}\,mm\,h^{-1}}$ (±25–30 %) for 20 years and 530–$830\,\mathrm{MJ\,ha^{-1}\,mm\,h^{-1}}$ (±20–25 %) for 30 years, and remains relatively wide, i.e. 580–$770\,\mathrm{MJ\,ha^{-1}\,mm\,h^{-1}}$ (±15 %), even for an 80-year record. Note that the temporal variability of the erosivity index is considerably larger than in the case of the annual total or maxima. For instance, the standard deviation of the annual erosivity index is more than double that of the annual precipitation total and more than 40 % larger than for annual 1 h precipitation maxima (estimated for the same station). The coverage probability (red line in Fig. 6, left panel) of the 90 % confidence interval is around 75 % for the 10-year record, and from 82 % for the 15-year record it increases only slowly, with an increasing record length of up to 87 % for the 80-year record.

The width of the confidence interval as well as the coverage probability are influenced by the variability of the at-

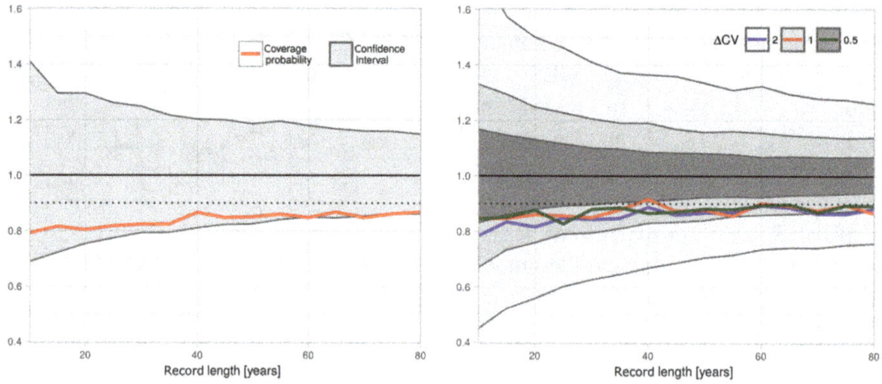

Figure 6. The average confidence intervals (relative to the long-term mean R factor, $669\,\mathrm{MJ\,ha^{-1}\,mm\,h^{-1}}$; grey area) and the coverage probability (thick lines) for different record lengths based on station C2TREB01 – Třeboň (left panel) and simulated data (right panel). The dotted line corresponds to the coverage probability of 90 %.

site erosivity index. This is demonstrated in Fig. 6 (right panel) for synthetic data generated from a Gamma distribution with parameters estimated from the C2TREB01 record (denoted $\Delta\mathrm{CV} = 1$) and with parameters modified such that the coefficient of variation of the modified distribution is half and double that of the C2TREB01 record ($\Delta\mathrm{CV} = 0.5$ and $\Delta\mathrm{CV} = 2$, respectively), and the mean R factor remains constant. Note that the Gamma distribution was not rejected by the Anderson–Darling test at the 0.05 significance level at most of the stations, including C2TREB01. It is clear that the confidence intervals as well as the coverage probability for the erosivity index simulated with parameters estimated from C2TREB01 (i.e. $\Delta\mathrm{CV} = 1$) correspond reasonably to those based on observed data. It is also clear (and expected) that the width of the confidence interval increases with the coefficient of variation. For instance, for a 15-year record and doubling of the coefficient of variation, it ranges from 0.5 to 1.57. For increasing record lengths the coverage probability increases and the width of the confidence interval decreases. Note that the confidence interval for the erosivity index with a large coefficient of variation remains relatively large (> 50 %), even for 80 years of data. The coverage probability, on the other hand, decreases only slightly with the coefficient of variation.

4.4 Spatial and temporal coverage

Figure 7 shows the average RMSE, coverage probability, and relative width of the 90 % confidence interval for the validation data in the cases of the GLS model (solid line), "areal-average" model (dotted line), and "areal-average" model considering only stations with an altitude of less than 600 m (dashed line). In the case of the GLS model, the RMSE for small numbers of stations and years is relatively large ($\approx 300\,\mathrm{MJ\,ha^{-1}\,mm\,h^{-1}}$). However, it quickly drops to $\approx 100\,\mathrm{MJ\,ha^{-1}\,mm\,h^{-1}}$ for 25 stations and then decreases almost linearly to $\approx 50\,\mathrm{MJ\,ha^{-1}\,mm\,h^{-1}}$ for 100 stations. The RMSE for both "areal-average" models depends on the number of stations only up to 25 stations and remains almost constant for larger numbers of stations ($\approx 110\,\mathrm{MJ\,ha^{-1}\,mm\,h^{-1}}$ when validated against all stations and $\approx 100\,\mathrm{MJ\,ha^{-1}\,mm\,h^{-1}}$ when only stations below 600 m are used). The RMSE depends only slightly on the number of years used for estimation of the R factor in the case of all models (GLS and both "areal-average" models). With respect to RMSE, the "areal-average" model is beneficial only when fewer than 20 stations are available. The RMSE for individual stations might be considerably larger. The 90 % quantile RMSE (not shown) is around $600\,\mathrm{MJ\,ha^{-1}\,mm\,h^{-1}}$ for both "areal-average" models (mostly independently of the number of stations) and from $2000\,\mathrm{MJ\,ha^{-1}\,mm\,h^{-1}}$ (10 stations) to $450\,\mathrm{MJ\,ha^{-1}\,mm\,h^{-1}}$ (30 stations) and $250\,\mathrm{MJ\,ha^{-1}\,mm\,h^{-1}}$ (100 stations) for the GLS model with 15 years of data.

The width of the 90 % confidence interval (Fig. 7, middle panel) for the GLS model drops from 120 % (5 stations) to 50 % (25 stations) and 30 % (100 stations). This value corresponds well to the width of the confidence interval for the full record of the C2TREB01 (Třeboň) station (see Fig. 6). The confidence interval for both "areal-average" models is approximately half that for the GLS model. The impact of the record length is small.

The coverage probability (Fig. 7, right panel) for the GLS model increases from 70 to 90 % for 5 and 10 stations, respectively. The coverage probability further increases only when more than 20 years of data are used. For shorter records, the coverage probability decreases to $\approx 80\,\%$ for 100 stations. This is a consequence of faster reduction of the confidence interval width when compared to the decrease in the RMSE. Similarly, the coverage probability decreases for both "areal-average" models, since the RMSE is almost constant for more than 25 stations, while the width of the confidence interval decreases. As with the RMSE, the coverage probability might be considerably lower for individual stations, especially in the case of both "areal-average" models (for a number of stations close to 0), while the coverage prob-

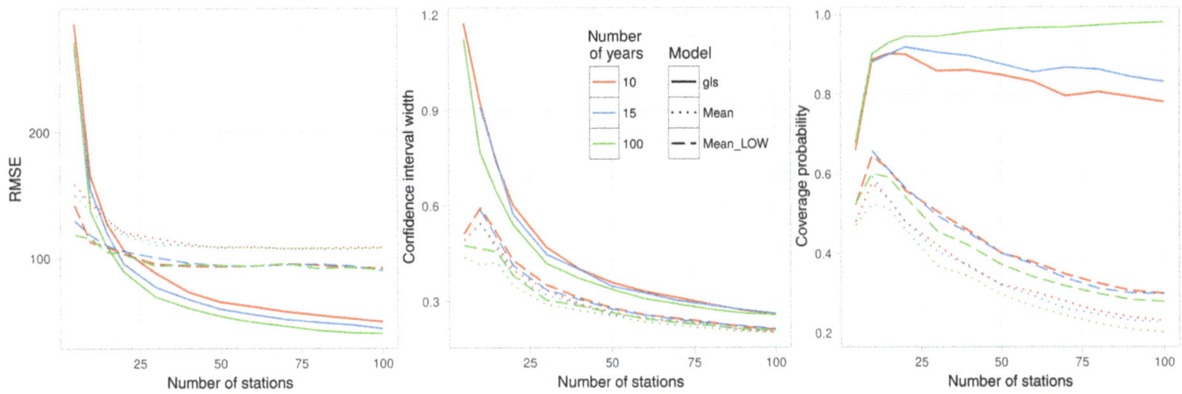

Figure 7. The RMSE, width of the confidence interval, the coverage probability for a GLS model (solid lines), the "areal-average" model considering all stations (dotted lines), and only those below 600 m (dashed lines) based on simulated data with various record lengths and number of stations.

ability is larger than 70 % for most of the stations in the case of the GLS model.

The results for all evaluated characteristics indicate that appropriate spatial coverage is more important than the length of the record, at least in situations when other relevant information to build a spatial model is available. However, at least 15–20 years of data should be considered (if possible) to provide reasonable coverage probabilities.

4.5 Comparison of different sources of uncertainty

Using the "validation data" and the GLS model, we calculated the coefficient of variation (CV) associated with formulation of the kinetic energy, spatial interpolation, and spatial and temporal coverage (Table 2). In addition, the CV was also calculated for estimates of the R factor based on different record lengths for the C2TREB01 (Třeboň) station and for the set of 96 stations considered for spatial interpolation. For the latter, the estimated CV was 23 % on average (9–43 % for all stations). This value can be interpreted as an indicator of natural variability of the R factor based on a 15-year record. Almost the same value (21 %) is estimated for the C2TREB01 (Třeboň) station and 15 years of data. The contribution of the kinetic energy formulation (≈ 13 %) and spatial interpolation (≈ 9 %) is about half of this value. As expected, the CV of the estimates for the C2TREB01 (Třeboň) station decreases with increasing record length (38, 16, and 12 % for 5, 50, and 80 years, respectively).

The same applies for the GLS model, for which, in addition, the CV also decreases with an increasing number of stations. When comparing corresponding record lengths, the CV is considerably smaller for the GLS model than for individual stations, providing a sufficient number of stations is considered in the model. For instance, for the R factor based on 15 years of data, the average CV is 31, 10, 7, and 6.4 % for 10, 30, 50, and 100 stations, respectively, while for the station data and the same record length the average CV was 23 %. In addition, considering 100 stations, the CV is only 13 (5) % for 5 (50) years. From Table 2 it is evident that not only does the average CV decrease, but that the same holds also for the stations with maximum variation. For instance, using 100 stations and 15 years, the maximum CV in the "validation set" is 12 % for a GLS model and 43 % for the station data. As the CV decreases with more stations and longer records, the relative importance of the expression

Table 2. The coefficient of variation (CV) for the R factor estimated by considering different formulas for kinetic energy, spatial interpolation methods, record length, and spatial coverage. In the second column, the average CV for all considered stations is given; the last two columns indicate the range of the CV from all stations in the "validation dataset".

Source	Coefficient of variation (%)	
Kinetic energy	12.7	(11.8–14.4)
Spatial interpolation	9.4	(1.5–29.7)
Number of stations (GLS model, 15 years)		
10 stations	31.0	(19.6–69.1)
30 stations	10.3	(7.6–21.4)
50 stations	7.6	(6.0–13.9)
Number of years (GLS model, 100 stations)		
5 years	13.0	(10.9–22.3)
15 years	6.4	(4.9–11.8)
50 years	4.9	(3.4–9.5)
Number of years (C2TREB01 – Třeboň)		
5 years	37.6	–
15 years	21.6	–
50 years	16.1	–
80years	11.8	–
Natural variability (96 stations, 15 years)		
	23.3	(9.1–43.0)

for the kinetic energy and spatial interpolation increases. The assessment of the sources of uncertainty could be done more formally using an analysis of variance (ANOVA) model (e.g. Yip et al., 2011) or a slightly more flexible linear mixed-effects model (see e.g. Hanel and Buishand, 2015).

5 Summary and conclusions

In the present paper we estimated the rainfall erosivity factor (R factor) for the area of the Czech Republic. The at-site values of the R factor based on a 15-year record for 96 stations were considered in several spatial models in order to provide estimates of the R factor for the whole area of the Czech Republic.

The spatial interpolation models included inverse distance weighting, simple and ordinary kriging, simple and ordinary cokriging, regression kriging with parameters estimated by the method of moments, and the GLS models. Several covariates have been considered to explain the spatial variation of the R factor over the area.

In addition, uncertainty due to kinetic energy formulation, spatial models, and spatial and temporal coverage was assessed by direct comparison of different methods and simulation studies.

The most important findings can be summarized as follows:

- the average R factor in the period 1989–2003 for the considered stations is 640 MJ ha^{-1} mm h^{-1}, with values for the individual stations between 320 and 1520 MJ ha^{-1} mm h^{-1};
- the at-site R factor is considerably correlated with average precipitation, coefficient of variation of monthly precipitation, the mean fraction of precipitation above the 95 % quantile of monthly precipitation, and longitude, while the correlation with altitude and latitude is weak;
- from the considered spatial models, a GLS model with altitude, latitude, mean precipitation, and the mean fraction of precipitation above the 95 % quantile of monthly precipitation provided the best performance according to three of four cross-validation indices;
- with respect to the cross-validation statistics, the spatial interpolation models that included long-term rainfall characteristics performed considerably better than those based on local interpolation and/or geographical information only;
- the resulting map based on the GLS model does not change considerably when stations with the largest R factor are excluded;
- when the number of stations and years available for interpolation is small, the relative contribution of the uncertainty due to the kinetic energy estimate and the spatial interpolation method is small compared to that due to the choice of the stations and time period;
- although the RMSE and confidence interval width decrease and coverage probability in general increases with record length and number of stations, reasonable estimates of the R factor may be obtained from relatively short records (e.g. 15–20 years) provided a sufficient number of stations are available and appropriate covariates can be found;
- the confidence intervals around the R factor estimates remain relatively wide even for long records, especially at locations with large natural variability of the annual erosivity index, and are considerably wider than those for annual rainfall total or precipitation maxima;
- the spatial model should in general be preferred over the areal-averaging of R factor estimates from individual stations unless only very short records for a small number of stations are available or appropriate covariates cannot be used.

6 Data availability

The original 10 min precipitation data as well as the gridded precipitation characteristics cannot be published due to licence of the Czech Hydrometeorological Institute. The R factor for 96 available stations calculated using 15 models for rainfall kinetic energy (Appendix A).

Table A1. Coefficients for calculation of rainfall kinetic energy in Eq. (A2). The acronyms used throughout the paper are given in the first column.

	e_{max}	a	b	
LP	28.9	0.54	0.059	Laws and Parsons (1943)
CA	28.0	0.76	0.090	Carter et al. (1974)
BF	29.0	0.72	0.050	Brown and Foster (1987)
MG	29.0	0.72	0.082	McGregor et al. (1995)
KI	29.3	0.28	0.018	Kinnell (1981)
RWa	26.4	0.67	0.035	Rosewell (1986)
RWb	28.1	0.60	0.040	Rosewell (1986)
MIa	24.6	0.46	0.037	McIsaac (1990)
MIb	29.2	0.51	0.011	McIsaac (1990)
MIc	28.8	0.45	0.033	McIsaac (1990)
MId	25.1	0.40	0.045	McIsaac (1990)
MIe	26.8	0.29	0.049	McIsaac (1990)
CT	35.9	0.56	0.034	Coutinho and Tomás (1995)

Appendix A: Considered formulations of rainfall kinetic energy

For the application of the Universal Soil Loss Equation, Wischmeier and Smith (1978) derived a logarithmic relationship between rainfall intensity and kinetic energy of the form (converted to metric units)

$$e_i = 210 + 89 \log (r_i/10), \quad \text{(A1)}$$

where (as in Eq. 2), r_i is the rainfall intensity (mm h^{-1}) during time interval i. Note that slightly different coefficients are provided by Renard et al. (1997). The logarithmic relationship implies that there is no upper limit to kinetic energy, whereas research has suggested that a maximum value does exist (see van Dijk et al., 2002, for references). Therefore, Wischmeier and Smith (1978) considered constant rainfall kinetic energy for intensities greater than 76 mm h^{-1}. Other authors used a relationship of the form

$$e_i = e_{\max}\left[1 - a \exp(-b r_i)\right], \quad \text{(A2)}$$

where $e_{\max}$ denotes the maximum kinetic energy contents and a and b are empirical constants. Many different combinations of the parameters $e_{\max}$, a, and b have been published. van Dijk et al. (2002) therefore proposed a relationship given in Eq. (2) as one providing estimates that are close to the average of many formulas for calculation of e_i. In the present paper, in addition to formulas given in Eqs. (2) and (A1), which are further referred to as "van Dijk" and "USLE", respectively, we calculated the e_i considering a set of coefficients for Eq. (A2) given in Table A1.

Appendix B: A resampling scheme for the assessment of temporal variability

Here we describe a nested bootstrap procedure in which samples of required length l (e.g. 10, 20, 80 years) are repeatedly drawn from the original (observed) record and resampled to obtain the R factor estimate with the corresponding confidence interval. Finally, the probability that the "true R factor" (here the estimate based on the full record) lies within the confidence interval is estimated.

The resampling is performed in the following steps:

1. choose the number of bootstrap samples for derivation of the confidence intervals (n_{CI}) and the number of bootstrap samples for the assessment of the coverage probability (n_{CP}); in our study we set $n_{\mathrm{CI}} = n_{\mathrm{CP}} = 500$;
2. draw a sample of length l with replacement from the original series of annual erosivity and denote this sample s;
3. draw a sample of length l with replacement from s and use it to calculate the average erosivity (i.e. the R factor);
4. repeat the previous step n_{CI} times;
5. calculate the 90 % confidence interval from the n_{CI} estimates of the R factor from step 3 and check whether this interval includes the true R factor;
6. repeat steps 2–5 n_{CP} times;
7. calculate the coverage probability associated with the record length l as the proportion of cases when the confidence interval from step 5 included the true R factor;
8. repeat the whole process for different record lengths l.

Please note that the described procedure provides n_{CP} estimates of the confidence intervals for specific l, and only their average is presented in the paper. Setting $n_{\mathrm{CP}} = 1$ might be sufficient in the situation that only the confidence intervals would be of interest.

Appendix C: A resampling scheme for the assessment of spatial and temporal coverage

The following scheme describes a nested bootstrap procedure for assessment of the RMSE, coverage probability, and the width of the 90 % confidence interval for a GLS model considering different lengths of the precipitation data and different numbers of stations. The assessment is based on resampling of a synthetic dataset of 100 years for 120 stations (denoted the "full dataset") and validated against the independent "validation dataset" (see Sect. 3.3.2). The procedure is summarized as follows:

1. draw a sample of n_{yr} years for n_{sta} stations from the "full dataset" and calculate the R factor for each station;
2. fit a GLS model of the form $R \sim \mathrm{NAVY} + r_{\mathrm{mea}} + Y$ using the sample from the previous step;
3. simulate data for the given n_{sta} stations from the fitted model (see e.g. Pinheiro and Bates, 2000);
4. refit the model and use this refitted model to predict the R factor for the "validation dataset";
5. repeat the previous two steps 500 times;
6. calculate the 90 % confidence interval around the estimated R factor and the RMSE for each station of the "validation dataset" from the 500 samples obtained in steps 2–5;
7. repeat the previous steps (1–6) 500 times;
8. repeat the whole procedure for different n_{yr} and n_{sta}.

The RMSE, confidence interval, and coverage probability for the "areal-average" model used for comparison were derived by replacing the estimates from the refitted model in step 4 by the areal average of the simulated data from step 3.

Acknowledgements. The research has been conducted within the framework of the project "Erosion runoff – increased risk of the residents and the water quality exposure in the context of the expected climate change" (VG20122015092) sponsored by the Ministry of the Interior of the Czech Republic. Data have been kindly provided by the Czech Hydrometeorological Institute. All calculations and plotting were done in R.

Edited by: M. Mikos

References

Angulo-Martínez, M., López-Vicente, M., Vicente-Serrano, S. M., and Beguería, S.: Mapping rainfall erosivity at a regional scale: a comparison of interpolation methods in the Ebro Basin (NE Spain), Hydrol. Earth Syst. Sci., 13, 1907–1920, doi:10.5194/hess-13-1907-2009, 2009.

Bonilla, C. A. and Vidal, K. L.: Rainfall erosivity in central Chile, J. Hydrol., 410, 126–133, 2011.

Borrelli, P., Diodato, N., and Panagos, P.: Rainfall erosivity in Italy: a national scale spatio-temporal assessment, Int. J. Digit. Earth, 9, 835–850, 2016.

Brádka, J.: Srážky na území ČSSR při jednotlivých typech povětrnostní situace, Hydrometeorological Institute, Prague, 1972.

Brázdil, R.: Vliv atlantského oceánu a středozemního moře na srážkové poměry léta na území ČSSR, J. E. Purkyne University, Brno, 1980.

Brown, L. and Foster, G.: Storm erosivity using idealized intensity distributions, T. ASAE, 30, 0379–0386, 1987.

Capra, A., Porto, P., and La Spada, C.: Long-term variation of rainfall erosivity in Calabria (Southern Italy), Theor. Appl. Climatol., doi:10.1007/s00704-015-1697-2, in press, 2015.

Carter, C., Greer, J., Braud, H., and Floyd, J.: Raindrop characteristics in South Central United States, T. ASAE, 17, 1033–1037, 1974.

Catari, G., Latron, J., and Gallart, F.: Assessing the sources of uncertainty associated with the calculation of rainfall kinetic energy and erosivity – application to the Upper Llobregat Basin, NE Spain, Hydrol. Earth Syst. Sci., 15, 679–688, doi:10.5194/hess-15-679-2011, 2011.

Coutinho, M. A. and Tomás, P. P.: Characterization of raindrop size distributions at the Vale Formoso Experimental Erosion Center, Catena, 25, 187–197, 1995.

Goovaerts, P.: Geostatistics for natural resources evaluation, in: Applied geostatistics series, Oxford University Press, Incorporated, Oxford, 1997.

Goovaerts, P.: Using elevation to aid the geostatistical mapping of rainfall erosivity, Catena, 34, 227–242, 1999.

Hanel, M. and Buishand, T. A.: Assessment of the sources of variation in changes of precipitation characteristics over the Rhine basin using a linear mixed-effects model, J. Climate, 28, 6903–6919, 2015.

Hanel, M. and Máca, P.: Spatial variability and interdependence of rain event characteristics in the Czech Republic, Hydrol. Process., 28, 2929–2944, 2014.

Hanel, M., Pavlásková, A., and Kyselý, J.: Trends in characteristics of sub-daily heavy precipitation and rainfall erosivity in the Czech Republic, Int. J. Climatol., 30, 1833–1845, 2016.

Haskard, A.: Anisotropic Matérn spatial covariance model: REML estimation and properties, University of Adelaide, Adelaide, 2007.

Hengl, T., Heuvelink, G., and Stein, A.: A generic framework for spatial prediction of soil variables based on regression-kriging, Geoderma, 120, 75–93, 2004.

Hengl, T., Heuvelink, G., and Rossiter, D.: About regression-kriging: From equations to case studies, Comput. Geosci., 33, 1301–1315, 2007.

Janeček, M., Kubátová, E., and Tippl, M.: Revised determination of the rainfall-runoff erosivity factor *R* for application of USLE in the Czech Republic, Soil Water Res., 1, 65–71, 2006.

Janeček, M., Bečvář, M., Bohuslávek, J., Dufková, J., Dumbrovský, M., Dostál, T., Hula, J., Kadlec, V., Krása, J., Kubátová, E., Novorný, I., Podhrázská, J., Tippl, M., Toman, F., Vopravil, J., and Vrána, K.: Guidelines on protection of agricultural land against soil erosion, Research Institute of Agricultural Engineering, Prague, 2007.

Janeček, M., Dostál, T., Kozlovsky Dufková, J., Dumbrovský, M., Hula, J., Kadlec, V., Kovář, P., Krása, J., Kubátová, E., Kobzová, D., Kudrnáčová, M., Novotný, I., Podhrázská, J., Pražan, J., Procházková, E., Středová, H., Toman, F., Vopravil, J., and Vlasák, J.: Guidelines on protection of agricultural land against soil erosion, Czech University of Life Sciences, Prague, 2012a.

Janeček, M., Květoň, V., Kubátová, E., and Kobzová, D.: Differentiation and regionalization of rainfall erosivity factor values in the Czech Republic, Soil Water Res., 7, 1–9, 2012b.

Janeček, M., Květoň, V., Kubátová, E., Kobzová, D., Vošmerová, M., and Chlupsová, J.: Values of rainfall erosivity factor for the Czech Republic, J. Hydrol. Hydromech., 61, 97–102, 2013.

Kinnell, P.: Rainfall intensity-kinetic energy relationships for soil loss prediction, Soil Sci. Soc. Am. J., 45, 153–155, 1981.

Kitanidis, P.: Generalized covariance functions in estimation, Math. Geol., 25, 525–540, 1993.

Krása, J., Středová, H., Dostál, T., and Novotný, I.: Rainfall erosivity research on the territory of the Czech Republic, in: Mendel a bioklimatologie, http://www.cbks.cz/SbornikBrno14/Krasa.pdf (last access: 20 October 2016), 2014.

Krása, J., Středová, H., Štěpánek, P., Hanel, M., Dostál, T., and Novotný, I.: Recent and future rainfall erosivity on the territory of the Czech Republic, EGU General Assembly Conference Abstracts, vol. 17, p. 7714, 2015.

Květoň, V., Zahradníček, J., and Žák, M.: Quality control and digitising of pluviographic measurements in the Czech Hydrometeorological Institute, Meteorologické zprávy, 57, 47–52, 2004.

Laws, J. O. and Parsons, D. A.: The relation of raindrop-size to intensity, Eos T. Am. Geophys. U., 24, 452–460, 1943.

Lee, M.-H. and Lin, H.-H.: Evaluation of Annual Rainfall Erosivity Index Based on Daily, Monthly, and Annual Precipitation Data of Rainfall Station Network in Southern Taiwan, Int. J. Distrib. Sensor Netw., 11, 1–15, doi:10.1155/2015/214708, 2014.

Lu, H. and Yu, B.: Spatial and seasonal distribution of rainfall erosivity in Australia, Soil Res., 40, 887–901, 2002.

McGregor, K., Bingner, R., Bowie, A., and Foster, G.: Erosivity index values for northern Mississippi, T. ASAE, 38, 1039–1047, 1995.

McIsaac, G.: Apparent geographic and atmospheric influences on raindrop sizes and rainfall kinetic energy, J. Soil Water Conserv., 45, 663–666, 1990.

Meddi, M., Toumi, S., and Assani, A. A.: Spatial and temporal variability of the rainfall erosivity factor in Northern Algeria, Arabian J. Geosci., 9, 1–13, 2016.

Meusburger, K., Steel, A., Panagos, P., Montanarella, L., and Alewell, C.: Spatial and temporal variability of rainfall erosivity factor for Switzerland, Hydrol. Earth Syst. Sci., 16, 167–177, doi:10.5194/hess-16-167-2012, 2012.

Minasny, B. and McBratney, A. B.: The Matérn function as a general model for soil variograms, Geoderma, 128, 192–207, 2005.

Minasny, B. and McBratney, A. B.: Spatial prediction of soil properties using EBLUP with the Matern covariance function, Geoderma, 140, 324–336, 2007.

Oliveira, P. T. S., Wendland, E., and Nearing, M. A.: Rainfall erosivity in Brazil: A review, Catena, 100, 139–147, 2013.

Panagos, P., Ballabio, C., Borrelli, P., Meusburger, K., Klik, A., Rousseva, S., Tadić, M. P., Michaelides, S., Hrabalíková, M., Olsen, P., Aalto, J., Lakatos, M., Rymszewicz, A., Dumitrescu, A., Beguería, S., and Alewell, C.: Rainfall erosivity in Europe, Sci. Total Environ., 511, 801–814, 2015.

Panagos, P., Ballabio, C., Borrelli, P., and Meusburger, K.: Spatio-temporal analysis of rainfall erosivity and erosivity density in Greece, Catena, 137, 161–172, 2016a.

Panagos, P., Borrelli, P., Spinoni, J., Ballabio, C., Meusburger, K., Beguería, S., Klik, A., Michaelides, S., Petan, S., Hrabalíková, M., Olsen, P., Aalto, J., Lakatos, M., Rymszewicz, A., Dumitrescu, A., Tadić, M. P., Diodato, N., Kostalova, J., Rousseva, S., Banasik, K., and Alewell, C.: Monthly Rainfall Erosivity: Conversion Factors for Different Time Resolutions and Regional Assessments, Water, 8, 119, doi:10.3390/w8040119, 2016b.

Pardo-Iguzquiza, E. and Chica-Olmo, M.: Geostatistics with the Matern semivariogram model: A library of computer programs for inference, kriging and simulation, Comput. Geosci., 34, 1073–1079, 2008.

Pinheiro, J. C. and Bates, D. M.: Linear Mixed-Effects Models: Basic Concepts and Examples, Springer, New York, NY, USA, 2000.

Porto, P.: Exploring the effect of different time resolutions to calculate the rainfall erosivity factor *R* in Calabria, southern Italy, Hydrol. Process., 30, 1551–1562, 2016.

Renard, K. G., Foster, G. R., Weesies, G. A., McCool, D., and Yoder, D.: Predicting soil erosion by water: a guide to conservation planning with the revised universal soil loss equation (RUSLE), Agriculture Handbook 703, US Government Printing Office, Washington, D.C., USA, 1997.

Rosewell, C. J.: Rainfall kinetic energy in eastern Australia, J. Clim. Appl. Meteorol., 25, 1695–1701, 1986.

Štekl, J., Brázdil, R., Kakos, V., Jež, J., Tolasz, R., and Sokol, Z.: Extreme daily precipitation on the territory of the Czech Republic in the period 1879–2000 and their synoptic causes, National Climatic Programme of the Czech Republic, 31, 140, 2001.

Svoboda, V., Hanel, M., Máca, P., and Kyselý, J.: Characteristics of rainfall events in RCM simulations for the Czech Republic, Hydrol. Earth Syst. Sci. Discuss., doi:10.5194/hess-2016-283, 2016a.

Svoboda, V., Hanel, M., Máca, P., and Kyselý, J.: Projected changes of rainfall event characteristics for the Czech Republic, J. Hydrol. Hydromech., 64, 415–425, doi:10.1515/johh-2016-0036, 2016b.

Tolasz, R., Brázdil, R., Bulíř, O., Dobrovolný, P., Dubrovský, M., Hájková, L., Halásová, O., Hostýnek, J., Janouch, M., Kohut, M., Kvętoņ, V., Lepka, Z., Lipina, P., Macková, J., Metelka, L., Míková, T., Mrkvica, Z., Močný, M., Nekovář, J., Němec, L., Pokorný, J., Reitschläger, D., Richterová, D., Rožnovský, J., Řepka, M., Semerádová, D., Sosna, V., Stříž, M., Šercl, P., Škáchová, H., Štěpánek, P., Štěpánková, P., Trnka, M., Valeriánová, A., Valter, J., Vaníček, K., Vavruška, F., Voženílek, V., Vráblík, T., Vysoudil, M., Zahradníček, J., Zusková, I., Žák, M., and Žalud, Z.: Altas podnebí Česka (Climate atlas of the Czech Republic), Český hydrometeorologický ústav, Universita Palackého, Palackého, 2007.

Van der Knijff, J., Jones, R., and Montanarella, L.: Soil erosion risk assessment in Europe, European Soil Bureau, Join Research Centre, Ispra, Italy, 2000.

van Dijk, A., Bruijnzeel, L., and Rosewell, C.: Rainfall intensity-kinetic energy relationships: a critical literature appraisal, J. Hydrol., 261, 1–23, 2002.

Verstraeten, G., Poesen, J., Demarée, G., and Salles, C.: Long-term (105 years) variability in rain erosivity as derived from 10-min rainfall depth data for Ukkel (Brussels, Belgium): Implications for assessing soil erosion rates, J. Geophys. Res.-Atmos., 111, 1–11, 2006.

Willmott, C. and Matsuura, K.: Advantages of the mean absolute error (MAE) over the root mean square error (RMSE) in assessing average model performance, Clim. Res., 30, 79–82, 2005.

Wischmeier, W. H. and Smith, D. D.: Predicting rainfall erosion losses – A guide to conservation planning, Agriculture Handbook 537, US Department of Agriculture, Maryland, USA, 1978.

Yin, S., Xie, Y., Nearing, M., and Wang, C.: Estimation of rainfall erosivity using 5- to 60-minute fixed-interval rainfall data from China, Catena, 70, 306–312, 2007.

Yip, S., Ferro, C. A., Stephenson, D. B., and Hawkins, E.: A simple, coherent framework for partitioning uncertainty in climate predictions, J. Climate, 24, 4634–4643, 2011.

11

Characterising the space–time structure of rainfall in the Sahel with a view to estimating IDAF curves

G. Panthou, T. Vischel, T. Lebel, G. Quantin, and G. Molinié
LTHE – UMR5564, Univ. Grenoble, IRD, CNRS, Grenoble, France

Correspondence to: T. Vischel (theo.vischel@ujf-grenoble.fr)

Abstract. Intensity–duration–area–frequency (IDAF) curves are increasingly demanded for characterising the severity of storms and for designing hydraulic structures. Their computation requires inferring areal rainfall distributions over the range of space scales and timescales that are the most relevant for hydrological studies at catchment scale. In this study, IDAF curves are computed for the first time in West Africa, based on the data provided by the AMMA-CATCH Niger network, composed of 30 recording rain gauges having operated since 1990 over a 16 000 km^2 area in south-western Niger. The IDAF curves are obtained by separately considering the time (intensity–duration–frequency, IDF) and space (areal reduction factor, ARF) components of the extreme rainfall distribution. Annual maximum intensities are extracted for resolutions between 1 and 24 h in time and from point (rain gauge) to 2500 km^2 in space. The IDF model used is based on the concept of scale invariance (simple scaling) which allows the normalisation of the different temporal resolutions of maxima series to which a global generalised extreme value (GEV) is fitted. This parsimonious framework allows one to use the concept of dynamic scaling to describe the ARF. The results show that coupling a simple scaling in space and time with a dynamical scaling that relates to space and time allows one to satisfactorily model the effect of space–time aggregation on the distribution of extreme rainfall.

1 Introduction

Torrential rain and floods have long been a major issue for hydrologists. For one, defining and computing their probability of occurrence is a scientific challenge per se, largely because it is a scale-dependent exercise. Second, and equally important, is the fact that they cause heavy environmental, societal, and economical damages – including human casualties – thus being a major concern for populations and decision makers.

The request of providing both an objective assessment of the probability of occurrence of high-impact rainfall and a tool for civil engineering structure design has found an answer through the calculation of intensity–duration–frequency (IDF) curves. These curves, generally computed from raingauge data, are intended to characterise the evolution of extreme rainfall distributions at a point when the duration of rainfall accumulation changes. However, rainfall at point location is not of the greatest interest when it comes to the hydrological and socio-economic impacts of extreme events, since it is essentially the convolution of the rainfall intensities over a catchment that characterises the severity of storms and creates the real threat.

This is why intensity–duration–area–frequency (IDAF) curves were conceived as a spatial extension of the IDF curves. Generally established by combining IDF curves and areal reduction factors (ARFs), they provide an estimation of extreme areal rainfall quantiles over a range of timescales and spatial scales.

Theoretical studies on IDF and ARFs have been an active research topic over the past 20 years or so (Koutsoyiannis et al., 1998; Menabde et al., 1999; De Michele et al., 2001, 2011, among others). IDF practical studies are also numerous but focused on regions where long series of sub-daily rainfall are available (e.g. Borga et al., 2005; Gerold and Watkins, 2005; Nhat et al., 2007; Bara et al., 2009; Ben-Zvi, 2009; Overeem et al., 2009; Awadallah, 2011; Ariff et al., 2012). On the other hand, when ARFs are computed from rain-gauge networks (Bell, 1976; Asquith and Famiglietti, 2000; Allen and DeGaetano, 2005), it requires a fair density of rain gauges in order to obtain accurate estimates of areal rainfall. The computation of IDAF curves must therefore deal with two major data requirements: (i) a high-density network of rain gauges and (ii) an array of long sub-daily rainfall series. In addition to these requirements, scientists face the challenge of producing coherent ARF and IDF models, if they wish for their IDAF model to be statistically consistent. This explains why there are so few studies dealing with the implementation of an IDAF model over a given region (e.g. De Michele et al., 2002, 2011; Castro et al., 2004; Ceresetti, 2011).

In fact, in some regions of the world there are virtually no IDF, ARF, and IDAF models that have ever been conceived because of data limitations. This is especially the case in many tropical regions, such as West Africa; in fact, one reason being that a harsh environment and resource scarcity have made it very challenging to operate recording rain-gauge networks. The few IDF studies available in the region (Oyebande, 1982; Mohymont et al., 2004; Oyegoke and Oyebande, 2008; Soro et al., 2010) are essentially empirical with no theoretical background that allows for the upgrade of their results in order to produce IDAF curves. ARF studies are still fewer, the most noticeable being an attempt by Rodier and Ribstein (1988) and Ribstein and Rodier (1994) at computing ARF values for a return period of 10 years, with no explicit inference of the areal rainfall distributions. All in all, there has never been any IDAF model derived for West Africa or sub-regions of West Africa. Yet, flood management – for which IDAF curves are a very useful tool – is now a major concern for West African countries. As a matter of fact and despite that West Africa is known for having experienced a major lasting drought over 1970–2000, numerous severe floods and exceptional inundations have struck the region over the last 2 decades (Tarhule, 2005; Descroix et al., 2012; Samimi et al., 2012). Moreover, flood damages in the region have been constantly increasing since 1950 (Di-Baldassarre et al., 2010).

While operational networks of the West African National Weather Services do not allow the establishment of IDAF curves in a consistent way – because they do not provide any long-term sub-daily rainfall series – there are other sources of data that can be used for that purpose. Among them are the 5 min rainfall series of the long-term AMMA-CATCH observing system covering a 16 000 km^2 area in south-western Niger from 1990 to present (Fig. 1). In this study we will make use of 30 series providing continuous 5 min rainfall records from 1990 to 2012.

This unique data set enables us to characterise the relationship between extreme rainfall distributions computed at various spatio-temporal scales and to propose IDAF curves for this characteristic Sahelian region.

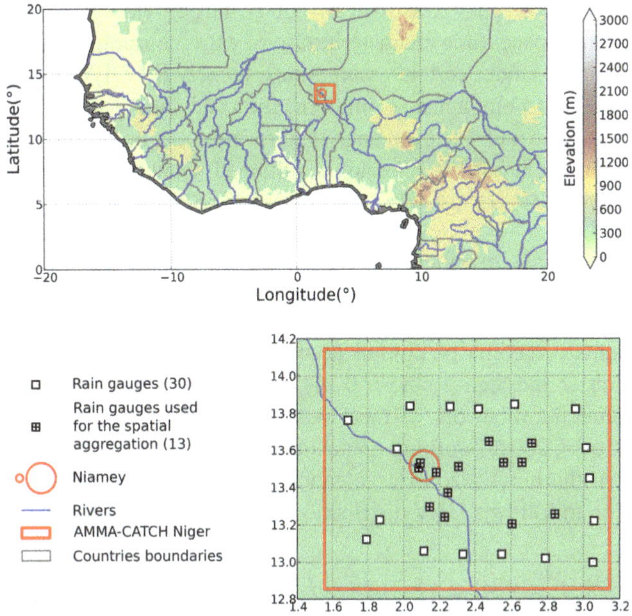

Figure 1. Study area. The background maps displays the elevation (metres).

2 IDAF curves in a generalised extreme value distribution and scale invariance framework

IDAF curves provide an estimate of areal rain rates – averaged over a given duration D and a given area A – for a given frequency of occurrence (currently expressed in term of return period T_r). In practice, IDAF curves are generally obtained by aggregating a temporal component and a spatial component represented respectively by the IDF computed at a point ($A = 0$) and by the ARF computed for a range of durations. In this framework, the most general formulation of an IDAF equation is as follows:

$$\mathrm{IDAF}(D, A, T_r) = \mathrm{IDF}(D, T_r) \times \mathrm{ARF}(D, A, T_r). \tag{1}$$

Assessing IDAF curves requires (i) inferring appropriate statistical distributions of rainfall to estimate the return periods and (ii) describing the statistical links between the distributions obtained at different space scales and timescales.

Several recent studies have confirmed that the generalised extreme value (GEV) distribution (Coles, 2001) provides a suitable framework to describe the distribution of extreme rainfall at a point (e.g. Overeem et al., 2008; Panthou et al., 2012; Papalexiou and Koutsoyiannis, 2013). Also, many authors have shown that rainfall displays scale invariance properties (Schertzer and Lovejoy, 1987; Gupta and Waymire, 1990; Burlando and Rosso, 1996; Bendjoudi et al., 1997; Veneziano et al., 2006), both in space and time. The temporal scaling properties give access to a direct analytical formulation of IDF curves (Menabde et al., 1999; Borga et al., 2005; Ceresetti, 2011), while the spatial scaling properties allow one to upscale IDF curves into IDAF curves (De Michele et al., 2002, 2011; Castro et al., 2004), thus providing an integrated space–time characterisation of extreme rainfall distributions. Under certain assumptions, namely the GEV distribution of point annual rainfall maxima and simple scaling in both time and space, an analytical formulation of the various components of Eq. (1) may be obtained, as will be detailed below.

2.1 GEV distribution

Let us define $I(D, A)$ as a random variable representing the annual maxima of rainfall accumulated over a given duration D and area A, and $i(D, A)$ a sample of $I(D, A)$. In the general framework of the block maxima sampling scheme (Coles, 2001), working on annual maxima generally ensures that the block size is large enough for the maxima distribution to follow a GEV distribution (Coles, 2001), written as

$$G(i;\mu,\sigma,\xi)=\exp\left\{-\left[1+\xi\left(\frac{i-\mu}{\sigma}\right)\right]^{-\frac{1}{\xi}}\right\} \quad (2)$$

for $1+\xi\left(\dfrac{i-\mu}{\sigma}\right)>0$,

where i is a generic notation for any value associated with a realisation of $I(D, A)$, μ being the location parameter, $\sigma > 0$ the scale parameter, and ξ the shape parameter of the GEV distribution. The shape parameter describes the behaviour of the distribution tail: a positive (negative) shape corresponds to a heavy-tailed (bounded) distribution. When ξ is equal to 0, the GEV reduces to the Gumbel distribution (light-tailed distribution):

$$G(i;\mu,\sigma)=\exp\left\{-\exp\left[-\left(\frac{i-\mu}{\sigma}\right)\right]\right\}. \quad (3)$$

2.2 Simple scaling in time and analytical formulation of IDF curves

The simple-scaling framework has been extensively used for deriving IDF curves (Menabde et al., 1999; Yu et al., 2004; Borga et al., 2005; Nhat et al., 2007; Bara et al., 2009; Ceresetti, 2011). An analytical formulation of the ARF was also given by Veneziano and Langousis (2005) in a multi-scaling framework. However, simple scaling is much more tractable than the multi-scaling approach and is more robust in terms of parameter inference; thus, this is the approach chosen here.

The annual maximum point rainfall random variable $\{I(D,0)\}$ follows a simple-scaling relation for a given duration D (with respect to a reference duration D_{ref}) if

$$I(D,0)\overset{d}{=}\lambda^{\eta}\times I(D_{\text{ref}},0), \quad (4)$$

where λ is a scale ratio ($\lambda = D/D_{\text{ref}}$), η is a scale exponent, and $\overset{d}{=}$ denotes an equality in distribution. Note that, for every duration D for which Eq. (4) holds, the normalised random variable $\{I(D,0)/D^{\eta}\}$ has the same statistical distribution as the normalised reference distribution $\{I(D_{\text{ref}},0)/D^{\eta}_{\text{ref}}\}$; this property will be used in the optimisation procedure in Sect. 4.2. Equation (4) implies (Gupta and Waymire, 1990)

$$E[I(D,0)]=\lambda^{\eta}\times E[I(D_{\text{ref}},0)] \quad (5)$$

and, more generally, a scaling of all the moments that can be written as

$$E[I^{q}(D,0)]=\lambda^{k(q)}\times E\left[I^{q}(D_{\text{ref}},0)\right] \quad (6)$$

or

$$\ln\{E[I^{q}(D,0)]\}=k(q)\ln(\lambda)+\ln\left\{E\left[I^{q}(D_{\text{ref}},0)\right]\right\}. \quad (7)$$

The notion of simple scaling is related to how $k(q)$ evolves with q. When this evolution is linear

$$k(q)=\eta q \quad (8)$$

and simple scaling holds (as opposed to multi-scaling if this relation in non-linear).

Checking whether the simple-scaling hypothesis is admissible over a given range of durations is thus equivalent to verifying through the data set whether the two following conditions are fulfilled (Gupta and Waymire, 1990):

- Eq. (7): log–log linearity between the statistical moments of any given order q
- Eq. (8): linearity between $k(q)$ and q.

Figure 2a illustrates these two conditions.

2.3 Spatial scaling, dynamical scaling, and ARF model

In its most general sense, the ARF is the ratio between areal rainfall and point rainfall, either for a given observed rain event or in a statistical sense. Here we are interested in deriving a statistical ARF that can be used for obtaining an analytical formulation of IDAF curves (which implies that the ARF does not depend on the return period considered); thus, this ARF denotes the ratio between the areal distribution and the point distribution of the annual rainfall maxima:

$$I(D,A)\overset{d}{=}\text{ARF}(D,A)\times I(D,0). \quad (9)$$

Note that Eq. (9) implies the following relationship:

$$\text{ARF}(D,A)=\frac{E[I(D,A)]}{E[I(D,0)]}. \quad (10)$$

In this study, the ARF model proposed by De Michele et al. (2001) is used. This model is based on two assumptions (which will have to be verified; see Sect. 5):

1. The studied rainfall variable is characterised by a simple-scaling relationship both in time and space.

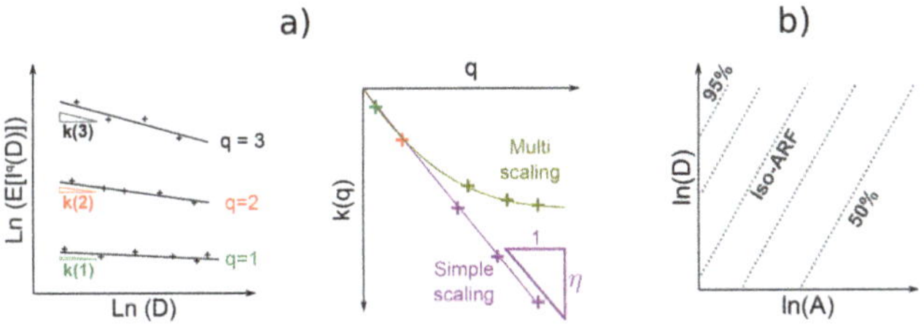

Figure 2. Visual model checks: **(a)** simple scaling; **(b)** ARF.

2. Time and spatial scales are linked by a so-called dynamic-scaling property written as

$$\left(\frac{D}{D_{\text{ref}}}\right) = \left(\frac{A}{A_{\text{ref}}}\right)^{z}, \tag{11}$$

where z is the dynamic-scaling exponent.

When these assumptions are verified, De Michele et al. (2001) show that the ARF can be written as

$$\text{ARF}(D,A) = \left[1+\omega\left(\frac{A^{a}}{D^{b}}\right)\right]^{\eta/b} = \left[1+\omega\left(\frac{A^{z}}{D}\right)^{b}\right]^{\eta/b}, \tag{12}$$

where η is the scaling exponent characterising the temporal simple scaling, a and b are two positive constant-scaling exponents linked by the relation $z = a/b$, and ω is a factor of normalisation.

This ARF formulation implies that iso-ARFs are lines in the plane $\{\ln(A), \ln(D)\}$ as shown in Fig. 2b (De Michele et al., 2001).

2.4 GEV simple-scaling IDAF model

By assuming that the maximum annual rainfall is GEV-distributed and that the scaling relations in time and space (Sects. 2.2 and 2.3) are verified, then the IDAF model is

$$I(D,A) \stackrel{d}{=} I(D_{\text{ref}},0) \times \lambda^{\eta} \times \text{ARF}(D,A). \tag{13}$$

As shown in the Appendix A, the compatibility of the simple-scaling and GEV frameworks is defined by the following equations:

$$I(D,A) \sim \text{GEV}\{\mu(D,A), \sigma(D,A), \xi(D,A)\} \tag{14}$$

$$\mu(D,A) = \mu_{\text{ref}} \times \lambda^{\eta} \times \text{ARF}(D,A) \tag{15}$$

$$\sigma(D,A) = \sigma_{\text{ref}} \times \lambda^{\eta} \times \text{ARF}(D,A) \tag{16}$$

$$\xi(D,A) = \xi_{\text{ref}}, \tag{17}$$

where μ_{ref}, σ_{ref} , and ξ_{ref} correspond to the GEV parameters computed for the arbitrary reference duration D_{ref} at a point, and $\lambda = D/D_{\text{ref}}$.

3 Data and implementation

Rainfall observing systems usually do not provide direct measurements at all the space scales and timescales required for an IDAF study; thus, it is necessary to derive from the raw data set an elaborated data set that allows one to verify the various assumptions found in the theoretical framework defined in Sect. 2.

Accordingly, this section describes both the rainfall samples initially available on our Sahelian region of south-western Niger and the process used to obtain the final data set from which the IDAF curves were computed. This process consists of two major steps:

1. space–time aggregation of the 5 min point rain rates in order to obtain the average rain rates for various space (A) and time (D) resolutions

2. extraction of extreme rainfall samples for each of the above resolutions.

3.1 The rainfall data set: AMMA-CATCH Niger records

The AMMA-CATCH Niger observing system was set up at the end of the 1980s as part of the long-term monitoring component of the HAPEX-Sahel experiment (Lebel et al., 1992). Since then, it has continuously operated a large array of meteorological and hydrological instruments, providing a unique set of high-resolution hydrometeorological data, covering a 16 000 km^2 area in south-western Niger. For the purpose of this research, a subset of thirty 5 min rainfall series was selected (Fig. 1), covering the entire 1990–2012 period. At each station, all years with more than 25 % of missing data have been removed in order to limit any sampling effect due to missing data in a particular year. After this quality control, all stations remain with at least 20 years of valid data, constituting our raw data sample.

To estimate areal rainfall intensities, this study makes use of the dynamical kriging interpolation method proposed by Vischel et al. (2011). Rain fields are produced over the domain of study at a time resolution of 5 min and a spatial resolution of 1 km^2. Dynamical kriging takes advantage of the time structure of 5 min rain fields to complement the purely spatial information provided by the gauge network. Instead of using a 3-D variogram (the inference of the space–time cross-covariance being notoriously non-robust), the method relies on the construction of Lagrangian rain fields which display a stable spatial structure represented by a nested exponential variogram. Dynamical kriging is an exact interpolator in the sense that the measured point values are replicated exactly; this interpolator is then used to produce 5 min rainfall grids, with a grid mesh of 1 km^2.

3.2 Space–time rainfall aggregation

The starting elements of the space–time aggregation process are the discretized fields of rain accumulated over a time increment $\Delta t = 5$ min and averaged over a square pixel of side length $\Delta xy = 1$ km. In the following, these rain fields are denoted as $r^*(x^*, y^*, t^*)$, where t^* is the ending time of the 5 min time step, and $\{x^*, y^*\}$ is the centre of the 1 km^2 pixel.

3.2.1 Spatial aggregation of 5 min rain fields

Let A be a surface over which the rainfall intensity is averaged. In this study, A is a square of side $Nx \times \Delta xy$ km $=$ $Ny \times \Delta xy$ km (corresponding to $Nx \times Ny$ pixels of 1 km^2). From the 5 min rain fields we can compute series of space averaged 5 min rain field accumulations, r_{A}^*, as

$$r_{\mathrm{A}}^*(x^*, y^*, t^*) = \frac{1}{A} \sum_{m=0}^{Nx-1} \sum_{n=0}^{Ny-1} r^* \quad (18)$$

$$\left\{x^* + \left(m - \frac{Nx-1}{2}\right)\Delta xy, y^* + \left(n - \frac{Ny-1}{2}\right)\Delta xy, t^*\right\}.$$

From these spatially averaged rain fields, spatial rainfall series have been extracted at rain-gauge locations. For each rain-gauge location (located at $\{x, y\}$), the nearest spatial rainfall series, r_{A}^* (located at $\{x^*, y^*\}$), is extracted. Figure 3a illustrates this approach the black circle of the right panel represents a rain-gauge located at $\{x, y\}$. In total, 12 scales of spatial aggregations have been retained to build the rainfall series: 1 km^2 (the pixel on which the station is located is selected) then 4, 9, 16, 25, 49, 100, 225, 400, 900, 1600, and 2500 km^2.

To limit border effects, the spatial aggregation is performed only in areas where the spatial distribution of stations is more or less isotropic. Each of the 30 measurement stations are considered individually; a circle centred on the station is plotted and divided in eight cardinal sectors (each sector has an angle of 45°). Only rain gauges having at least one other rain gauge present in at least seven of the eight sectors are retained for spatial aggregation (see Fig. 3a); the distance of the other gauges from the centre station is not taken into account for the selection, only the presence or absence of a rain gauge in the sector. Only 13 gauges (out of 30) satisfy this criterion (Fig. 1). They are referred to in the following as a central rain gauge (CR) because their localisation is used as a central point, from which the 12 areas of aggregation are delimited. In total there are thus 156 areal series (12 areas of aggregation centred on 13 different locations).

3.2.2 Time aggregation of 5 min point series and 5 min spatial series

A time aggregation procedure is applied to the 30 point 5 min rainfall series and to the 156 spatial rainfall series.

Let D be a given duration of Nt 5 min time steps ($D = Nt \times \Delta t$). The time aggregation is done by using a moving

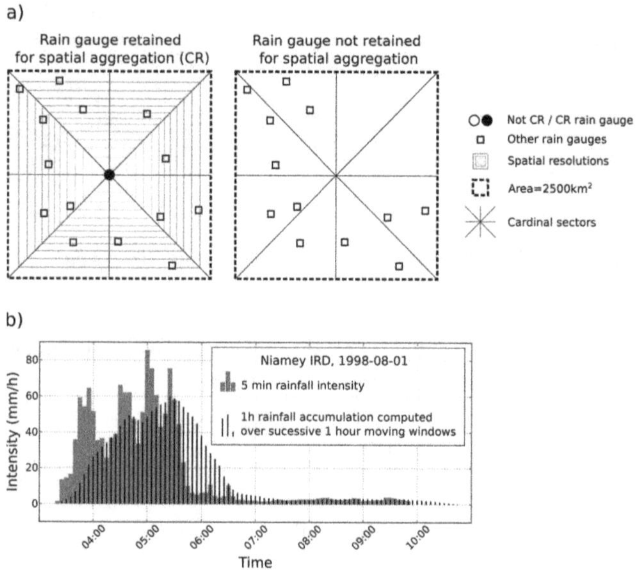

Figure 3. Space and time aggregation procedures. **(a)** Illustration of the procedure leading to select (right case) or to reject (left case) a gauge for becoming a centre for spatial aggregation; **(b)** time aggregation at a point: comparing a hyetograph of 5 min rainfall to a hyetograph of 1 h rainfall.

time window of length D over which the 5 min rainfall intensity is averaged (this moving window procedure is carried out in order to make sure that we will be able to extract the maximum for each duration considered). The time aggregation can be written as

$$r_{D,0}^*(x, y, t^*) = \frac{1}{D} \sum_{p=0}^{Nt-1} r_0^*(x, y, t^* - p \times \Delta t) \quad (19)$$

in the case of 5 min point series located at $\{x, y\}$ ($A = 0$), and

$$r_{D,A}^*(x^*, y^*, t^*) = \frac{1}{D} \sum_{p=0}^{Nt-1} r_{\mathrm{A}}^*(x^*, y^*, t^* - p \times \Delta t) \quad (20)$$

for a given surface A in the case of 5 min spatial rainfall series located at $\{x^*, y^*\}$.

Thus, $Nt = 12$ for $D = 1$ h, $Nt = 24$ for $D = 2$ h and so forth. This procedure is illustrated in Fig. 3b. The 11 different time resolutions considered in this study range from 1 to 24 h (1, 2, 3, 4, 6, 8, 10, 12, 15, 18, and 24 h) and are all obtained from the original 5 min series.

3.2.3 Extraction of extreme rainfall: annual block maxima

The use of GEV distribution to model the extreme rainfall series requires using the block maxima procedure to extract rainfall extremes. It consists of defining annual blocks of observations separately for each of the 11 different time resolutions considered and to take the maxima within each block.

A sample of 23 (1990–2012) annual maximum rainfall values $\{i(D,A)\}$ is thus obtained for each spatial aggregation and duration.

In summary

- There are 13 reference locations.
- Around each of the 13 reference locations, 12 areas of increasing size 1, 4, 9, 16, 25, 49, 100, 225, 400, 900, 1600, and 2500 km^2 are defined.
- For each of these 156 (13 × 12) areas, 11 time series of 23 (1990–2012) annual maximum values are constructed, corresponding to 11 different durations of rainfall accumulation 1, 2, 3, 4, 6, 8, 10, 12, 15, 18, 24 h.

4 Inferences of the individual components of the model

The proposed model has seven parameters: the temporal scale exponent (η), the three ARF parameters (a, b, ω), and the three GEV parameters (μ_{ref}, σ_{ref}, ξ_{ref}). After having tested different optimisation procedures (most notably a global maximum likelihood estimation and the two-step method proposed by Koutsoyiannis et al., 1998), a three-step method was finally retained, since it gave the best results in the evaluation of the IDAF model (see Sect. 5). These three steps are explained in the following paragraphs (Sects. 4.1 to 4.3). For each step, an illustration based on the result obtained for the Niamey Aéroport station is given in Fig. 4.

4.1 Temporal simple scaling: estimation of η

The temporal scaling of the IDAF model is described by the η parameter. The inference of η is achieved in two steps. The first one consists of computing $k(q)$ for different moments q through a linear regression between the logarithm of the statistical moments of order q ($E[I^q]$) and the durations D (see Fig. 4a, left panel); next, η is obtained by a linear regression between $k(q)$ and q (see Fig. 4a, right panel). At the Niamey Aéroport station, the value obtained for η is equal to -0.91.

4.2 GEV parameters: $\mu_{\text{ref}}, \sigma_{\text{ref}}, \xi_{\text{ref}}$

The GEV parameters μ_{ref}, σ_{ref}, ξ_{ref} are estimated on the point samples, using the property that all normalised samples $\{i(D,0)/D^{\eta}\}$ must come from the same distribution if simple scaling holds. All normalised samples are pooled in one single sample on which the GEV parameters are estimated (this methodology corresponds to the second step of the two-step method proposed by Koutsoyiannis et al., 1998). Figure 4b illustrates this process at the Niamey Aéroport rain gauge: initial samples are displayed on the left panel normalised samples are plotted on the right panel. The fitted GEV and the estimated GEV parameters are also given in this figure.

In comparison with fitting the GEV parameters separately to each sample constituted for each duration, this method aims at limiting sampling effects by fitting the GEV parameters on a single sample gathering all rainfall durations. The maximum likelihood and the L-moments methods were tested for estimating the GEV parameters. The estimation provided by these methods gave similar results, probably due to the large sample size. The results of the L-moments method are presented here, while this method is generally considered better than the Maximum Likelihood Estimation (MLE) for the estimation of high quantiles when the length of the series is short (Hosking and Wallis, 1997).

4.3 Spatial scaling

The estimation of a, b, and ω was carried out by minimising the mean square difference between the empirical ARF (Eq. 10) and the model ARF (Eq. 12), as originally proposed by De Michele et al. (2001). Other scores (mean and max. absolute error, bias, etc.) and variables (difference between the observed and model mean areal rainfall) have been tried but gave poorer results in validation (Sect. 5). Figure 4c shows the comparison between the empirical ARF and the model ARF at the Niamey Aéroport station and the parameters obtained for the theoretical ARF model.

4.4 Regional model

The point parameter inference ($\mu_{\text{ref}}, \sigma_{\text{ref}}, \xi_{\text{ref}}$ and η) has been performed on each of the 30 point rainfall series, which thus provide 30 IDF models. The complete IDAF model has been fitted to each of the 13 rain gauges CR. The obtained IDF and IDAF parameters display neither any coherent spatial pattern nor any trend over the domain, as may be seen in Fig. 5. Sampling effects due to the small area and the short length of the series may explain this point, since a trend has been observed on a larger domain at the regional scale for daily rainfall (Panthou et al., 2012).

Assuming a spatial homogeneity of rainfall distribution (no spatial pattern), annual maxima series have been pooled together to obtain regional samples. The regional samples were used to fit the IDAF model over the domain in order to limit sampling effects. The point regional sample pools together the 30 rainfall samples directly provided by the 30 rain gauges $i(D,0)$; the 12 areal regional samples obtained for each of the 12 spatial resolutions $\{i(D,A); A = 1,\ldots,2500\,\text{km}^2\}$ result from pooling together the 13 individual series (CRs) computed as explained in Sect. 3.

Table 1 presents the parameters obtained for the global IDAF model. The obtained GEV parameters are $\mu_{\text{ref}} = 40.6\,\text{mm h}^{-1}$, $\sigma_{\text{ref}} = 10.8\,\text{mm h}^{-1}$, and $\xi_{\text{ref}} = 0.1$. When upscaled to the daily duration $\mu(24\,\text{h}) = 2.29\,\text{mm h}^{-1}$ ($55.0\,\text{mm day}^{-1}$) and $\sigma(24\,\text{h}) = 0.61\,\text{mm h}^{-1}$ ($14.6\,\text{mm day}^{-1}$). It is worth noting that these latter values are coherent with those

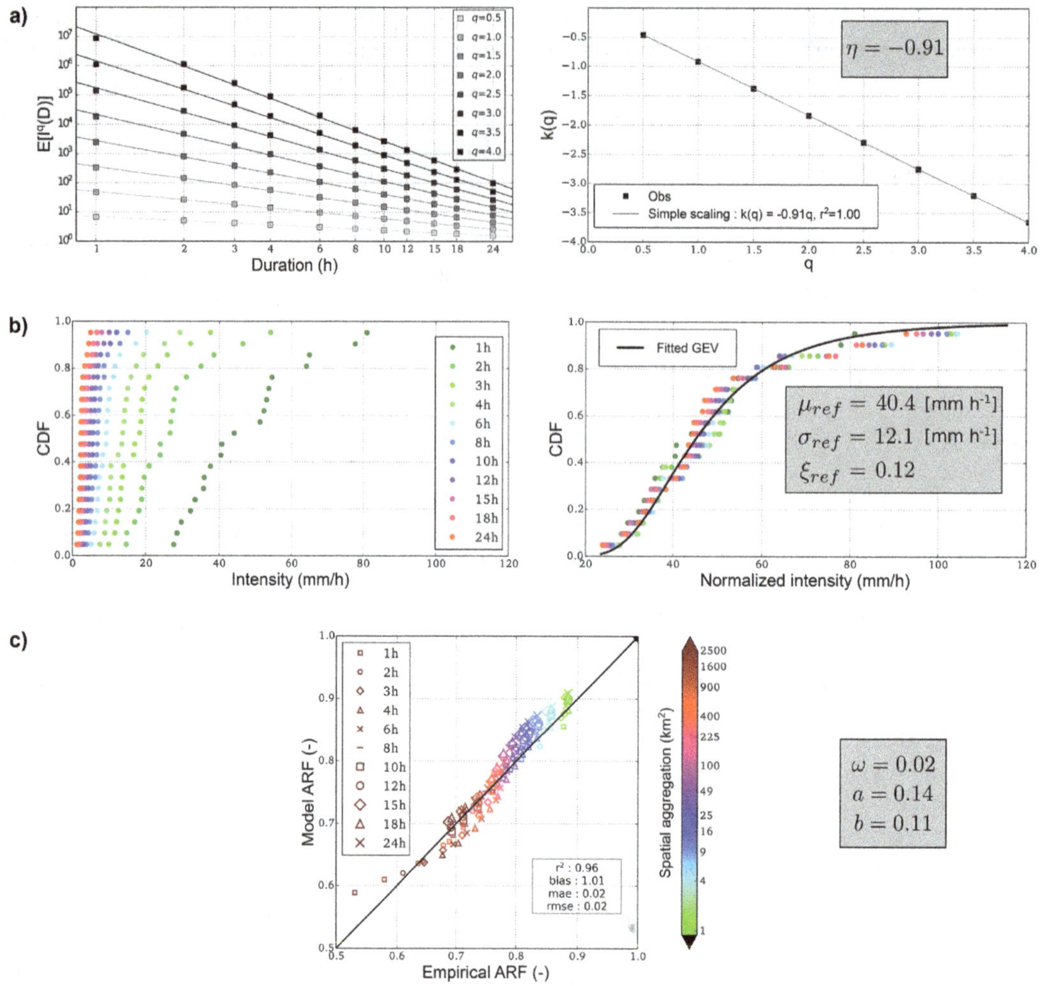

Figure 4. Example of IDAF model inference at the Niamey Aéroport rain gauge: **(a)** checking of the temporal simple-scaling conditions (left: linear relationship between the logarithm of the statistical moments of order q and the durations D, right: linear relationship between $k(q)$ and q) and estimation of the temporal simple-scaling exponent; **(b)** left: empirical cumulative distribution of annual maxima, right: global fitting of the GEV parameters; **(c)** comparison between empirical and modelled ARF.

Table 1. Obtained parameters for the global IDAF model (a) and corresponding GEV parameter values for the different durations D for the point scale A=0 (b).

(a) μ_{ref}	σ_{ref}	ξ	η	a	b	z	ω
40.60	10.81	0.10	−0.90	0.165	0.156	1.06	0.026

(b) GEV parameter	1 h	2 h	3 h	4 h	6 h	8 h	10 h	12 h	15 h	18 h	24 h
μ (mm h^{-1})	40.60	21.68	15.02	11.58	8.02	6.19	5.05	4.29	3.50	2.97	2.29
σ (mm h^{-1})	10.81	5.77	4.00	3.08	2.14	1.65	1.35	1.14	0.93	0.79	0.61
ξ (–)						0.10					

obtained for a much larger area in this region by Panthou et al. (2012, 2013), working on the data of 126 daily rain gauges covering the period 1950–1990. Note also that the temporal scale exponent (η) is large (0.9), which means that the intensity strongly decreases as the duration increases. This is not surprising given the strong convective nature of rainfall in this region. Similar values of the temporal scaling exponents are obtained in regions where strong convective systems occur (Mohymont et al., 2004; Van-de Vyver and Demarée, 2010; Ceresetti et al., 2011), while lower values are obtained in regions where extreme rainfall is generated by different kinds of meteorological systems (for example

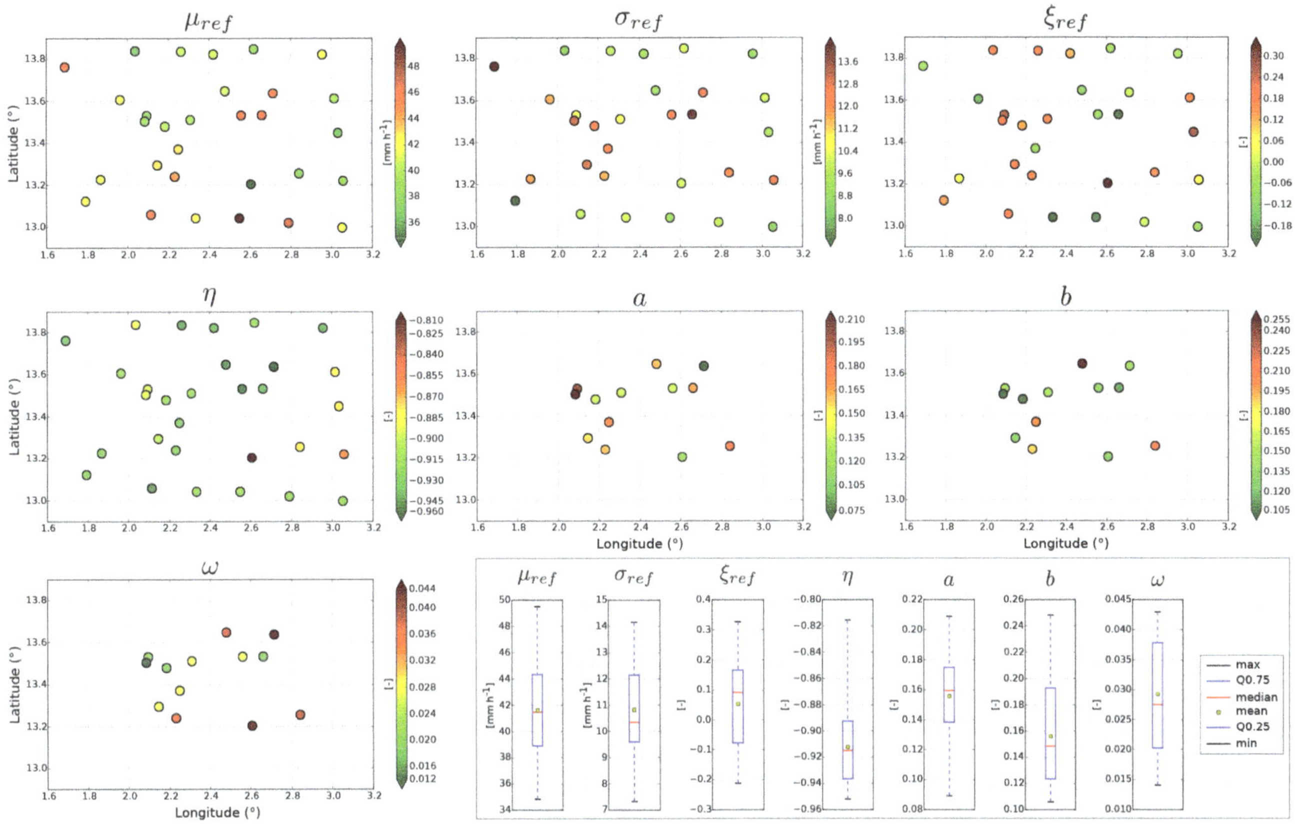

Figure 5. Map of the obtained IDF parameters (η, μ_{ref}, σ_{ref}, and ξ_{ref} fitted on the 30 rain gauge samples) and IDAF parameters (η, μ_{ref}, σ_{ref}, ξ_{ref}, a, b, and ω fitted on the 13 CR samples). The grey box shows box plots of the parameter values obtained at the different rain gauges.

in many mid-latitude regions; see e.g. Menabde et al., 1999; Borga et al., 2005; Nhat et al., 2007). The dynamic-scaling exponent is roughly equal to 1 which means that increasing the surface by a given factor leads to a similar ARF change than increasing the duration by the same factor (keeping in mind that this rule applies only to the range of time–space resolutions explored here).

5 IDAF model evaluation

The evaluation of the IDAF model is carried out in two successive stages. First each component used to build the final model (temporal simple scaling, ARF model, and GEV distribution) is checked individually; next, the global goodness of fit (GOF) is tested using the Anderson–Darling (AD) and the Kolmogorov–Smirnov (KS) tests.

In Fig. 6 two series of graphs are plotted in order to verify whether the simple-scaling hypothesis holds for the time dimension. On the left are the plots of $\ln(E[I^q])$ vs. $\ln(D)$ designed to check the log–log linearity between these two variables (Eq. 7); on the right are the plots of q vs. $k(q)$ aimed at checking the linearity between these two variables (Eq. 8). At all three spatial scales, there is a clear linearity of the plots, meaning that the two conditions for accepting the temporal simple-scaling hypothesis are fulfilled. Note that the graphs shown are those obtained on the regional samples for three different spatial scales only (point scale, 100 and 2500 km^2), but the quality of the fitting is similar for all the other spatial scales.

Simple scaling in space and dynamical scaling (e.g. the relationship between time and spatial scaling) are checked in Fig. 7. This figure compares the empirical ARFs (Eq. 10) computed on the regional samples and the ARFs obtained with the model (Eq. 12) for all the space scales and timescales pooled together. With a determination coefficient (r^2) of 0.98, and a very small root mean square error, it appears that the model restitutes very well the empirical ARF at all space scales and timescales, except at the hourly time step and for the three largest surfaces (900, 1600, and 2500 km^2), for which the model significantly underestimates the observed reduction factor. At such space scales and timescales the finite size of the convective systems generating the rain fields creates a significant external intermittency (see Ali et al., 2003, on the distinction between internal and external intermittency). It thus seems that the simple-scaling framework holds only as long as the influence of the external intermittency is negligible or weak. Consequently, it is likely that the overestimation of the ARF by the simple scaling-

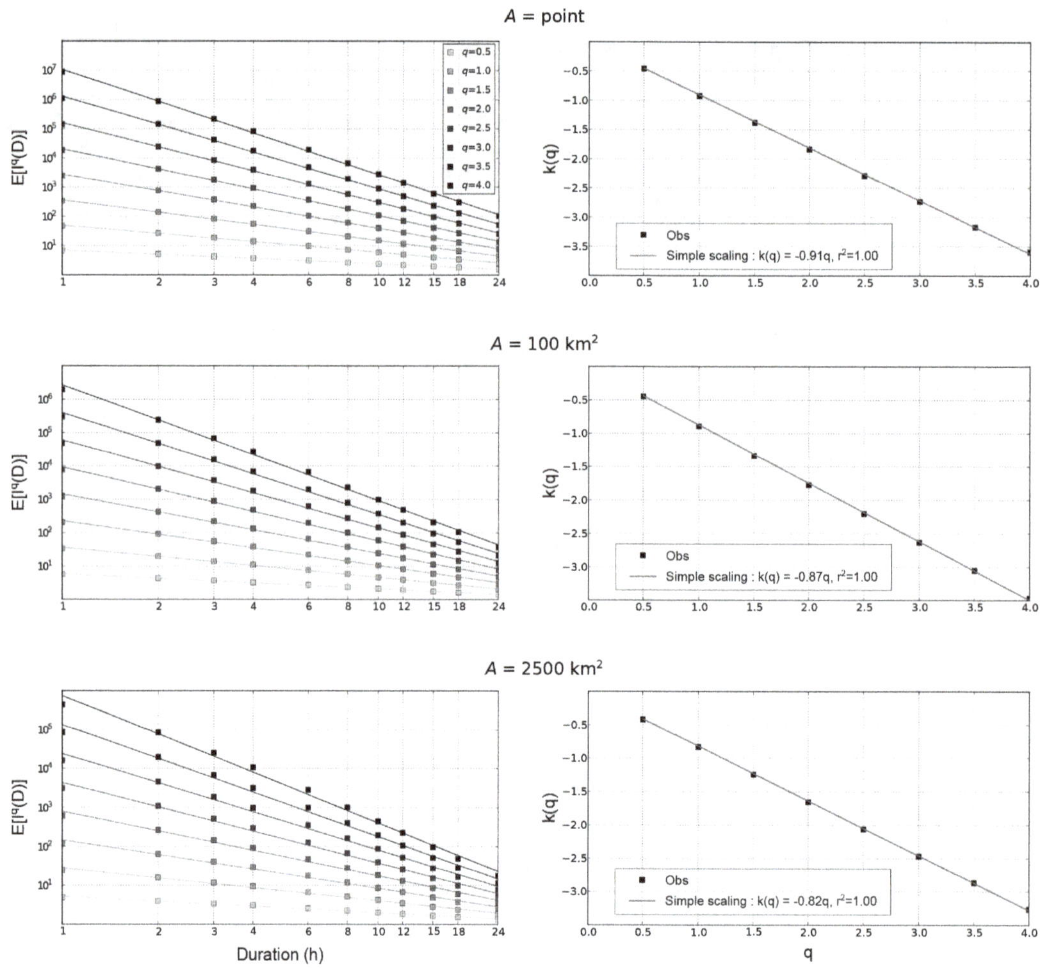

Figure 6. Checking of the temporal simple-scaling conditions for the regional samples defined by the 30 available rain gauges for point resolution (top), and the 13 CRs (see Sect. 3.2.1) for resolutions at $100\,km^2$ (middle) and $2500\,km^2$ (bottom).

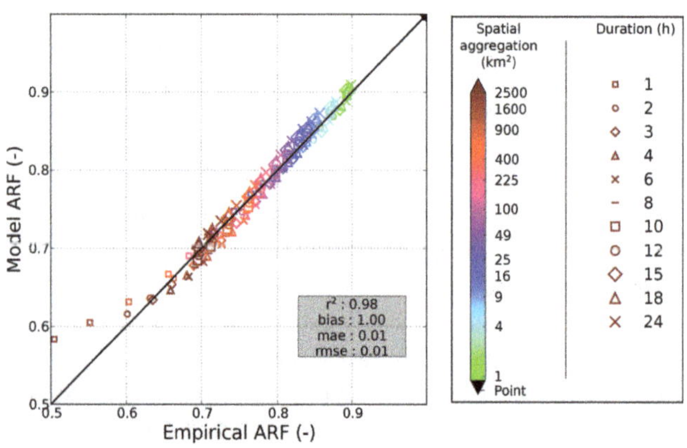

Figure 7. Comparison between empirical ARF (obtained with the regional samples: 30 rain gauges for point resolution and 13 CRs for other spatial resolutions) and the modelled ARF IDAF model.

based model would be observed for larger space scales and timescales than the ones the AMMA-CATCH data set allows one to explore.

Figure 8 illustrates that the global model is also able to reproduce very correctly the mean areal rainfall intensity over the whole time–space domain explored here, except again for the hourly time step and the largest surfaces.

As the IDAF model is primarily designed to estimate high quantiles, its ability to represent the mean is not a sufficient skill. It is thus of primary importance to evaluate its ability to also represent correctly high return levels and extreme quantiles. This was realised by visually inspecting return level plots and by using GOF statistical tests computed in a cross-validation mode (all the stations are used to calibrate the model except one which is used to validate the model prediction). These tests are used to quantitatively assess how well the theoretical GEV distribution based on the IDAF model fits the empirical Cumulative Distribution Functions (CDF) of the observed annual maxima for each spatio-temporal resolution. Each test provides a statistic and its corresponding

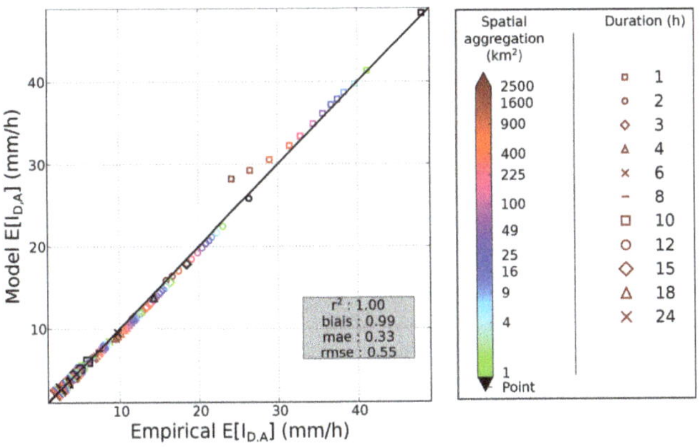

Figure 8. Comparison between empirical mean areal rainfall intensity (obtained with the regional samples: 30 rain gauges for point resolution and 13 CRs for other spatial resolutions) and global IDAF model for different spatio-temporal aggregations.

p value. The p value is used as an acceptation/rejection criterion by fixing a threshold of non-exceedance (here 1, 5, and 10 %).

The return level plots displayed in Fig. 9 for two reference locations and three time steps allow a visual inspection of the capacity of the global IDAF to fit the empirical samples. The p values of the two GOF tests are given in the inset caption. As could be expected, there is a significant dispersion of the results obtained on individual samples. The difficulty of reproducing correctly the empirical distribution when combining the smallest time steps with the largest areas is confirmed. While similar graphs were plotted for the other 11 reference locations, it is obviously difficult to obtain a relevant global evaluation from the visual examination of such plots.

Figure 10 aims at tackling this limitation by representing this information in a more synthetic way. In this figure the percentage of individual series for which the IDAF model is rejected by the AD GOF test is mapped for each duration and spatial aggregation for two levels of significance (1 and 10 %). Here it is worth remembering that we have 30 individual series for the point scale, and 13 different individual series for each of the 12 spatial scales, meaning that, for a given time step, the percentage of rejections/acceptations are computed from a total of 186 ($30 + 12 \times 13$) test values. Here again, the limits of the model for small time steps and large areas are clearly visible; one can also notice a larger number of rejections for small areas and the highest durations (duration higher than 12 h and area smaller than 25 km^2). Apart from that, the number of rejections of the null hypothesis remains low. The KS test (not shown) displays similar results with fewer rejections of the null hypothesis. It thus appears fair to conclude that, over the range of space scales and timescales covered by the AMMA-CATCH network, a simple scaling approach allows for computing realistic ARFs, the limit of validity being reached for areas roughly larger than 1000 km^2 at the hourly time step.

6 Discussion and conclusions

Up to now the rarity of rainfall measurements at high space–time resolution in tropical Africa has not allowed comprehensive studies on the scaling properties of rain fields in that region to be carry out. From 1990 the recording rain-gauge network of the AMMA-CATCH observing system has sampled rainfall in a typical Sahelian region of West Africa at a time resolution of 5 min and a space resolution of 20 km, over an area slightly larger than $1° \times 1°$. This data set was used here for characterising the space–time structure of extreme rainfall distribution, the first time such an attempt has been made in this region where rainfall is notoriously highly variable.

Simple scaling was shown to hold for both the time and the space dimensions over a space–time domain ranging from 1 to 24 h and from the point scale to 2500 km^2; it was further shown that dynamical scaling relates the timescales to the space scales, leading to propose a global IDAF model valid over this space–time domain, under the assumption that extreme rainfall values are GEV distributed.

Different optimisation procedures were explored in order to infer the seven parameters of this global IDAF model. A three-step procedure was finally retained, the global IDAF model being fitted to a global sample built from all the different samples available for a given space scales and timescale. This model has been evaluated through different graphical methods and scores. These scores show that the ARFs yielded by the IDAF model fit significantly well (in a statistical sense) with the observed ARFs over our space–time domain, except for the part of the domain combining the smallest timescales with the largest space scales. This limitation is likely related to the larger influence of the external intermittency of the rain fields at such space scales and timescales.

Despite the growing accuracy of rainfall remote-sensing devices, this study demonstrates that dense rain-gauge networks operating in a consistent way over long periods of time are still keys to the statistical modelling of extreme rainfall. In the numerous regions where rainfall is undersampled by operational networks and where satellite monitoring is not accurate enough to provide meaningful values of high rainfall at small space scales and timescales, dense networks covering a limited area may provide the information necessary for complementing the operational networks and satellite monitoring. In West Africa, south of the Niger site, AMMA-CATCH has been operating another site of similar size in a Sudanian climate since 1997 (Ouémé Catchment, Benin), providing ground for a similar study in a more humid tropical climate.

As mentioned in the discussion of Sect. 5, there is however a limitation of these two research networks, linked to their spatial coverage. Extending the area sampled by these networks to something of the order of $2° \times 2°$ would indeed allow for studying more finely the effect of the limited size of the convective systems onto the statistical properties of

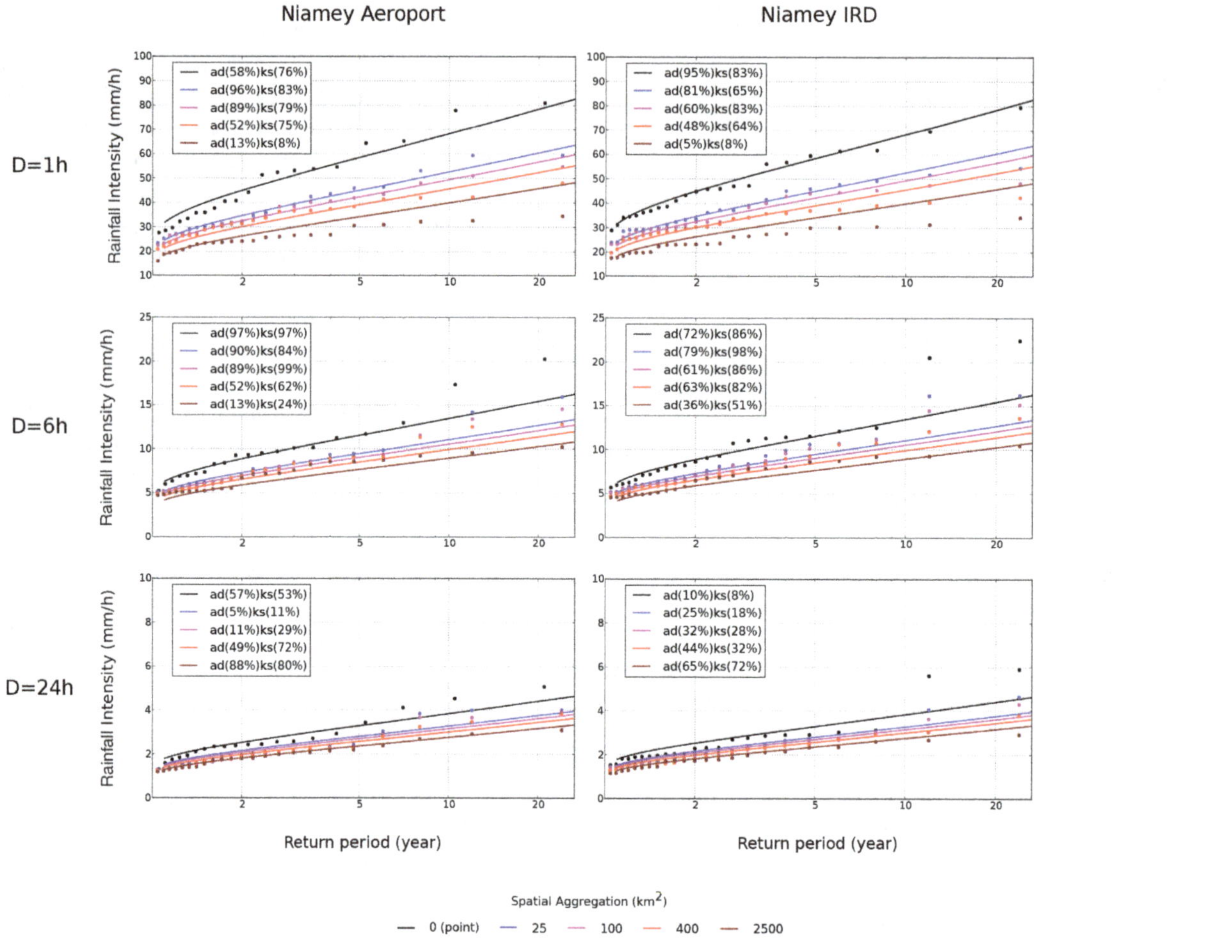

Figure 9. Empirical return level plot obtained at two rain gauges in comparison with the global IDAF model for different durations (1 h, 6 h, and 24 h from top to bottom) and different spatial aggregations (from point to 2500 km^2).

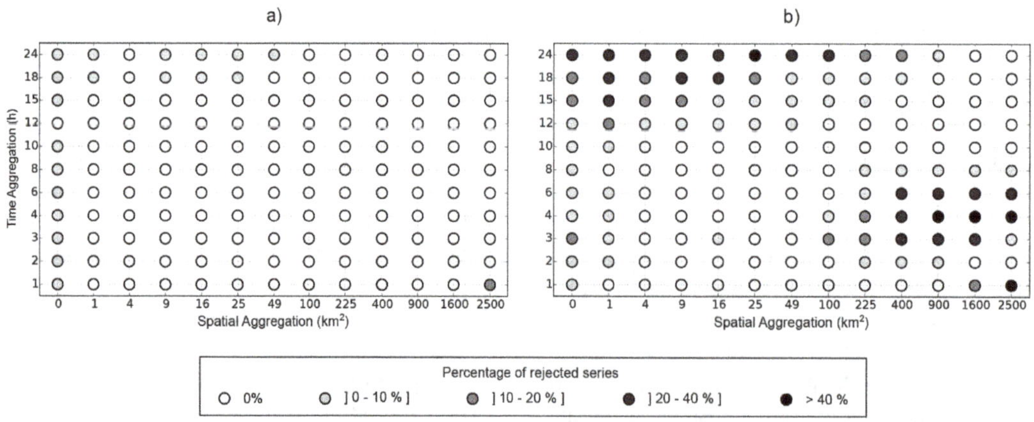

Figure 10. Anderson–Darling GOF test (cross-validation): percentage of rejected series for 1 % **(a)** and 10 % **(b)** significance level for the global IDAF model. Note that there are 30 series for the point scale (0 km^2) and 13 for the other spatial aggregation (CR).

the associated rain fields. However, this means enlarging the area by a factor of 4, making it much more costly and difficult from a logistical point of view to survey properly. For the years to come, AMMA-CATCH remains committed to operating both the Niger and the Benin sites for documenting possible evolutions of the rainfall regimes at fine space scales and timescales in the context of global change as well as for verifying whether the scaling relationships proposed here still hold for quantiles at higher time periods. As a matter of fact, one strong hypothesis of the model proposed here is that the ARF is independent of the return period. This hypothesis seems verified for return periods smaller than the length of our time series, but it is not possible to infer whether this really holds for higher return periods. Therefore, developing an IDAF model able to account for a possible evolution of the ARF with the return period level is a path that has to be explored, copulas being a candidate for such a development (e.g. Singh and Zhang, 2007; Ariff et al., 2012).

It is also envisioned to test other IDAF model formulations based on alternative approaches for modelling the scale relationships, among which the method proposed by Overeem et al. (2010) seems of particular interest.

Appendix A

A brief explanation of the transition between the simple-scaling framework used to described the space–time scaling of maximum annual rainfall (Eq. 13) and the GEV model used to model the statistical distribution of these maxima (Eqs. 15 to 17) is given here.

The random variables $I(D_{\text{ref}},0)$ and $I(D,A)$ are modelled by a GEV model:

$$\text{Prob}\{I(D_{\text{ref}},0)\leq i(D_{\text{ref}},0)\}= \exp\left\{-\left[1+\xi(D_{\text{ref}},0)\left(\frac{i(D_{\text{ref}},0)-\mu(D_{\text{ref}},0)}{\sigma(D_{\text{ref}},0)}\right)\right]^{-\frac{1}{\xi(D_{\text{ref}},0)}}\right\} \tag{A1}$$

and

$$\text{Prob}\{(I(D,A)\leq i(D,A)\} = \exp\left\{-\left[1+\xi(D,A)\left(\frac{i(D,A)-\mu(D,A)}{\sigma(D,A)}\right)\right]^{-\frac{1}{\xi(D,A)}}\right\}. \tag{A2}$$

Letting $c=\lambda^{\eta}\times \text{ARF}(D,A)$, then Eq. (13) becomes

$$I(D,A)\overset{d}{=}I(D_{\text{ref}},0)\times c \tag{A3}$$

and if Eq. (13) holds, then

$$\begin{aligned}\text{Prob}\{I(D,A)\leq i(D,A)\} &= \text{Prob}\{I(D_{\text{ref}},0)\times c\leq i(D_{\text{ref}},0)\times c\}\\ &= \text{Prob}\{I(D_{\text{ref}},0)\leq i(D_{\text{ref}},0)\}\end{aligned} \tag{A4}$$

and

$$\text{Prob}\{I(D,A)\leq i(D,A)\} = \exp\left\{-\left[1+\xi(D_{\text{ref}},0)\left(\frac{i(D_{\text{ref}},0)-\mu(D_{\text{ref}},0)}{\sigma(D_{\text{ref}},0)}\right)\right]^{-\frac{1}{\xi(D_{\text{ref}},0)}}\right\}. \tag{A5}$$

By replacing $I(D_{\text{ref}},0)$ by $I(D,A)/c$ we obtain

$$\text{Prob}\{I(D,A)\leq i(D,A)\} = \exp\left\{-\left[1+\xi(D_{\text{ref}},0)\left(\frac{\frac{I(D,A)}{c}-\mu(D_{\text{ref}},0)}{\sigma(D_{\text{ref}},0)}\right)\right]^{-\frac{1}{\xi(D_{\text{ref}},0)}}\right\} \tag{A6}$$

or

$$\text{Prob}\{I(D,A)\leq i(D,A)\} = \exp\left\{-\left[1+\xi(D_{\text{ref}},0)\left(\frac{I(D,A)-\mu(D_{\text{ref}},0)\times c}{\sigma(D_{\text{ref}},0)\times c}\right)\right]^{-\frac{1}{\xi(D_{\text{ref}},0)}}\right\}. \tag{A7}$$

The equality between Eq. (A2) and Eq. (A7) gives

$$\mu(D,A)=\mu_{\text{ref}}\times c \tag{A8}$$

$$\sigma(D,A)=\sigma_{\text{ref}}\times c \tag{A9}$$

$$\xi(D,A)=\xi_{\text{ref}}. \tag{A10}$$

Equations (A8)–(A10) correspond to Eqs. (15)–(17) in the main text.

Acknowledgements. This research was partly funded by the LABEX OSUG@2020 (ANR10 LABX56), partly by the French project ESCAPE (ANR10-CEPL-005), and partly by IRD and INSU through the support to the AMMA-CATCH observing system. Gérémy Panthou's Ph.D. grant is funded by SOFRECO.

Edited by: P. Molnar

References

Ali, A., Lebel, T., and Amani, A.: Invariance in the Spatial Structure of Sahelian Rain Fields at Climatological Scales, J. Hydrometeorol., 4, 996–1011, 2003.

Allen, R. and DeGaetano, A.: Areal Reduction Factors for Two eastern United States Regions with High Rain-Gauge Density, J. Hydrol. Eng., 10, 327–335, 2005.

Ariff, N., Jemain, A., Ibrahim, K., and Wan Zin, W.: IDF relationships using bivariate copula for storm events in Peninsular Malaysia, J. Hydrol., 470–471, 158–171, 2012.

Asquith, W. H. and Famiglietti, J. S.: Precipitation areal-reduction factor estimation using an annual-maxima centered approach, J. Hydrol., 230, 55–69, 2000.

Awadallah, A.: Developing Intensity-Duration-Frequency Curves in Scarce Data Region: An Approach using Regional Analysis and Satellite Data, Engineering, 03, 215–226, 2011.

Bara, M., Kohnová, S., Gaál, L., Szolgay, J., and Hlavcoá, K.: Estimation of IDF curves of extreme rainfall by simple scaling in Slovakia, Contributions to Geophysics and Geodesy, 39, 187–206, 2009.

Bell, F.: The areal reduction factor in rainfall frequency estimation, Tech. rep. 35, Institute of hydrology, Natural Environment Research Council, UK, 1976.

Ben-Zvi, A.: Rainfall intensity-duration-frequency relationships derived from large partial duration series, J. Hydrol., 367, 104–114, 2009.

Bendjoudi, H., Hubert, P., Schertzer, D., and Lovejoy, S.: Interprétation multifractale des courbes intensité-duréé-fréquence des précipitations, C. R. Acad. Sci. Paris, Sciences de la terre et des planétes/Earth & Planetary Sciences, 325, 323–326, 1997.

Borga, M., Vezzani, C., and Fontana, G. D.: Regional rainfall depth-duration-frequency equations for an Alpine region, Natural Hazards, 36, 221–235, 2005.

Burlando, P. and Rosso, R.: Scaling and muitiscaling models of depth-duration-frequency curves for storm precipitation, J. Hydrol., 187, 45–64, 1996.

Castro, J. J., Cârsteanu, A. A., and Flores, C. G.: Intensity-duration-area-frequency functions for precipitation in a multifractal framework, Physica A: Statistical Mechanics and its Applications, 338, 206–210, 2004.

Ceresetti, D.: Structure spatio-temporelle des fortes précipitations: Applicatio á la région Cévennes Vivarais, Ph.D. thesis, Université de Grenoble, 2011.

Ceresetti, D., Anquetin, S., Molinié, G., Leblois, E., and Creutin, J.: Multiscale Evaluation of Extreme Rainfall Event Predictions Using Severity Diagrams, Weather Forecast., 27, 174–188, 2011.

Coles, S.: An introduction to statistical modeling of extreme values, Springer, London; New York, 2001.

De Michele, C., Kottegoda, N. T., and Rosso, R.: The derivation of areal reduction factor of storm rainfall from its scaling properties, Water Resour. Res., 37, 3247–3252, 2001.

De Michele, C., Kottegoda, N., and Rosso, R.: IDAF (intensity-duration-area frequency) curves of extreme storm rainfall: a scaling approach, Water Sci. Technol., 45, 83–90, 2002.

De Michele, C., Zenoni, E., Pecora, S., and Rosso, R.: Analytical derivation of rain intensity-duration-area-frequency relationships from event maxima, J. Hydrol., 399, 385–393, 2011.

Descroix, L., Genthon, P., Amogu, O., Rajot, J. L., Sighomnou, D., and Vauclin, M.: Change in Sahelian Rivers hydrograph: The case of recent red floods of the Niger River in the Niamey region, Global Planet. Change, 98–99, 18–30, 2012.

Di-Baldassarre, G., Montanari, A., Lins, H., Koutsoyiannis, D., Brandimarte, L., and Böschl, G.: Flood fatalities in Africa: From diagnosis to mitigation, Geophys. Res. Lett., 37, 1–5, 2010.

Gerold, L. and Watkins, D.: Short Duration Rainfall Frequency Analysis in Michigan Using Scale-Invariance Assumptions, J. Hydrol. Eng., 10, 450–457, 2005.

Gupta, V. and Waymire, E.: Multiscaling properties of spatial rainfall and river flow distributions, J. Geophys. Res., 95, 1999–2009, 1990.

Hosking, J. and Wallis, J. R.: Regional frequency analysis: an approach based on L-moments, Cambridge University Press, Cambridge UK, 1997.

Koutsoyiannis, D., Kozonis, D., and Manetas, A.: A mathematical framework for studying rainfall intensity-duration-frequency relationships, J. Hydrol., 206, 118–135, 1998.

Lebel, T., Sauvageot, H., Hoepffner, M., Desbois, M., Guillot, B., and Hubert, P.: Rainfall estimation in the Sahel: the EPSAT-NIGER experiment, Hydrol. Sci., 37, 201–215, 1992.

Menabde, M., Seed, A., and Pegram, G.: A simple scaling model for extreme rainfall, Water Resour. Res., 35, 335–339, 1999.

Mohymont, B., Demarée, G. R., and Faka, D. N.: Establishment of IDF-curves for precipitation in the tropical area of Central Africa – comparison of techniques and results, Nat. Hazards Earth Syst. Sci., 4, 375–387, doi:10.5194/nhess-4-375-2004, 2004.

Nhat, L., Tachikawa, Y., Sayama, T., and Takara, K.: A simple scaling charateristics of rainfall in time and space to derive intensity duration frequency relationships, Annu. J. Hydraul. Eng., 51, 73–78, 2007.

Overeem, A., Buishand, A., and Holleman, I.: Rainfall depth-duration-frequency curves and their uncertainties, J. Hydrol., 348, 124–134, 2008.

Overeem, A., Buishand, T. A., and Holleman, I.: Extreme rainfall analysis and estimation of depth-duration-frequency curves using weather radar, Water Resour. Res., 45, W10424, doi:10.1029/2009WR007869, 2009.

Overeem, A., Buishand, T. A., Holleman, I., and Uijlenhoet, R.: Extreme value modeling of areal rainfall from weather radar, Water Resour. Res., 46, W09514, doi:10.1029/2009WR008517, 2010.

Oyebande, L.: Deriving rainfall intensity-duration-frequency relationships and estimates for regions with inadequate data, Hydrol. Sci. J., 27, 353–367, 1982.

Oyegoke, S. and Oyebande, L.: A new technique for analysis of extreme rainfall for Nigeria, Environ. Res. J., 2, 7–14, 2008.

Panthou, G., Vischel, T., Lebel, T., Blanchet, J., Quantin, G., and Ali, A.: Extreme rainfall in West Africa: A regional modeling, Water Resour. Res., 48, 1–19, 2012.

Panthou, G., Vischel, T., Lebel, T., Quantin, G., Pugin, A.-C., Blanchet, J., and Ali, A.: From pointwise testing to a regional vision: An integrated statistical approach to detect nonstationarity in extreme daily rainfall. Application to the Sahelian region, J. Geophys. Res.-Atmos., 118, 8222–8237, 2013.

Papalexiou, S. and Koutsoyiannis, D.: Battle of extreme value distributions: A global survey on extreme daily rainfall, Water Resour. Res., 49, 187–201, 2013.

Ribstein, P. and Rodier, J.: La predetermination des crues sur des petits bassins saheliens inferieurs à 10 km^2, Tech. rep., ORSTOM, Laboratoire d'hydrologie, Montpellier, 1994.

Rodier, J. and Ribstein, P.: Estimation des caracteristiques de la crue decennale pour les petits basins versants du Sahel couvrant de 1 a 10 km^2, Tech. rep., ORSTOM, Laboratoire d'hydrologie, Montpellier, 1988.

Samimi, C., Fink, A. H., and Paeth, H.: The 2007 flood in the Sahel: causes, characteristics and its presentation in the media and FEWS NET, Nat. Hazards Earth Syst. Sci., 12, 313–325, doi:10.5194/nhess-12-313-2012, 2012.

Schertzer, D. and Lovejoy, S.: Physical Modeling and Analysis of Rain and Clouds by Anisotropic Scaling Multiplicative Processes, J. Geophys. Res., 92, 9693–9714, 1987.

Singh, V. and Zhang, L.: IDF Curves Using the Frank Archimedean Copula, J. Hydrol. Eng., 12, 651–662, 2007.

Soro, G., Goula, B. T. A., Kouassi, F., and Srohourou, B.: Update of Intensity-Duration-Frequency curves for precipitation of short durations in tropical area of West Africa (cote d'Ivoire), J. Appl. Sci., 10, 704–715, 2010.

Tarhule, A.: Damaging Rainfall and Flooding: The Other Sahel Hazards, Climatic Change, 72, 355–377, 2005.

Van-de Vyver, H. and Demarée, G. R.: Construction of Intensity-Duration-Frequency (IDF) curves for precipitation at Lubumbashi, Congo, under the hypothesis of inadequate data, Hydrological Sciences Journal – Journal des Sciences Hydrologiques, 55, 555–564, 2010.

Veneziano, D. and Langousis, A.: The areal reduction factor: A multifractal analysis, Water Resour. Res, 41, W07008, doi:10.1029/2004WR003765, 2005.

Veneziano, D., Furcolo, P., and Iacobellis, V.: Imperfect scaling of time and space-time rainfall, J. Hydrol., 322, 105–119, 2006.

Vischel, T., Quantin, G., Lebel, T., Viarre, J., Gosset, M., Cazenave, F., and Panthou, G.: Generation of High-Resolution Rain Fields in West Africa: Evaluation of Dynamic Interpolation Methods, J. Hydrometeorol., 12, 1465–1482, 2011.

Yu, P., Yang, T., and Lin, C.: Regional rainfall intensity formulas based on scaling property of rainfall, J. Hydrol., 295, 108–123, 2004.

12

Numerical modeling of flow and transport in the Bari industrial area by means of rough walled parallel plate and random walk models

Claudia Cherubini[1,2], Nicola Pastore[3], Dimitra Rapti[4], and Concetta I. Giasi[3]

[1]Department of Physics & Earth Sciences, University of Ferrara, Via Saragat 1-, 44122, Ferrara, Italy
[2]School of Civil Engineering, University of Queensland, St Lucia, Brisbane 4072, Australia
[3]DICATECh, Department of Civil, Environmental, Building Engineering, and Chemistry, Politecnico di Bari, Bari, Italy
[4]New Energies And environment Company (NEA) Via Saragat 1-, 44122 Ferrara, Italy

Correspondence: Claudia Cherubini (claudia.cherubini@unife.it) and Nicola Pastore (nicola.pastore@poliba.it)

Abstract. Modeling fluid flow and solute transport dynamics in fractured karst aquifers is one of the most challenging tasks in hydrogeology.

The present study investigates the hotspots of groundwater contamination in the industrial area of Modugno (Bari – southern Italy), where the limestone aquifer has a fractured and karstic nature.

A rough walled parallel plate model coupled with a geostatistical analysis to infer the values of the equivalent aperture has been implemented and calibrated on the basis of piezometric data. Using the random walk theory, the steady-state distribution of hypothetical contamination with the source at the hotspot has been carried out, reproducing a pollution scenario which is compatible with the observed one. From an analysis of the flow and transport pattern it is possible to infer that the anticline affecting the Calcare di Bari formation in direction ENE–WSW influences the direction of flow as well as the propagation of the contaminant.

The results also show that the presence of nonlinear flow influences advection, in that it leads to a delay in solute transport with respect to the linear flow assumption. This is due to the non-constant distribution of solutes according to different pathways for fractured media which is related to the flow rate.

1 Introduction

The characterization and the description of phenomena that involve fractured aquifers, especially when considered in relation to water resource exploitation, are important issues because fractured aquifers serve as the primary source of drinking water for many areas of the world. In fractured rock aquifers, groundwater is stored in the fractures, joints, bedding planes, and cavities of the rock mass. The ability of a fracture to transmit water as well as contaminants depends primarily on the size of the opening, or the fracture aperture.

The parallel plate model is widely used to simulate flow in a fracture due to its simplicity in idealizing a fracture. Many researchers (Baker, 1955; Huitt, 1956; Snow, 1965, 1970; Gale, 1977) have used flow between smooth parallel plates as a model for flow in fractures. The solution to the Navier–Stokes equation for flow between parallel plates, known as plane Poiseuille flow, has been known to fluid mechanicians since the nineteenth century.

Witherspoon et al. (1980) and Elliott and Brown (1988) suggested that a factor should be introduced into the parallel plate theory to take account of the effects of joint surface properties.

Zimmermann and Bodvarsson (1996) discussed the problem of fluid flow through a rock fracture within the context of fluid mechanics. The derivation of the "cubic law" was given as the solution to the Navier–Stokes equations for flow between smooth, parallel plates. They analyzed the various geometric and kinematic conditions that are necessary in order for the Navier–Stokes equations to be replaced by the more tractable lubrication or HeleShaw equations and reviewed various analytical and numerical results pertaining to the problem of relating the effective hydraulic aperture to the statistics of the aperture distribution.

Some researchers proposed a variable aperture model, stating that it is better adapted to describing flow and transport channeling effects than a parallel plate model (Neretnieks et al., 1982; Bourke, 1987; Pyrak-Nolte, 1988; Tsang and Tsang, 1989; Tsang et al., 2001) where fracture apertures can be described by normal (Lee et al., 2003), lognormal (e.g., Keller, 1998; Keller et al., 1999), or gamma distributions (Tsang and Tsang, 1987), or a self-affine scale invariance (Plouraboué et al., 1995).

Neuzil and Tracy (1981) presented a model for flow in fractures where the flow is envisioned as occurring in a set of parallel plate openings with different apertures whose distribution was lognormal and used a modified Poiseuille equation.

They showed that the flow conformed to the cubic law and also that the maximum flow occurs through the largest apertures, thereby emphasizing that flow occurs through preferred paths. Thus, in their analysis, the flow depended on the tail of the frequency distribution.

Tsang and Tsang (1987) proposed a theoretical approach to interpret flow in a tightly fractured medium in terms of flow through a system of statistically equivalent one-dimensional channels of variable aperture. The channels were statistically equivalent in the sense that the apertures along each flow channel are generated from the same aperture density distribution and spatial correlation length.

Oron and Berkowitz (1998) have examined the validity of applying the "local cubic law" (LCL) to flow in a fracture which is bounded by impermeable rock surfaces. A two-dimensional order-of-magnitude analysis of the Navier–Stokes equations yields three conditions for the applicability of LCL flow, as a leading-order approximation in a local fracture segment with parallel or nonparallel walls. These conditions demonstrate that the "cubic law" is valid provided that aperture is measured not on a point-by-point basis, but rather as an average over a certain length. Experimental work by Plouraboué et al. (2000) in self-affine rough fractures with various translations of the opposing fracture surfaces indicated that heterogeneity in the flow field caused deviations from the parallel plate model for fracture flow. Some researchers often find it convenient to represent aperture fields in terms of equivalent aperture in the parallel plate model (Zheng et al., 2008).

Brush and Thomson (2003) developed three-dimensional flow models to simulate fluid flow through various random synthetic rough walled fractures created by combining random fields of aperture and the mean wall topography or midsurface, which quantifies undulation about the fracture plane.

The total flow rates from three-dimensional Stokes simulations were within 10 % of LCL simulations with geometric corrections for all synthetic fractures. Differences between the Navier–Stokes (NS) and Stokes simulations clearly demonstrated that inertial forces can significantly influence the internal flow field within a fracture and the total flow rate across a fracture.

Klimczak et al. (2010) carried out flow simulations through fracture networks using the discrete fracture network model (DFN) where flow was modeled through fracture networks with the same spatial distribution of fractures for correlated and uncorrelated fracture length-to-aperture relationships. Results indicate that flow rates are significantly higher for correlated DFNs. Furthermore, the length-to-aperture relations lead to power-law distributions of network hydraulic conductivity which greatly influence equivalent permeability tensor values. These results confirm the importance of the correlated square root relationship of displacement to length scaling for total flow through natural opening-mode fractures and, hence, emphasize the role of these correlations for flow modeling.

Wang et al. (2015) developed and tested a modified LCL (MLCL) taking into account local tortuosity and roughness, and works across a low range of local Reynolds numbers. The MLCL was based on (1) modifying the aperture field by orienting it with the flow direction and (2) correcting for local roughness changes associated with local flow expansion/contraction. In order to test the MLCL, they compared it with direct numerical simulations with the Navier–Stokes equations using real and synthetic three-dimensional rough walled fractures, previously corrected forms of the LCL, and experimental flow tests. The MCL proved to be more accurate than previous modifications of the LCL.

The continuous time random walk (CTRW) approach provides a versatile framework for modeling (non-Fickian) solute transport in fractured media.

Berkowitz et al. (2001) examined a set of analytical solutions based on the CTRW approach to analyze breakthrough data from tracer tests to account for non-Fickian (or scale-dependent) dispersion behavior that cannot be properly quantified by using the advection–dispersion equation.

Cortis and Birkholzer (2008) developed a macroscopic model based on the CTRW framework, to characterize the interaction between the fractured and porous rock domains by using a probability distribution function of residence times. They presented a parametric study of how CTRW parameters evolve, describing transport as a function of the hydraulic conductivity ratio between fractured and porous domains.

Srinivasan et al. (2010) presented a particle-based algorithm that treats a particle trajectory as a subordinated stochastic process that is described by a set of Langevin equations, which represent a CTRW. They used convolution-based particle tracking (CBPT) to increase the computational efficiency and accuracy of these particle-based simulations. The combined CTRW–CBPT approach allows us to convert any particle tracking legacy code into a simulator capable of handling non-Fickian transport.

Dentz et al. (2015) developed a general CTRW approach for transport under radial flow conditions starting from the random walk equations for the quantification of non-local solute transport induced by heterogeneous flow distributions and by mobile–immobile mass transfer processes. They observed power-law tails of the solute breakthrough for broad distributions of particle transit times and particle trapping times. The combined model displayed an intermediate regime, in which the solute breakthrough is dominated by the particle transit times in the mobile zones, and a late time regime that is governed by the distribution of particle trapping times in immobile zones.

The present study is aimed at analyzing the scenario of groundwater contamination of the industrial area of Modugno (Bari – southern Italy) where the limestone aquifer has a fractured and karstic nature.

Previous studies carried out in the same aquifer have applied different conceptual models to model fluid flow and contaminant transport.

Cherubini (2008) applied the discrete feature approach (Diersch, 2002) where the three-dimensional geometry of the subsurface domain describing the matrix structure was combined by interconnected two-dimensional and one-dimensional discrete feature elements in two dimensions in order to simulate, respectively, fractures and karstic cavities in the Bari limestone aquifer. The fracture distribution was inferred from a nonparametric geostatistical analysis (indicator Kriging) of fracture frequency data which had been derived by Rock Quality Designation (RQD) (Priest and Hudson, 1976) data of the contaminated area of the ex Gasometer.

Cherubini et al. (2008) compared the flow modeling results of the previous work with those from a new hydrogeological reconstruction of the heterogeneities in the same aquifer by means of multiple realizations conditioned to borehole data (RQD population), in order to obtain a three-dimensional distribution of fracture frequency, cavities, and terra rossa lenses.

Cherubini and Pastore (2010) applied the nested sequential indicator simulation algorithm to represent the geological architecture of the Bari limestone aquifer which provided reliable prediction of fluid flow. According to phenol transport, the presence of preferential pathways was detected.

Cherubini et al. (2013)a realized a three-dimensional flow model of the Bari limestone aquifer supported by a detailed local-scale geologic model realized by means of sequential indicator simulation (SIS) of lithofacies unit sequences. In this study, a lumped parameter approach was used and calibrated on the groundwater discharge and global hydraulic gradient where fluid flow in fractures was represented by the cubic law, and the Darcy–Weisbach equation was used to estimate the resistance term in the karst network.

Masciopinto et al. (2010) adopted a conceptual model consisting of a three-dimensional parallel set of horizontal planar fractures in between rock layers, each fracture having a variable aperture generated by a stationary random field conditioned to the data derived from pumping-tracer tests. The particle tracking solution was combined with the PHREEQC-2 results to study two-dimensional laminar/non-laminar flow and reactive transport with biodegradation in each fracture of the conceptual model.

Masciopinto and Palmiotta (2013) derived new equations of the fracture aperture as functions of a tortuosity factor to simulate fluid flow and pollutant transport in fractured aquifers. MODFLOW/MT3DMS water velocity predictions were compared with those obtained using a specific software application which solves flow and transport problems in a three-dimensional set of parallel fissures. The results of a pumping/tracer test carried out in a fractured limestone aquifer in Bari (southern Italy) have been used to calibrate advective–dispersive tracer fluxes given by the applied models. Successful simulations of flow and transport in the fractured limestone aquifer were achieved by accommodating the new tortuosity factor in models whose importance lies in the possibility of switching from a discrete to a continuum model by taking into account the effective tracer velocity during flow and transport simulations in fractures.

Masciopinto and Visino (2017) carried out filtration tests on a set of 16 parallel limestone slabs with thicknesses of about 1 cm where rough surfaces and variable fracture apertures had been artificially created. The experimental filtration results suggest that model simulations of perturbed virus transport in fractured soils need to also consider pulse-like sources and sinks of viruses. This behavior cannot be simulated using conventional model equations without including a new kinetic model approach.

The present work focuses on the investigation of the hotspots of aquifer contamination in order to infer the location of the sources.

A rough walled parallel plate model has been implemented and calibrated on the basis of piezometric data and has coupled a geostatistical analysis to infer the values of the equivalent aperture.

The current study introduces a novel approach for simulating flow and transport in fractured aquifers by means of combining a rough walled parallel plate model for the flow simulation coupled with inverse modeling and geostatistical analysis to infer the values of the equivalent aperture together with the random walk theory to reproduce the scenario of contamination.

2 Geological and hydrogeological framework

It is well known that hydraulic properties and consequently fluid circulation and contaminant propagation in carbonate rocks are strongly influenced by the degree of rock fracturing and, in general, the presence of mechanical discontinuities, like faults, joints, or other tectonic elements such as syncline or anticline axes (Caine et al., 1996; Caine and Forster, 1999; Antonellini et al., 2014; Billi et al., 2003). Also, the deformation mechanisms are mainly controlled by the physico-chemical properties of rocks, which are, in turn, the result of different composition, depositional setting, and diagenetic evolution (Zhang and Spiers, 2005; Rustichelli et al., 2012).

From the geological point of view, the investigated area is located in the Murge Plateau corresponding to a broad antiformal structure oriented WNW–ESE and represents the bulging foreland of the Pliocene–Pleistocene Southern Apennines orogenic belt (Pieri et al., 1997; Doglioni, 1994; Forster and Evans, 1991; Korneva et al., 2014; Parise and Pascali, 2003).

The stratigraphy of the Murge area consists of a Variscan crystalline basement topped by 6–7 km thick Mesozoic sedimentary cover (represented by the Calcare di Bari formation) followed by relatively thin and discontinuous Cenozoic and Quaternary deposits (Calcareniti di Gravina formation). Figure 1 shows the simplified geological map of the area of Bari.

2.1 Calcare di Bari formation (Cretaceous)

The Calcare di Bari succession consists of biopeloidal and peloidal wackestones/packstones alternating with stromatolithic bindstones with frequent intercalations of dolomitic limestones and grey dolostones. The formation shows a thickness of about 470 m. Most of the Calcare di Bari formation shows facies features related to peritidal environments; only the upper part suggests a relatively more distal and deeper environment belonging to an external platform setting (Pieri et al., 2012; Fig. 1).

This succession appears stratified and fissured and, where it does not show tectonic discontinuities, it is characterized by a subhorizontal or slightly inclined lying position. This formation is subjected to the complex and relevant karstic phenomena that locally lead to the formation of cavities of different shapes and sizes, partially or completely filled by "terra rossa" deposits. The degree of fracturing degree affecting the Calcare di Bari formation is quite variable and mainly depends on the geological and structural (faults, anticline axis, etc.) evolution of the area, including faulting and folding. Also, the distribution of the local measurement of the RQD index is confirmed by the variability of the electrical resistivity along geoelectrical profiles (with lengths from 500 to 1000 m) and from the propagations of the P and S waves (seismic measurements; length of about 1000 m).

On the basis of borehole and in situ surveys, carried out by private companies, the following was observed.

- The fracturing degree of the Calcare di Bari formation is quite variable and it is expressed by the RQD values that vary between 16 % and 25 % (maximum borehole depths: 30 m). Based on the classification system of Deere and Deere (1988) the rock mass is of "very poor rock quality" (RQD < 25 %).
- Medium values of the Rock Mass Rating (RMR) of about 36, indicate, following the Bieniawski classification (1989), very poor rock mass (class IV).
- In addition, profiles of the electrical resistivity (depth < 30 m) allow us to emphasize the presence of very variable electric resistivity values with variations between 100 (low fracture carbonates rocks) and 1700 Ohm m for very fractured formations, with local values on the order of 3–4000 Ohm m in the case of underground cavities.
- Similarly, the velocities of seismic waves P and S have average values on the order of 1300 and 800 m s^{-1}, respectively, in highly fractured limestones, and 2300 (P) and 1400 (S) m s^{-1} for compact formations.

2.2 Calcareniti di Gravina formation (lower Pleistocene)

This unit unconformably lies on the Calcare di Bari formation. Its thickness varies from a few meters to 20 m and its depositional environments are related to offshore settings. The lower boundary is transgressive and is locally marked by reddish residual deposits and/or by brackish silty deposits passing upward to shallow water calcarenite rich in bioclasts.

As regards the structural features of these deposits, it is possible to observe that the anticline affecting the Cretaceous succession of the Calcare di Bari formation with an ENE–WSW axial direction (Fig. 1) causes a partial diversion of the water courses, whose path seems to be also influenced by some NE–SW fault (NE of Modugno). The former phenomenon is due to the antithetically dipping flanks of the gentle fold, while the latter effect is likely a consequence of the denser fracturing along the shear zone and hence the increased erodibility of the local outcropping limestone enhancing the water flow concentration.

In general, the limestone bedrock hosts a wide and thick aquifer due to a diffuse rock fracturing and the karstic phenomena.

Moreover, the irregular spatial distribution of the fractures and karstic channels makes the Bari aquifer very anisotropic. The average hydraulic conductivity of this aquifer is generally estimated to be 10^{-3} to 10^{-4} m s^{-1}.

The groundwater flows toward the sea, under a low gradient, in different subparallel fractured layers separated by compact (i.e., not fractured) rock blocks.

In proximity to the coast, the carbonate (Mesozoic) stratum contains freshwater flowing in phreatic conditions and floating on underlying saltwater of continental intrusion. The location of the transition zone between freshwater and salt water has thickness and position variable and changes over time depending on the distribution of the hydrostatic pressures of the system.

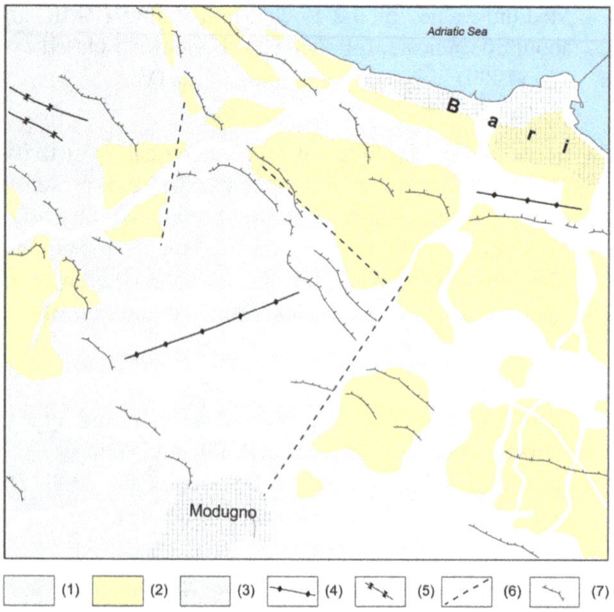

Figure 1. Simplified geological map of the area of Bari: (1) Calcare di Bari formation (Cretaceous); (2) Calcareniti di Gravina formation (Lower Pleistocene); (3) hydrographic network; (4) anticlinal axis; (5) syncline axis; (6) fault (uncertain); (7): escarpment.

3 Hydrologic and hydrogeologic water budget

The effective infiltration has been estimated by means of the hydrologic and hydrogeologic water budget of the subtended basin. Climatic data registered in the thermopluviometric stations present in the area have been elaborated and the average rainfall module and the monthly evapotranspiration have been calculated for the 3 decades 1974–2005.

Twelve climatic stations have been considered (Bari – hydrographic station, Bari – observatory station, Bitonto, Grumo Appula, Adelfia, Casamassima, Mercadante, Ruvo di Puglia, Corato, Altamura, Santeramo, Gioia del Colle) and for each station the monthly rainfall and evapotranspiration map has been realized by means of the *inverse distance weighting* algorithm. The latter has been estimated by means of the Thornthwaite method, applying a crop coefficient of 0.40.

The hydrologic and hydrogeologic basins have been defined on the basis of literature data and the regional thematic cartography.

The lithotypes in the study area are principally limestones and calcarenites with secondary permeability, characterized by a high transmissivity. The zones in proximity to tectonic structures create preferential flow paths, but at the same time generate a dismemberment of the aquifer that could not be able to feed the flow downstream. Because of that it proves to be difficult to carry out a zonation of recharge areas, and therefore a constant runoff coefficient of 0.10 has been considered for the whole basin. In Fig. 2 the map of the (a) annual precipitation and (b) estimated evapotranspiration evaluated for the hydrological basin of the study area is shown.

4 Well-performance tests: step-drawdown tests

Ninety-eight long-term step-drawdown hydraulic tests have been analyzed in the study area.

A step-drawdown test is a single-well test in which the well is pumped at a low constant discharge rate until the drawdown within the well stabilizes.

Step-drawdown tests can be used to evaluate the characteristics of the well and its immediate environment. Unlike the aquifer test, it is not designed to produce reliable information concerning the aquifer, even though it is possible to estimate the transmissivity of the immediate surroundings of the catchment. This test determines the critical flow rate of the well, as well as the various head losses and drawdowns as functions of pumping rates and times. Finally, it is designed to estimate the well efficiency, to set an exploitation pumping rate, and to specify the depth of installation of the pump.

The total drawdown at a pumping well is given by

$$s = (A_1 + A_2) \cdot Q + B \cdot Q^2, \tag{1}$$

where s (L) represents the registered drawdown, Q ($\mathrm{L^3\,T^{-1}}$) the pumped flow rate, A_1 ($\mathrm{T\,L^{-2}}$) the linear aquifer loss coefficient, and A_2 ($\mathrm{T\,L^{-2}}$), e, and B ($\mathrm{T^2\,L^{-5}}$) are, respectively, the linear and nonlinear well-loss coefficients.

This equation can be made explicit in terms of aquifer transmissivity T ($\mathrm{L^2\,T^{-1}}$), the transmissivity of the damage zone T_{SKIN} ($\mathrm{L^2\,T^{-1}}$), and the nonlinear term β ($\mathrm{T^2\,L^{-4}}$) (Cherubini and Pastore, 2011):

$$s = \left[\frac{1}{T2\pi}\ln\left(\frac{R}{r_{\mathrm{w}}}\right) + \frac{1}{2\pi}\left(\frac{1}{T_{\mathrm{SKIN}}} - \frac{1}{T}\right)\ln\left(\frac{r_{\mathrm{SKIN}}}{r_{\mathrm{w}}}\right)\right]Q + \left[\frac{\beta}{4\pi^2}\left(\frac{1}{r_{\mathrm{w}}} - \frac{1}{R}\right)\right]Q^2, \tag{2}$$

where r_{w} (L) represents the well radius, r_{SKIN} (L) the radius of the damage zone, and R (L) the radius of influence of the well.

The total drawdown is formed from three components: the hydraulic component of the aquifer assuming a valid Thiem function, a skin function presented by Cooley and Cunningham (1979) assuming that the transmissivity and the radius of the damage zone are, respectively, equal to $T_{\mathrm{SKIN}} = T/2$ e $r_{\mathrm{SKIN}} = 2r_{\mathrm{w}}$, and a contribution related to nonlinear losses introduced by Wu (2001).

The radius of influence of the well is obtained by means of the Sichart equation:

$$R = 3000 \cdot s \cdot \sqrt{K}. \tag{3}$$

In Fig. 3 is reported the statistical distribution of the estimated transmissivity values along the study area.

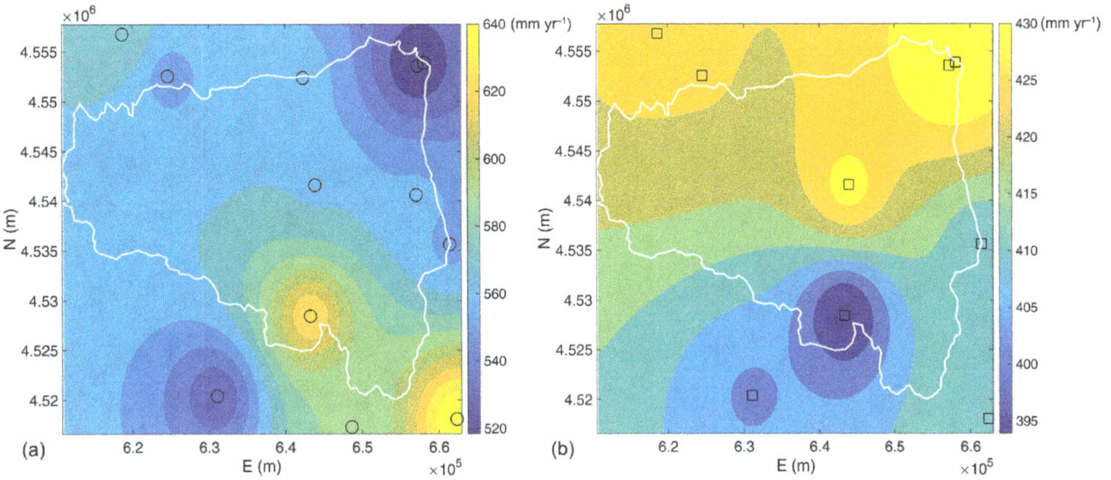

Figure 2. Map of **(a)** annual precipitation and **(b)** estimated evapotranspiration evaluated for the hydrological basin of the study area.

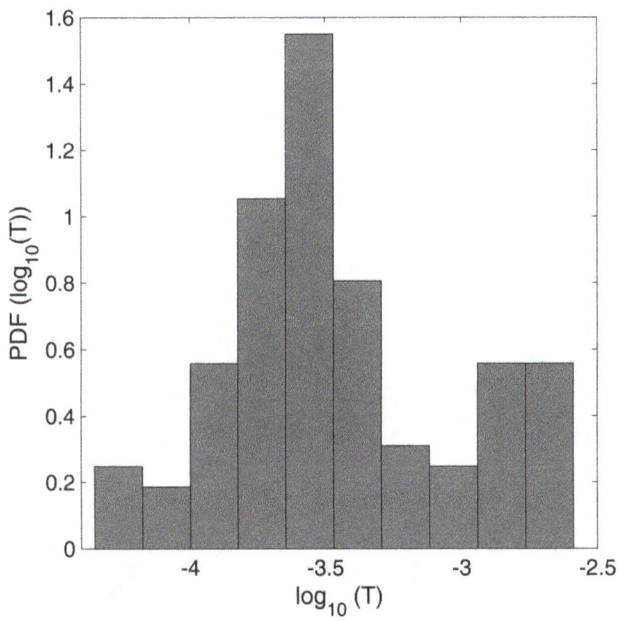

Figure 3. Statistical distribution of $\log_{10}(T)$.

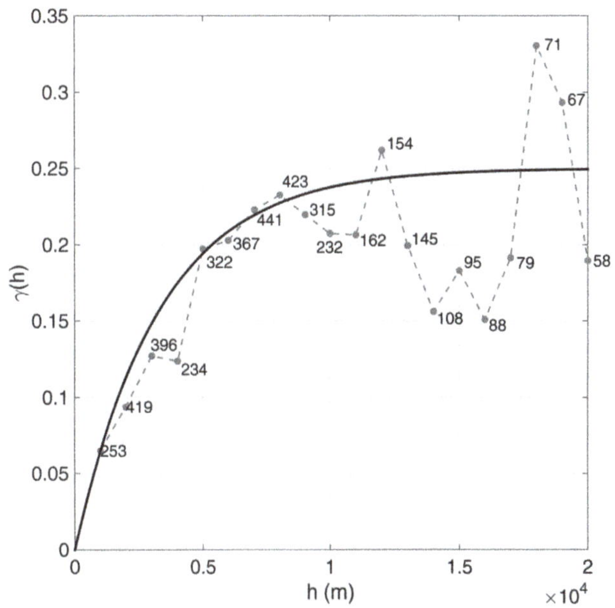

Figure 4. Omnidirectional experimental variogram fitted with an exponential model; sill = 1.2, range = 10000 m.

5 Linear model of regionalization of transmissivity

The geostatistical analysis has been carried out on the $\log_{10}$ transmissivity values using the S-GemS open-source code (Remy, 2004).

The experimental variogram, which provides a description of how the data are related (correlated) with distance, has been calculated (Fig. 4). Because the Kriging algorithm requires a positive definite model of spatial variability, the experimental variogram cannot be used directly. Instead, a model must be fitted to the data to approximately describe the spatial continuity of the data. An exponential model has been used to fit the experimental variogram described by the function:

$$\gamma(h) = C\left[1 - \exp\left(-\frac{h}{a}\right)\right], \quad (4)$$

where C represents the variance (sill), h (L) the lag, and a (L) the correlation length (range). In our case C assumes a value of 1.2 and a a value of 10000 m.

Figure 5 shows the ordinary Kriging interpolation of $\log_{10}(T)$.

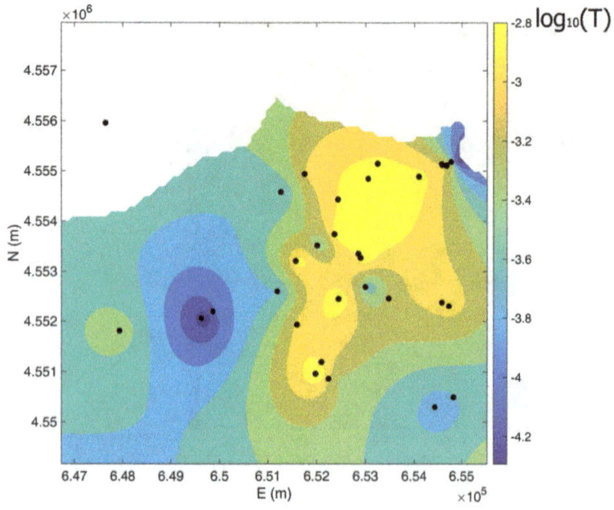

Figure 5. Ordinary Kriging interpolation of $\log_{10}(T)$.

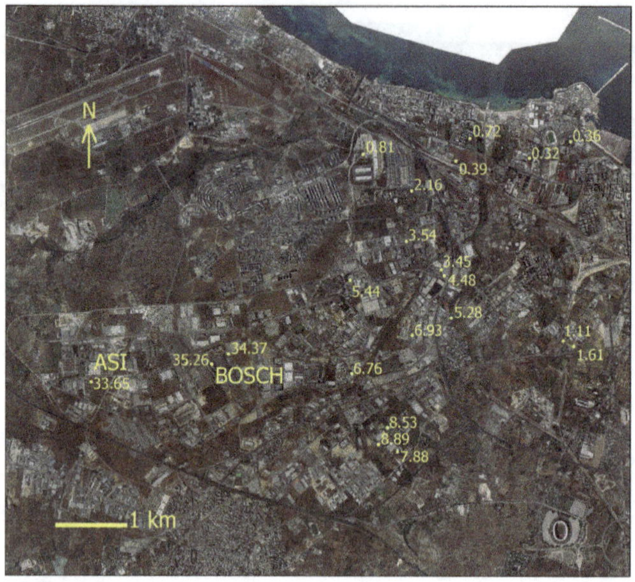

Figure 6. Measured piezometric heads (m, slm) from the February 2012 monitoring campaign.

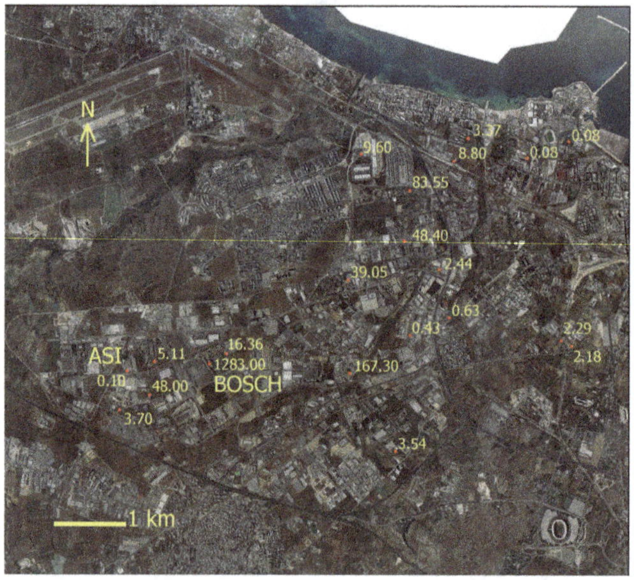

Figure 7. Location of the detected contamination by PCE ($\mu g\,L^{-1}$).

6 Analysis of piezometric data

Figure 6 shows the spatial distribution of hydraulic heads on the basis of the 2012 measurement campaign. A global trend in the direction of groundwater flow from SW to NE is evident. A relevant aspect is the presence of high hydraulic head values in proximity to ASI and Bosch wells.

Possible explanations for the increase in hydraulic gradient are (1) lower transmissivity as shown through the step drawdown tests; (2) the transition from a more permeable outcropping lithotype to a less permeable one resulting in a decrease in the effective infiltration; (3) the presence of sinkholes and fissures at the surface giving rise to a point source recharge; and (4) hydraulic disconnection due to lower interconnectivity of the fracture system.

The aquifer transmissivity in that zone is on the order of $10^{-5}\,m^2\,s^{-1}$. The trend observed in the hydraulic gradient confirms the increase in the aquifer transmissivity from upstream to downstream: in fact, the tests carried out in proximity to the coast have returned a transmissivity value of 10^{-} to $10^{-3}\,m^2\,s^{-1}$.

7 Analysis of the scenario of contamination for the study area

The various monitoring campaigns carried out have shown a contamination by chlorinated aliphatic hydrocarbons which, unlike petroleum products, are denser than water and can exist as dense non-aqueous phase liquids (DNAPLs).

The presence of two hotspot areas has been detected, located upstream of the groundwater flow, coherently with the state of contamination detected downstream.

Figure 7 shows the location of the detected contamination ($\mu g\,L^{-1}$).

The pollution indicator has been chosen on the basis of the toxicologic and cancirogenic parameters, the solubility, the sorption coefficient, and the maximum detected contaminant concentration. On the basis of the results of this screening, the tetrachloroethylene (PCE) has the highest concentration, as well as low values of reference dose factors (RDFs) and slope factors (SFs).

8 Parallel rough walled fracture model

The simplest model of flow through rock fractures is the parallel plate model (Huitt, 1956; Snow, 1965) which conceptualizes the fractured medium as made by a set of smooth parallel plates with the same hydraulic aperture b (L) that are separated by a uniform distance. This is actually the only geometrical fracture model for which an exact calculation of the hydraulic conductivity is possible.

Natural fractures present rough walls and complex geometries. Nonlinear flow may occur through rough walled rock fractures; as a consequence, the inertial effect dominates the flow dynamics, giving rise to a deviation from Darcy's law. Fluid flow through a set of natural fracture planes can be expressed using the Darcy–Weisbach equation:

$$\frac{\mathrm{d}h}{\mathrm{d}x} = -\frac{f}{D}\frac{v^2}{2g}, \tag{5}$$

where D (L) represents the hydraulic diameter ($2b$ for the parallel plate model), f is the Darcy–Weisbach coefficient, h (L) is the hydraulic head, x (L) is the distance, and v ($\mathrm{L\,T^{-1}}$) is the average velocity in fracture calculated as

$$v = \frac{q}{n_\mathrm{f} b}, \tag{6}$$

where $q(\mathrm{L^2\,T^{-1}})$ is the volumetric flow rate per unit length of fracture and n_f (–) is the number of fractures.

The Darcy–Weisbach equation can be rewritten in terms of volumetric flow per unit length:

$$\overline{q} = -\left[n_\mathrm{f} b \frac{\sqrt{\frac{4b}{f} g}}{\sqrt{|\nabla h|}} \right] \nabla h. \tag{7}$$

The term in square brackets represents the equivalent hydraulic transmissivity $T_\mathrm{eq}(f, \nabla h)$ of the n_f rough walled fractures.

The Darcy–Weisbach coefficient or friction factor depends on the flow regime. In the case of a smooth walled fracture and linear flow regime, f is equal to

$$f = \frac{96}{Re}, \tag{8}$$

where Re represents the Reynolds number:

$$Re = \frac{\rho v D}{\mu}. \tag{9}$$

Substituting Eq. (8) into Eq. (7), the cubic law (Witherspoon et al., 1980) where q is proportional to the cubic power of the fracture aperture is obtained:

$$q = n_\mathrm{f} \frac{\rho g}{\mu} \frac{b^3}{12}. \tag{10}$$

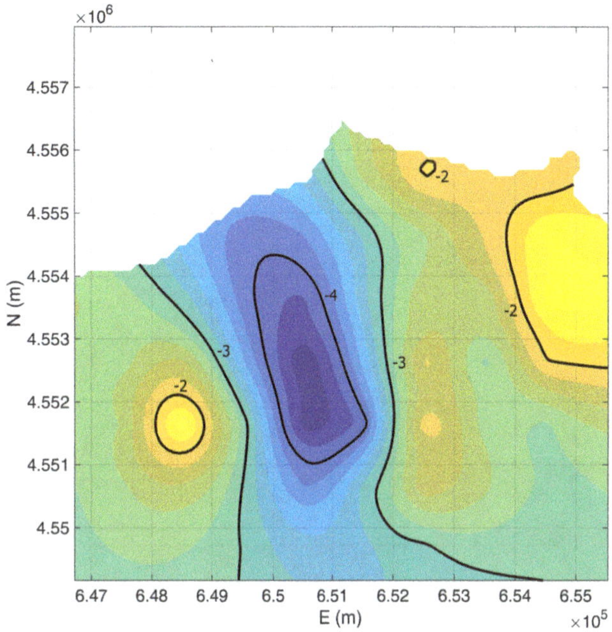

Figure 8. Map of log10 of aquifer transmissivity determined by means of the inverse modeling algorithm.

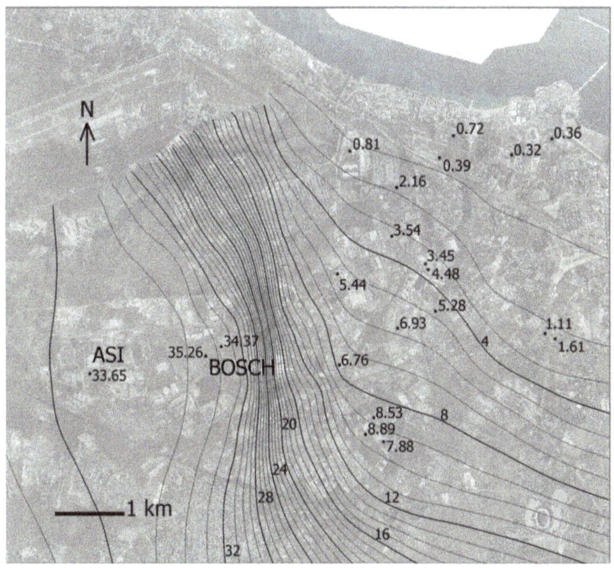

Figure 9. Map of simulated hydraulic heads.

The cubic law is not always adequate to represent the flow process in natural fractures, so a deviation from linearity can be observed.

The friction factor depends on the flow regime described by the Reynolds number and can be represented by the following relationship found by Nazridoust at al. (2006):

$$f = \frac{123}{Re}\left(1 + 0.12 Re^{0.687}\right). \tag{11}$$

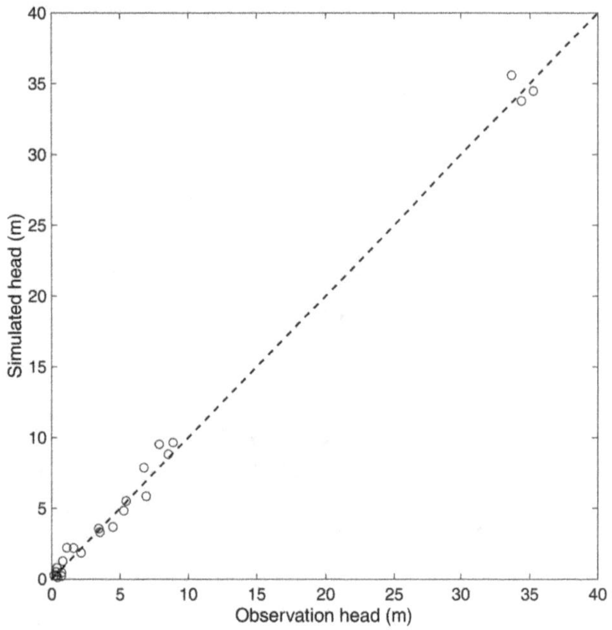

Figure 10. Graph of the calibration.

9 Inverse flow modeling

Inverse modeling is a technique used to estimate unknown model parameters using as input data punctual values of the state variables (hydraulic head flow). Generally, in real problems the number of parameters to estimate (n) is higher than the number of measured values (m). For example, this is the case with mapping hydraulic transmissivity values varying continuously in space.

For underdetermined inverse problems of this kind the objective function (L) can be written in this way:

$$L(\boldsymbol{y},\boldsymbol{s}) = L_{\text{fitness}}(\boldsymbol{y},\boldsymbol{s}) + L_{\text{penalty}}(\boldsymbol{s}), \tag{12}$$

where $\boldsymbol{s}$ represents the vector of measured values of state variables (e.g. hydraulic transmissivity), and $\boldsymbol{y}$ represents the vector of parameter values.

The fitness function responds to maximum likelihood criteria between the observed and simulated values and can be written as

$$L_{\text{fitness}}(\boldsymbol{y},\boldsymbol{s}) = (\boldsymbol{y} - \boldsymbol{h}(\boldsymbol{s}))^T \mathbf{R}^{-1} (\boldsymbol{y} - \boldsymbol{h}(\boldsymbol{s})), \tag{13}$$

where h represents the model that, starting from the parameter vector, estimates the state variable, and $\mathbf{R}$ is the measurement error covariance matrix. Generally this function can be reduced to the square root of the sum of the squared difference between the measured and simulated RMSE:

$$L_{\text{fitness}} = \frac{\|\boldsymbol{y} - \boldsymbol{h}(\boldsymbol{s})\|^2}{\Delta H^2}, \tag{14}$$

where ΔH represents a parameter of accuracy of observed data.

The penalty function is used to discriminate the solutions with values of the fitness function comparable by means of geostatistical criteria (Kitanidis, 1995):

$$L_{\text{penalty}} = (\boldsymbol{s} - \boldsymbol{X\beta})^T \mathbf{Q}^{-1} (\boldsymbol{s} - \boldsymbol{X\beta}). \tag{15}$$

where $\mathbf{Q}$ represents the spatial covariance matrix, $\boldsymbol{X}$ is a unit vector, and β is the mean of the values of the parameters. The penalty function can be rewritten by eliminating β:

$$L_{\text{penalty}} = \boldsymbol{s}^T \boldsymbol{G}\boldsymbol{s}\boldsymbol{G} \equiv \mathbf{Q}^{-1} - \frac{\mathbf{Q}^{-1}\boldsymbol{X}\boldsymbol{X}^T\mathbf{Q}^{-1}}{\boldsymbol{X}^T\mathbf{Q}^{-1}\boldsymbol{X}}. \tag{16}$$

The common assumption is that the spatial distribution of the parameters follows the geostatistical distribution defined by the variogram. Under this hypothesis the covariance matrix present in the penalty function can be defined as

$$Q_{ij} = 2\gamma\left(\left|x_i - x_j\right|\right) i, j = 1, \ldots, n. \tag{17}$$

10 Solute transport modeling

Solute transport in fracture neglecting the effect of matrix diffusion and the chemical reactions can be described by the following advection–dispersion equation:

$$\frac{\partial c}{\partial t} + \overline{v} \cdot \nabla c = \nabla \cdot (\mathbf{D}\nabla c), \tag{18}$$

where c ($\mathrm{M\,L^{-3}}$) is the concentration of a solute and $\mathbf{D}$ ($\mathrm{L^2\,T^{-1}}$) is the symmetric dispersion tensor with the following components:

$$\begin{aligned} Dxx &= \left(\alpha_{\mathrm{L}} v_x^2 + \alpha_{\mathrm{T}} v_y^2\right) / |, v| \\ Dyy &= \left(\alpha_{\mathrm{T}} v_x^2 + \alpha_{\mathrm{L}} v_y^2\right) / |v|, \\ Dxy &= (\alpha_{\mathrm{L}} - \alpha_{\mathrm{T}}) v_x v_y / |v|, \end{aligned} \tag{19}$$

where α_{L} (L) and α_{T} (L) are the longitudinal and transverse dispersivities, respectively.

For pure advective transport the particle moves along the flow lines. In order to represent dispersion phenomena, the random walk method adds a random displacement to each particle, independently of the other particles, in addition to advective displacement.

For a given time step Δt, considering the tensorial nature of the dispersion and the spatially variable velocity field, each particle moves according to

$$\begin{aligned} x_p(t+\Delta t) &= x_p(t) + v'_x \Delta t + Z_1\sqrt{2D_L\Delta t}\frac{v_x}{|v|} \\ &\quad - Z_2\sqrt{2D_T\Delta t}\frac{v_y}{|v|}, \\ y_p(t+\Delta t) &= y_p(t) + v'_y \Delta t + Z_1\sqrt{2D_L\Delta t}\frac{v_y}{|v|} \\ &\quad + Z_2\sqrt{2D_T\Delta t}\frac{v_x}{|v|}, \end{aligned} \tag{20}$$

with

$$v'_x = v_x + \frac{\partial D_{xx}}{\partial x} + \frac{\partial D_{xy}}{\partial y},$$
$$v'_y = v_y + \frac{\partial D_{xy}}{\partial x} + \frac{\partial D_{yy}}{\partial y},$$
$$D_L = \alpha_L |v|,$$
$$D_T = \alpha_T |v|, \quad (21)$$

where Z_1 and Z_2 are two normally distributed random variables. For steady-state flow and for a source constant intensity, the assumption that the particles N released in time interval $(t_1, t_1 + \Delta t)$ follow exactly the same random trajectories of the particles N released during the previous interval $(t_1, t_1 - \Delta t)$ is possible. Under this assumption only N particles are needed to simulate the location of the particles at a previous time step.

11 Results and discussion

11.1 Flow modeling

The MODFLOW numerical code coupled with the inverse model approach presented in the previous section has been used to model groundwater flow.

The numerical simulations have been carried out on a two-dimensional domain of 968.7 km^2. The domain has been discretized by means of a structured grid of 100 m in size.

In correspondence to the coastline, a first type of boundary condition has been imposed ($h = 0$ m), and along the detected watershed a second type of boundary condition ($q = 0\,\mathrm{m^2\,s^{-1}}$); the recharge from upstream is simulated by means of a first type of boundary condition where the hydraulic heads are equal to the detected regional values $h = 32$–41 m (Piano di Tutela delle Acque Regione Puglia, Tav. 6.2 http://old.regione.puglia.it/index.php?page=documenti&id=29&opz=getdoc, last access: February 2018).

A second type of boundary condition on the whole simulation domain has been imposed that concerns the mean effective infiltration calculated from the hydrologic budget $q = 0.037\,\mathrm{m^3\,d^{-1}}$.

The algorithm of inverse modeling has been applied to carry out the estimation of the spatial distribution of the equivalent transmissivity (Fig. 8) on the basis of the observed hydraulic head (vector $\boldsymbol{y}$), the regionalization model (matrix $\mathbf{Q}$) described by the variogram of the logarithm of the hydraulic transmissivity determined in the previous section.

The inverse model algorithm follows those steps. (1) Start from a conditional simulation of the log of T_{eq} determined by means of the hydraulic tests conducted in the area. (2) A set of pilot points are chosen in the area using a regularly spaced criterion and the value of T_{eq} has been determined for each pilot point (vector $\boldsymbol{s}$). (3) By means of the ordinary Kriging interpolation of the pilot points the map of T_{eq} is obtained and represents the input datum of the flow numerical model. (4) The hydraulic head has been determined using the flow numerical model (vector $\boldsymbol{h}$) and the values of the objective function have been determined using Eq. (12). (5) The values of T_{eq} are updated for each pilot point.

Using the Levenberg–Marquardt algorithm, the values of T_{eq} for each pilot point are updated as long as the objective function is minimized.

Figure 9 shows the results obtained from the flow model in steady-state condition, calibrated with the measurement campaign of February 2012 (Table 1).

Table 2 shows the data of model calibration and Fig. 10 shows the graph of the calibration.

The outcomes of the calibration are satisfactory. The comparison between the simulated and observed data has given a mean absolute residual equal to 0.57 m, an RMSE equal to 4.57 m, and a correlation coefficient r^2 equal to 0.997. In the following figures and tables are shown the results for the flow model.

The simulated hydraulic head distribution together with the equivalent transmissivity map put into evidence how the anticline affecting the Calcare di Bari formation in direc-

Table 1. Comparison between the observed and simulated hydraulic heads with related residuals, relative to the measurement campaign of February 2012.

Name	Obs. head (m)	Computed head (m)	Residual head (m)
P10	4.480	3.682	−0.798
L1-S	5.278	4.835	−0.443
P11	1.611	2.205	0.594
P19	1.110	2.217	1.107
P14	0.321	0.515	0.194
L2-S	0.722	0.466	−0.256
P4	0.386	0.801	0.415
L3-S	2.163	1.870	−0.293
P3	5.441	5.519	0.078
P16	3.536	3.315	−0.221
L4-S	3.450	3.567	0.117
P18	6.926	5.851	−1.075
L5-S	33.649	35.587	1.938
L8-S	8.532	8.809	0.277
L7-S	7.880	9.516	1.636
L6-S	8.892	9.651	0.759
P13	0.807	1.281	0.474
L9	0.705	0.236	−0.469
L10	0.167	0.276	0.109
L11	0.317	0.279	−0.038
L12	0.360	0.245	−0.115
L13	0.418	0.144	−0.274
L14-S	34.370	33.776	−0.594
L15-S	35.260	34.477	−0.783
P2	6.760	7.865	1.105

Table 2. Data of model calibration.

Mean residual	−0.138
Mean absolute residual	0.566
Root mean squared residual	0.743
Sum of squared weighted residual	4.571

tion ENE–WSW influences the flow directions. Furthermore, they highlight how the hydraulic circulation is more active along the coast coherently with a higher degree of fracturing and karst phenomena.

Once the equivalent hydraulic transmissivity map has been obtained, and assuming a value of the number of sets of fractures n_f, the spatial distribution of the mean equivalent aperture and the velocity field can be obtained.

Assuming the cubic law to be valid, the mean equivalent aperture can be obtained as

$$b_\mathrm{eq} = \sqrt[3]{12 \frac{T_\mathrm{eq}}{n_\mathrm{f}} \frac{\mu}{\rho g}}. \tag{22}$$

The velocity field results:

$$v_x = -\frac{\rho g}{\mu} \frac{b_\mathrm{eq}^2}{12} h_x,$$
$$v_y = -\frac{\rho g}{\mu} \frac{b_\mathrm{eq}^2}{12} h_y, \tag{23}$$

whereas assuming the Darcy–Weisbach equation to be valid, the mean equivalent aperture and the flow field can be obtained by means of the following iterative steps starting from the values of b_eq, v_x, and v_y previously evaluated:

$$Re^k = \frac{\rho \left|v^k\right| 2b_\mathrm{eq}^k}{\mu},$$
$$f^{k+1} = \frac{123}{Re^k}\left(1 + 0.12\left(Re^{k0.687}\right)\right),$$
$$b_\mathrm{eq}^{k+1} = \sqrt[3]{\frac{T_\mathrm{eq}}{n_\mathrm{f}^2} \frac{f^{k+1}}{4g} |\nabla h|},$$
$$v_x^{k+1} = -\frac{\sqrt{\frac{4b_\mathrm{eq}^{k+1}}{f^{k+1}} g}}{\sqrt{|\nabla h|}} h_x,$$
$$v_y^{k+1} = -\frac{\sqrt{\frac{4b_\mathrm{eq}^{k+1}}{f^{k+1}} g}}{\sqrt{|\nabla h|}} h_y. \tag{24}$$

Figure 11 shows the relative percentage of error in the flow velocity magnitude for different numbers of fractures.

As the number of fractures increases, the velocity magnitude decreases; therefore, the friction factor reaches a value of $96/Re$. Anyway, the percentage of error in the flow velocity magnitude seems not to be negligible. In fact, for $n_\mathrm{f} = 28$ a minimum value of 8 % is obtained, reaching a value of 28 % for $n_\mathrm{f} = 4$. These results show that under natural hydraulic gradient conditions in fractured limestone the nonlinearity of the flow cannot be negligible. It is clear that under a forced hydraulic gradient due to anthropic stresses the equivalent transmissivity decreases dramatically, with a value less than 40 % of Darcian-like hydraulic transmissivity (Cherubini et al., 2012).

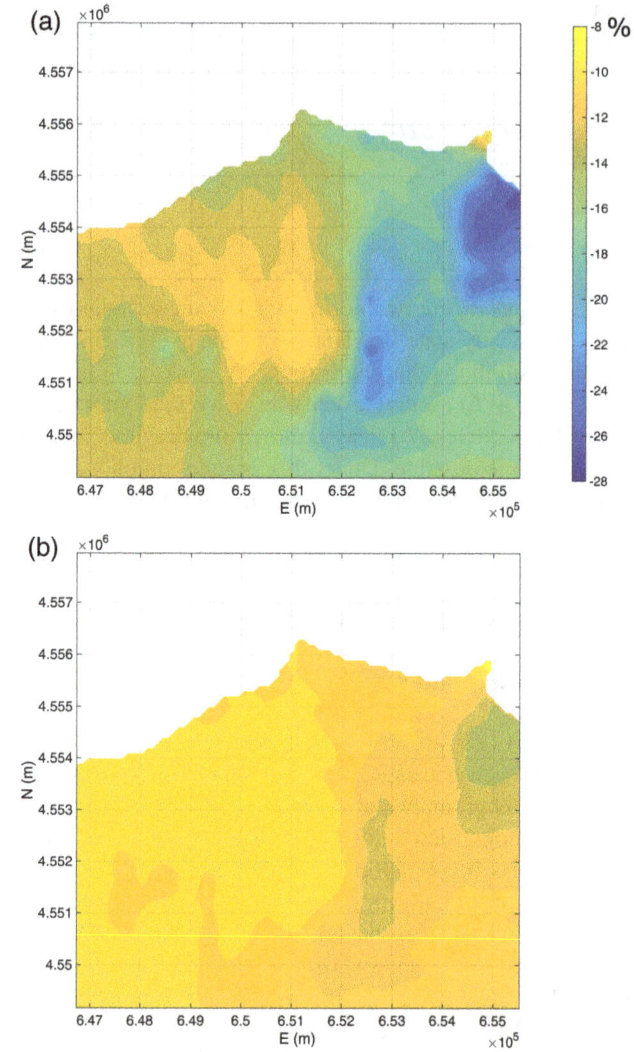

Figure 11. Relative percentage of error in the flow velocity magnitude for different numbers of fractures: **(a)** $n_\mathrm{f} = 4$; **(b)** $n_\mathrm{f} = 28$.

11.2 Analysis of the scenario of contamination

A particle tracking transport method has been applied for the simulation of contaminant transport. A punctual source contamination has been imposed in correspondence to the detected hotspot equal to a PCE concentration of 1283 $\mu g\,L^{-1}$. This localization is coherent with the soil contamination detected in the area.

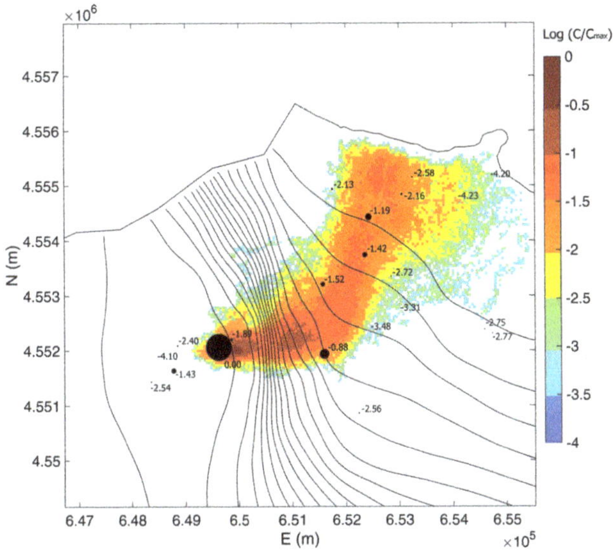

Figure 12. Steady-state distribution of hypothetical contamination using the random walk model with the source contamination localized in correspondence to the hotspot of the contamination considering a number of fracture of $n_f = 20$ and a longitudinal and transversal dispersion coefficient equal to $\alpha_L = 70$ m and $\alpha_T = 7$ m.

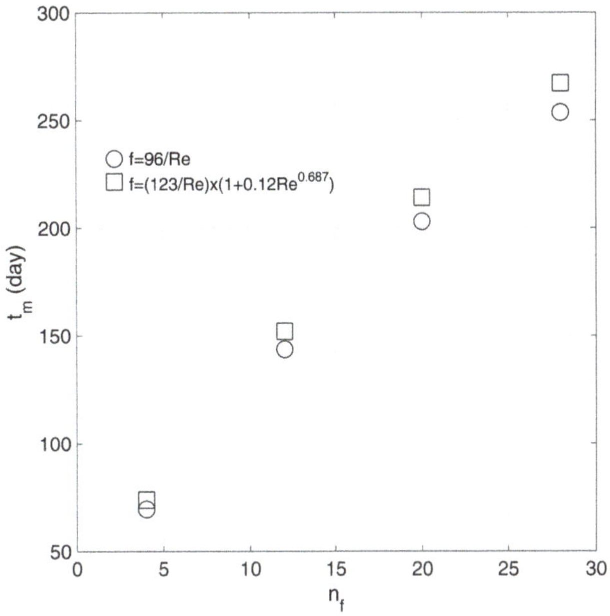

Figure 14. Mean travel time t_m at varying numbers of fractures n_f for linear and nonlinear models.

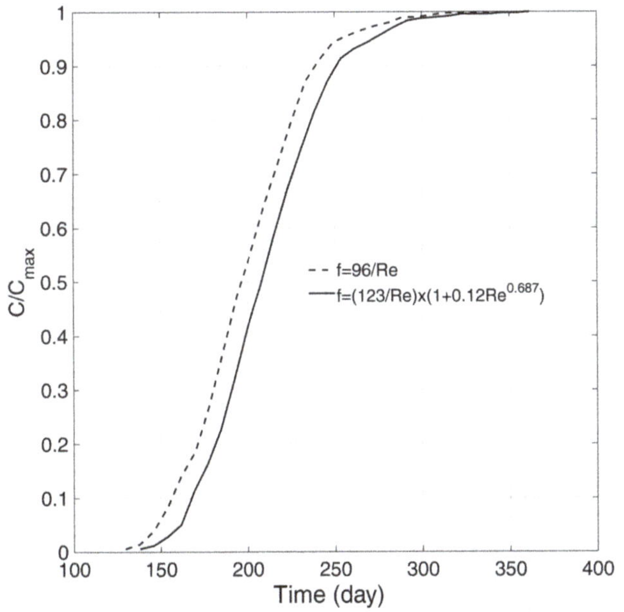

Figure 13. Breakthrough curves of hypothetical continuous contamination released in correspondence to the hotspot, determined for linear and nonlinear flow models, evaluated at the downstream boundary.

In order to solve the advective transport equation a numerical Lagrangian particle-based random walk method is implemented. For each time step a constant number of 500 particles have been released into the domain from the source. According to the Rauch et al. (2005) assumption reported in the previous section, only 500 particles are needed to simulate the steady-state distribution of the hypothetical contamination. Even if the source of contamination has been considered punctual, the obtained simulation scenario proves to be compatible with the observed one, and therefore it is possible to assume that the sources of contamination are located in correspondence to the detected hotspot (Fig. 12).

Figure 13 shows the breakthrough curves of hypothetical continuous contamination released in correspondence to the hotspot, determined for linear and nonlinear flow models, evaluated at the downstream boundary for $n_f = 20$.

Figure 14 shows the mean travel time at a varying number of fractures for the linear and nonlinear models. With an increasing number of fractures, the travel time increases in a linear way, because the cross-sectional area increases as well. The figures highlight that travel time for the nonlinear model is higher than the linear assumption. In a particular way, the percentages of error are in the range of 6.22 %–5.34 % passing from $n_f = 4$ ($Re = 0.02$–10.60) to $n_f = 28$ ($Re = 0.002$–1.51). This is coherent with what was detected by Cherubini et al. (2012, 2013b, 2014), who carried out hydraulic and tracer tests on an artificially created fractured rock sample and found a pronounced mobile–immobile zone interaction leading to a non-equilibrium behavior of solute transport.

The existence of a non-Darcian flow regime was shown to influence the velocity field by giving rise to a delay in solute migration with respect to the values that could be obtained under the assumption of a linear flow field. Furthermore, the presence of inertial effects was shown to enhance non-equilibrium behavior. In a particular manner, they found that the percentage of error in the travel time with respect to the linear flow assumption varied in the range of 5.90 %–40.75 %, corresponding to a range of *Re* of 29.48–52.16. These results highlight the fact that as the scale of observation increases, the error in the mean travel time with respect to the linear flow model becomes more relevant. In fact, at field scale also for *Re*, just above the unit ($n_f = 28$), the error is equal to 5.34 %, comparable with the error of 5.90 % found at laboratory scale for *Re* equal to 29.48. This means that under anthropic stresses multiple pumpings or injections give rise to a higher flow velocity and then higher *Re*, leading to a dramatic delay on contaminant transport. Therefore, nonlinear flow must be considered in order to have a more accurate estimation of the breakthrough curve and mean travel time of contaminated scenarios.

In fracture networks, the presence of nonlinear flow plays an important role in the distribution of the solutes according to the different pathways. In fact, the energy spent to cross the path should be proportional to the resistance to flow associated with the single pathway, which in a nonlinear flow regime is not constant but depends on the flow rate. This means that by changing the boundary conditions, the resistance to flow varies and consequently the distribution of solute in the main and secondary pathways also changes, giving rise to a different behavior of solute transport (Cherubini et al., 2014).

This concept has to be taken into account in case of cleanup of the aquifer using for example the Pump & Treat system. The multiple pumping and reinjection of the treated groundwater give rise to a higher flow velocity in the aquifer, resulting in a much greater hydraulic gradient. In this case nonlinear flow behavior has to be taken into account in order to obtain more accurate cleanup strategies.

12 Conclusions

The present study is aimed at analyzing the scenario of groundwater contamination (by investigating the hotspots) of the industrial area of Modugno (Bari – southern Italy) where the limestone aquifer has a fractured and karstic nature.

The presence of hotspot areas has been detected, located upstream of the groundwater flow, coherently with the state of contamination detected downstream and the soil contamination.

A rough walled parallel plate model has been implemented and calibrated on the basis of piezometric data and has coupled a geostatistical analysis to infer the values of the equivalent aperture. Using the random walk theory, the steady-state distribution of hypothetical contamination with the source contamination at the hotspot has been carried out.

The flow and transport models have reproduced the flow pattern well and have given a pollution scenario that is compatible with the observed one.

From an analysis of the flow and transport pattern it is possible to infer that the anticline affecting the Calcare di Bari formation in direction ENE–WSW influences the direction of flow as well as the propagation of the contaminant.

The results also show that the presence of nonlinear flow influences advection, in that it leads to a delay in solute transport with respect to the linear flow assumption. Moreover, the distribution of solute according to different pathways is not constant, but is related to the flow rate.

This is due to the non-proportionality between the energy spent to cross the path and the resistance to flow for fractured media, which affects the distribution of the solutes according to the different pathways.

The obtained results represent the fundamental basis for a detailed study of the contaminant propagation in correspondence to the hotspot area in order to find the best cleanup strategies and optimize any anthropic intervention on the industrial site.

Future developments of the current study will be to implement a transient model and to include the density-dependent flow in the simulations.

Author contributions. CC and NP dealt with data analysis, modeling and interpretation, DR dealt with the geology part and interpretation of results, and CIG dealt with interpretation of results and theoretical analysis.

Competing interests. The authors declare that they have no conflict of interest.

Acknowledgements. The authors would like to acknowledge the editor and reviewers for their valuable comments and suggestions that helped improve the manuscript.

Edited by: Philippe Ackerer

References

Antonellini, M., Cilona, A., Tondi, E., Zambrano, M., and Agosta, F.: Fluid-flow numerical experiments of faulted porous carbonates, Northwest Sicily (Italy), Mar. Petrol. Geol., 55, 186–201, 2014.

Baker, W. J.: Flow in fissured formations, Proceedings of 4th Worm Petroleum Congress, Carlo Colombo, Rome, 379–393, 1955.

Berkowitz, B., Kosakowski, G., Margolin, G., and Scher, H.: Application of Continuous Time Random Walk Theory to Tracer Test Measurements in Fractured and Heterogeneous Porous Media, Ground Water, 39, 593–603, 2001.

Bieniawski, Z. T.: Engineering Rock Mass Classifications, John Wiley & Sons, New York, 251, 1989.

Billi, A., Salvini, F., and Storti, F.: The damage zone-fault core transition in carbonate rocks: implications for fault growth, structure and permeability, J. Struct. Geol., 25, 1779–1794, 2003.

Brush, D. J. and Thomson, N. R.: Fluid flow in synthetic rough-walled fractures: Navier-Stokes, Stokes, and local cubic law simulations, Water Resour. Res., 39, 1085, https://doi.org/10.1029/2002WR001346, 2003.

Bourke, P. J.: Channeling of flow through fractures in rock, in Proceedings of GEOVAL-1987 International Symposium, 1, Swedish Nuclear Power Inspectorate (SKI), Stockholm, 167–177, 1987.

Caine, J. S. and Forster, C. B.: Fault zone architecture and fluid flow: Insights from field data and numerical modeling, in: Faults and sub-surface fluid flow in the shallow crust, edited by: Haneberg, W. C., Mozley, P. S., Moore, C., and Goodwin, L. B., AGU Geophysical Monograph, 113, 101–127, 1999.

Caine, J. S., Evans, J. P., and Forster, C.: Fault zone architecture and permeability structure, Geology, 24, 1025–1028, 1996.

Cherubini, C.: A modeling approach for the study of contamination in a fractured aquifer, Geotechnical and geological engineering, Springer, the Netherlands, 26, 519–533, 2008.

Cherubini, C. and Pastore, N.: Modeling contaminant propagation in a fractured and karstic aquifer, Fresen. Environ. Bull., 19, 1788–1794, 2010.

Cherubini, C. and Pastore, N.: Critical stress scenarios for a coastal aquifer in southeastern Italy, Nat. Hazards Earth Syst. Sci., 11, 1381–1393, https://doi.org/10.5194/nhess-11-1381-2011, 2011.

Cherubini, C., Pastore, N., and Francani, V.: Different approaches for the characterization of a fractured karst aquifer, World scientific and engineering academy and society, Transactions on fluid mechanics, Wisconsin-Usa, 1, 29–35, 2008.

Cherubini, C., Giasi, C. I., and Pastore, N.: Bench scale laboratory tests to analyze non-linear flow in fractured media, Hydrol. Earth Syst. Sci., 16, 2511–2522, https://doi.org/10.5194/hess-16-2511-2012, 2012.

Cherubini, C., Giasi, C., and Pastore, N.: Fluid flow modeling of a coastal fractured karstic aquifer by means of a lumped parameter approach, Environ. Earth Sci., 70, 2055–2060, 2013a.

Cherubini, C., Giasi, C. I., and Pastore, N.: Evidence of non-Darcy flow and non-Fickian transport in fractured media at laboratory scale, Hydrol. Earth Syst. Sci., 17, 2599–2611, https://doi.org/10.5194/hess-17-2599-2013, 2013b.

Cherubini, C., Giasi, C. I., and Pastore, N.: On the reliability of analytical models to predict solute transport in a fracture network, Hydrol. Earth Syst. Sci., 18, 2359–2374, https://doi.org/10.5194/hess-18-2359-2014, 2014.

Cooley, R. L. and Cunningham, A. B.: Consideration of energy loss in theory of flow to wells, J. Hydrol., 43, 161–184, 1979.

Cortis, A. and Birkholzer, J.: Continuous time random walk analysis of solute transport in fractured porous media, Water Resour. Res., 44, W06414, https://doi.org/10.1029/2007WR006596, 2008.

Deere, D. U. and Deere, D. W.: The rock quality designation (RQD) index in practice, Rock classification systems for engineering purposes, edited by: Kirkaldie, L., ASTM Special Publication Philadelphia, Am. Soc. Test. Mater., 984, 91–101, 1988.

Dentz, M., Kang, P. K., and Le Borgne, T.: Continuous Time Random Walks for Non-Local Radial Solute Transport, Adv. Water Resour., 82, 16–26, 2015.

Diersch, H. J. G.: FEFLOW finite element subsurface flow and transport simulation system, User's manual/Reference manual/White papers, Release 5.1, WASY Ltd, Berlin, 2002.

Doglioni, C.: The Puglia uplift (SE Italy) An anomaly in the foreland of the Apenninic subduction due to buckling of a thick continental lithosphere, Tectonics, 13, 1309–1321, 1994.

Elliott, G. M. and Brown, E. T.: Laboratory measurement of the thermo-hydro-mechanical properties of rock, Q. J. Eng. Geol. Hydroge., 21, 299–314, 1988.

Forster, C. B. and Evans, J. P.: Hydrogeology of thrust faults and crystalline thrust sheets: results of combines field and modelling studies, Geophys. Res. Lett., 18, 979–982, 1991.

Gale, J. A.: Numerical Field and Laboratory Study of Flow in Rocks with Deformable Fractures, Sci. Ser. 72, Inland Waters Dir., Water Resources Branch, Ottawa, Ontario, Canada, 1977.

Huitt, J. L.: Fluid Flow in Simulated Fractures, Amer. Inst. Chem. Eng. J., 2, 259–264, 1956.

Lee, J., Kang, J. M., and Choe, J.: Experimental analysis on the effects of variable apertures on tracer transport, Water Resour. Res., 39, 1015, https://doi.org/10.1029/2001WR001246, 2003.

Keller, A.: High resolution, non-destructive measurement and characterization of fracture apertures, Int. J. Rock Mech. Min., 35, 1037–1050, 1998.

Keller, A. A., Roberts, P. V., and Blunt, M. J.: Effect of fracture aperture variations on the dispersion of contaminants, Water Resour. Res., 35, 55–63, 1999.

Kitanidis, P.: Quasi?linear geostatistical theory for inversing, Water Resour. Res., 31, 2411–2419, https://doi.org/10.1029/95WR01945, 1995.

Klimczak, C., Schultz, R. A., Parashar, R., and Reeves, D. M.: Cubic law with aperture-length correlation: implications for network scale fluid flow, Hydrogeol. J., 18, 851–862, 2010.

Korneva, I., Tondi, E., Agosta, F., Rusctichelli, A., Spina, V., Bitonte, R., and Di Cuia, R.: Structural properties of fractured and faulted Cretaceous platform carbonates, Murge Plateau (Southern Italy), Mar. Petrol. Geol., 57, 312–326, 2014.

Masciopinto, C. and Palmiotta, D.: Flow and Transport in Fractured Aquifers: New Conceptual Models Based on Field Measurements, Transport Porous Med., 96, 117–133, 2013.

Masciopinto, C. and Visino, F.: Strong release of viruses in fracture flow in response to a perturbation in ionic strength: Filtration/retention tests and modeling, Water Res., 126, 240–251, 2017.

Masciopinto, C., Volpe, A., Palmiotta, D., and Cherubini, C.: A combined PHREEQC-2/parallel fracture model for the simulation of laminar/non-laminar flow and contaminant transport with reactions, J. Contam. Hydrol., 117, 94–108, 2010.

Nazridoust, K., Ahmadi, G., and Smith, D. H.: A new friction factor correlation for laminar, single-phase flows through rock fractures, J. Hydrol., 329, 315–328, https://doi.org/10.1016/j.jhydrol.2006.02.032, 2006.

Neretnieks, I., Eriksen, T., and Tahtinen, P.: Tracer movement in a single fissure in granitic rock: Some experimental results and their interpretation, Water Resour. Res., 18, 849–858, 1982.

Neuzil, C. E. and Tracy, J. V.: Flow Through Fractures, Water Resour. Res., 17, 191–199, 1981.

Oron, A. P. and Berkowitz, B.: Flow in rock fractures: The local cubic law assumption reexamined, Water Resour. Res., 34, 2811–2825, 1998.

Parise, M. and Pascali, V.: Surface and subsurface environmental degradation in the karst of Apulia (southern Italy), Environ. Geol., 44, 247–256, 2003.

Pieri, P., Festa, V., Moretti, M., and Tropeano, M.: Quaternary tectonic activity of the Murge area (Apulian foreland-Southern Italy), Ann. Geofis., 40, 1395–1404, 1997.

Pieri, P., Sabato, L., Spalluto, L., and Tropeano, M.: Note Illustrative della Carta Geologica d'Italia alla scala 1 : 50.000 foglio 438, Bari, Nuova Carta Geologica d'Italia alla scala 1 : 50.000 – Progetto CARG, ISPRA-Servizio Geologico d'Italia, ISBN 2035-8008, January 2012.

Plouraboué, F., Kurowski, P. J., Hulin, P., Roux, S., and Schmittbuhl, J.: Aperture of rough crack, Phys. Rev., 51, 1675–1685, 1995.

Plouraboué, F., Kurowski, P., Boffa, J. M., Hulin, J. P., and Roux, S.: Experimental study of the transport properties of rough self-affine fractures, J. Contam. Hydrol., 46, 295–318, 2000.

Priest, S. D. and Hudson, J. A.: Discontinuity spacings in rock, Int. J. Rock Mech. Min., 13, 135–148, 1976.

Pyrak-Nolte, L. J., Cook, N. G. W., and Nolte, D.: Fluid percolation through single fractures, Geophys. Res. Lett., 15, 1247–1250, 1988.

Rauch, R., Schäfer, W., and Wagner, C.: Solute Transport Modelling. An introduction to Models and Solution Strategies, Gebrüder Borntraeger Verlagsbuchhandlung, Berlin Tuggart, 2005.

Remy, N.: Geostatistical Earth Modeling Software: User's Manual, Stanford University, 15–18, available at: http://sgems.sourceforge.net/ (last access: July 2018), 2004.

Rustichelli, A., Tondi, E., Agosta, F., Cilona, A., and Giorgioni, M.: Development and distribution of bed-parallel compaction bands and pressure solution seams in carbonates (Bolognano Formation, Majella Mountain, Italy), J. Struct. Geol., 37, 181–199, 2012.

Snow, D. T.: A Parallel Plate Model of Fractured Permeable Media, PhD Dissertation, University of California, 1965.

Snow, D. T.: The Frequency and Apertures of Fractures in Rocks, Int. J. Rock Mech. Min., 7, 23–40, 1970.

Srinivasan, G., Tartakovsky, D. M., Dentz, M., Viswanathan, H., Berkowitz, B., and Robinson, B. A.: Random walk particle tracking simulations of non-Fickian transport in heterogeneous media, J. Comput. Phys., 229, 4304–4314, 2010.

Tsang, Y. W. and Tsang, C. F.: Channel model of flow through fractured media, Water Resour. Res., 23, 467–479, 1987.

Tsang, Y. W. and Tsang, C.-F.: Flow channeling in a single fracture as two-dimensional strongly heterogeneous permeable medium, Water Resour. Res., 25, 2076–2080, 1989.

Tsang, C.-F., Tsang, Y. W., Birkhölzer, J., and Moreno, L.: Dynamic channeling of flow and transport in saturated and unsaturated heterogeneous media, in Flow and Transport Through Unsaturated Fractured Rock, 2nd ed., Geophys. Monograph, 42, 33–44, AGU, Washington, D.C., 2001.

Wang, L., Cardenas, M. B., Slottke, D. T., Ketcham, R. A., and Sharp Jr., J. M.: Modification of the Local Cubic Law of fracture flow for weak inertia, tortuosity and roughness, Water Resour. Res., 51, 2064–2080, 2015.

Witherspoon, P. A., Wang, J. S. Y., Iwai, K., and Gale, J. E.: Validity of the cubic law for fluid flow in a deformable rock fracture, Water Resour. Res., 16, 1016–1034, 1980.

Wu, Y. S.: NonDarcy Displacement of Immiscible Fluids in Porous Media, Water Resour. Res., 37, 2943–2950, https://doi.org/10.1029/2001WR000389, 2001.

Zhang, X. and Spiers, C. J.: Compaction of granular calcite by pressure solution at room temperature and effects of pore fluid chemistry, Int. J. Rock Mech. Min., 42, 950–960, 2005.

Snow, D. T.: A Parallel Plate Model of Fractured Permeable Media, PhD Dissertation, University of California, 1965.

Snow, D. T.: The Frequency and Apertures of Fractures in Rocks, Int. J. Rock Mech. Min., 7, 23–40, 1970.

Srinivasan, G., Tartakovsky, D. M., Dentz, M., Viswanathan, H., Berkowitz, B., and Robinson, B. A.: Random walk particle tracking simulations of non-Fickian transport in heterogeneous media, J. Comput. Phys., 229, 4304–4314, 2010.

Tsang, Y. W. and Tsang, C. F.: Channel model of flow through fractured media, Water Resour. Res., 23, 467–479, 1987.

Tsang, Y. W. and Tsang, C.-F.: Flow channeling in a single fracture as two-dimensional strongly heterogeneous permeable medium, Water Resour. Res., 25, 2076–2080, 1989.

Tsang, C.-F., Tsang, Y. W., Birkhölzer, J., and Moreno, L.: Dynamic channeling of flow and transport in saturated and unsaturated heterogeneous media, in Flow and Transport Through Unsaturated Fractured Rock, 2nd ed., Geophys. Monograph, 42, 33–44, AGU, Washington, D.C., 2001.

Wang, L., Cardenas, M. B., Slottke, D. T., Ketcham, R. A., and Sharp Jr., J. M.: Modification of the Local Cubic Law of fracture flow for weak inertia, tortuosity and roughness, Water Resour. Res., 51, 2064–2080, 2015.

Witherspoon, P. A., Wang, J. S. Y., Iwai, K., and Gale, J. E.: Validity of the cubic law for fluid flow in a deformable rock fracture, Water Resour. Res., 16, 1016–1034, 1980.

Wu, Y. S.: NonDarcy Displacement of Immiscible Fluids in Porous Media, Water Resour. Res., 37, 2943–2950, https://doi.org/10.1029/2001WR000389, 2001.

Zhang, X. and Spiers, C. J.: Compaction of granular calcite by pressure solution at room temperature and effects of pore fluid chemistry, Int. J. Rock Mech. Min., 42, 950–960, 2005.

13

Leaf-scale experiments reveal an important omission in the Penman–Monteith equation

Stanislaus J. Schymanski and Dani Or

Department of Environmental Sciences, ETH Zurich, 8092 Zurich, Switzerland

Correspondence to: Stanislaus J. Schymanski (stan.schymanski@env.ethz.ch)

Abstract. The Penman–Monteith (PM) equation is commonly considered the most advanced physically based approach to computing transpiration rates from plants considering stomatal conductance and atmospheric drivers. It has been widely evaluated at the canopy scale, where aerodynamic and canopy resistance to water vapour are difficult to estimate directly, leading to various empirical corrections when scaling from leaf to canopy. Here, we evaluated the PM equation directly at the leaf scale, using a detailed leaf energy balance model and direct measurements in a controlled, insulated wind tunnel using artificial leaves with fixed and predefined stomatal conductance. Experimental results were consistent with a detailed leaf energy balance model; however, the results revealed systematic deviations from PM-predicted fluxes, which pointed to fundamental problems with the PM equation. Detailed analysis of the derivation by Monteith (1965) and subsequent amendments revealed two errors: one in neglecting two-sided exchange of sensible heat by a planar leaf, and the other related to the representation of hypostomatous leaves, which are very common in temperate climates. The omission of two-sided sensible heat flux led to bias in simulated latent heat flux by the PM equation, which was as high as 50 % of the observed flux in some experiments. Furthermore, we found that the neglect of feedbacks between leaf temperature and radiative energy exchange can lead to additional bias in both latent and sensible heat fluxes. A corrected set of analytical solutions for leaf temperature as well as latent and sensible heat flux is presented, and comparison with the original PM equation indicates a major improvement in reproducing experimental results at the leaf scale. The errors in the original PM equation and its failure to reproduce experimental results at the leaf scale (for which it was originally derived) propagate into inaccurate sensitivities of transpiration and sensible heat fluxes to changes in atmospheric conditions, such as those associated with climate change (even with reasonable present-day performance after calibration). The new formulation presented here rectifies some of the shortcomings of the PM equation and could provide a more robust starting point for canopy representation and climate change studies.

1 Introduction

A vast number of current global land surface models, hydrological models and inverse approaches to deduce evaporation from remote sensing data employ the analytical solution for the latent heat flux from plant leaves derived by Monteith (1965), based on an earlier formulation for a wet surface by Penman (1948), see e.g. Overgaard et al. (2006) and Dolman et al. (2014). This so-called Penman–Monteith equation (henceforth referred to as the PM equation), which introduced stomatal resistance into Penman's formalism, found widespread use in the prediction of latent heat flux based on estimates of leaf and canopy resistance to water vapour. Whereas the PM equation is generally believed to provide an adequate physical description of transpiration from an individual leaf, it is commonly applied at the canopy scale, where aerodynamic and bulk stomatal resistance are difficult to estimate and are usually deduced empirically from measurements of transpiration and an inverted PM equation (Raupach and Finnigan, 1988) or from observed surface temperatures (Tanner and Fuchs, 1968). The scaling up from leaf to canopy is commonly done in an ad hoc manner by replacing the leaf-scale resistances with their assumed canopy-scale counterparts, often without any additional physics involved. The leaf-canopy scaling and use of data at daily or monthly scales has led to various empirical corrections to the PM equation (Allen, 1986), which may have obscured more fundamental issues with the derivations by Monteith (1965). On the other hand, Langensiepen et al. (2009) proposed a detailed leaf-scale parametrisation of the PM equation and averaging over the canopy and time that yielded reasonable agreement with sap-flow-derived canopy transpiration estimates, without empirical corrections to the PM equation.

A number of authors have focused on biases introduced by the simplifications inherent in the PM equation, such as the linearisation of the saturation vapour pressure curve and the neglect of dependency of net irradiance on surface temperature, and proposed various approaches to reduce such biases (Paw U and Gao, 1988; McArthur, 1990; Milly, 1991; Widmoser, 2009). Interestingly, even 50 years after its derivation, we have not found a rigorous test of the PM equation at the leaf scale, whereas our analysis of the derivations by Monteith (1965) and later amendments revealed two errors in considering the effect of stomata and the two-sided exchange of sensible heat. Therefore, the objectives of the present study are to

1. develop an experimental set-up allowing direct and independent measurement of all components of the energy balance of a single leaf and the relevant boundary conditions,
2. compare different analytical and numerical leaf energy balance and gas exchange models with experimental results and
3. derive an improved analytical representation of latent and sensible heat fluxes at the leaf scale.

The study is structured as follows. We first present a physically based, explicit leaf energy balance and gas exchange model, to serve as a reference for the physical processes. The explicit model is then used to rederive the Penman and Penman–Monteith (PM) equations while highlighting all simplifying assumptions inherent in these formulations. Subsequently, we will derive a general analytical formulation based on the approach by Penman (1952) and analyse consistency between the various analytical solutions and the explicit leaf energy and gas exchange model. In the next step, we will present an experimental set-up allowing us to measure all components of the leaf energy balance under fully controlled conditions, using artificial leaves with known stomatal conductance. Experimental results will be compared with the explicit numerical model and the different analytical solutions, assessing potential bias.

2 Materials and methods

The detailed derivations are described in the Appendix, while the experimental methods are discussed in detail in a technical note submitted to HESS (Schymanski et al., 2017). Here, we only summarise the key points and concepts necessary to understand the flow of the paper. All symbols used in this paper are listed and described in the Appendix, Table A1. Additionally, all derivations, data and code to reproduce the results are provided online and can be accessed at https://github.com/schymans/Schymanski_leaf-scale_2016/tree/1.0.0 and doi:10.5281/zenodo.241259 (Schymanski, 2017). Data analysis was performed using the free software SageMath, the Sage Mathematics Software System (version 7.3), The Sage Developers, 2016 (http://www.sagemath.org).

2.1 Explicit leaf energy balance and gas exchange model

The detailed leaf energy balance model used here is based on derivations published previously (Schymanski et al., 2013; Schymanski and Or, 2015, 2016), and is reproduced here after reorganisation of equations for consistency with the present paper.

The leaf energy balance is determined by the dominant energy fluxes between the leaf and its surroundings, including radiative, sensible and latent energy exchange (linked to mass exchange). These are illustrated in Fig. 1. Focusing on steady-state conditions, the energy balance can be written as

$$R_{\mathrm{s}} = E_{\mathrm{l}} + H_{\mathrm{l}} + R_{\mathrm{ll}}, \tag{1}$$

where R_{s} ($\mathrm{W\,m^{-2}}$) is absorbed short-wave radiation, E_{l} ($\mathrm{W\,m^{-2}}$) is the latent heat flux away from the leaf, H_{l} ($\mathrm{W\,m^{-2}}$) is the sensible heat flux away from the leaf and R_{ll} ($\mathrm{W\,m^{-2}}$) is the net emitted long-wave radiation, i.e. the emitted minus the absorbed radiation. In the above, extensive variables are defined per unit leaf area. Following our previous work (Schymanski et al., 2013), this study considers spatially homogeneous planar leaves, i.e. homogenous illumination and a negligible temperature gradient between the two sides of the leaf. The net long-wave emission is represented by the difference between blackbody radiation at leaf temperature (T_{l}, K) and that at the temperature of the surrounding objects (T_{w}, commonly represented by air temperature, T_{a}, both in K) (Monteith and Unsworth, 2007):

$$R_{\mathrm{ll}} = a_{\mathrm{sH}} \epsilon_{\mathrm{l}} \sigma (T_{\mathrm{l}}^4 - T_{\mathrm{w}}^4), \tag{2}$$

where a_{sH} is the fraction of projected leaf area exchanging radiative and sensible heat (2 for a planar leaf, 1 for a soil surface), ϵ_{l} is the leaf's long-wave emissivity (≈ 1) and σ ($\mathrm{W\,K^{-4}\,m^{-2}}$) is the Stefan–Boltzmann constant. Total convective heat transport away from the leaf is represented as

$$H_{\mathrm{l}} = a_{\mathrm{sH}} h_{\mathrm{c}} (T_{\mathrm{l}} - T_{\mathrm{a}}), \tag{3}$$

where h_{c} ($\mathrm{W\,K^{-1}\,m^{-2}}$) is the average one-sided convective heat transfer coefficient, determined by properties of the leaf boundary layer.

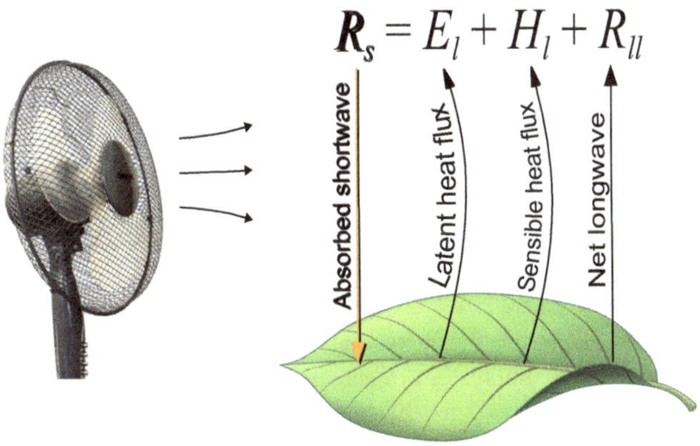

Flux	Equation	Driver
Latent heat	$E_l = L\, M_w\, g_{tw} (C_{wl} - C_{wa})$	Vapour conc.
Sensible heat	$H_l = 2h_c (T_l - T_a)$	Temperature
Longwave	$R_{ll} = 2\sigma (T_l^4 - T_a^4)$	Temperature

$(C_{wl} - C_{wa}) = f(RH, T_l, T_a)$

$g_{tv}, h_c \sim sqrt(wind)$

Figure 1. Components of the leaf energy balance and their thermodynamic drivers. Bent arrows indicate fluxes that are directly affected by wind speed. Table at bottom illustrates the drivers for each flux (temperature differences for sensible and radiative heat exchange, water vapour concentration differences for mass exchange and hence latent heat flux). Additional equations below the table illustrate that the driver for latent heat flux is also related to temperature differences and that the transfer coefficients for both latent and sensible heat flux depend on wind. L: latent heat of vaporisation, M_w: molecular mass of water, g_tw: total leaf conductance to water vapour, C_wl: concentration of water vapour in leaf-internal air, C_wa: concentration of water vapour in free air stream, h_c: one-sided heat transfer coefficient, T_l: leaf temperature, T_a: air temperature, σ: Stefan–Boltzmann constant, RH: relative humidity of the free air stream, g_bw: leaf boundary layer conductance to water vapour.

Latent heat flux (E_l, W m^{-2}) is directly related to the transpiration rate ($E_\mathrm{l,mol}$, mol m^{-2} s^{-1}) by

$$E_\mathrm{l} = E_\mathrm{l,mol} M_\mathrm{w} \lambda_E, \qquad (4)$$

where M_w (kg mol^{-1}) is the molar mass of water and λ_E (J kg^{-1}) the latent heat of vaporisation. $E_\mathrm{l,mol}$ (mol m^{-2} s^{-1}) was computed in molar units as a function of the concentration of water vapour within the leaf (C_wl, mol m^{-3}) and in the free air (C_wa, mol m^{-3}) (Incropera et al., 2006, Eq. 6.8):

$$E_\mathrm{l,mol} = g_\mathrm{tw}(C_\mathrm{wl} - C_\mathrm{wa}), \qquad (5)$$

where g_tw (m s^{-1}) is the total leaf conductance for water vapour, dependent on stomatal (g_sw) and the boundary layer conductance (g_bw) in the following way:

$$g_\mathrm{tw} = \frac{1}{\frac{1}{g_\mathrm{sw}} + \frac{1}{g_\mathrm{bw}}}. \qquad (6)$$

Both the one-sided leaf convective heat transfer coefficient (h_c) and boundary layer conductance to water vapour (g_bw, m s^{-1}) relate to the same physical principles of diffusion and boundary layer dynamics; i.e. both depend on leaf size (L_l, m), wind speed (v_w, m s^{-1}) and the level of turbulence in the air stream (critical Reynolds number, N_{Re_c}), expressed in the dimensionless Nusselt and Lewis numbers (N_{Nu_L} and N_{Le}, respectively). The relation of h_c to g_bw additionally depends on whether stomata are present on one side of the leaf only ($a_\mathrm{s} = 1$) or both sides of the leaf ($a_\mathrm{s} = 2$). The relevant equations to compute all of these variables as a function of air temperature, pressure and vapour pressure (T_a, P_a and P_wa, respectively), wind speed (v_w), turbulence and leaf properties are given in the Appendix, Sects. B1–B4.

Figure 2 illustrates the use of measurements and the different equations to compute the leaf energy balance components. Leaf temperature (T_l) needs to be computed by iteration, using the leaf energy balance model, due to the non-linearities in Eqs. (2) and (B5). Note that a direct measurement of T_l (e.g. using infrared sensors) would enable direct computation of R_ll and H_l, and finally E_l from the energy balance as $E_\mathrm{l} = R_\mathrm{s} - R_\mathrm{ll} - H_\mathrm{l}$ without any iterations. This illustrates that the use of any of the analytical solutions explained below is not necessary if T_l is known and questions the approach proposed by Tanner and Fuchs (1968), where observed leaf or surface temperature is inserted into the Penman–Monteith equation to estimate transpiration rate.

2.2 Generalisation of Penman's analytical approach

The PM equation derived by Monteith (1965) was based on the analytical solution for evaporation from a wet surface by Penman (1948). The key point of Penman's analytical solution is to express evaporation as a function of the surface–air vapour pressure difference and sensible heat flux as a function of surface–air temperature difference. Here, we will follow the succinct derivation presented in the Appendix of Penman (1952) and use our notation for a leaf to obtain a general solution applicable either to a transpiring leaf or an evaporating surface. In the first step, we will introduce general transfer coefficients for latent heat (c_E, W m^{-2} Pa^{-1}) and sensible heat (c_H, W m^{-2} K^{-1}), satisfying the following equations:

$$E_\mathrm{l} = c_E(P_\mathrm{wl} - P_\mathrm{wa}) \qquad (7)$$

and

$$H_\mathrm{l} = c_H(T_\mathrm{l} - T_\mathrm{a}). \qquad (8)$$

(Please refer to Appendix B3 for a discussion of the meaning of Eq. 7 compared to Eq. 5, and conversion of transfer coefficients.)

Equations (7), (8) and the leaf energy balance equation (Eq. 1) form a system of three equations with four unknowns: E_l, H_l, T_l and P_wl. In order to eliminate T_l, Penman assumed that the ratio of the vapour pressure difference between the

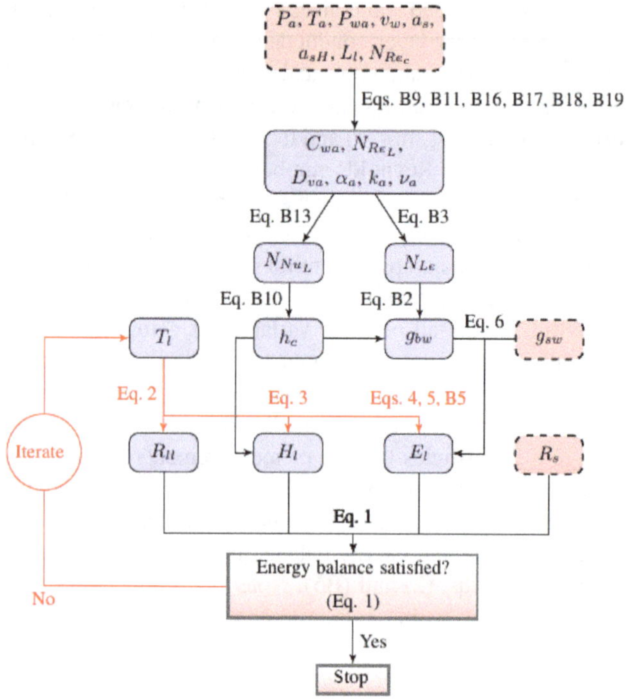

Figure 2. Flow chart of computation procedure for different leaf energy balance components. Dashed pink boxes with rounded corners indicate external input, while solid blue rounded boxes indicate computed variables. Note the central role of leaf temperature, which needs to be computed by iteration against the leaf energy balance.

surface and the saturation vapour pressure at air temperature (P_{was}) to the temperature difference between the surface and the air can be approximated by the slope of the saturation vapour pressure curve at air temperature (Δ_{eTa}, Pa K^{-1}):

$$\Delta_{\text{eTa}} = \frac{P_{\text{wl}} - P_{\text{was}}}{T_{\text{l}} - T_{\text{a}}}. \tag{9}$$

This gives four equations (Eqs. 1, 7, 8 and 9) that can be solved for the four unknowns E_{l}, H_{l}, T_{l} and P_{wl}:

$$E_{\text{l}} = \frac{\Delta_{\text{eTa}} c_E (R_{\text{s}} - R_{\text{ll}}) + c_E c_H (P_{\text{was}} - P_{\text{wa}})}{\Delta_{\text{eTa}} c_E + c_H}, \tag{10}$$

$$H_{\text{l}} = \frac{c_H (R_{\text{s}} - R_{\text{ll}}) + c_E c_H (P_{\text{wa}} - P_{\text{was}})}{\Delta_{\text{eTa}} c_E + c_H}, \tag{11}$$

$$T_{\text{l}} = T_{\text{a}} + \frac{(R_{\text{s}} - R_{\text{ll}}) + c_E (P_{\text{wa}} - P_{\text{was}})}{\Delta_{\text{eTa}} c_E + c_H} \tag{12}$$

and

$$P_{\text{wl}} = \frac{\Delta_{\text{eTa}} (R_{\text{s}} - R_{\text{ll}} + P_{\text{wa}} c_E) + P_{\text{was}} c_H}{\Delta_{\text{eTa}} c_E + c_H}. \tag{13}$$

In the original formulations by Penman and Monteith, the term $R_{\text{s}} - R_{\text{ll}}$ is referred to as net available energy, and for a ground surface, it is represented by net radiation minus ground heat flux ($R_{\text{N}} - G$). For a leaf, there is no ground heat flux, and $R_{\text{N}} = R_{\text{s}} - R_{\text{ll}}$. In most applications of the analytical solutions, R_{ll} is not explicitly calculated, but it is assumed that R_{N} is known, neglecting the dependence of R_{ll} on the leaf temperature. This neglect can be alleviated by linearising the equation for R_{ll} (Leuning et al., 1989), which was also done in Sect. 2.4, where we rederive Eqs. (10)–(12) based on a linearised equation for R_{ll}, eliminating the need for separate estimation of R_{N}.

To solve Eqs. (10)–(13), one only needs information about c_H and c_E, appropriate for a leaf or an evaporating surface, whichever is the system of interest. For a planar leaf, $c_H = a_{\text{sH}} h_{\text{c}}$ with $a_{\text{sH}} = 2$ as the leaf exchanges sensible heat on both sides, whereas for a soil surface, $a_{\text{sH}} = 1$. Comparison of Eqs. (4) and (7) with the common representation of $E_{\text{l,mol}}$ as a function of total leaf conductance to water vapour (g_{tw}) and water vapour mole fractions (Eq. B6) suggests that

$$c_E = M_{\text{w}} \lambda_E g_{\text{tw,mol}} / P_{\text{a}}, \tag{14}$$

where $g_{\text{tw,mol}}$ has an aerodynamic component related to g_{bw} (and hence h_{c}) and a surface-specific component, related to g_{sw}, as described in Appendix B1. Since planar leaves can have stomata on one or both sides, the relation between h_{c} and g_{bw} is not universal; i.e. a_{s} in Eq. (B2) can be equal to 1 or 2, whereas for a soil surface $a_{\text{s}} = 1$.

2.3 Inconsistencies in the PM equation

From the general form (Eqs. 10–12), we can recover various analytical forms used for latent heat flux (e.g. Penman, 1948, 1952; Monteith, 1965), with the appropriate substitutions for c_E and c_H. This is shown in detail in the Appendix, Sect. B8, where we also illustrate some inconsistencies in the published derivations. Here, we will discuss errors in the derivation of the PM equation, when it is intended for the simulation of leaf transpiration. The derivation is based on the Penman equation for a wet surface (Penman, 1948), which can be recovered from the above general solution by substituting $c_E = f_{\text{u}}$ and $c_H = \gamma_{\text{v}} f_{\text{u}}$ into Eq. (10) (Fig. 3a):

$$E_{\text{w}} = \frac{\Delta_{\text{eTa}} (R_{\text{s}} - R_{\text{ll}}) + f_{\text{u}} \gamma_{\text{v}} (P_{\text{was}} - P_{\text{wa}})}{\Delta_{\text{eTa}} + \gamma_{\text{v}}}, \tag{15}$$

where f_{u} (W Pa^{-1} m^{-2}) is often referred to as the "wind function".

Monteith (1965) rederived the Penman equation for wet surface evaporation (Eq. 15) using a different set of arguments and arrived at an equivalent equation (Eq. 8 in Monteith, 1965):

$$E_{\text{w}} = \frac{\Delta_{\text{eTa}} (R_{\text{s}} - R_{\text{ll}}) + \rho_{\text{a}} c_{\text{pa}} (P_{\text{was}} - P_{\text{wa}}) / r_{\text{a}}}{\Delta_{\text{eTa}} + \gamma_{\text{v}}}, \tag{16}$$

where r_{a} (s m^{-1}) is the leaf boundary layer resistance to sensible heat flux. Equation (16) is consistent with Eq. (15) if Penman's wind function (f_{u}) is replaced by

$$f_{\text{u}} = \frac{\rho_{\text{a}} c_{\text{pa}}}{\gamma_{\text{v}} r_{\text{a}}}. \tag{17}$$

Monteith pointed out that the ratio between the conductance to sensible heat and the conductance to water vapour transfer, expressed in the psychrometric constant (γ_v, Pa K^{-1}) would be affected by stomatal resistance (r_s, s m^{-1}) and hence proposed to replace the psychrometric constant by γ_v^*:

$$\gamma_v^* = \gamma_v(1 + \frac{r_s}{r_a}), \tag{18}$$

leading to the so-called Penman–Monteith equation for transpiration:

$$E_l = \frac{\Delta_{eTa}(R_s - R_{ll}) + \rho_a c_{pa}(P_{was} - P_{wa})/r_a}{\Delta_{eTa} + \gamma_v\left(1 + \frac{r_s}{r_a}\right)}. \tag{19}$$

To test whether Eq. (19) is physically consistent for a planar leaf, we deduce it from our more general Eq. (10), using suitable definitions for c_E , c_H , r_a and r_s. Equation (19), with γ_v defined in Eq. (B46), could be recovered by substituting $c_E = \epsilon\lambda_E\rho_a/(P_a(r_s + r_v))$ and $c_H = c_{pa}\rho_a/r_a$ into Eq. (10), with subsequent substitution of $r_v = r_a$ (implicit in Eq. 17, considering that $f_u = c_E$). Note, however, that r_a in Monteith's derivation is defined as one-sided resistance to sensible heat exchange (Monteith and Unsworth, 2013, p. 231), neglecting the fact that planar leaves exchange sensible heat on both sides. We suppose that this omission is related to the original Penman derivation, developed for a soil surface, which exchanges latent and sensible heat across one interface, and hence is not appropriate for a leaf. To alleviate this constraint, one could define r_a and r_s as total (two-sided) leaf resistances, but in this case, the simplification $r_v \approx r_a$ is not valid for hypostomatous leaves, as r_v would then be twice the value of r_a. This is illustrated in Fig. 3c, where sensible heat flux is released from both sides of the leaf, while latent heat flux is only released from the abaxial side, implying that $a_{sh} = 2$ and $a_s = 1$.

Monteith and Unsworth (2013) acknowledged that a hypostomatous leaf could exchange sensible heat on two sides, but latent heat on one side only, and proposed to represent this fact by further modifying γ_v^* to

$$\gamma_v^* = n_{MU}\gamma_v(1 + r_s/r_a), \tag{20}$$

where $n_{MU} = 1$ for leaves with stomata on both sides and $n_{MU} = 2$ for leaves with stomata on one side, i.e. $n_{MU} = a_{sh}/a_s$ in our notation. Insertion of Eq. (20) into Eq. (16) yields what we will call the Monteith–Unsworth (MU) equation, which only differs from the Penman–Monteith equation by the additional factor n_{MU}:

$$E_l = \frac{\Delta_{eTa}(R_s - R_{ll}) + \rho_a c_{pa}(P_{was} - P_{wa})/r_a}{\Delta_{eTa} + \gamma_v n_{MU}\left(1 + \frac{r_s}{r_a}\right)}. \tag{21}$$

However, this was done by specifying r_s and r_a as one-sided resistances when inserting them into the term for γ_v in Eq. (16), which was already based on the approximation $r_v \approx r_a$, which is not valid for hypostomatous leaves, as explained above. If we replace r_a by $r_a = r_a/a_{sh}$ in Eq. (16) before substitution of Eq. (20), we obtain a corrected MU equation:

$$E_l = \frac{\Delta_{eTa}(R_s - R_{ll}) + \rho_a c_{pa}(P_{was} - P_{wa})a_{sh}/r_a}{\Delta_{eTa} + \gamma_v a_{sh}/a_s\left(1 + \frac{r_s}{r_a}\right)}, \tag{22}$$

which only differs from Eq. (21) by the factor a_{sh} ($= 2$) in the nominator. Equations (19) and (22) are only equivalent to each other if $a_{sh} = 1 = a_s$, implying that Eq. (19) is not applicable for any planar leaves. For symmetrical amphistomatous leaves, $a_{sh} = 2 = a_s$, in which case the classic PM equation is only missing a factor of 2 in the nominator, as pointed out by Jarvis and McNaughton (1986, Eq. A9).

2.4 Analytical solution including radiative feedback

The above analytical solutions eliminated the non-linearity problem of the saturation vapour pressure curve, but they do not consider the dependency of the long-wave component of the leaf energy balance (R_{ll}) on leaf temperature (T_l), as expressed in Eq. (2). Therefore, the above analytical equations are commonly used in conjunction with fixed value of R_{ll}, either taken from observations or the assumption that $R_{ll} = 0$. Here, we replace the non-linear Eq. (2) by its tangent at $T_l = T_a$, which is given by

$$R_{ll} = 4a_{sh}\epsilon_l\sigma T_a^3 T_l - a_{sh}\epsilon_l\sigma(T_w^4 + 3T_a^4). \tag{23}$$

Note that the common approximation of $T_w = T_a$ simplifies the above equation to $R_{ll} = 4a_{sh}\epsilon_l\sigma(T_a^3 T_l - T_a^4)$. The linearisation introduces a bias of less than -20 W m^{-2} in the calculation of R_{ll} for leaf temperatures ± 20 K of air temperature, compared to Eq. (2) (see Fig. A3).

We can now use a similar procedure as in Sect. 2.2, but this time it is aimed at eliminating P_{wl} using the Penman assumption, rather than eliminating T_l. We first eliminate c_E from Eq. (7) by introducing the psychrometric constant as

$$\gamma_v = c_H/c_E, \tag{24}$$

and introduce it into Eq. (8) to obtain

$$H_l = \gamma_v c_E(T_l - T_a). \tag{25}$$

Then, we insert the Penman assumption (Eq. 9) to eliminate P_{wl} and obtain

$$E_l = \frac{c_H\left(\Delta_{eTa}(T_l - T_a) + P_{was} - P_{wa}\right)}{\gamma_v}. \tag{26}$$

We can now insert the linearised Eqs. (23), (26) and (8) into the energy balance equation (Eq. 1), and solve for leaf tem-

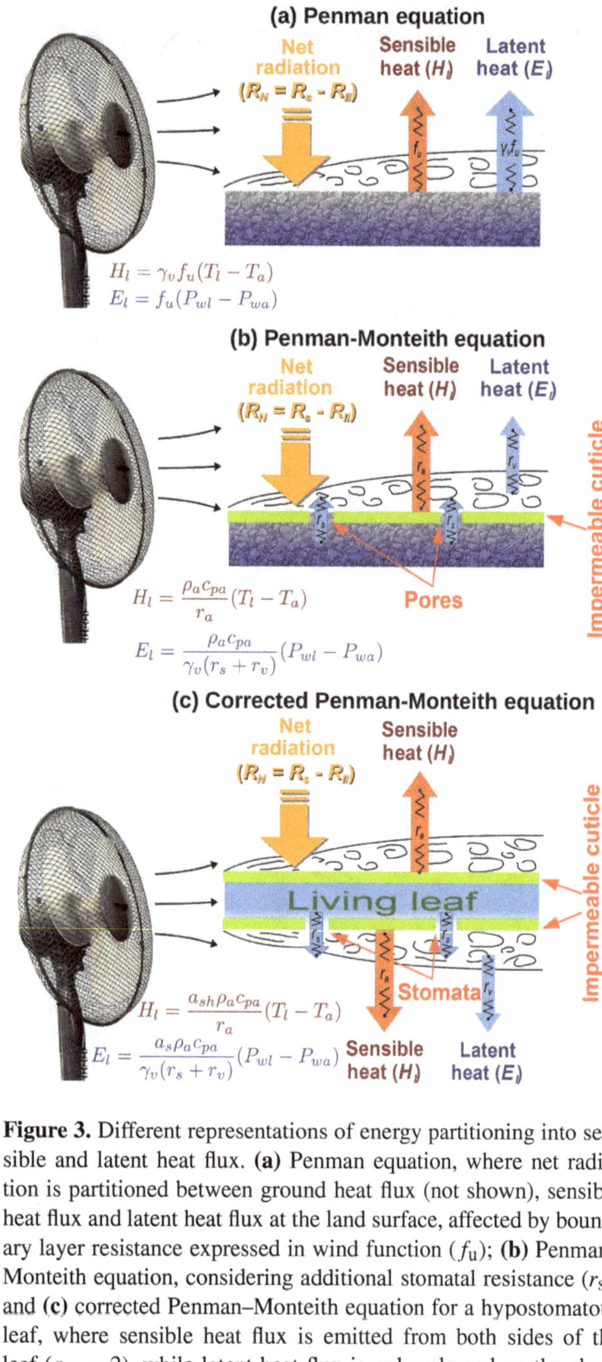

Figure 3. Different representations of energy partitioning into sensible and latent heat flux. **(a)** Penman equation, where net radiation is partitioned between ground heat flux (not shown), sensible heat flux and latent heat flux at the land surface, affected by boundary layer resistance expressed in wind function (f_u); **(b)** Penman–Monteith equation, considering additional stomatal resistance (r_s); and **(c)** corrected Penman–Monteith equation for a hypostomatous leaf, where sensible heat flux is emitted from both sides of the leaf ($a_{sh} = 2$), while latent heat flux is only released on the abaxial (lower) side of the leaf ($a_s = 1$).

perature (T_l) to obtain

$$T_l = \Big(R_s + c_H T_a + c_E (\Delta_{eTa} T_a + P_{wa} - P_{was}) \tag{27}$$

$$+ a_{sh}\epsilon_l \sigma \left(3T_a^4 + T_w^4\right)\Big) \frac{1}{c_H + c_E \Delta_{eTa} + 4a_{sh}\epsilon_l \sigma T_a^3}, \tag{28}$$

where the temperature of the surroundings is commonly assumed to equal air temperature ($T_w = T_a$). Equation (27) can be reinserted into Eqs. (8), (26) and (23) to obtain analytical expressions for H_l, E_l and R_{ll}, respectively, which satisfy the energy balance (Eq. 1). Alternatively, the value of T_l obtained from Eq. (27) for specific conditions could be used to calculate any of the energy balance components using the fundamental equations described in Fig. 2. However, in this case, bias in T_l due to simplifying assumptions included in the derivation of Eq. (27) could result in an unclosed leaf energy balance ($R_s - R_{ll} - H_l - E_l \neq 0$).

2.5 Comparisons of numerical and analytical models with observations

Variations in leaf temperature and leaf energy balance components were simulated using a detailed numerical model (Sect. 2.1), and various analytical solutions, including the Penman–Monteith equation (PM; Eq. 19), the Monteith–Unsworth equation (MU; Eq. 21), our corrected Monteith–Unsworth equation (MUc; Eq. 22) and the analytical solution using linearised net long-wave balance (Rlin, based on Eq. 27). All the above models require similar forcing data, i.e. irradiance or net radiation, air temperature and humidity, wind speed or aerodynamic resistance and stomatal resistance. To compare the adequacy of the different models for capturing the key physical processes, we have used identical environmental forcing in all models, consisting of absorbed short-wave radiation (R_s), air temperature (T_a) and vapour pressure (P_{wa}), wind speed (v_w), stomatal conductance (g_{sw}) and characteristic length of the leaf (L_l). Wind speed and L_l were used to calculate the one-sided convective heat transfer coefficient (h_c, Eq. B10), which is then used to calculate the leaf boundary layer conductance for water vapour (g_{bw}, Eq. B2) in the numerical model. The value of h_c is converted to r_a in the PM equation using Eq. (B51), whereas in our new analytical models $c_H = a_{sh} h_c$. In those models that do not consider feedbacks between leaf temperature and net radiation, i.e. PM, MU and MUc, we assumed that net radiation equals the absorbed short-wave radiation, i.e. $R_N = R_s$ (p. 79 in Monteith and Unsworth, 2013). For verification of the results using experimental data, we designed a new experimental set-up, as described below. The forcing corresponding to each experimental data point was used in the different models as described above, producing a simulation data point by each model for each observation data point. Independent calculation of plausible ranges of stomatal resistance or conductance ($r_s = 1/g_{sw}$) is described below, and from within these plausible ranges, we chose values that led to best possible reproduction of the data by the numerical model.

3 Experimental set-up

To separate the physical aspects of leaf energy and gas exchange from complex biological control, we used artificial leaves with laser-perforated surfaces representing fixed stom-

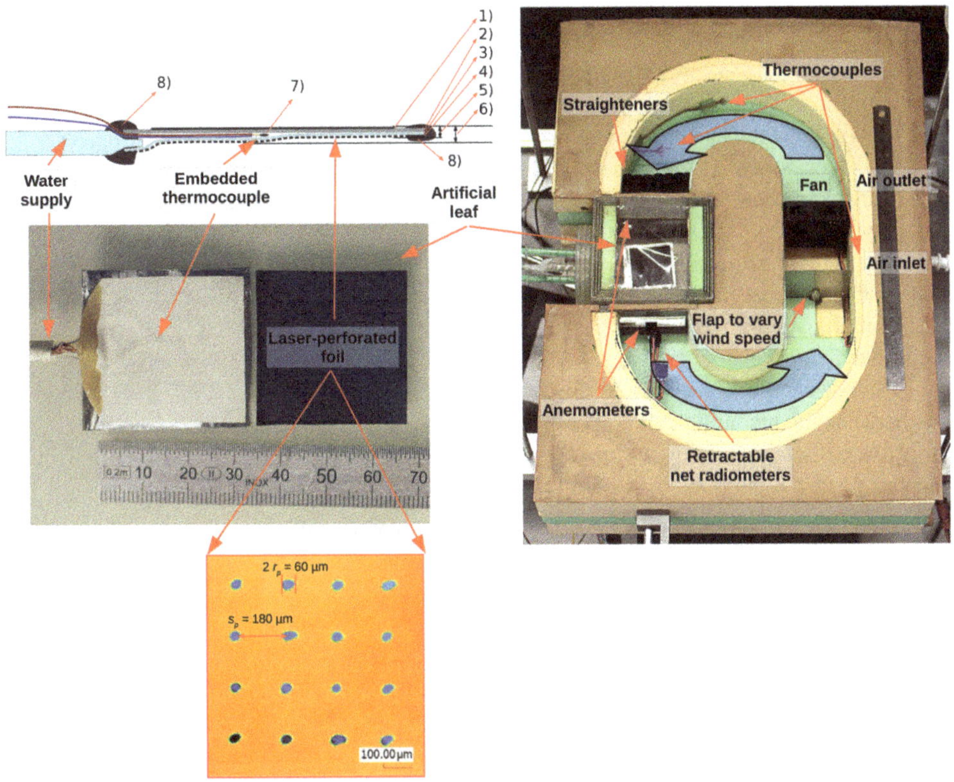

Figure 4. Artificial leaf and wind tunnel. (a): cross section of artificial leaf; (b): leaf image before full assembly; (c): topography of laser-perforated foil with 60 µm pore diameter and 180 µm spacing; (d): wind tunnel. The following are indicated in Panel (a): (1) black aluminium tape (0.05 mm thick); (2) aluminium tape (0.08 mm); (3) absorbent filter paper (0.1–0.2 mm); (4) laser-perforated foil (0.01–0.05 mm); (5) min. leaf thickness: 0.3–0.4 mm; (6) max. leaf thickness: 0.35–0.65 mm; (7) thermocouple; (8) glue.

atal apertures and continuous water supply monitored by micro-flow sensors (Fig. 4a–c). We further constructed a specialised insulated leaf wind tunnel (Fig. 4d) permitting full control of atmospheric conditions including air temperature, humidity, irradiance and wind speed and allowing direct measurement of all leaf energy balance components independently, including net radiation latent and sensible heat flux. A detailed documentation of the leaf wind tunnel and the artificial leaves along with the relevant thermodynamic calculations has been submitted as a technical note to HESS (Schymanski et al., 2016).

3.1 Artificial leaves

The artificial leaves were constructed of a core made of porous filter paper (Whatman no. 41), glued onto aluminium tape and connected to a water supply by a thin tube, flattened at one end and tightly glued between the aluminium foil and the filter paper, using Araldite epoxy resin (Fig. 4a–c). Along with the water supply tube, a thin copper–constantan thermocouple (TG-TI-40) was placed between the filter paper and the adhesive aluminium tape. The water supply was connected to a high-resolution liquid flow metre (SLI-0430, Sensirion AG, Staefa, Switzerland) and a water supply with a water table 1–3 cm below the position of the leaf, to ensure that the liquid flow did not exceed the transpiration rate while maintaining minimum head loss along the flow path.

Different laser perforations were performed by Ralph Beglinger (Lasergraph AG, Würenlingen, Switzerland), Robert Voss (ETH Zurich, Switzerland) and Rolf Brönnimann (EMPA, Zurich, Switzerland), and the geometry of laser perforations was measured using a confocal laser scanning microscope (CLSM VK-X200, Keyence, Osaka, Japan). See Fig. A2 for examples.

The stomatal conductance resulting from a particular perforation size and density was computed following the derivations presented by Lehmann and Or (2015), assuming that the stomatal conductance results from two resistances in a series: the throat resistance (r_{sp}), resulting from the width of the perforation and the thickness of the perforated foil, and the vapour shell resistance (r_{vs}), resulting from the size and spacing of the stomata, which can be understood as the resistance related to distribution of the point-source water vapour over the entire one-sided leaf boundary layer. We hereby neglect any internal resistance (termed "end correction" by Lehmann and Or, 2015), as we assume that the wet filter paper has direct contact with the perforated foil. The relevant equations are described in Appendix B10.

3.2 Leaf wind tunnel

Leaf energy and gas exchange were measured in a thermally insulated wind tunnel with full control over energy and mass exchange (Fig. 4). The wind tunnel is circular, with two straight sections of 25 cm length each, a fan in one of the straight sections and a transparent window and leaf holder in the opposite straight channel. The fan circulates the wind as indicated by the arrows in Fig. 4d, subjecting it to controlled wind conditions. The wind tunnel features an air inlet just before the fan and an air outlet just after the fan, where the air is assumed to be well mixed across the tunnel cross section. In this way, leaf gas exchange can be deduced from the concentration difference between the incoming and outgoing air and the controlled flow rate of air into the wind tunnel. For this purpose, the incoming air was supplied by a humidifier providing prescribed vapour pressure and flow rate.

The sensible heat flux (H_l) was deduced from the chamber energy balance by computing the amount of heat exchanged with the surroundings through the exchange of air and subtracting the amount of heat added by the fan. Since the fan was placed inside the chamber, the amount of heat it injected was assumed to be equal to its power consumption, which was kept constant by a programmable power controller, while wind speed was varied by adjusting the position of a wing in the flow path (Fig. 4) and monitored using a miniature thermal flow sensor. A stack of 3 cm long plastic straws in the flow path was used to reduce spiralling of the air flow caused by the rotating fan. The main wind tunnel was built out of foamed insulation material, while the leaf chamber itself had two layers of polypropylene foil on each side (above and below the leaf) to permit the transmission of short-wave and long-wave radiation while minimising conductive heat transfer (see position of the artificial leaf in Fig. 4). We used retractable miniature net radiation sensors to periodically measure the net radiative load on the leaf. Copper–constantan thermocouples were placed in the air stream upstream and downstream of the leaf chamber, lightly inserted into the wind tunnel wall on the inside and the outside of the chamber, and in the duct through which air was supplied to the wind tunnel by an external humidifier, providing a flow rate of up to 10 L min^{-1} and controlled air temperature and dew point.

The leaf wind tunnel was used to measure steady-state conditions under given forcing (air temperature, humidity, wind speed and irradiance). Sensible heat exchange between the leaf and the surrounding air was computed from total chamber heat exchange, using monitored flow rate and temperature of incoming and outgoing air (Fig. 5). The relevant thermodynamic calculations are presented in a separate technical note (Schymanski et al., 2016).

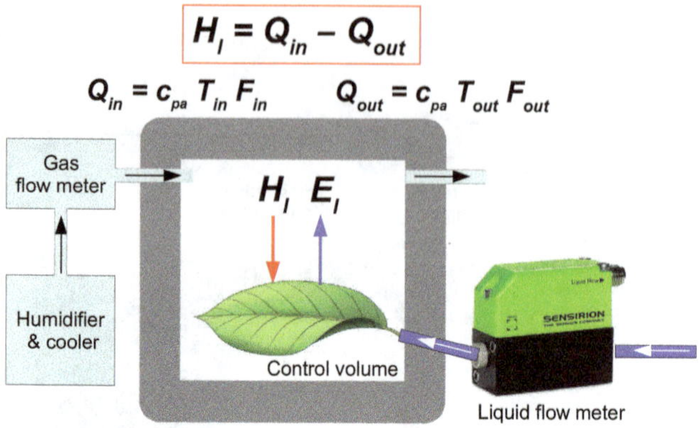

Figure 5. Simplified energy balance of insulated wind tunnel. Latent heat flux (E_l) is calculated from liquid flow rate into leaf, sensible heat flux (H_l) is calculated from difference in heat content of incoming and outgoing air (c_{pa}: heat capacity of air; T_{in}, T_{out}: air temperatures of incoming and outgoing air; F_{in}, F_{out}: incoming and outgoing air flow rates).

Table 1. Perforation characteristics and resulting stomatal conductances, computed using Eqs. (B55) and (B56), following the procedure described in the Appendix, Sect. B10. Foil thickness is 25 μm.

Pore density mm^{-2}	Pore area μm^{-2}	Pore radius μm	g_{sw} m s^{-1}
27.3–38.2	710–1572	15–22	0.022–0.046
7.1–7.8	890–1886	16–24	0.006–0.012

g_{sw}: calculated stomatal conductance.

4 Results

4.1 Correspondence between experimental results and numerical model

Experiments were performed for various artificial leaves with different stomatal conductances under varying air humidity or varying wind speed, in the absence of short-wave radiation. Stomatal conductance was deduced form confocal laser scanning microscope (CLSM) images of the perforated foils, as described above. The ranges of stomatal geometries and deduced conductances for the two different leaves presented here are given in Table 1. A more detailed analysis of correspondence between experimental results obtained for a larger variety of artificial leaves and the numerical model are presented in a technical note (Schymanski et al., 2016). Here, we only present selected experiments that highlight systematic differences between the various analytical solutions, the numerical model and observations.

The numerical model reproduced observed sensible and latent heat fluxes very accurately (Fig. 6) using stomatal conductance values within the narrow ranges deduced from CLSM images (Table 1) with no other forms of calibration.

The experimental conditions and stomatal conductances are given in the figure captions. Observed net radiation (R_{nleaf}) was a little bit below the sum of observed latent and sensible heat fluxes ($E_{\mathrm{l}} + H_{\mathrm{l}}$, Fig. 6), suggesting that the energy balance was not entirely closed. Simulated net radiation ($R_{\mathrm{nleaf}} = R_{\mathrm{s}} - R_{\mathrm{ll}}$) was in between the observed R_{nleaf} and the observed $E_{\mathrm{l}} + H_{\mathrm{l}}$. Consistent with Fig. 1, the net radiative exchange was not sensitive to wind speed in Fig. 6a, while E_{l} and H_{l} responded strongly (in opposite directions).

4.2 Performance of analytical leaf gas exchange and energy balance models

The analytical models generally underestimated latent heat flux, but the model based on linearised R_{ll} (Rlin) showed very little bias and closely reproduced the observed latent and sensible heat fluxes, as it permitted calculation of the net long-wave component (in contrast with PM, MU and MUc expressions that assumed $R_{\mathrm{ll}} = 0$). The calculations based on the Penman–Monteith equation significantly underestimated latent heat flux, especially at high stomatal conductances (PM values almost 50 % lower than the observed values in Fig. 7). The Monteith–Unsworth (MU) equation produced an even stronger underestimation of latent heat flux in our results, whereas our corrected Monteith–Unsworth (MUc) equation was a lot closer to the observed heat fluxes than either the MU or the PM equations. However, only Eq. (27) (Rlin) was able to capture the asymmetry between latent and sensible heat fluxes caused by net absorption of long-wave radiation, as all the other calculations were based on the assumption of zero radiative exchange ($R_{\mathrm{ll}} = 0$), i.e. $H_{\mathrm{l}} = -E_{\mathrm{l}}$. Our results suggest that the omission of the long-wave radiative feedback (MUc) resulted in a much smaller effect than the omission of two-sided sensible heat exchange (PM, MU), compared to the most comprehensive analytical solution (Rlin) and observations.

Since we were not able to systematically assess the effects of irradiance and air temperature in our lab experiments, we conducted a numerical experiment where we compared simulations by the numerical model with simulations by the best analytical model and the PM equation. The results shown in Fig. 8 suggest that our new analytical solution (Eq. 27) behaves very similarly to the numerical model, whereas the PM equation misrepresents the sensitivities of latent and sensible heat fluxes to both irradiance and air temperature.

5 Discussion

> This age values usefulness more highly than correctness, and the making of money more highly than both. In fact, there is definitely something suspect about an examiner who would bother at all with whether an idea is correct or not. (Raupach and Finnigan, 1988)

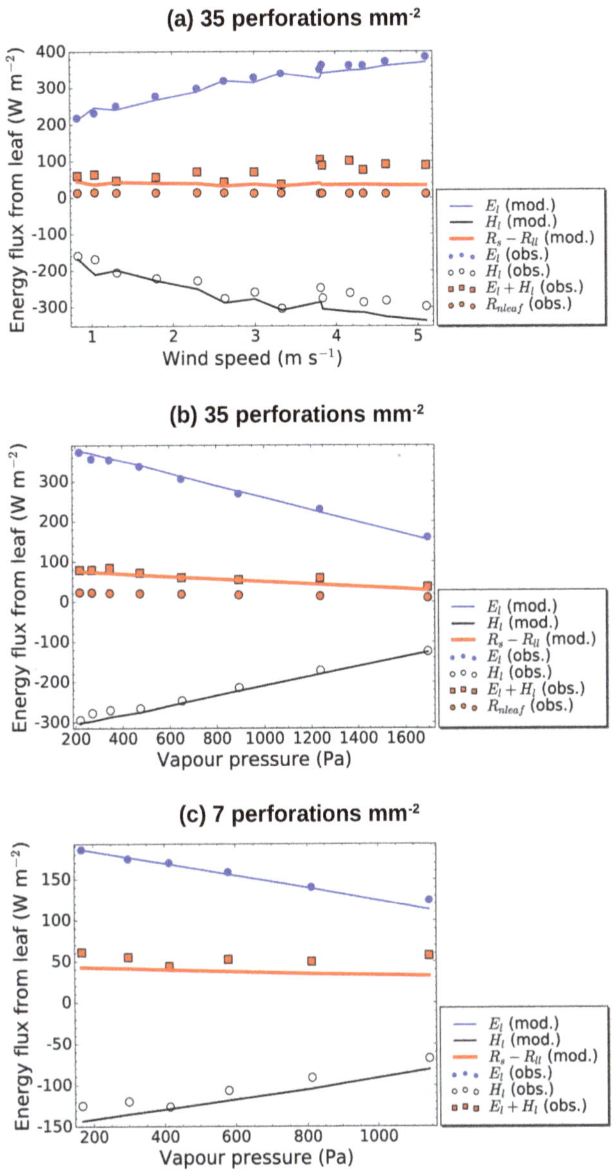

Figure 6. Numerical simulations vs. observed fluxes of sensible, latent and radiative heat in response to varying wind speed and vapour pressure. Numerical model results (lines) are based on observed boundary conditions representative of observations (dots). The boundary conditions are summarised as follows: **(a)** $g_{\mathrm{sw}} = 0.042\,\mathrm{m\,s^{-1}}$; $R_{\mathrm{s}} = 0$; $T_{\mathrm{a}} = 295.0$–$296.5\,\mathrm{K}$; $P_{\mathrm{wa}} = 1187$–$1278\,\mathrm{Pa}$; **(b)** $g_{\mathrm{sw}} = 0.035\,\mathrm{m\,s^{-1}}$; $R_{\mathrm{s}} = 0$; $T_{\mathrm{a}} = 295.7$–$296.0\,\mathrm{K}$; $v_{\mathrm{w}} = 1.0\,\mathrm{m\,s^{-1}}$; **(c)** $g_{\mathrm{sw}} = 0.0065\,\mathrm{m\,s^{-1}}$; $R_{\mathrm{s}} = 0$; $T_{\mathrm{a}} = 296.1$–$296.7\,\mathrm{K}$; $v_{\mathrm{w}} = 0.7\,\mathrm{m\,s^{-1}}$. g_{sw}: stomatal conductance; T_{a}: air temperature; P_{wa}: vapour pressure; E_{l}: latent heat flux; H_{l}: sensible heat flux; R_{s}: absorbed short-wave radiation; R_{ll}: net emitted long-wave radiation; $R_{\mathrm{nleaf}} = R_{\mathrm{s}} - R_{\mathrm{ll}}$.

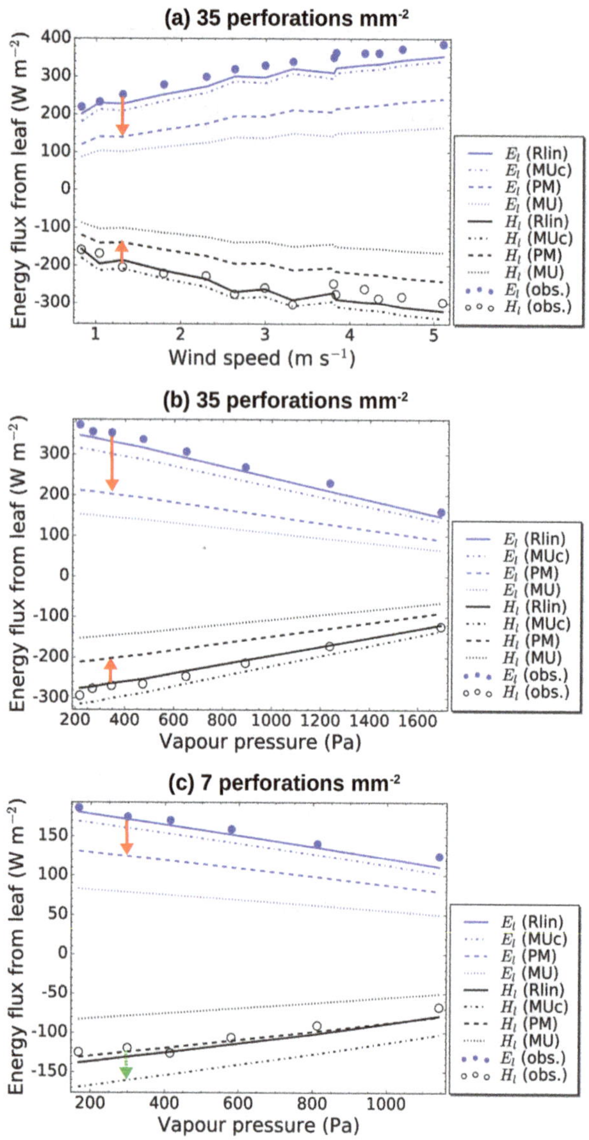

Figure 7. Analytical simulations vs. observed fluxes of sensible and latent heat in response to varying wind and vapour pressure. Numerical model results (lines) are based on observed boundary conditions representative of observations (dots). Conditions are the same as in Fig. 6. E_l: latent heat flux; H_l: sensible heat flux; Rlin: based on linearised long-wave balance (Eq. 27); MUc: corrected Monteith–Unsworth equation (Eq. 22); PM: Penman–Monteith equation (Eq. 19); MU: Monteith–Unsworth equation (Eq. 21). Red arrows indicate the magnitudes of biases in the PM equation; the dashed green arrow marks the maximum bias in sensible heat flux in the MUc equation for the experimental conditions.

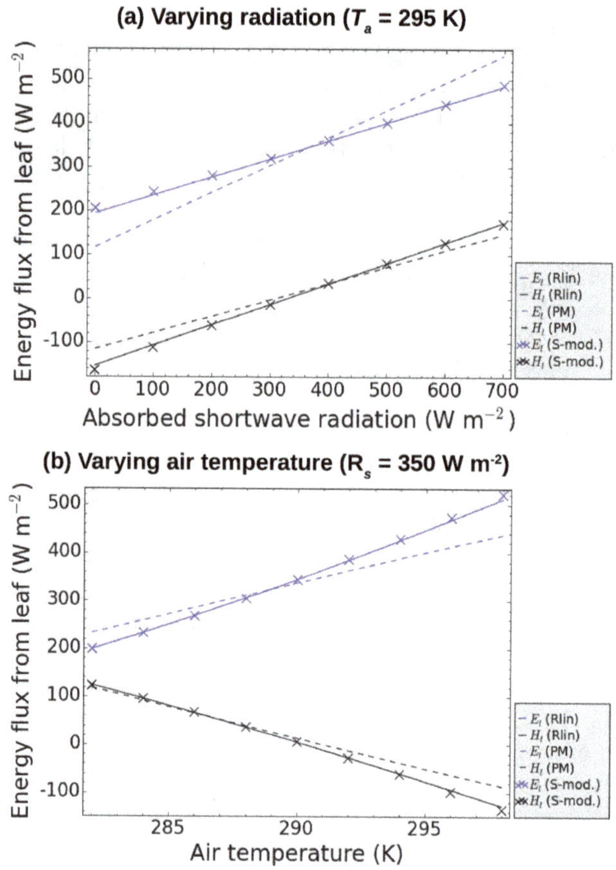

Figure 8. Numerical vs. analytical simulations of sensible and latent heat in response to varying irradiance and air temperature. Crosses represent numerical solution of leaf energy balance model (S-mod.), solid lines our new analytical solution based on linearised long-wave balance (Rlin; Eq. 27) and dashed lines the Penman–Monteith equation (PM; Eq. 19). Simulation conditions: $g_{sw} = 0.045\ \mathrm{m\,s^{-1}}$; $P_{wa} = 1300\ \mathrm{Pa}$; $v_w = 1\ \mathrm{m\,s^{-1}}$. g_{sw}: stomatal conductance; T_a: air temperature; P_{wa}: vapour pressure; E_l: latent heat flux; H_l: sensible heat flux; R_s: absorbed short-wave radiation.

The widespread use of the PM equation is mainly due to its simplicity and usefulness, the latter of which is contingent on its ability to accurately represent the sensitivity of evapotranspiration to atmospheric variables and surface properties (boundary layer and bulk stomatal conductances).

In our rederivation and subsequent analyses, we have identified two errors in the PM equation and in the "corrected" MU formulation by Monteith and Unsworth (2013). Both formulations are based on evaporation from a soil surface, which exchanges sensible and radiative heat only on one side, whereas planar leaves have two sides exposed to the surrounding air. Failure to recognise this omission has led to a second error in the MU formulation, where an additional reduction to transpiration was introduced to represent leaves that exchange water vapour from one side only. For a leaf, the energy for transpiration in darkness is mainly supplied by sensible heat flux (on both sides), which increases with increasing wind speed. In contrast, the energy for evaporation from a soil surface in darkness is supplied by sensible heat on the evaporating surface only (and by soil heat flux from below). The neglect of the additional exchange of sensible heat on the second side of the leaf in the PM and MU models led to significant underestimation of the observed transpiration rates in our experiments, where the sensible heat flux is the main source of energy for transpiration (in the absence of short-wave radiation). Note, however, that the bias is not constant and not always negative. As illustrated in Fig. 8, the negative bias decreases with increasing irradiance or air temperature, goes to 0 at a certain combination of temperature and irradiance and then becomes positive at higher values of irradiance and/or temperature. This is because under conditions where the leaf temperature is lower than ambient, sensible heat flux is a source of energy for transpiration, whereas under conditions where the leaf is warmer than the air, sensible heat flux competes for energy with transpiration. The omission of sensible heat exchange by the second leaf surface has therefore the most drastic effects when leaf temperature most strongly deviates from air temperature. It may also be noteworthy in this context that the expression for aerodynamic resistance (r_a) given by Monteith (1965, Eq. 14) has been pointed out by other authors to result in heat transfer 2.5 times higher than expected if interpreted as a one-sided resistance (Parlange et al., 1971). This may have arisen from the confusion about one-sided vs. two-sided energy exchange. Our experimental results clearly illustrate that the inconsistencies we found in the PM and MU equations are not just semantic, but actually lead to very significant biases in simulated transpiration rates for known stomatal resistance, which would alternatively lead to biases in deduced resistance for known transpiration rates. The results further illustrate that our correction for two-sided leaves improves reproduction of leaf-scale measurements tremendously (MUc vs. PM in Fig. 7), but additional consideration of the surface temperature–long-wave emission feedback (Eq. 27 and Rlin in Fig. 7) is also important to accurately capture the characteristics of the leaf energy balance, particularly the sensible heat flux.

Although the upscaling of a physically based leaf-scale model to a canopy or land surface is fraught with various challenges, including characterisation of the stomatal or canopy conductance, canopy-scale boundary layer conductance, consideration of canopy storage and distinction between radiative and aerodynamic surface temperatures (Monteith, 1965; Tanner and Fuchs, 1968; Jarvis and McNaughton, 1986; Raupach, 1995; Mallick et al., 2013, 2016), we believe that care must be taken to start off with the correct leaf-scale model. In this context, we wish to point out that Monteith (1965) referred to a single leaf when deriving the PM equation (evident in the abstract and from Page 208 onwards in his paper), and that the use of the PM equation at canopy scale is commonly motivated in the context of a big leaf analogy or by aggregation of many representative leaves, implying that the physics valid for a leaf is also valid for a canopy (e.g. Lhomme et al., 2012; Verhoef and Allen, 2000). Therefore, the omission of the second side of a leaf when Monteith (1965) used the Penman equation as a basis for his derivation is likely relevant when representing canopy fluxes using the PM equation.

In the present study, we have developed an experimental set-up allowing to control all relevant boundary conditions at the leaf scale, including stomatal conductance, and measuring (to our knowledge, for the first time) all components of the leaf energy balance. In contrast to previous tests of the PM equation, which were conducted at the canopy scale, where boundary layer and canopy conductances could not be measured directly, we have been able to greatly constrain model parametrisation by independent measurements of stomatal conductance. This has led to the discovery that the PM equation, in its original formulation and common use, does not accurately represent leaf-scale processes. Our newly derived analytical solutions (Eqs. 27 and 22) are able to not only more accurately reproduce leaf-scale sensible and latent heat fluxes, but they also allow direct calculation of leaf temperature, which could be used as an additional diagnostic variable at the canopy scale and potentially improve remote-sensing-based evaporation products.

Given the widespread and successful use of the PM equation, the question arises whether common practice, which relies on parameterisation by fitting resistance terms that provide a match with observations, somehow compensates for the errors we identified in the present study. The answer is "yes and no". As shown in Fig. 8, there are certain conditions for which the PM equation and the corrected solutions yield very similar results, and one could easily obtain a much closer match to the experimental results by fitting significantly larger values for r_a and r_s in the PM equation than those estimated from diffusive resistance of perforated surfaces (the laser perforations in our artificial leaves). This may also explain the lack of general bias in PM-estimated transpiration rates when applied at the canopy scale (e.g. Langensiepen et al., 2009). However, the sensitivity of latent and sensible heat flux to changing atmospheric conditions (e.g. short-wave irradiance and air temperature) deduced from the PM equation would clearly be different than the trends produced by the corrected equations and numerical simulations (Fig. 8). This suggests that use of the PM equation for projections under future climate change scenarios could lead to a bias in the results. This potential source of bias could be reduced using the corrected equation presented in this study (irrespective of the estimated resistance values fitted for a canopy).

6 Conclusions

In this study, we revisit the governing equations for the exchange of water vapour and energy between a planar leaf and a surrounding air stream under forced convection. We derived general analytical solutions for steady-state sensible and latent heat fluxes from a leaf and the corresponding leaf temperature (Eqs. 10–12) based on the approach by Penman (1952). The general equation permits comparison between different analytical solutions available in the literature by substituting appropriate formulations of the sensible and latent heat transfer coefficients. Our analysis reveals that the Penman–Monteith equation (Eq. 19), even with its modification by Monteith and Unsworth (2013) (Eq. 21) is not accurate for a typical planar leaf due to omission of the radiative and sensible heat fluxes from one side of a leaf. We demonstrated how our general solution can be used to obtain a more consistent representation of leaf energy and gas exchange, in agreement with leaf-scale experimental data (using artificial leaves). We propose that the same approach could prove useful to derive a more accurate canopy-scale representation of latent and sensible heat fluxes, considering their coupling with radiative exchange and ground heat flux. The new generalised leaf-scale equations offer a promise for more consistent responses of latent and sensible heat fluxes to changes in atmospheric forcing in future climates than the responses predicted by the original PM equation (due to the omissions therein).

Appendix A: Additional tables and figures

Table A1. Table of symbols and standard values used in this paper. All area-related variables are expressed per unit leaf area.

Variable	Description (value)	Units
A_p	Cross-sectional pore area	$\mathrm{m^2}$
a_s	Fraction of one-sided leaf area covered by stomata (1 if stomata are on one side only, 2 if they are on both sides)	1
a_sh	Fraction of projected area exchanging sensible heat with the air (2)	1
α_a	Thermal diffusivity of dry air	$\mathrm{m^2\,s^{-1}}$
β_B	Bowen ratio (sensible / latent heat flux)	1
c_E	Latent heat transfer coefficient	$\mathrm{J\,Pa^{-1}\,m^{-2}\,s^{-1}}$
c_H	Sensible heat transfer coefficient	$\mathrm{J\,K^{-1}\,m^{-2}\,s^{-1}}$
c_pa	Specific heat of dry air (1010)	$\mathrm{J\,K^{-1}\,kg^{-1}}$
C_wa	Concentration of water in the free air	$\mathrm{mol\,m^{-3}}$
C_wl	Concentration of water in the leaf air space	$\mathrm{mol\,m^{-3}}$
d_p	Pore depth	m
D_va	Binary diffusion coefficient of water vapour in air	$\mathrm{m^2\,s^{-1}}$
Δ_eTa	Slope of saturation vapour pressure at air temperature	$\mathrm{Pa\,K^{-1}}$
E_l	Latent heat flux from leaf	$\mathrm{J\,m^{-2}\,s^{-1}}$
$E_\mathrm{l,mol}$	Transpiration rate in molar units	$\mathrm{mol\,m^{-2}\,s^{-1}}$
E_w	Latent heat flux from a wet surface	$\mathrm{J\,m^{-2}\,s^{-1}}$
ϵ	Water-to-air molecular weight ratio (0.622)	1
ϵ_l	Long-wave emissivity of the leaf surface (1.0)	1
f_u	Wind function in Penman approach, $f(u)$ adapted to energetic units	$\mathrm{J\,Pa^{-1}\,m^{-2}\,s^{-1}}$
g	Gravitational acceleration (9.81)	$\mathrm{m\,s^{-2}}$
g_bw	Boundary layer conductance to water vapour	$\mathrm{m\,s^{-1}}$
$g_\mathrm{bw,mol}$	Boundary layer conductance to water vapour	$\mathrm{mol\,m^{-2}\,s^{-1}}$
g_sw	Stomatal conductance to water vapour	$\mathrm{m\,s^{-1}}$
$g_\mathrm{sw,mol}$	Stomatal conductance to water vapour	$\mathrm{mol\,m^{-2}\,s^{-1}}$
g_tw	Total leaf conductance to water vapour	$\mathrm{m\,s^{-1}}$
$g_\mathrm{tw,mol}$	Total leaf layer conductance to water vapour	$\mathrm{mol\,m^{-2}\,s^{-1}}$
γ_v	Psychrometric constant	$\mathrm{Pa\,K^{-1}}$
h_c	Average one-sided convective transfer coefficient	$\mathrm{J\,K^{-1}\,m^{-2}\,s^{-1}}$
H_l	Sensible heat flux from leaf	$\mathrm{J\,m^{-2}\,s^{-1}}$
k_a	Thermal conductivity of dry air	$\mathrm{J\,K^{-1}\,m^{-1}\,s^{-1}}$
k_dv	Ratio D_va / V_m	$\mathrm{mol\,m^{-1}\,s^{-1}}$
L_l	Characteristic length scale for convection (size of leaf)	m
λ_E	Latent heat of evaporation (2.45e6)	$\mathrm{J\,kg^{-1}}$
$M_\mathrm{N_2}$	Molar mass of nitrogen (0.028)	$\mathrm{kg\,mol^{-1}}$
$M_\mathrm{O_2}$	Molar mass of oxygen (0.032)	$\mathrm{kg\,mol^{-1}}$
M_w	Molar mass of water (0.018)	$\mathrm{kg\,mol^{-1}}$
N_{Gr_L}	Grashof number	1
N_{Le}	Lewis number	1
n_MU	$n = 2$ for hypostomatous, $n = 1$ for amphistomatous leaves	1
N_{Nu_L}	Nusselt number	1
n_p	Pore density	$\mathrm{m^{-2}}$
N_{Pr}	Prandtl number (0.71)	1
N_{Re_c}	Critical Reynolds number for the onset of turbulence	1
N_{Re_L}	Reynolds number	1
N_{Sh_L}	Sherwood number	1
ν_a	Kinematic viscosity of dry air	$\mathrm{m^2\,s^{-1}}$
P_a	Air pressure	Pa
$P_\mathrm{N_2}$	Partial pressure of nitrogen in the atmosphere	Pa
$P_\mathrm{O_2}$	Partial pressure of oxygen in the atmosphere	Pa

Table A1. Continued.

Variable	Description (value)	Units
P_{wa}	Vapour pressure in the atmosphere	Pa
P_{was}	Saturation vapour pressure at air temperature	Pa
P_{wl}	Vapour pressure inside the leaf	Pa
r_a	One-sided boundary layer resistance to heat transfer (r_H in Monteith and Unsworth (2013, p. 231))	$s\,m^{-1}$
r_{bw}	Boundary layer resistance to water vapour, inverse of g_{bw}	$s\,m^{-1}$
R_{ll}	Long-wave radiation away from leaf	$J\,m^{-2}\,s^{-1}$
R_{mol}	Molar gas constant (8.314472)	$J\,K^{-1}\,mol^{-1}$
r_p	Pore radius (for ellipsoidal pores, half the pore width)	m
R_s	Solar short-wave flux	$J\,m^{-2}\,s^{-1}$
r_s	Stomatal resistance to water vapour (Monteith and Unsworth, 2013, p. 231)	$s\,m^{-1}$
r_{sp}	Diffusive resistance of a stomatal pore	$s\,m^2\,mol^{-1}$
r_{sw}	Stomatal resistance to water vapour, inverse of g_{sw}	$s\,m^{-1}$
r_{tw}	Total leaf resistance to water vapour, $r_{bv} + r_{sv}$	$s\,m^{-1}$
r_v	Leaf BL resistance to water vapour; Monteith and Unsworth, 2013, Eq. 13.16	$s\,m^{-1}$
r_{vs}	Diffusive resistance of a stomatal vapour shell	$s\,m^2\,mol^{-1}$
ρ_a	Density of dry air	$kg\,m^{-3}$
ρ_{al}	Density of air at the leaf surface	$kg\,m^{-3}$
S	Factor representing stomatal resistance in Penman (1952)	1
s_p	Spacing between stomata	m
σ	Stefan–Boltzmann constant ($5.67e-8$)	$J\,K^{-4}\,m^{-2}\,s^{-1}$
T_a	Air temperature	K
T_l	Leaf temperature	K
T_w	Radiative temperature of objects surrounding the leaf	K
V_m	Molar volume of air	$m^3\,mol^{-1}$
v_w	Wind velocity	$m\,s^{-1}$

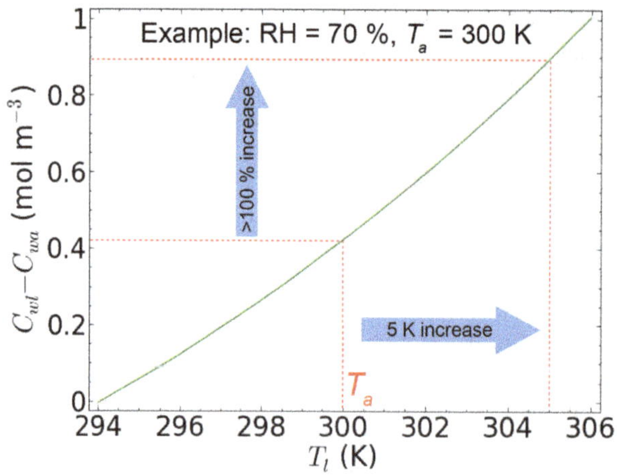

Figure A1. Dependence of the leaf–air water vapour concentration difference ($C_{wl} - C_{wa}$) on leaf temperature (T_l). In this example, (70 % relative humidity, 300 K air temperature (T_a), the water vapour concentration difference doubles for an increase in leaf temperature by 5 K relative to air temperature or drops to 0 for a decrease in leaf temperature by 6 K. Plot obtained by inserting Eq. (B5) into Eq. (B4). C_{wa} was obtained substituting T_a for T_l and multiplying Eq. (B5) by the assumed relative humidity of 0.7.

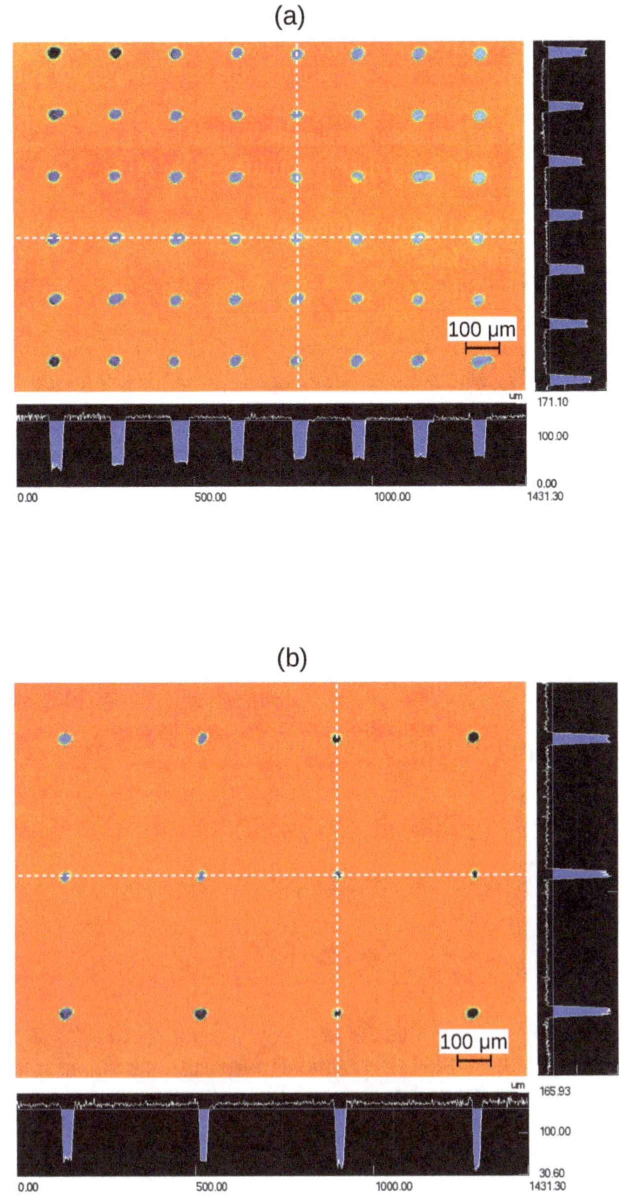

Figure A2. Example confocal laser scanning microscope (CLSM) images of perforated foils summarised in Table 1. **(a)** 35 perforations per mm^2, **(b)** 7.8 perforations per mm^2. Colours represent surface elevation, black bars at the bottom and on the right of each picture show topographic profiles of transects crossing perforations (white dashed lines in main images), with the detection thresholds (10 µm below surface) marked as blue-filled areas.

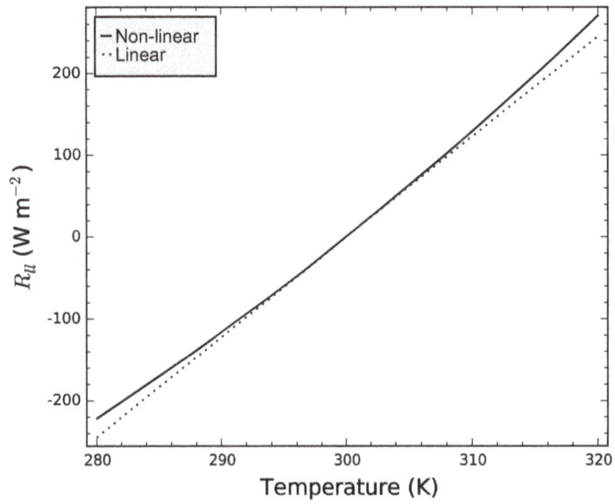

Figure A3. Net long-wave radiation away from leaf as a function of leaf temperature. Solid line represents Eq. (2), while the dotted line represents the linearised function (Eq. 23). Calculations are based on 300 K air and wall temperature (T_a and T_w, respectively).

Appendix B: Mathematical derivations

B1 Boundary layer conductance to water vapour

The total leaf conductance to water vapour is determined by the boundary layer and stomatal conductances and equal to 1 over the sum of their respective resistances ($g_{\text{tw}} = 1/(r_{\text{sw}} + r_{\text{bw}})$. The boundary layer conductance for water vapour is equivalent to the mass transfer coefficient for a wet surface (Incropera et al., 2006, Eq. 7.41):

$$g_{\text{bw}} = N_{Sh_L} D_{\text{va}}/L_{\text{l}}, \tag{B1}$$

where N_{Sh_L} is the dimensionless Sherwood number and D_{va} is the diffusivity of water vapour in air. If the convection coefficient for heat is known, the one for mass (g_{bw}) can readily be calculated from the relation (Incropera et al., 2006, Eq. 6.60)

$$g_{\text{bw}} = \frac{a_{\text{s}} h_{\text{c}}}{\rho_{\text{a}} c_{\text{pa}} N_{Le}^{1-n}}, \tag{B2}$$

where a_{s} is the fraction of one-sided transpiring surface area in relation to the surface area for sensible heat exchange, c_{pa} is the constant-pressure heat capacity of air, n is an empirical constant ($n = 1/3$ for general purposes) and N_{Le} is the dimensionless Lewis number, defined as (Incropera et al., 2006, Eq. 6.57)

$$N_{Le} = \alpha_{\text{a}}/D_{\text{va}}, \tag{B3}$$

where α_{a} is the thermal diffusivity of air. The value of a_{s} was set to 1 for leaves with stomata on one side only, and to 2 for stomata on both sides. Other values could be used for leaves only partly covered by stomata.

B2 Effect of leaf temperature on the leaf–air vapour concentration gradient

The concentration difference in Eq. (5) is a function of the temperature and the vapour pressure differences between the leaf and the free air. Assuming that water vapour behaves like an ideal gas, we can express its concentration as

$$C_{\text{wl}} = \frac{P_{\text{wl}}}{R_{\text{mol}} T_{\text{l}}}, \tag{B4}$$

where P_{wl} is the vapour pressure inside the leaf, R_{mol} is the universal gas constant and T_{l} is leaf temperature. A similar relation holds for the vapour concentration in free air, $C_{\text{wa}} = P_{\text{wa}}/(R_{\text{mol}} T_{\text{l}})$. In this study, the vapour pressure inside the leaf is assumed to be the saturation vapour pressure at leaf temperature, which is computed using the Clausius–Clapeyron relation (Hartmann, 1994, Eq. B.3):

$$P_{\text{wl}} = 611 \exp\left(\frac{\lambda_E M_{\text{w}}}{R_{\text{mol}}}\left(\frac{1}{273} - \frac{1}{T_{\text{l}}}\right)\right), \tag{B5}$$

where λ_E is the latent heat of vaporisation and M_{w} is the molar mass of water.

Note that the dependence of the leaf–air water concentration difference ($C_{\text{wl}} - C_{\text{wa}}$) in Eq. (B4) is very sensitive to leaf temperature. For example, if the leaf temperature increases by 5 K relative to air temperature, $C_{\text{wl}} - C_{\text{wa}}$ would double, while if leaf temperature decreased by 6 K, $C_{\text{wl}} - C_{\text{wa}}$ would go to 0 at 70 % relative humidity (Fig. A1).

B3 Concentration or vapour pressure gradient driving transpiration?

Note that $E_{\text{l,mol}}$ is commonly expressed as a function of the vapour pressure difference between the free air (P_{wa}) and the leaf (P_{wl}), in which the conductance ($g_{\text{tw,mol}}$) is expressed in molar units (mol m^{-2} s^{-1}):

$$E_{\text{l,mol}} = g_{\text{tw,mol}} \frac{P_{\text{wl}} - P_{\text{wa}}}{P_{\text{a}}}. \tag{B6}$$

For $P_{\text{wl}} = P_{\text{wa}}$, Eq. (5) can still give a flux, whereas Eq. (B6) gives zero flux. This is because the concentrations of vapour in air (mol m^{-3}) can differ due to differences in temperature, even if the partial vapour pressures are the same (see Eq. B4). Therefore, the relation between g_{tw} and $g_{\text{v,mol}}$ has an asymptote at the equivalent temperature. It can be obtained by combining Eqs. (5) and (B6) and solving for $g_{\text{tw,mol}}$:

$$g_{\text{tw,mol}} = g_{\text{tw}} \frac{P_{\text{a}}(P_{\text{wa}} T_{\text{l}} - P_{\text{wl}} T_{\text{a}})}{(P_{\text{wa}} - P_{\text{wl}}) R_{\text{mol}} T_{\text{a}} T_{\text{l}})}. \tag{B7}$$

For $T_{\text{l}} = T_{\text{a}}$, the relation simplifies to

$$g_{\text{tw,mol}} = g_{\text{tw}} \frac{P_{\text{a}}}{R_{\text{mol}} T_{\text{a}}}, \tag{B8}$$

which, for typical values of P_{a} and T_{a}, amounts to $g_{\text{tw,mol}} \approx$ 40 mol m$^{-3} g_{\text{tw}}$. For all practical purposes, we found that Eqs. (5) and (B6) with $g_{\text{tw,mol}} = g_{\text{tw}} \frac{P_{\text{a}}}{R_{\text{mol}} T_{\text{a}}}$ give similar results when plotted as functions of leaf temperature.

B4 Model closure

Given climatic forcing as P_{a}, T_{a}, R_{s}, P_{wa} and v_{w}, and leaf-specific parameters a_{s}, a_{sH}, L_{l} and g_{sw}, we need to compute C_{wa}, h_{c}, g_{bw} and a series of other derived variables, as described below.

The vapour concentration in the free air can be computed from vapour pressure analogously to Eq. (B4):

$$C_{\text{wa}} = \frac{P_{\text{wa}}}{R_{\text{mol}} T_{\text{a}}}. \tag{B9}$$

The heat transfer coefficient (h_{c}) for a flat plate can be determined using the non-dimensional Nusselt number (N_{Nu_L}):

$$h_{\text{c}} = k_{\text{a}} \frac{N_{Nu_L}}{L_{\text{l}}}, \tag{B10}$$

where k_a is the thermal conductivity of the air in the boundary layer and L_l is a characteristic length scale of the leaf.

For sufficiently high wind speeds, inertial forces drive the convective heat transport (forced convection) and the relevant dimensionless number is the Reynolds number (N_{Re_L}), which defines the balance between inertial and viscous forces (Incropera et al., 2006, Eq. 6.41):

$$N_{Re_L} = \frac{v_w L_l}{\nu_a}, \tag{B11}$$

where v_w is the wind velocity ($m\,s^{-1}$), ν_a is the kinematic viscosity of the air and L_l is taken as the length of the leaf in wind direction.

In the absence of wind, buoyancy forces, driven by the density gradient between the air at the surface of the leaf and the free air dominate convective heat exchange (free or natural convection). The relevant dimensionless number here is the Grashof number (N_{Gr_L}), which defines the balance between buoyancy and viscous forces (Incropera et al., 2006, Eqs. 9.3 and 9.65):

$$N_{Gr_L} = \frac{g\left(\frac{\rho_a - \rho_{al}}{\rho_{al}}\right) L_l^3}{\nu_a^2}, \tag{B12}$$

where g is gravity, while ρ_a and ρ_{al} are the densities of the gas in the atmosphere and at the leaf surface, respectively.

For $N_{Gr_L} \ll N^2_{Re_L}$, forced convection is dominant and free convection can be neglected, whereas for $N_{Gr_L} \gg N^2_{Re_L}$ free convection is dominant and forced convection can be neglected (Incropera et al., 2006, p. 565). For simplicity, the analysis is limited to forced conditions, which are satisfied by considering wind speeds greater than 0.5 m s^{-1} for 5×5 cm leaves.

The average Nusselt number under forced convection was calculated as a function of the average Reynolds number and a critical Reynolds number (N_{Re_c}) that determines the onset of turbulence and depends on the level of turbulence in the free air stream or leaf surface properties (Incropera et al., 2006, p. 412):

$$N_{Nu_L} = (0.037 N_{Re_L}^{4/5} - C_1) N_{Pr}^{1/3}, \tag{B13}$$

with

$$C_1 = 0.037 C_2^{4/5} - 0.664 C_2^{1/2} \tag{B14}$$

and

$$C_2 = \frac{N_{Re_L} + N_{Re_c} - |N_{Re_c} - N_{Re_L}|}{2}. \tag{B15}$$

Equation (B15) was introduced to make Eq. (B13) valid for all Reynolds numbers, and following considerations explained in our previous work (Schymanski et al., 2013), we chose $N_{Re_c} = 3000$ in the present simulations.

In order to simulate steady-state leaf temperatures and the leaf energy balance terms using the above equations, it is necessary to calculate ρ_a, D_{va}, α_a, k_a and ν_a, while L_l, Re_c and g_{sv} are input parameters, and P_{wa} and v_w (vapour pressure and wind speed) are part of the environmental forcing. D_{va}, α_a, k_a and ν_a were parameterised as functions of air temperature (T_a) only, by fitting linear curves to published data (Monteith and Unsworth, 2007, Table A.3):

$$D_{va} = (1.49 \times 10^{-7}) T_a - 1.96 \times 10^{-5} \tag{B16}$$

$$\alpha_a = (1.32 \times 10^{-7}) T_a - 1.73 \times 10^{-5} \tag{B17}$$

$$k_a = (6.84 \times 10^{-5}) T_a + 5.62 \times 10^{-3} \tag{B18}$$

$$\nu_a = (9 \times 10^{-8}) T_a - 1.13 \times 10^{-5}. \tag{B19}$$

Assuming that air and water vapour behave like an ideal gas, and that dry air is composed of 79 % N_2 and 21 % O_2, we calculated air density as a function of temperature, vapour pressure and the partial pressures of the other two components using the ideal gas law:

$$\rho_a = \frac{n_a M_a}{V_a} = M_a \frac{P_a}{R_{mol} T_a}, \tag{B20}$$

where n_a is the amount of matter (mol), M_a is the molar mass ($kg\,mol^{-1}$), P_a the pressure, T_a the temperature and R_{mol} the molar universal gas constant. This equation was used for each component, i.e. water vapour, N_2 and O_2, where the partial pressures of N_2 and O_2 are calculated from atmospheric pressure minus vapour pressure, yielding

$$\rho_a = \frac{M_w P_{wa} + M_{N_2} P_{N_2} + M_{O_2} P_{O_2}}{R_{mol} T_a}, \tag{B21}$$

where M_{N_2} and M_{O_2} are the molar masses of nitrogen and oxygen, respectively, while P_{N_2} and P_{O_2} are their partial pressures, calculated as

$$P_{N_2} = 0.79 (P_a - P_{wa}) \tag{B22}$$

and

$$P_{O_2} = 0.21 (P_a - P_{wa}). \tag{B23}$$

B5 Analytical solutions by Penman

In order to obtain analytical expressions for the different leaf energy balance components, one would need to solve the leaf energy balance equation for leaf temperature first. However, due to the non-linearities of the blackbody radiation and the saturation vapour pressure equations, an analytical solution has not been found yet. Penman (1948) proposed a workaround, which we reproduced below, adapted to our notation and to a wet leaf, while Penman's formulations referred to a wet soil surface. He formulated evaporation from a wet surface as a diffusive process driven by the vapour pressure difference near the wet surface and in the free air:

$$E_w = f_u (P_{wl} - P_{wa}), \tag{B24}$$

where E_w (J s^{-1} m^{-2}) is the latent heat flux from a wet surface and f_u is commonly referred to as the wind function. Penman then defined the Bowen ratio (Eq. 10 in Penman, 1948) as

$$\beta_B = H_l/E_w = \gamma_v \frac{T_l - T_a}{P_{wl} - P_{wa}}, \tag{B25}$$

where H_l is the sensible heat flux and γ_v is the psychrometric constant, referring to the ratio between the transfer coefficients for sensible heat and that for water vapour.

In order to eliminate T_l, Penman introduced a term for the ratio of the vapour pressure difference between the surface and the saturation vapour pressure at air temperature (P_{was}) to the temperature difference between the surface and the air:

$$\Delta_{eTa} = \frac{P_{wl} - P_{was}}{T_l - T_a}, \tag{B26}$$

and he proposed to approximate this term by the slope of the saturation vapour pressure curve evaluated at air temperature, which can be obtained by substitution of T_a for T_l and differentiation of Eq. (B5) with respect to T_a:

$$\Delta_{eTa} = \frac{611\lambda_E M_w \exp\left(\frac{\lambda_E M_w}{R_{mol}}\left(\frac{1}{273} - \frac{1}{T_a}\right)\right)}{R_{mol} T_a^2}. \tag{B27}$$

For further discussion of the meaning of this assumption, please refer to Mallick et al. (2014).

Substitution of Eq. (9) in (B25) (Eq. 15 in Bowen, 1926) yields

$$\beta_B = \frac{\gamma_v}{\Delta_{eTa}} \frac{(P_{wl} - P_{was})}{(P_{wl} - P_{wa})}. \tag{B28}$$

Substituting E_w for E_l in the energy balance equation (Eq. 1), inserting $H_l = \beta_B E_w$ (Eq. B25) and solving for E_w gives

$$E_w = \frac{R_s - R_{ll}}{\beta_B + 1}. \tag{B29}$$

Substitution of Eq. (B28) into (B29), equating with Eq. (B24) and solving for P_{wl} gives

$$P_{wl} = \frac{f_u(\Delta_{eTa} P_{wa} + \gamma_v P_{was}) + \Delta_{eTa}(R_s - R_{ll})}{f_u(\Delta_{eTa} + \gamma_v)}. \tag{B30}$$

Now, insertion of Eq. (B30) into Eq. (B24) gives the so-called Penman equation:

$$E_w = \frac{\Delta_{eTa}(R_s - R_{ll}) + f_u \gamma_v (P_{was} - P_{wa})}{\Delta_{eTa} + \gamma_v}. \tag{B31}$$

Equation (15) is equivalent to Eq. (16) in Penman (1948), but Eq. (17) in Penman (1948), which should be equivalent to Eq. (B30), has P_{wl} (e_s in Penman's notation) on both sides, so it seems to contain an error. In his derivations, Penman expressed $R_s - R_{ll}$ as "net radiant energy available at surface" and pointed out that the above two equations can be used to estimate E_l and T_l from air conditions only. This neglects the fact that R_{ll} is also a function of the leaf temperature. To estimate surface temperature, Eq. (B30) can be inserted into Eq. (9) and solved for T_l, yielding

$$T_l = \frac{R_s - R_{ll} + f_u(\gamma_v T_a + \Delta_{eTa} T_a + P_{wa} - P_{was})}{f_u(\gamma_v + \Delta_{eTa})}. \tag{B32}$$

B5.1 Introduction of stomatal resistance by Penman (1952)

To account for stomatal resistance to vapour diffusion, Penman (1952) introduced an additional multiplicator (S) in Eq. (B24) (Penman, 1952, Appendix 13):

$$E_l = f_u S (P_{wl} - P_{wa}), \tag{B33}$$

where $S = 1$ for a wet surface (leading to Eq. B24) and $S < 1$ in the presence of significant stomatal resistance.

In accordance with Eqs. (B24) and (B25), H_l can be expressed as (Penman, 1952, Appendix 13)

$$H_l = \gamma_v f_u (T_l - T_a). \tag{B34}$$

Substitution of Penman's simplifying assumption ($T_l - T_a = (P_{wl} - P_{was})/\Delta_{eT}$, Eq. (9) is the first step to eliminating T_l:

$$H_l = \frac{\gamma_v f_u (P_{wl} - P_{was})}{\Delta_{eTa}}. \tag{B35}$$

A series of algebraic manipulations involving Eqs. (B33), (B35) and (1) and the resulting Eq. (B36) is given in Penman (1952, Appendix 13). When solving Eqs. (B33), (B35) and (1) for E_l, H_l and P_{wl}, we obtained

$$E_l = \frac{S\Delta_{eTa}(R_s - R_{ll}) + S\gamma_v f_u (P_{was} - P_{wa})}{S\Delta_{eT} + \gamma_v} \tag{B36}$$

$$H_l = \frac{\gamma_v (R_s - R_{ll}) + S\gamma_v f_u (P_{wa} - P_{was})}{S\Delta_{eTa} + \gamma_v} \tag{B37}$$

$$P_{wl} = \frac{(\Delta_{eTa}/f_u)(R_s - R_{ll}) + (S\Delta_{eTa} P_{wa} + \gamma_v P_{was})}{S\Delta_{eTa} + \gamma_v}. \tag{B38}$$

B5.2 Analytical solutions for leaf temperature, f_u, γ_v and S

Equation (B38) can be inserted into Eq. (9) and solved for leaf temperature to yield

$$T_l = T_a + \frac{R_s - R_{ll} - S f_u (P_{was} - P_{wa})}{f_u (S\Delta_{eT} + \gamma_v)}. \tag{B39}$$

Penman (1952) proposed to obtain values of f_u and S for a plant canopy empirically and described ways to do this. However, for a single leaf, f_u and S could also be obtained analytically from our detailed mass and heat transfer model.

Comparison of Eq. (B33) with (B6) (after substituting Eq. 4) reveals that S is equivalent to

$$S = \frac{M_w g_{tw,mol} \lambda_E}{P_a f_u}, \tag{B40}$$

where f_u was defined by Penman (1948) as the transfer coefficient for wet surface evaporation, i.e. a function of the boundary layer conductance only.

To find a solution for f_u, we first formulate E_w as transpiration from a leaf where $g_{tw} = g_{bw}$, using Eqs. (4), (B6) and (B8):

$$E_w = \frac{\lambda_E M_w g_{bw}}{R_{mol} T_a}(P_{wl} - P_{wa}). \tag{B41}$$

Comparison of Eq. (B41) with (B24) gives f_u as a function of g_{bw}:

$$f_u = g_{bw}\frac{\lambda_E M_w}{R_{mol} T_a}. \tag{B42}$$

Comparison of Eq. (B34) and (3) reveals that

$$\gamma_v = \frac{a_{sh} h_c}{f_u}, \tag{B43}$$

and insertion of Eqs. (B42) and (B2) give γ_v as a function of a_{sh} and a_s:

$$\gamma_v = a_{sh}/a_s \frac{N_{Le}^{\frac{2}{3}} R_{mol} T_a \rho_a c_{pa}}{\lambda_E M_w}. \tag{B44}$$

Now, we can insert Eqs. (B42), (B8) and (6) into (B40) to obtain S as a function of g_{sw} and g_{bw}:

$$S = \frac{g_{sw}}{g_{bw} + g_{sw}}. \tag{B45}$$

The above equation illustrates that S is not just a function of stomatal conductance, but also the leaf boundary layer conductance, explaining why Penman (1952) found that S depends on wind speed.

B6 Psychrometric constant in the Penman–Monteith equation

Monteith and Unsworth (2013) provide a definition of γ_v as

$$\gamma_v = \frac{c_{pa} P_a}{\lambda_E \epsilon}, \tag{B46}$$

where ϵ is the ratio of molecular weights of water vapour and air (given by Monteith and Unsworth, 2013 as 0.622). The molar mass of air is $M_a = \rho_a V_a / n_a$, while according to the ideal gas law, $V_a/n_a = R_{mol} T_a / P_a$, which yields for $\epsilon = M_w/M_a$:

$$\epsilon = \frac{M_w P_a}{R_{mol} T_a \rho_a}. \tag{B47}$$

Inserting Eqs. (B21), (B22) and (B23) into the above, T_a cancels out, and at standard atmospheric pressure of 101 325 Pa, we obtain values for ϵ between 0.624 and 0.631 for vapour pressure ranging from 0 to 3000 Pa, compared to the value of 0.622 mentioned by Monteith and Unsworth (2013).

B7 Meaning of resistances in PM equation

As opposed to the formulations in Sect. 2.1, where sensible and latent heat transfer coefficients (h_c and g_{tw}, respectively) translate leaf–air differences in temperature or vapour concentration to fluxes, resistances in the PM equation are defined in the context of the following two equations (Monteith and Unsworth, 2013, Eqs. 13.16 and 13.20):

$$E_l = \frac{a_s \lambda_E \rho_a \epsilon}{P_a (r_v + r_s)}(P_{wl} - P_{wa}) \tag{B48}$$

and

$$H_l = \frac{a_{sh} \rho_a c_{pa}}{r_a}(T_l - T_a), \tag{B49}$$

where r_v and r_s are the one-sided leaf boundary layer and stomatal resistances to water vapour, respectively, and r_a is the one-sided leaf boundary layer resistance to sensible heat transfer. Note that we introduced a_s, a_{sh} and r_s in Eqs. (B48) and (B49) based on the description on p. 231 in Monteith and Unsworth (2013), where the authors also assumed that $r_v \approx r_a$.[1]

Comparison of Eq. (B48) (after substitution of Eq. B47) with our fundamental diffusion equation (Eq. 5, after substitution of Eqs. B4 and B9 and insertion into Eq. 4) reveals that under isothermal conditions ($T_l = T_a$)

$$r_v = a_s / g_{bw}, \tag{B50}$$

while comparison of Eq. (B49) with Eq. (3) reveals that

$$r_a = \frac{\rho_a c_{pa}}{h_c}. \tag{B51}$$

B8 Comparison of our general analytical solution with original Penman and Penman–Monteith equations

From the general form (Eq. 10), we can recover most of the above analytical solutions by appropriate substitutions for c_E and c_H, but closer inspection of the necessary substitutions reveals some inconsistencies.

The Penman equation for a wet surface (Eq. 15) can be recovered by substituting $c_E = f_u$ and $c_H = \gamma_v f_u$ into Eq. (10) (Fig. 3a), while additional substitution of Eq. (17) leads to recovery of Eq. (16), the Penman equation as reformulated by Monteith (1965). The formulation for leaf transpiration derived by Penman (1952) (Eq. B36) is obtained by substituting $c_E = S f_u$ (deduced from Eq. B33) and $c_H = \gamma_v f_u$ (from Eq. B34). These substitutions are consistent with the formulations of latent and sensible heat flux given in Eqs. (B34) and (B24) or (B33), as long as f_u and r_a refer to the total resistances of a leaf to latent and sensible heat flux, respectively, as Eq. (17) in conjunction with $c_H = \gamma_v f_u$ implies that

$$c_H = (\rho_a c_{pa})/r_a. \tag{B52}$$

[1] Division of Eq. (B51) by (B50) and substitution of Eqs. (B2), (B3), (B17) and (B16) reveals that $r_a/r_v = N_{Le}^{-2/3} = 1.082$.

Similarly, the Penman–Monteith equation (Eq. 19 with γ_v defined in Eq. B46) could be recovered by substituting $c_E = \epsilon\lambda_E \rho_a/(P_a(r_s + r_v))$ and $c_H = c_{pa}\rho_a/r_a$, with subsequent substitution of $r_v = r_a$. Note, however, that these substitutions are not consistent with Eqs. (B48) and (B49), as the factors a_s and a_{sh} (referring to the number of leaf faces exchanging latent and sensible heat flux, respectively) are missing (Fig. 3b cf. 3c). This is because the PM equation was derived with a soil surface in mind, which exchanges latent and sensible heat only on one side, and hence is not appropriate for a leaf. To alleviate this constraint, one could define r_a and r_s as total (two-sided) leaf resistances, but in this case, the simplification $r_v \approx r_a$ is not valid for hypostomatous leaves, as r_a would then be only half of r_v. This is illustrated in Fig. 3c, where sensible heat flux is released from both sides of the leaf, while latent heat flux is only released from the abaxial side, implying that $a_{sh} = 2$ and $a_s = 1$.

Monteith and Unsworth (2013) acknowledged that a hypostomatous leaf could exchange sensible heat on two sides, but latent heat on one side only and introduced the parameter $n_{MU} = a_{sh}/a_s$ to account for this (Eq. 21). Using our general equation, it should be possible to reproduce the MU equation (Eq. 21) by substituting $c_E = a_s\epsilon\lambda_E\rho_a/(P_a(r_s+r_v))$ (deduced from Eq. B48) and $c_H = a_{sh}c_{pa}\rho_a/r_a$ (deduced from Eq. B49) into Eq. (10). However, the result of this substitution, as presented in Eq. (B53), is not the same as Eq. (21) after substitution of Eq. (B46) and $n_{MU} = a_{sh}/a_s$, which would result in Eq. (B54):

$$E_l = \frac{a_s\epsilon\lambda_E\left(\Delta_{eTa}r_a\left(R_s - R_{ll}\right) + a_{sh}c_{pa}\rho_a\left(P_{was} - P_{wa}\right)\right)}{P_a a_{sh}c_{pa}\left(r_s + r_a\right) + \Delta_{eTa}a_s\epsilon\lambda_E} \quad \text{(B53)}$$

vs.

$$E_l = \frac{a_s\epsilon\lambda_E\left(\Delta_{eTa}r_a\left(R_s - R_{ll}\right) + c_{pa}\rho_a\left(P_{was} - P_{wa}\right)\right)}{P_a a_{sh}c_{pa}\left(r_s + r_a\right) + \Delta_{eTa}a_s\epsilon\lambda_E}. \quad \text{(B54)}$$

Note the missing a_{sh} in the nominator of Eq. (B54), as pointed out in the main text.

B9 Surface temperature dependence of net radiation

In the main text, Eq. (2) was linearised by taking its derivative with respect to T_l, defining this derivative as the slope of a linear function of temperature with an intercept chosen to make this function intersect with Eq. (2) at $T_l = T_a$. The result is given in Eq. (23) and plotted in Fig. (A3).

B10 Calculation of stomatal conductance from pore dimensions

At least three confocal laser scanning images of each perforated foil were analysed and average pore area (A_p, m^2), pore radius (r_p, m), number of pores per surface area (n_p, m^{-2}) and average distance to nearest neighbour (s_p, m) was computed for each image. The resulting stomatal conductance was computed following the derivations presented by Lehmann and Or (2015), assuming that the stomatal conductance results from two resistances in a series: the throat resistance (r_{sp}), dependent on the areas of the pores and the thickness of the perforated foil (d_p), and the vapour shell resistance (r_{vs}), dependent on the size and spacing of the stomata, which can be understood as the resistance related to distribution of the point-source water vapour over the entire one-sided leaf boundary layer. We hereby neglect any internal resistance (termed "end correction" by Lehmann and Or, 2015), as we assume that the wet filter paper has direct contact with the perforated foil. The throat resistance was computed (Eq. 1 in Lehmann and Or, 2015) as

$$r_{sp} = \frac{d_p}{A_p k_{dv} n_p}, \quad \text{(B55)}$$

where k_{dv} is the ratio of the vapour diffusion coefficient and the molar volume of air (D_{va}/V_m), and $A_p = \pi r_p^2$. For the vapour shell resistance, we use the formulation originally proposed by Bange (1953):

$$r_{vs} = \left(\frac{1}{4r_p} - \frac{1}{\pi s_p}\right)\frac{1}{k_{dv}n_p}, \quad \text{(B56)}$$

where s_p (m) is the spacing between stomata, inferred from the images as $s_p = 1/\sqrt{n_p}$. Stomatal conductance (g_{sw}) for each image was then calculated following Eq. (B8), i.e. $g_{sw} = g_{sw,mol}R_{mol}T_a/P_a$, after substituting $g_{sw,mol} = 1/(r_{sp} + r_{vs})$.

Author contributions. Stanislaus J. Schymanski performed the mathematical derivations, designed and carried out the experiments and wrote the paper. Dani Or was involved in the design of the experimental set-up, interpretation of the results and writing the paper.

Competing interests. The authors declare that they have no conflict of interest.

Acknowledgements. The authors are very grateful to Dani Breitenstein for his assistance in designing and constructing the wind tunnel, to Stefan Meier and Joni Dehaspe for assistance in constructing artificial leaves and to Hans Wunderli for assistance in the lab. We also wish to acknowledge technical advice by Roland Kuenzli (DMP AG, Fehraltorf, Switzerland), laser perforation services by Ralph Beglinger (Lasergraph AG, Würenlingen, Switzerland), Robert Voss (ETH Zurich, Switzerland) and Rolf Brönnimann (EMPA, Zurich, Switzerland), helpful feedback from Tim Reichenau and Claire Zimmermann (Univ. Köln, Germany) and constructive reviewer comments by Stefan Dekker and an anonymous reviewer. The project benefited from funding by the Swiss National Science Foundation (project 200021 135077).

Edited by: P. Gentine

References

Allen, R.: A Penman for All Seasons, J. Irrig. Drain. Eng.-ASCE, 112, 348–368, doi:10.1061/(ASCE)0733-9437(1986)112:4(348), 1986.

Bange, G. G. J.: On the quantitative explanation of stomatal transpiration, Acta Bot. Neerl., 2, 255–296, 1953.

Bowen, I. S.: The Ratio of Heat Losses by Conduction and by Evaporation from any Water Surface, Phys. Rev., 27, 779–787, doi:10.1103/PhysRev.27.779, 1926.

Dolman, A. J., Miralles, D. G., and de Jeu, R. A.: Fifty years since Monteith's 1965 seminal paper: the emergence of global ecohydrology: the emergence global ecohydrology, Ecohydrology, 7, 897–902, doi:10.1002/eco.1505, 2014.

Hartmann, D. L.: Global physical climatology, Academic Press, 1994.

Incropera, F. P., DeWitt, D. P., Bergman, T. L., and Lavine, A. S.: Fundamentals of Heat and Mass Transfer, John Wiley & Sons, 6th Edn., 2006.

Jarvis, A. J. and McNaughton, K. G.: Stomatal Control of Transpiration: Scaling Up from Leaf to Region, Adv. Ecol. Res., 15, 1–49, 1986.

Langensiepen, M., Fuchs, M., Bergamaschi, H., Moreshet, S., Cohen, Y., Wolff, P., Jutzi, S. C., Cohen, S., Rosa, L. M. G., Li, Y., and Fricke, T.: Quantifying the uncertainties of transpiration calculations with the Penman–Monteith equation under different climate and optimum water supply conditions, Agr. Forest Meteorol., 149, 1063–1072, doi:10.1016/j.agrformet.2009.01.001, 2009.

Lehmann, P. and Or, D.: Effects of stomata clustering on leaf gas exchange, New Phytol., 207, 1015–1025, doi:10.1111/nph.13442, 2015.

Leuning, R., Grace, J., Monteith, J. L., Milford, J. R., Unsworth, M. H., and Fowler, D.: Leaf Energy Balances: Developments and Applications [and Discussion], Philos. T. R. Soc. B, 324, 191–206, doi:10.1098/rstb.1989.0044, 1989.

Lhomme, J. P., Montes, C., Jacob, F., and Prévot, L.: Evaporation from Heterogeneous and Sparse Canopies: On the Formulations Related to Multi-Source Representations, Bound.-Lay. Meteorol., 144, 243–262, doi:10.1007/s10546-012-9713-x, 2012.

Mallick, K., Jarvis, A., Fisher, J. B., Tu, K. P., Boegh, E., and Niyogi, D.: Latent Heat Flux and Canopy Conductance Based on Penman–Monteith, Priestley–Taylor Equation, and Bouchet's Complementary Hypothesis, J. Hydrometeorol., 14, 419–442, doi:10.1175/JHM-D-12-0117.1, 2013.

Mallick, K., Jarvis, A. J., Boegh, E., Fisher, J. B., Drewry, D. T., Tu, K. P., Hook, S. J., Hulley, G., Ardö, J., Beringer, J., Arain, A., and Niyogi, D.: A Surface Temperature Initiated Closure (STIC) for surface energy balance fluxes, Remote Sens. Environ., 141, 243–261, doi:10.1016/j.rse.2013.10.022, 2014.

Mallick, K., Trebs, I., Boegh, E., Giustarini, L., Schlerf, M., Drewry, D. T., Hoffmann, L., von Randow, C., Kruijt, B., Araùjo, A., Saleska, S., Ehleringer, J. R., Domingues, T. F., Ometto, J. P. H. B., Nobre, A. D., de Moraes, O. L. L., Hayek, M., Munger, J. W., and Wofsy, S. C.: Canopy-scale biophysical controls of transpiration and evaporation in the Amazon Basin, Hydrol. Earth Syst. Sci., 20, 4237–4264, doi:10.5194/hess-20-4237-2016, 2016.

McArthur, A. J.: An accurate solution to the Penman equation, Agr. Forest Meteorol., 51, 87–92, doi:10.1016/0168-1923(90)90043-6, 1990.

Milly, P. C. D.: A refinement of the combination equations for evaporation, Surv. Geophys., 12, 145–154, doi:10.1007/BF01903416, 1991.

Monteith, J. and Unsworth, M.: Principles of Environmental Physics, Academic Press, Oxford, UK, 3rd Edn., 2007.

Monteith, J. L.: Evaporation and environment, Sym. Soc. Exp. Biol., 19, 205–234, doi:10.1002/qj.49710745102, 1965.

Monteith, J. L. and Unsworth, M. H.: Principles of environmental physics: plants, animals, and the atmosphere, 4th Edn., Elsevier/Academic Press, Amsterdam, Boston, 2013.

Overgaard, J., Rosbjerg, D., and Butts, M. B.: Land-surface modelling in hydrological perspective – a review, Biogeosciences, 3, 229–241, doi:10.5194/bg-3-229-2006, 2006.

Parlange, J.-Y., Waggoner, P. E., and Heichel, G. H.: Boundary Layer Resistance and Temperature Distribution on Still and Flapping Leaves, Plant Physiol., 48, 437–442, 1971.

Paw U, K. T. and Gao, W.: Applications of solutions to non-linear energy budget equations, Agr. Forest Meteorol., 43, 121–145, doi:10.1016/0168-1923(88)90087-1, 1988.

Penman, H. L.: Natural Evaporation from Open Water, Bare Soil and Grass, P. R. Soc. A, 193, 120–145, doi:10.1098/rspa.1948.0037, 1948.

Penman, H. L.: The physical bases of irrigation control, International Horticultural Congress, 2, 913–924, 1952.

Raupach, M. R.: Vegetation-atmosphere interaction and surface conductance at leaf, canopy and regional scales, Agr. Forest Meteorol., 73, 151–179, doi:10.1016/0168-1923(94)05071-D, 1995.

Raupach, M. R. and Finnigan, J. J.: "Single-layer models of evaporation from plant canopies are incorrect but useful, whereas multilayer models are correct but useless": Discuss, Funct. Plant Biol., 15, 705–716, doi:10.1071/PP9880705, 1988.

Schymanski, S.: Data and code to reproduce computational results and plots, doi:10.5281/zenodo.241259, 2017.

Schymanski, S. J. and Or, D.: Wind effects on leaf transpiration challenge the concept of "potential evaporation", Proc. IAHS, 371, 99–107, doi:10.5194/piahs-371-99-2015, 2015.

Schymanski, S. J. and Or, D.: Wind increases leaf water use efficiency, Plant Cell Environ., 39, 1448–1459, doi:10.1111/pce.12700, 2016.

Schymanski, S. J., Or, D., and Zwieniecki, M.: Stomatal Control and Leaf Thermal and Hydraulic Capacitances under Rapid Environmental Fluctuations, PLoS ONE, 8, e54231, doi:10.1371/journal.pone.0054231, 2013.

Schymanski, S. J., Breitenstein, D., and Or, D.: Technical note: An experimental setup to measure latent and sensible heat fluxes from (artificial) plant leaves, Hydrol. Earth Syst. Sci. Discuss., doi:10.5194/hess-2016-643, in review, 2017.

Schymanski, S. J., Breitenstein, D., and Or, D.: Technical note: An experimental setup to measure latent and sensible heat fluxes from (artificial) plant leaves, Hydrol. Earth Syst. Sci. Discuss., doi:10.5194/hess-2016-643, in review, 2017.

Tanner, C. B. and Fuchs, M.: Evaporation from unsaturated surfaces: A generalized combination method, J. Geophys. Res., 73, 1299–1304, doi:10.1029/JB073i004p01299, 1968.

Verhoef, A. and Allen, S. J.: A SVAT scheme describing energy and CO_2 fluxes for multi-component vegetation: calibration and test for a Sahelian savannah, Ecol. Model., 127, 245–267, doi:10.1016/S0304-3800(99)00213-6, 2000.

Widmoser, P.: A discussion on and alternative to the Penman–Monteith equation, Agr. Water Manage., 96, 711–721, doi:10.1016/j.agwat.2008.10.003, 2009.

14

On the sensitivity of urban hydrodynamic modelling to rainfall spatial and temporal resolution

G. Bruni[1], R. Reinoso[2], N. C. van de Giesen[1], F. H. L. R. Clemens[1,3], and J. A. E. ten Veldhuis[1]

[1]Department of Water Management, Faculty of Civil Engineering and Geosciences, Delft University of Technology, Delft, the Netherlands
[2]Department of Geoscience and Remote Sensing, Faculty of Civil Engineering and Geosciences, Delft University of Technology, Delft, the Netherlands
[3]Deltares, Delft, the Netherlands

Correspondence to: G. Bruni (g.bruni@tudelft.nl)

Abstract. Cities are increasingly vulnerable to floods generated by intense rainfall, because of urbanisation of flood-prone areas and ongoing urban densification. Accurate information of convective storm characteristics at high spatial and temporal resolution is a crucial input for urban hydrological models to be able to simulate fast runoff processes and enhance flood prediction in cities. In this paper, a detailed study of the sensitivity of urban hydrodynamic response to high resolution radar rainfall was conducted. Rainfall rates derived from X-band dual polarimetric weather radar were used as input into a detailed hydrodynamic sewer model for an urban catchment in the city of Rotterdam, the Netherlands. The aim was to characterise how the effect of space and time aggregation on rainfall structure affects hydrodynamic modelling of urban catchments, for resolutions ranging from 100 to 2000 m and from 1 to 10 min. Dimensionless parameters were derived to compare results between different storm conditions and to describe the effect of rainfall spatial resolution in relation to storm characteristics and hydrodynamic model properties: rainfall sampling number (rainfall resolution vs. storm size), catchment sampling number (rainfall resolution vs. catchment size), runoff and sewer sampling number (rainfall resolution vs. runoff and sewer model resolution respectively).

Results show that for rainfall resolution lower than half the catchment size, rainfall volumes mean and standard deviations decrease as a result of smoothing of rainfall gradients. Moreover, deviations in maximum water depths, from 10 to 30 % depending on the storm, occurred for rainfall resolution close to storm size, as a result of rainfall aggregation. Model results also showed that modelled runoff peaks are more sensitive to rainfall resolution than maximum in-sewer water depths as flow routing has a damping effect on in-sewer water level variations. Temporal resolution aggregation of rainfall inputs led to increase in de-correlation lengths and resulted in time shift in modelled flow peaks by several minutes. Sensitivity to temporal resolution of rainfall inputs was low compared to spatial resolution, for the storms analysed in this study.

1 Introduction

Rainfall is a key input to hydrological models and a crucial issue for hydrologists is to find the importance of the spatial structure of rainfall in relation to flood generation (Segond et al., 2007). Many studies conducted in large natural catchments have shown that spatial variability of rainfall is important in determining both timing and volume of rainfall transformed into runoff (Obled et al., 1994) and thus timing of simulated basin response and magnitude of the response peak (Dawdy and Bergman, 1969; Krajewski et al., 1991; Seliga et al., 1992). It has been suggested, with much less evidence, that this is also true for small catchments with shorter response times, such as urban catchments (Blanchet et al., 1992; Obled et al., 1994). Urban catchments are characterised by a high percentage of imperviousness, which leads to a high proportion of the rainfall producing runoff.

It is therefore expected that the effect of spatial rainfall variability on water flows is greater in urban catchments than in rural ones, where local variation of rainfall input is smoothed and delayed within the soil as a result of infiltration in pervious areas (Obled et al., 1994, among others). Previous studies have shown that urban catchments, characterised by a fast hydrological response due to both low interception and infiltration, are highly sensitive to small-scale spatial and temporal variability of the precipitation field (Bell and Moore, 2000; Einfalt et al., 2004; Gires et al., 2013) In the past, a lot of studies have addressed requirements and approaches for flood modelling (Schmitt et al., 2004; Balmforth and Dibben, 2006; Parker et al., 2011; Pathirana et al., 2011; Priest et al., 2011; Neal et al., 2012; Ozdemir et al., 2013). More recently, studies have shown the impact of rainfall variability on hydrodynamic models outputs (Gires et al., 2012; Liguori et al., 2012; Vieux and Imgarten, 2012).

As resolutions of available data and models have increased, rainfall variability information at high resolution has become a critical component to study hydrological response in urban drainage systems using hydrological models. Weather radars are more suitable for this purpose than rain gauge networks as they have better spatial coverage. Weather radars, such as S-band and C-band radars, are already used by meteorological institutes worldwide in order to (indirectly) measure and predict precipitation at national and regional scales. Nonetheless, several studies have shown that the spatial resolution of operational radar network measurements is insufficient to meet the scale of urban hydrodynamics (Berne et al., 2004; Emmanuel et al., 2011; Schellart et al., 2012). Because of their relatively low cost and small size, X-band radars are ideally suited for local rainfall estimation. These radars measure at high resolutions, both in space and time, and much closer to the ground than S- or C-band radars, which for operational purposes, cover large distances and thus point higher especially at locations several tens of kilometres away from the radar sites. X-band radars have been tested locally and show better performances in catching the rapidly changing characteristics of intense rainfall than rain gauges (Jensen and Pedersen, 2005). This is particularly the case when the distance between rain gauges is larger than 3 to 4 km (Wood et al., 2000).

The effects of radar spatial resolution on hydrological model outputs were addressed by Ogden and Julien (1994) by using length scales to characterise rainfall data and catchments, such as storm de-correlation length, grid size of rainfall data, characteristic catchment length and grid size of the distributed runoff model. In their study, Ogden and Julien aimed to explain variability in hydrological responses based on rainfall and catchment characteristics, for two catchments of 30 and 100 km^2, using fully distributed rainfall–runoff models. They recommended rainfall spatial resolution of 0.4 the square root of the watershed area, in order to avoid deviations in runoff flows. This corresponds to 1 km resolution for a 10 km^2 watershed, and 4 km resolution for a 100 km^2 watershed, as was also found by Segond et al. (2007). Several other studies on natural catchments also found that the influence of rainfall resolution is directly related to the spatial variability of the storm and of the catchment that transforms rainfall into runoff (Krajewski et al., 1991; Winchell et al., 1998; Koren et al., 1999, among others).

The purpose of this paper was to analyse the sensitivity of urban hydrodynamic model outputs to spatial and temporal resolutions of rainfall inputs derived from weather radar data at intra-urban scale. Sensitivity was analysed according to spatial characteristics of rainfall and urban catchment properties as well as model topology. Sensitivity was quantified using dimensionless parameters that describe relationships between rainfall resolution and spatial characteristics of the urban catchment, storm cells and model topology. Some of them were chosen according to their previous use by Odgen and Julien (1994). In this study rainfall estimates were used derived from dual-polarimetric X-band radar (IDRA), operated by Delft University of Technology (TU Delft) and located at CESAR, Cabauw Experimental Site for Atmospheric Research (Leijnse et al., 2010; Otto and Russchenberg, 2011). A detailed urban hydrodynamic model for a catchment in the city of Rotterdam was chosen as a pilot case. Catchment conditions are representative of urban districts in lowland areas, especially delta cities, where almost half of the world population lives. Lowland catchments are characterised by flat terrain, therefore the mechanism dominating sewer flow is different from sloped terrain, where flow is driven by gravitation. This study aims at analysing the sensitivity of this urban hydrodynamic model to changes in rainfall spatial and temporal resolution. The study's focus is on model uncertainty related to rainfall input; model performance is not tested here, since storms were virtually applied to the catchment, which did not allow a proper model validation based on water level and flow observations. However, model geometry was strictly checked and model parameters were estimated based on literature values and experts opinion, so that the model is considered to be a reliable representation of local pluvial response.

Results were used to address the following questions:

- Does small-scale precipitation variability affect hydrological response and can a highly detailed semi-distributed model properly describe such a response?
- Is high-resolution rainfall information required when a storm does not present pronounced space–time variability?
- Does sensitivity of small-sized urban catchments to spatial and temporal variability of precipitation depend on catchment scale?

The findings have relevance for the use of high-resolution radar data in flood forecasting and flood protection in cities, at intra-urban scale. It provides a contribution to the debate on radar spatial resolution requirements for urban drainage modelling of small-scale urban catchments at district level, i.e. up to 3 km^2.

The paper is organised as follows. Section 2 presents the case study, hydrodynamic modelling approach and provides an analysis and description of rainfall fields used to conduct the sensitivity analysis. In Sect. 3 scale lengths are defined and then used to obtain a set of dimensionless parameters that will characterise relationships between rainfall fields, spatial resolution of rainfall and catchment characteristics. In Sect. 4 results of the scale analysis are shown and discussed. Lastly, conclusions are presented in Sect. 5.

2 Presentation of the case study and data sets

2.1 Case study and model description

This paper focuses on the central district of Rotterdam, the Netherlands (Fig. 1). The district is densely populated and includes mainly residential areas with approximately 30 000 inhabitants, as well as businesses and shopping centres. The district has a size of 3.4 km^2. Two green areas are located in the southern part of the district, sized 6 and 24 ha. The southern border of the district is formed by the Meuse River. The district belongs to a polder area below sea level. As a result, the area is nearly flat and there is not a dominant flow direction. During rainfall, excess storm water needs to be pumped out into the river system or temporarily stored elsewhere. Meanwhile, net rainfall fills sewer systems and storage basins up to the level of external weirs, where overflows to surface water take place if rainfall continues. An underground storage facility with a capacity of 10 000 m^2 has been built in the district to reduce flood risk during heavy rainfall events.

A hydrodynamic urban drainage model has been built for the catchment area using SOBEK-Urban software (Deltares, 2014). Although fully distributed models best describe the effect of rainfall variability on a catchment, the use of a highly detailed semi-distributed model with runoff areas of approximately the same size or smaller than the highest rainfall input resolution of 100 m × 100 m, is a close alternative. The combined sewer system was modelled in 1-D and consists of around 3000 manhole nodes (most of them are with runoff) and 11 external weirs, which serve as outflow points. The model contains four pressurised pipes interconnecting parts of the sewer system. Two external pumping stations transport water to the waste water treatment plant and to the river. Rainfall–runoff processes are modelled in SOBEK-RR (Deltares, 2014). The main components in this model are surface water storage, evaporation, infiltration and delay of surface runoff before entering the sewer system. Surface water storage occurs when rainwater forms puddles. When the water level exceeds the given maximum street storage, runoff is generated. Infiltration is computed on pervious surfaces by the Horton equation. Runoff to the sewer system is computed as a function of net rainfall and runoff factors, which depend on length, roughness, slope and percentage of imperviousness of the areas. According to Dutch guidelines (Stichting RIONED, 2004), four different area types were used with different sets of runoff parameter values (Table 1): closed paved, open paved, roof flat and roof sloped (with slope larger than 4 %) areas. The open paved area type represents paved streets with bricks, which allow water to infiltrate and to be retained within the road surface. Green areas are not taken into account by the model, as they are assumed to be disconnected from the sewer system. The rainfall–runoff module is lumped and its basic unit is the "runoff area". Each runoff area contains different types of surface, the runoff of which enters the sewer system through the manhole nodes. Further details of the software package used in this study are provided in the Appendix.

Table 1. Surface characteristics of the central district catchment in Rotterdam used for hydrodynamic modelling: percentage, runoff coefficient and storage coefficient.

Type of area	Overall percentage (%)	Runoff coefficient (min^{-1})	Storage coefficient (mm)
Open paved flat	40	0.2	0.5
Closed paved flat	14	0.2	0.5
Roof flat	16	0.2	2
Roof sloped (slope larger than 4 %)	30	0.5	0

Table 2. Specification of the X-band radar of CESAR.

Dual polarimetric X-band radar	
Radar type	FMCW
Polarisation	Dual polarisation
Frequency	9.475 GHz
Highest range resolution	30 m
Min range	230 m
Max range	< 122 km
Max unambiguous radial velocity	19 $m\,s^{-1}$
Temporal resolution	1 min
Beamwidth	1.8°
Elevation	0.5°

2.2 Rainfall data

Rainfall data were obtained from CESAR (Leijnse et al., 2010) which provides data from a dual-polarimetric X-band radar collected at 30 m range resolution and a maximum unambiguous range of 15 km approximately. Other specifications on the new generation X-band radar device can be found in Table 2. Aggregations were made from radar rainfall rates at 30 m polar pixels based on reflectivity for values smaller than 30 dBZ, differential phase otherwise (Otto and

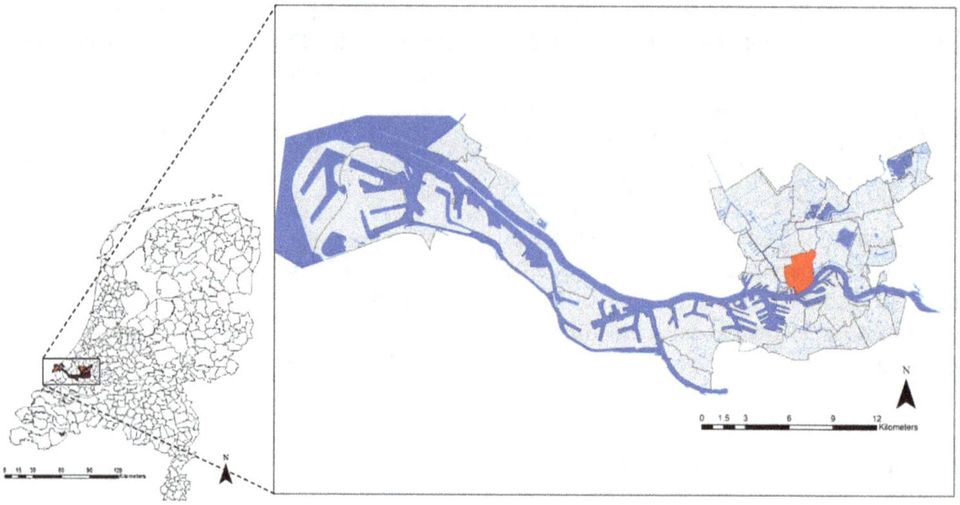

Figure 1. Localisation of Centrum district (in red in the right panel), situated in Rotterdam urban area (right panel and in red in the left panel), the Netherlands (left panel).

Russchenberg, 2011). The X-band radar has been operational intermittently since 27 June 2008. From the available data sets provided by CESAR, four rainfall storms could be selected for analysis based on a minimum mean rainfall volume of 3 mm over the area size of the studied catchment, the size of which is 3.4 km^2. Lower rainfall volumes produce insufficient runoff to allow proper hydrodynamic analysis. According to the classification adopted by Emmanuel et al. (2012), events are grouped as follows:

- Event 1 and Event 2: storm organised in rain bands;
- Event 3: storm less organised;
- Event 4: light rain.

In Event 1, a long-lived squall line was measured on 3 January 2012. The convective storm moved eastward with a velocity of 20 $m\,s^{-1}$ approximately. A squall line is a line of convective cells that forms along a cold front with a predominately trailing stratiform precipitation (Storm et al., 2007). Squall lines are typically associated with a moderate shear between 10 and 20 $m\,s^{-1}$ and strong updraft (Weisman and Rotunno, 2004). If winds increase rapidly with height ahead of a strong front, thunderstorms triggered along the boundary may organise into severe storms called supercell storms. The X-band radar was able to capture storm features associated with supercell. The overall duration of the event was short, 1 h in total, but the most intense peak lasted 10 min at the end of the storm, and with rainfall intensities higher than 100 $mm\,h^{-1}$. The most affected part of the catchment was the central and the northwestern part, while the southern part was affected by light rain. Event 2, occurring on 10 September 2011, can be characterised as a cluster of convective and organised storm cells that moved in a northeast direction. The storm moved northeastward with a velocity of 16 $m\,s^{-1}$ approximately. The storm system showed a convective spread area larger than the first event and with slower shift. The storm lasted 2 h, between 18:00–20:00 UTC, with the most intense part concentrated between 19:00 and 20:00 UTC. Intensities ranged between 30 and 60 $mm\,h^{-1}$, and the whole central part, from south to north of the catchment was affected, while east and west bands were less exposed. In Event 3, which occurred on 28 June 2011 from 22:00 to 24:00 UTC, mesoscale observations showed a non-organised squall line moving northeast, with a speed of 15 $m\,s^{-1}$ approximately and containing rainfall rate cores of at least 10 $mm\,h^{-1}$. Rainfall rate values of 50 $mm\,h^{-1}$ were founded over small areas during 22:00–23:00 UTC, travelling from southwest toward northeast and affecting all the catchment. Lastly, Event 4, which occurred on 29 October 2012, was a stratiform precipitation moving eastward at 13 $m\,s^{-1}$ approximately and showing uniform rainfall rates. Rainfall retrieval was based on reflectivity only, of about 8 $mm\,h^{-1}$. Storms motions and directions were estimated based on a centroid-based storm association algorithm, inspired by Johnson et al. (1998). For each event, total rainfall volumes in terms of minimum, maximum and mean value of all pixels affecting the area can be found in Fig. 2, as well as their standard deviation, giving a first insight into the variability of the event. Figure 3 presents radar images showing the maximum intensity minute of each one of the selected rainfall events, as well as the location of the catchment with respect to them and the main direction of the storms.

3 Methods

In this study, effects of radar spatial resolution on hydrological model outputs were analysed by means of length

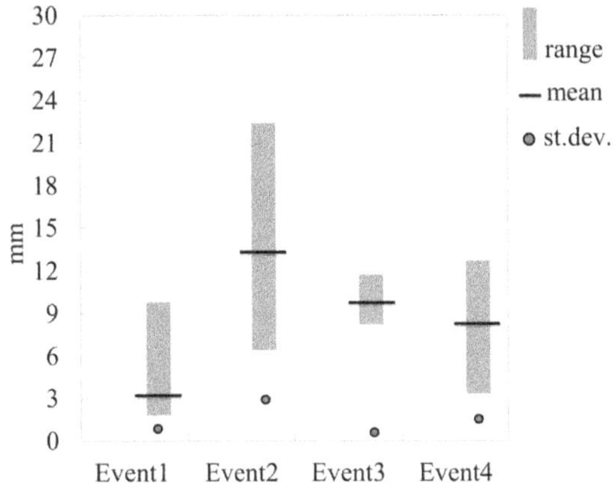

Figure 2. Characteristics of the four selected storm events: rainfall volume range (maximum and minimum for all 100 m × 100 m pixels over the catchment area), mean and standard deviations.

scales. Building upon the approach introduced by Ogden and Julien (1994), length scales were developed for urban catchments and adjusted and extended for application to hydrodynamic urban drainage models (Table 3, Fig. 4). A scale dependency between storm, catchment and model topology for small-scale urban catchments, was studied based on rainfall fields derived from polarimetric radar, using spatial resolutions of 100, 500, 1000 and 2000 m, obtained by upscaling the original resolution. The finest spatial resolution, namely 100 m, was chosen for being the highest resolution at which radar rainfall data were provided. The 1000 m resolution was selected for being the resolution at which most of the national weather radar networks work, the 500 m was chosen as an intermediate resolution between X-band radar and C-band national radar network resolutions. The 2000 m resolution was used to represent uniform rainfall conditions over the catchment. Results were analysed to investigate the effect of different spatial and temporal rainfall data resolutions on rainfall volumes, peak runoff and in-sewer water depths at locations inside the catchment, according to dimensionless parameters specified.

3.1 Scale lengths

3.1.1 Rainfall lengths

Rainfall length L_R was defined as the rainfall resolutions used as input into the hydrodynamic model to observe the response of the catchment. Rainfall data were spatially aggregated from the original resolution (30 m near the radar, 100 m elsewhere) to 500, 1000 and 2000 m. In this work storms where captured at distances from radar such that the finest grid resolution was 100 m × 100 m.

3.1.2 Storm and catchment lengths

To characterise storm size, de-correlation length of the storm L_D was defined as the distance from which rainfall rates are statistically independent. For each of the four storms under study, de-correlation lengths were determined as the range of the experimental anisotropic semi-variogram computed over the study area. The semi-variogram function was originally defined by Matherson (1963) as half the average squared difference between points separated by a distance h (Eq. 1). It is calculated as

$$\gamma(h) = \frac{1}{2m(h)} \sum_i [(Z(x_{i+h}) - Z(x_i))^2], \tag{1}$$

where $m(h)$ is the set of all pairwise Euclidean distances h, and Z are the rainfall values at spatial locations. Storm de-correlation length was defined as the range of the semi-variogram, i.e. the distance at which the sill is first reached; the sill is defined as the limit of the semi-variogram tending to infinite lag distances (see Fig. 5). Besides the value of the lag distance, in this paper the direction is also taken into account: we computed the anisotropic semi-variogram (Goovaerts, 2000; Haberlandt, 2007; Emmanuel et al., 2012), in four directions, spaced 45°. Since the limiting length is the minimum storm length, the minimum of the four ranges was taken as storm length for the study.

Storm de-correlation length was compared to pixel size of radar rainfall estimates L_R and to catchment length L_C, computed as the square root of the catchment size.

3.1.3 Model lengths

Characteristic lengths of the model topology are a result of modeller's choices based on available data, options of applied software and acceptable computational effort. Runoff length L_{RA} characterises the spatial resolution of the runoff model and was defined as the square root of the averaged runoff areas' size. Runoff length quantifies the size of the grid over which runoff is generated: if $L_{RA} \ll L_C$, the catchment is divided into sufficiently small elements to describe the spatial variability of the catchment characteristics. Moreover, spatial variability in rainfall rates can be properly captured by the runoff model if $L_R < L_{RA}$. If $L_R > L_{RA}$, rainfall rates can no longer be correctly attributed to associated runoff areas, which may distort the hydrological response pattern (Ogden and Julien, 1994).

Sewer length L_S characterises the inter-pipe distance; it is roughly the urban equivalent of drainage density for natural catchments. L_S was defined as the ratio between catchment size and the total length of the piped system. Similar to L_{RA}, the condition $L_R \ll L_S$ guarantees that the sewer pipe system routes the correct rainfall volume, previously transformed in runoff over the corresponding runoff area.

Table 3. Scale lengths related to catchment, runoff areas and sewer density, for the total catchment as well as length scale ranges for the 10 subcatchments.

Length scales (m)	Code	Event 1	Event 2	Event 3	Event 4
Storm de-correlation length	L_D	950	1000	1480	1600
Runoff length: mean (median)	L_{RA}		28 (23)		
Sewer length	L_S		43		
Catchment length	L_C		2.024		
Subcatchment runoff length (range)	(Sub) L_{RA}		21–59		
Subcatchment sewer length (range)	(Sub) L_S		33–78		
Subcatchment length (range)	(Sub) L_C		429–2.024		

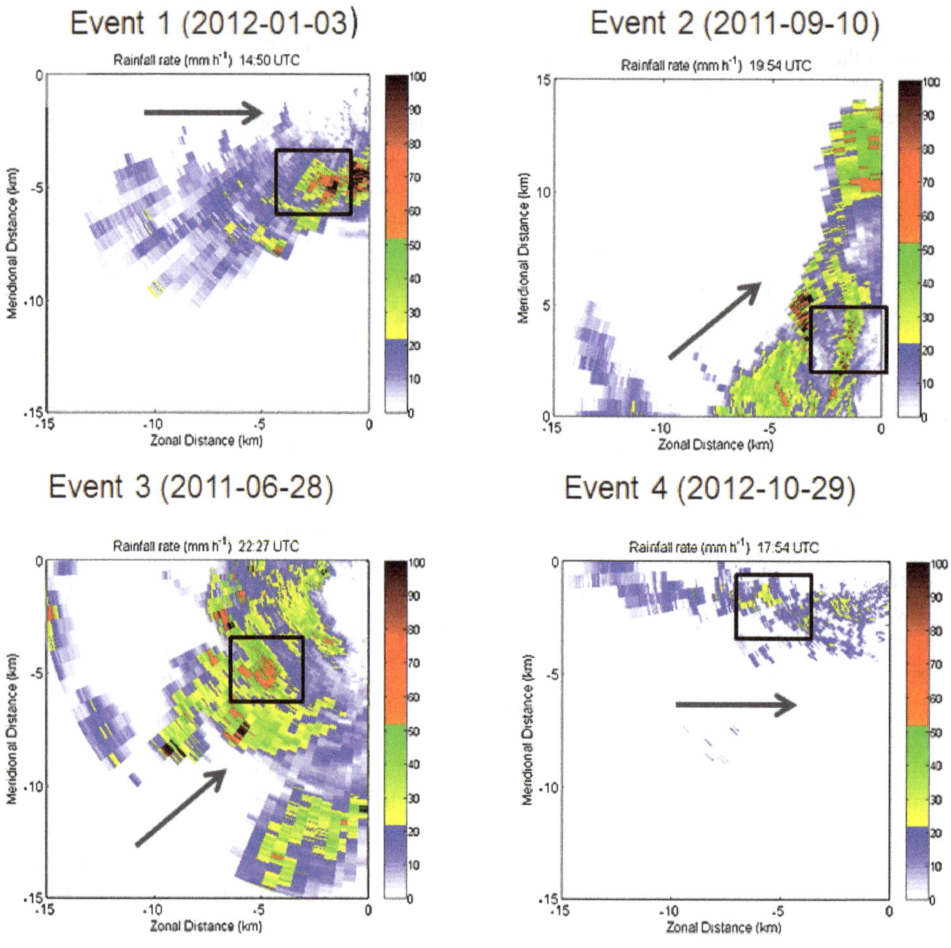

Figure 3. Plots of the maximum intensity time step for the four storm events, main direction of the storm (grey arrow), and virtual position of the catchment with respect to storm movement (black square). Zonal distances in east–west and north–south direction from X-band radar position. The latter is at (0, 0) and the maximum range is 15 km. Events 1, 3 and 4 were detected in the southwestern quadrant of the radar coverage, while Event 2 was detected in the northwestern quadrant.

3.1.4 Definition of subcatchments

The analysis involving model lengths was conducted at subcatchment scale to compare results for different model lengths: the district was divided into 11 subcatchments (Fig. 4). In lowland areas, drainage systems are often interlinked and looped and flow direction changes over the course of a storm event as the system first fills and then starts routing the storm water. This implies that flow directions and subcatchment boundaries are changeable and cannot be defined based on topography or network configuration. For this reason, in order to define subcatchment boundaries, we performed the following steps (according to a previous work of ten Veldhuis and Skovgård Olsen, 2012):

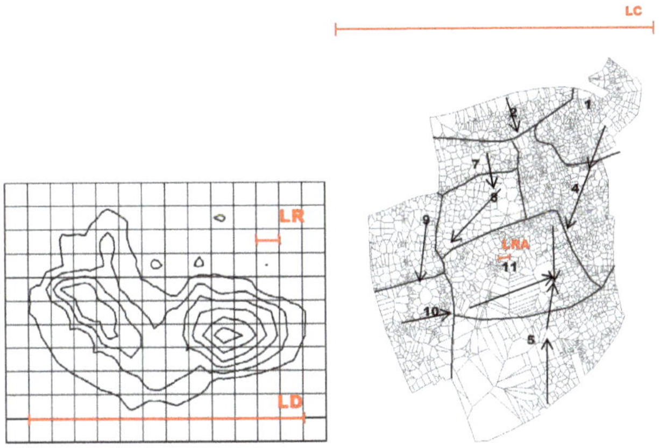

Figure 4. Storm de-correlation length (L_D) and rainfall resolution (L_R) in left panel. Catchment length (L_C), runoff length (L_{RA}) in right panel; the catchment is divided into 11 independent subcatchments. Red arrows represent main flow directions. Runoff areas are also displayed, their average size is reported in Table 2.

1. We ran simulations under long-lasting uniform storms;
2. We made sure no overflow towards surface water bodies occurred (in that case, a direction change would affect the sewer flow);
3. We detected sewer pipes with $Q = 0$;
4. We delineate subcatchments as if the latter were removed;
5. We compared flows at outlets of the 11 subcatchments in "looped" conditions (the original model) and "branched" conditions (model after the removal of cross boundary conduits); we found high agreement between the two results; therefore we accepted the catchments' delineation as a satisfactory approximation.

A visual inspection of the sewer network helped to understand the direction of flow: since no overflows occurred for the events used in this study, the system drains received water toward the main pumping station. Under this condition the main sewer conduits collect all water from peripheral conduits. We could therefore observe the flow direction in the main conduit.

3.2 Dimensionless parameters

Using the length scales, dimensionless parameters were computed to analyse relationships between spatial characteristics of rainfall, catchment and its hydrological response.

3.2.1 Rainfall sampling number (L_R/L_D)

"Rainfall sampling number" (L_R/L_D) was defined as the ratio between rainfall length (L_R) and storm de-correlation length (L_D) in order to study rainfall gradient smoothing in terms of the relationship between the estimated rainfall field and the storm inherent structure. This parameter is similar to the "storm smearing" effect defined by Ogden and Julien (1994); it accounts for the deformation of the storm structure caused by rainfall measurements of coarser resolution than the storm length. For instance, rainfall intensities in storm cells with sizes smaller than applied rainfall spatial resolution will be averaged out, leading to an underestimation in rainfall rates in the area affected by the storm cells and a overestimation in the area surrounding the cells.

In other words, as L_R tends to L_D, rain rates in high intensity regions tend to decrease, and conversely rainfall intensities in adjacent regions tend to increase. The overall effect is a reduction of rainfall gradients. Dimensionless rainfall sampling number quantifies this effect.

3.2.2 Catchment sampling number (L_R/L_C)

The second dimensionless parameter, "catchment sampling number" (L_R/L_C), also referred to as "watershed smearing" by Ogden and Julien (1994), was defined as the ratio between rainfall length (L_R) and catchment length (L_C). It accounts for rainfall transfer across catchment boundaries, as the rainfall spatial resolution approaches the size of the catchment. When the parameter exceeds 1, location of rainfall cells with respect to the catchment becomes uncertain and rainfall variability is not properly captured by the catchment. In other words, when dealing with small size storms, the position of the storm with respect to the catchment is no longer properly represented for rainfall resolutions approaching or exceeding catchment length. This affects the hydrological response: a storm moving near the boundaries of the catchment is averaged across the catchment boundary, so rainfall is artificially transferred outside the catchment. This effect is quantified by the catchment sampling number, relating the size of the catchment to the size of the radar pixel.

3.2.3 Runoff sampling number (L_R/L_{RA})

The third parameter is called "runoff sampling number" (L_R/L_{RA}), which is the ratio between rainfall length (L_R) and runoff area length (L_{RA}). This, similar to catchment sampling number, quantifies the correct assignment of rainfall values to the corresponding runoff area. The higher this ratio, the less precise is the rainfall assignment to the correct runoff area, but also the lower this ratio, the more unable is the model to capture rainfall variability, as the model resolution is coarser than the rainfall resolution. This parameter relates to the rainfall–runoff module of the model, which has rainfall as input and runoff discharge into one of the nodes of the sewer network as output. Runoff sampling number relates model input data resolution to runoff model resolution, and intends to measure the "smearing" of runoff flows induced by low rainfall resolution compared to runoff area resolution.

3.2.4 Sewer sampling number (L_R/L_S)

The fourth dimensionless parameter is the "sewer sampling number" (L_R/L_S), defined as the ratio between rainfall length (L_R), and intra-sewer length (L_S), which is computed as the average length of conduits in the system. The lower the sewer sampling number, the less sensitive is the drainage network to rainfall variability: a low "sewer sampling number" means that the inter-pipe distance is larger than the rainfall pixel size, so the sewer system cannot catch rainfall variability. Conversely, for higher sewer sampling numbers rainfall input is too coarse compared to the sewer network density and this may result in lack of accuracy of modelled water levels and sewer overflows. The "smearing effect" for sewer flows is related to the runoff smearing effect, quantified by the runoff sampling number, but they differ in this respect: the latter focuses on runoff model output, namely discharge towards the sewer network, while the sewer index represents the routing within the piped system and so it quantifies the smearing effect for in-sewer water levels. Water levels in pipes are affected by runoff discharge but also by upstream sewer inflows. As it is not possible to isolate the effect at the level of individual pipes, it is analysed at the outlet of each independent subcatchment.

3.2.5 Normalisation of model output results

To compare results between rainfall resolutions and between storms, model results were normalised with respect to results related to the highest rainfall spatial resolution: total rainfall volumes, runoff peaks and maximum in-sewer water depths were normalised according Eqs. (2), (3) and (4) respectively:

$$V_{\text{norm}}\left(L_{R_i}\right) = \frac{V(L_{R_i})}{V(L_{R100})} \tag{2}$$

$$Q_{\text{norm}}\left(L_{R_i}\right) = \frac{Q(L_{R_i})}{Q(L_{R100})} \tag{3}$$

$$\text{WD}_{\text{norm}}\left(L_{R_i}\right) = \frac{\text{WD}(L_{R_i})}{\text{WD}(L_{R100})}, \tag{4}$$

where L_{R_i} represents parameter values at the rainfall resolution under consideration (100, 500, 1000 or 2000 m) and L_{R100} represents values at 100 m rainfall resolution, used as a reference for normalisation.

3.3 Temporal resolution analysis

While the focus of this paper is on spatial scales, a preliminary investigation of the effect of temporal resolution on model outcomes was conducted to see how temporal resolution interrelates with spatial resolution. To this end, rainfall data were aggregated to 5 and 10 min temporal resolutions.

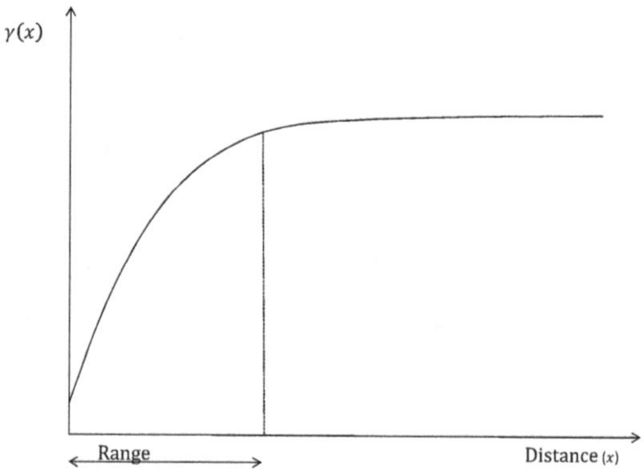

Figure 5. Sketch of semi-variogram: the range is the distance (x) from the origin beyond which the semi-variogram $\gamma(x)$ tends to infinity.

The temporal aggregation was performed by averaging out 5 (10) consecutive 1 min rainfall values at a time, obtaining temporal resolution of 5 (10) min. Semi-variograms were computed for these resolutions to study the relation between temporal resolution and the spatial structure of rainfall. The effect of the variation in rainfall temporal resolution on model outputs was quantified through the comparison of time to maximum water depths. Combined time–space resolutions were studied for Events 3 and 4: both events were simulated at two spatial and two temporal resolutions, namely 100 m, 1000 m, 5 min and 10 min, composing four different spatio-temporal rainfall scenarios.

4 Results and discussion

Results of length scale calculations are presented in Table 3, dimensionless parameter values are shown in Table 4. Storm de-correlation lengths vary between 950 and 1600 m for the four storm events. Subcatchment lengths vary from 429 to 2024 m, while runoff and sewer lengths in the hydrodynamic model vary between about 20 and 80 m, representing the model's high spatial resolution. Dimensionless parameter values show that rainfall sampling numbers vary from 0.06 for Event 4 to 0.11 for Event 1 at 100 m rainfall resolution and increase to 1.25 and 2.11 respectively at 2000 m rainfall resolution. Catchment sampling number increases from 0.05 to 0.99 for 100 and 2000 m, while runoff and sewer sampling numbers vary from 1.9 to 4.7 at 100 m resolution to 25.5 to 93.3 at 2000 m resolution, runoff sampling numbers being slightly higher than sewer sampling numbers.

Model results of the four storm events were compared against dimensionless parameters to identify trends and variability as a function of storm characteristic, radar resolution and model resolution.

Table 4. Dimensionless parameters values derived from scale length values, for the four different rainfall resolutions used in the study. Values presented for runoff sampling and sewer sampling numbers, represent value ranges for the 10 subcatchments (outlined in Fig. 4).

Rainfall resolution	Rainfall sampling number L_R/L_D				Catchment sampling number	Runoff sampling number	Sewer density sampling number
(m)	Event 1	Event 2	Event 3	Event 4	L_R/L_C	L_R/L_{RA}	L_R/L_S
100	0.11	0.10	0.07	0.06	0.05	2.6–4.7	1.9–3.8
500	0.53	0.50	0.34	0.31	0.25	13.1–23.3	6.4–19.1
1000	1.05	1.00	0.68	0.63	0.49	26.1–46.7	12.8–38.3
2000	2.11	2.00	1.35	1.25	0.99	52.3–93.3	25.5–76.5

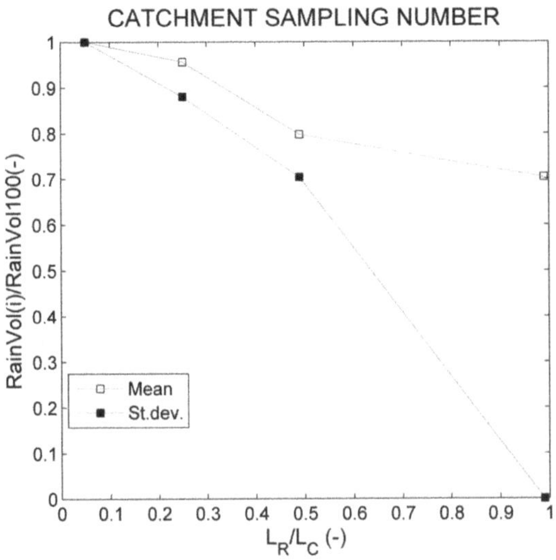

Figure 6. Normalised rainfall volumes versus catchment sampling number (L_R/L_C): mean and standard deviation of normalised rainfall volumes computed over the catchment, for the four events.

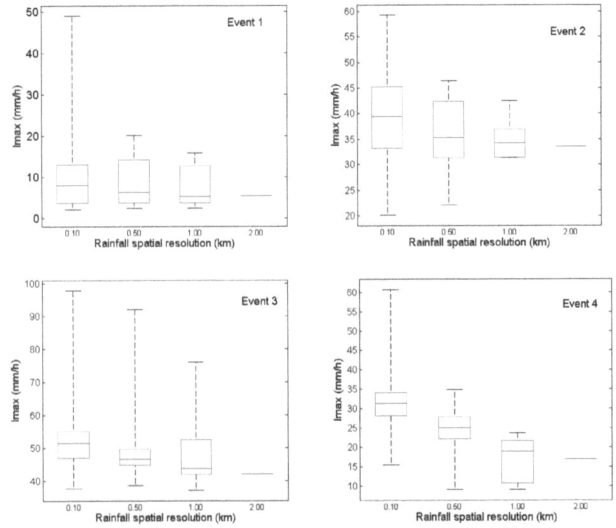

Figure 7. Boxplots of maximum rainfall intensity (mm h^{-1}) among all pixels covering the catchment area, for the four spatial resolutions (the 2000 m shows a unique value corresponding to rainfall uniformly distributed over the catchment), for the four events analysed.

4.1 Effect of spatial resolution

4.1.1 Total rainfall volumes versus catchment sampling number

Figure 6 shows mean and standard deviation of normalised rainfall volumes (according to Eq. 2) computed over the catchment, versus catchment sampling number. This result was obtained only analysing various rainfall resolutions and no hydrological modelling was used.

The results show that mean normalised rainfall volumes decrease by 5, 20 and 30 % with respect to the 100 m resolution reference, for L_R/L_C 0.2, 0.5 and 1 respectively. Standard deviations decrease by 2, 30 and 100 % respectively. Mean and standard deviation decrease progressively for catchment sampling number values above 0.2. This means that rainfall gradients decrease as rainfall values are smoothed at coarser resolution and that rainfall volumes decrease as smoothing of rainfall values at the catchment boundaries artificially transfers rainfall across the boundary. According to the findings of Ogden and Julien (1994), this effect, called by them "catchment smearing", occurs for catchment sampling numbers greater than 0.4. In contrast, results of the present study show that this effect already occurs at smaller sampling numbers, namely 0.2, and becomes stronger for values greater than 0.2. Figure 7 presents boxplots for maximum rainfall intensity values per pixel, over the studied catchment as a function of rainfall spatial resolution. The median of maximum intensity values shows a mild decrease for coarser rainfall resolutions. The smoothing effect is more pronounced for Event 3 and Event 4, where convective cells move closer to catchment boundaries. This results in storm cells being smoothed across catchment boundaries.

Event 1 is characterised by a 1 km wide storm line passing over the catchment very rapidly, resulting in steep rainfall gradients that are strongly smoothed when rainfall input resolution is reduced. When resolution is reduced from 100 to 500 m, spatial structure of the storm line is decomposed, leading to a reduction in maximum rainfall intensities (Fig. 7) in the area affected by the storm. As resolution is reduced from 1000 to 2000 m resolution, storm structure is lost and rainfall becomes uniform over the catchment. Storm cells in Event 2 are characterised by steeper spatial gradients in rainfall intensities compared to Event 1 and as a result maximum rainfall intensity values are more strongly affected by changes in rainfall resolution: the upper 25 % values decrease as a result of rainfall gradient smoothing, especially as resolution is reduced from 100 to 500 m. The lowest 25 % values increase as a result of gradient smoothing and storm structure decomposition, especially as resolution is reduced from 500 to 1000 m, where the variation between first and third quartile values is reduced from about 10 to 5 mm h^{-1}. Events 3 and 4 present a clear reduction of the median as a result of rainfall aggregation across the catchment boundary. The variation between first and third quartile values is larger at 1000 m resolution than at 100 and 500 m resolution. For Event 3, this is due to the non-organised structure of rainfall cells: local rainfall cells found at 100 m resolution are smoothed out at 500 m resolution, while at 1000 m resolution the most active convective area affects two out of nine pixels covering the catchment, i.e. the lowest 25 % values are relatively high. Event 4 is characterised by stratiform precipitation showing uniform rainfall rates. Upper quartile values decrease from 34 to 22 mm h^{-1}; lower quartile values reduce from 28 mm h^{-1} to about 10 mm h^{-1} at 1000 m resolution. This is a result of rainfall gradient smoothing and storm cells spreading southward due to spatial aggregation, while the core of the storm remains within the catchment boundaries. The strongest effect of rainfall coarsening in this case is found in a strong reduction of rainfall gradients. As a general conclusion, spatial aggregation leads to smoothing of rainfall gradients, while the effect on rainfall intensities' distribution strongly depends on spatial dimensions of storm cells and the movement of storm cells relative to the catchment boundaries.

4.1.2 Normalised maximum in-sewer water depths and runoff peaks versus rainfall resolution

Figure 8 summarises the effect of rainfall spatial resolution coarsening on semi-distributed hydrodynamic model outputs in terms of maximum computed water depths and maximum runoff flows in all nodes, per storm event. The in-sewer maximum water depths and runoff peaks at every node of the model are normalised using Eqs. (3) and (4). Results presented in the boxplots show that normalised runoff peaks are more strongly affected by changing spatial resolution of rainfall inputs compared to normalised maximum water depths. The largest effect of spatial aggregation is found for Event 4 (Fig. 8 last column), where upper and lower quartile values of runoff peaks are reduced by 40 to 60 % at 2000 m resolution with respect to the reference at 100 m resolution. Normalised maximum water depths are less strongly affected; upper quartile values remain almost unchanged, while lower quartile values decrease by up to 30 %. Event 4 has a pronounced spatial structure that is strongly affected by rainfall resolution coarsening and this directly translates into stronger changes in runoff volumes compared to the other events. Largest changes in normalised maximum water depths are found for Event 1, where upper and lower quartile values change by up to 40 % as a result of spatial redistribution of rainfall due to resolution coarsening. This event is characterised by small total rainfall volumes, resulting in small flows and water depth variations, which in turn translate into large relative differences.

Smaller changes in water depths compared to runoff flows are explained by the fact that water depths are influenced by rainfall–runoff inputs as well as by sewer routing and by storage being dominant over flow in drainage systems characterised by small gradients.

For Events 1, 2 and 3, changes in normalised water depths and runoff flows are of the same order of magnitude at 500 and 1000 m resolutions, which indicates that the effect of rainfall resolution coarsening from 100 to 500 m is not further amplified as resolution is reduced to 1000 m. When resolution is further reduced to 2000 m, corresponding to uniform rainfall over the catchment, values above 3rd quartile tend to increase as areas previously affected by low rainfall receive higher rainfall as a result of gradient smoothing.

4.1.3 Spatial structure of rainfall: anisotropic semi-variogram

Figure 9 shows experimental multi-directional spatial semi-variograms for each of the four storm events. For each storm and each time step, the semi-variogram was computed in four directions, from 0 to 180°, starting from north and going clockwise at an angle step of 45° (directions at 0 and 180° are the same, thus plots coincide). To obtain a unique semi-variogram representative of overall storm duration, for each direction, a weighted average of all semi-variograms was computed, assigning a higher weight to those of higher variance. This criterion was chosen to focus the study on more pronounced spatial rainfall structures, without losing information on the temporal evolution of the storm. Rainfall data used for the calculation are those estimated at the highest temporal and spatial resolution of IDRA radar, 1 min and 100 m respectively, in order to analyse rainfall structure at its most accurate description. The semi-variogram of Event 1 (Fig. 9 top left) presents a unique structure with a range of 1200 m in three out of four directions, while at 90° direction the range is smaller, reaching a de-correlation distance at 950 m. This is quite expected since Event 1 is a squall line moving from west to east, thus the gradient at 90° is steeper than at 180°. All four semi-variograms show a fast rise, al-

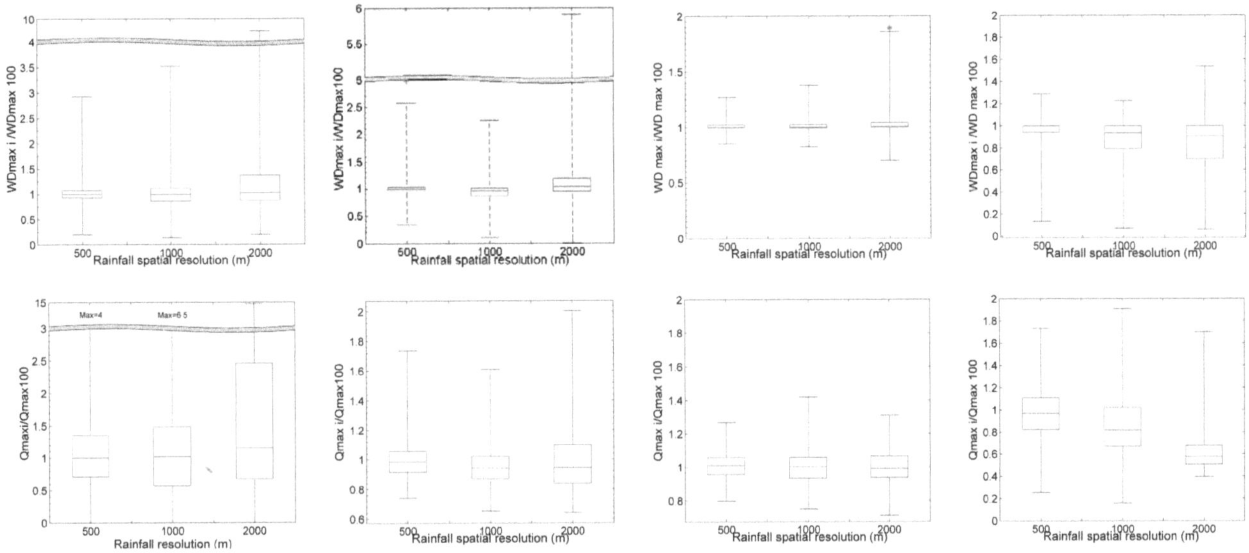

Figure 8. Boxplots of the normalised maximum water depths (top panel) and runoff peaks (bottom panel) computed for all nodes in the model, for Events 1 (left) to 4 (right).

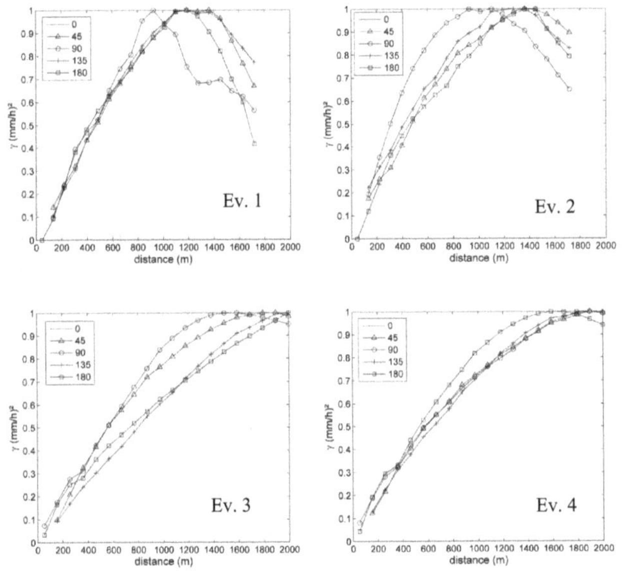

Figure 9. Instantaneous experimental multi-directional spatial semi-variogram of non-zero rainfall for each of the four storms.

though the shape of the one at 90° diverts considerably from the rest.

The same results are found for Event 2: the directional semi-variogram at 90° shows a faster rise compared to the other directions, thus the storm structure is clearly oriented. The de-correlation distance is 1000 m. No explanation was found to interpret the pronounced decrease in the semi-variograms of Events 1 and 2. We can only report that the

Table 5. Range derived from experimental semi-variograms for different temporal aggregations, for all four events.

Rainfall	Range (m)		
	$\Delta t = 1\,\mathrm{min}$	$\Delta t = 5\,\mathrm{min}$	$\Delta t = 10\,\mathrm{min}$
Event 1	950	960	970
Event 2	1000	1200	1450
Event 3	1480	> 2000	> 2000
Event 4	1600	1500	1500

same behaviour was found in storms belonging to the same rainfall group defined by Emmanuel et al. (2012).

Semi-variograms of Events 3 and 4 show a milder rise compared to Events 1 and 2. They are characterised by a different type of rainfall structure: Event 3 is a non-organised storm band, it seems to have a more defined structure in the 45 and 90° directions, the range of which is 1480 m (see also Table 5). The curves at 135 and 180° directions do not reach stability, meaning that the de-correlation distance exceeds the catchment size for which the semi-variogram was calculated. The rainfall structure of Event 4 shows a more isotropic behaviour. This is an expected result, since light rain storms are characterised by low and uniform rainfall rates. The de-correlation distance is 1600 m, highest among the four events, found in the 180° direction. The de-correlation distances found by means of this geostatistical approach were used to compute the rainfall sampling numbers discussed in the next section.

4.1.4 Normalised in-sewer maximum water depths versus rainfall sampling number

The rainfall sampling number is a measure for what Ogden and Julien (1994) referred to as "storm smearing": rainfall rates in convective regions tend to decrease while rain rates in low-intensity regions tend to increase as a result of spatial aggregation. The overall effect is thus a flattening of rainfall gradients. This happens when the resolution of the volume unit measured by the weather radar approaches or exceeds the rainfall de-correlation length, thus the rainfall sampling number exceeds 1. This effect is also due to "catchment smearing" addressed in Sect. 4.1.1. The effect of rainfall sampling number on in-sewer water depths was analysed for all four rainfall events. In-sewer depths were analysed at the outlets of the 10 subcatchments (Fig. 4) to study the effect of storm smearing in relation to catchment characteristics and in-sewer flow routing. Maximum water depth values were normalised with respect to values at 100 m resolution. Figure 10 shows normalised maximum water depths against rainfall sampling number, at the outlet nodes of the 10 subcatchments and the outlet node of the whole catchment (catchment number 11 in Fig. 10). For all events, deviations in normalised water depth increase for L_R/L_D increasing to 0.5 and 1. For Event 1, when L_R/L_D exceeds 1, deviations slightly reduce in 5 of out 11 catchments while slightly increasing for 6 subcatchments, depending on local re-distribution of rainfall. Subcatchment 2 shows highest deviation at $L_R/L_D = 0.5$, followed by a decrease for coarser resolutions. This is because the subcatchment is located at the boundary of the storm, where at 500 m spatial resolution rainfall gradients increase, while at 1000 m resolution gradients are reduced due to averaging over a larger region not affected by the storm. This directly affects the maximum water depth in underlying subcatchments. The opposite situation occurs in subcatchment 5, which is located in the southern part of the catchment with the closest node at 1.2 km from the convective region, beyond the de-correlation length. The storm only affects this southern region when rainfall data are aggregated to the 2000 m resolution, so the storm "virtually" extends from the northern part of the catchment to the whole catchment. A similar effect is noticed at the same subcatchment for Event 2. Results suggest that for most subcatchments, storm smearing occurs for L_R/L_D ratio above 0.5. Figure 9 shows that while storm smearing already affects water depths at values of L_R/L_D below 0.5, the effect becomes a lot stronger for values between 0.5 and 2. Results are in agreement with the findings of Ogden and Julien (1994), who found for their catchments that "storm smearing" occurred for $L_R/L_D > 0.8$. This implies that for the storm events used in this study, with de-correlation lengths of 0.95 to 1.6 km, the current resolution of operational weather radars (1000 m) is insufficient to have a proper estimation of intra-urban hydrodynamics.

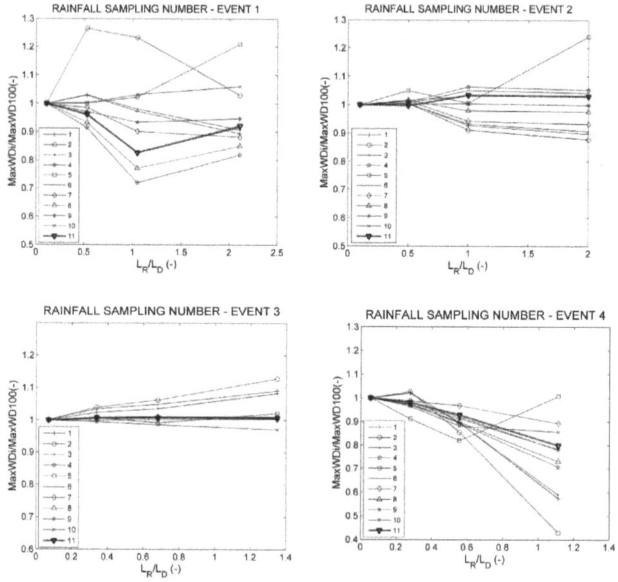

Figure 10. Normalised maximum in-sewer water depths versus rainfall sampling number (L_R/L_D): results at the outlet of the 10 subcatchments (numbered 1 to 10) and of the whole catchment (no. 11).

4.1.5 Normalised runoff peaks versus runoff sampling number

Normalised maximum runoff flows of all runoff areas were averaged within each of the 11 (sub)catchments for all four events and plotted versus corresponding runoff sampling numbers (Fig. 11) to study effects of rainfall smoothing on runoff inputs at (sub)catchment level. Deviations from 100 m simulation results remain between 0.9 and 1.1 for $L_R/L_{RA} < 20$, while higher deviations up to almost 50 % occur for $L_R/L_{RA} > 20$. At the original rainfall input resolution of 100 m, L_R/L_{RA} is below 10, so rainfall pixel size used to feed the urban hydrological model is up to 10 times larger than runoff model resolution. As L_R/L_{RA} grows larger, computed maximum runoff flows increasingly deviate as a result of rainfall smoothing and of catchment smearing, discussed in Sect. 4.1.1.

4.1.6 Normalised maximum water depths versus sewer sampling number

As presented in Sect. 3, sewer sampling number represents a measure of the ability of the sewer system to capture rainfall variability. For the model used in this study, intra-sewer pipe distances are quite small, ranging from 33 to 78 m: this means that there are 700 to 900 m of sewer pipes per 100 m × 1100 m of catchment area. The idea here is to analyse the combination effect of rainfall resolution and sewer model resolution. Figure 12 presents normalised maximum water depths as a function of sewer sampling numbers aver-

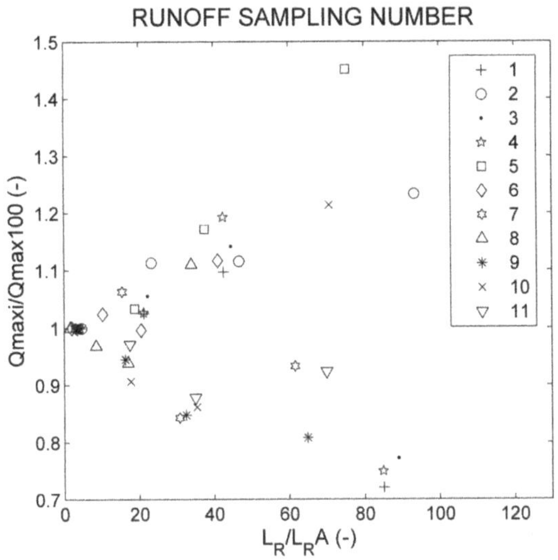

Figure 11. Normalised runoff peaks versus runoff sampling number (R_R/L_{RA}): results averaged over each of the 10 subcatchments (numbered 1 to 10) and over the whole catchment (no. 11).

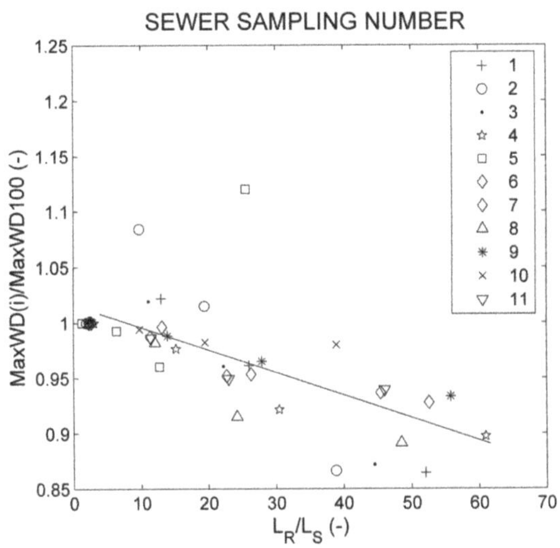

Figure 12. Normalised maximum water depths versus sewer sampling number (L_R/L_S): results at the outlet of the 10 catchments and of the whole catchment (no. 11).

aged per subcatchment, for all four events. Results show that maximum water depths tend to decrease for increasing sewer sampling numbers. In general, deviations from the reference case are smaller for in-sewer water depths, ranging from 0.87 to 1.13, than for runoff peaks, which are in the range 0.7–1.5. This is due to the smoothing effect of flow routing through the pipe system on in-sewer water depths.

4.2 Effects of temporal resolution

4.2.1 Changes in spatial structure of rainfall due to time aggregation

X-band radar images are obtained at 1 min temporal scale: the radar completes radar scans in 1 min. In order to analyse the effect of temporal resolution on rainfall spatial anisotropic semi-variograms, raw rainfall data were aggregated by averaging the original radar images to 5 and 10 min resolutions. The anisotropic experimental semi-variogram was then computed based on the aggregated values (Fig. 13). Anisotropic semi-variograms for these time resolutions were used to examine the interrelationship between temporal resolution and spatial structure of rainfall. Results show that the semi-variograms change in shape more strongly when aggregating from 1 min to 5 min compared to aggregating from 5 to 10 min resolutions. The range derived from the semi-variograms increases for lower temporal resolutions. This is especially clear for Event 3, where at 5 and 10 min the storm structure within the catchment boundaries is lost as the semi-variograms become monotonic in any of the four directions considered. In Event 4, the range expands until the catchment limits for three out of four directions, while in the 90° direction the semi-variogram range decreases. Events 1 and 2 seem less affected by changes in temporal resolution; the shape of the curves changes but the range expands only few tens of metres. Table 5 summarises ranges for all rainfall events as a function of time resolution.

4.2.2 Effect of temporal resolution on timing of maximum water depths

The effect of changes in rainfall temporal resolution on model outputs was quantified in terms of the time shift of maximum water depths with respect to the reference case. Figure 14 shows time shifts of maximum water depths between the reference simulation (100 m, 1 min) and 5 and 10 min simulations (both at 100 m spatial resolution) at the outlets of the 11 (sub)catchments, for Event 1. Results show that the timing of maximum water depths shifts by up to 4 min for aggregation to 5 min resolution and by up to 10 min for 10 min resolution. Time shifts were also calculated for all 3000 nodes of the catchment model (results not shown here). At 5 min resolution, the time shift of maximum water depths with respect to the reference case is less than 5 min for 99.4 % of all nodes. At 10 min resolution, time shifts of more than 5 min occur in 0.86 % of the nodes; time shifts in all other nodes are less than 5 min.

Figure 15 shows time shifts of maximum water depths with respect to two reference cases: 100 m, 1 min and 1000 m, 1 min. For Event 3, results show that the model is most sensitive to temporal aggregation to 10 min, at 100 m spatial resolution. The time delay of maximum water depths compared to the reference case is between 8 and 16 min. At

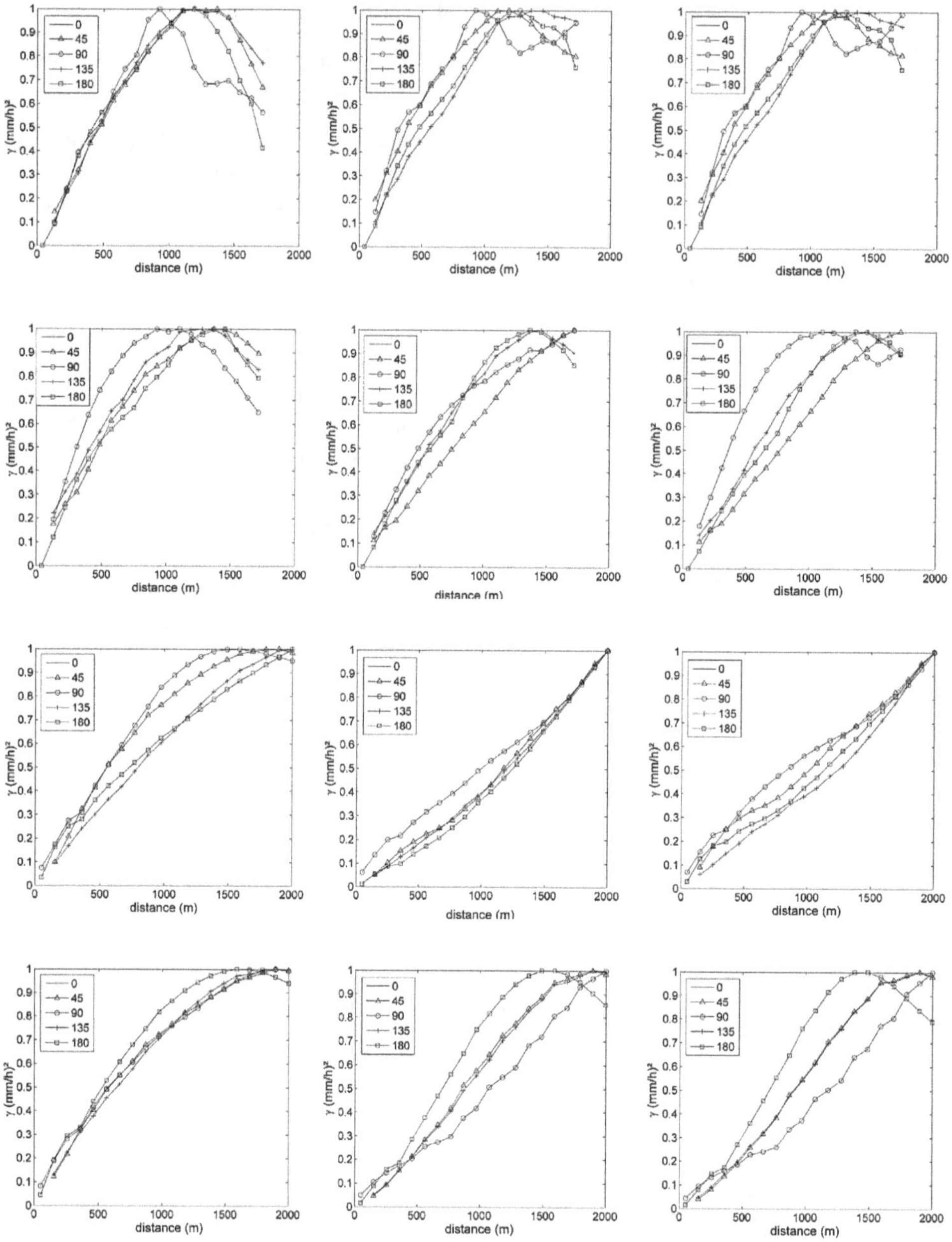

Figure 13. Anisotropic experimental semi-variograms for the four rainfall events (in rows) and different temporal resolutions: 1, 5 and 10 min (left, central and right column respectively).

the 1000 m resolution, the effect of temporal aggregation on timing of maximum water depths is comparatively smaller. The relatively high impact of 100 m and 10 min resolution simulation is explained by the change in rainfall structure induced by temporal aggregation. As shown in Fig. 13, third row, de-correlation length becomes larger than the catchment size. This effect already occurs at 5 min aggregation, but is more pronounced at 10 min aggregation. In both cases time aggregation results in enlargement of the area affected by convective storm cells, in a smoothing of rainfall peaks, and in a change in timing of rainfall peaks. This results in delay or anticipation of maximum water depths, depending on the relative position of a node with respect to the storm and also depending on the temporal position of rainfall peak values, and therefore on the temporal sampling process (for instance, if peak values are within the same 5 or 10 min sampling interval, time to peak will be hardly shifted). If peak values are averaged out with previous or following no-peak values, this will result in an anticipation or delay of sampled rainfall values and consequently anticipation/delay in hydrological response. A possible explanation for why this effect is noticeable only at 10 min is because the concentration time

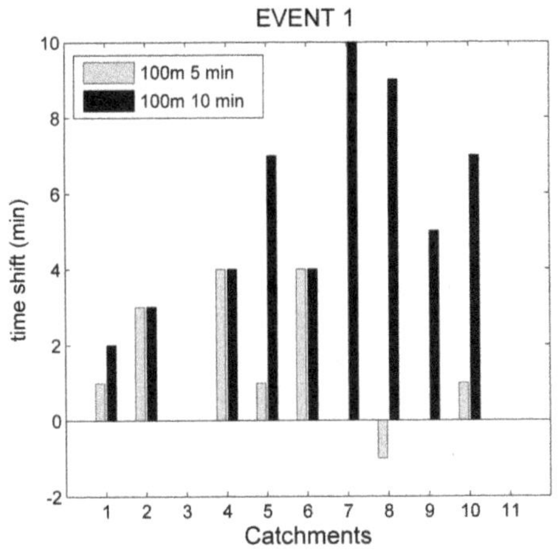

Figure 14. Time shift between maximum water depths of reference case (100 m spatial resolution, 1 min temporal resolution), and 5 and 10 min simulation, at the outlets of the 10 subcatchments and of the whole catchment (no. 11).

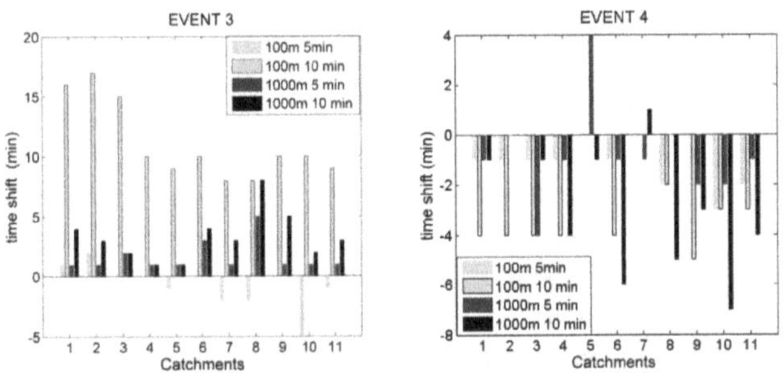

Figure 15. Differences in time to maximum water depth at the outlets of the 10 subcatchments and of the whole catchment (no. 11) for Events 3 and 4. Simulations at the highest spatial and temporal resolutions (100 m/1000 m 1 min respectively) are taken as reference.

of the 11 nodes is lower than 10 min. In order to notice an impact on model output, the time step of rainfall input must be smaller than the concentration time of the catchment at the outlet (Vaes et al., 2001) (being the concentration time the time rainfall needs to travel from the furthest place in the catchment to the chosen outlet of the sewer system). For Event 4, temporal aggregation results in anticipation of maximum water by 1 to 7 min at most of the catchment outlets.

Moreover, the effects of time aggregation on model performance have been analysed through the comparison in maximum water depths between simulations. Deviations in maximum water depths with respect to the reference case were below 0.05 m. This shows that the effect of rainfall spatial aggregation is much more important than that of temporal aggregation, in this specific case study and under these rainfall scenarios. Due to the low deviations found, results have not been reported here.

5 Conclusions

The sensitivity of an urban hydrodynamic model to spatial and temporal resolutions of weather radar data was investigated in this paper. Analyses are based on a densely populated urban catchment in Rotterdam, the Netherlands and four rainfall events that were derived from polarimetric X-band radar data. Rainfall and catchment properties were characterised using various length scales: catchment size and storm de-correlation length, which depend on the specific site and storm; rainfall data resolution, which depends on rainfall measurement resolution; and runoff resolution and sewer density, which are modeller's choices. Sensitivity of model outputs to rainfall spatial resolution was analysed in relation to: catchment size, through catchment sampling number (L_R/L_C); storm length, by means of rainfall sampling number (L_R/L_D); runoff resolution of the model, through runoff sampling number (L_R/L_{RA}); and sewer density, with the sewer sampling number (L_R/L_S). The first parameter is responsible for the uncertainty of rainfall location with respect to watershed boundaries; the second parameter describes smoothing of rain rate gradients; the third and fourth parameters describe the ability of the model (the runoff model and the sewer model respectively) to capture the rainfall structure. Storm length was been computed as the range of anisotropic experimental semi-variograms. Four rainfall spatial resolutions (100, 500, 1000 and 2000 m) and three temporal resolutions (1, 5 and 10 min) were analysed. Results obtained in this study show:

- As the ratio L_R/L_C increases (in this particular case for $L_R/L_C > 0.2$), there is a progressive decrease of both rainfall volume mean and standard deviation. Rainfall gradients decrease due to smoothing induced by rainfall resolution coarsening; mean rainfall over the catchment decreases as smoothed storm core cells extend beyond the catchment boundaries. The effect of spatial resolution coarsening on rainfall values strongly depends on the movement of storm cells relative to the catchment.

- As the ratio L_R/L_D increases (in this particular case for $L_R/L_D > 0.9$), "rainfall smearing" occurs, inducing deviations in maximum modelled in-sewer water depths. The magnitude of deviations depends on the spatial structure of the storm and variability in rainfall gradients which determines how much the rainfall field is destructured by resolution coarsening. Results are in line with what was found by Ogden and Julien (1994).

- As the ratio L_R/L_{RA} increases, deviations in runoff peaks occur. For $L_R/L_{RA} > 20$, deviations in runoff peaks are above 10 % with respect to the reference case (at 100 m rainfall resolution). This implies that, when operational weather radar products (1000 m spatial resolution) are used to feed a hydrodynamic model, runoff model outputs are not correctly represented by the model at runoff area resolutions lower than 50 m.

- As the ratio L_R/L_S increases, deviations from the reference case (100 m resolution) occur: these are smaller for in-sewer water depths, ranging from 0.87 to 1.13, than for runoff peaks, which are in the range 0.7–1.5. This is due to the smoothing effect of flow routing through the pipe system on in-sewer water depths.

Additionally, an analysis of the change in spatial structure of rainfall due to time aggregation was conducted. To this end, the impact on model results was quantified in terms of time shift of maximum water depths with respect to the reference case at 1 min temporal resolution. Experimental anisotropic semi-variograms temporal aggregations at 5 and 10 min show that rainfall field structure changes due to temporal resolution coarsening. Rainfall correlation length increases by several 100 m due to time aggregation (up to 45 % of original decorrelation length). For all rainfall events, smoothing of rainfall fields induced by temporal aggregation results in peak time shifts up to 6 min. Model outputs are most strongly affected when rainfall temporal aggregation leads to complete distortion of the rain field, which happened for one of the four events in this study.

This study was a first attempt to characterise how the effect of space and time aggregation on rainfall structure affects hydrodynamic modelling of urban catchments, for resolutions ranging from 100 to 2000 m and from 1 to 10 min. It was investigated how rainfall change in resolution is absorbed by the model, giving an indication of scale relationships between: storm structure, its representation, catchment size and model structure. In this study four storm events were used that could be derived from an experimental polarimetric X-band radar.

The findings of this study helped to provide initial insights into how small-scale precipitation variability affects hydrological response and to what extent an urban drainage model can properly describe such a response. The outcomes showed that critical thresholds are to be expected in terms of the relationship between rainfall resolution and model scales. This study points out that scale relationships are relevant in determining model output sensitivities. To give a more robust meaning to these sampling numbers, more storm events should be analysed and more catchments should be tested to confirm the findings of this study. Additionally, model sensitivity to rainfall input resolution should be analysed in relation to other sources of uncertainty, such as those related to model structure and model parameter estimation. This requires installation of a polarimetric radar in the city, which is planned for the near future. This will enable model validation according to locally observed rainfall and sewer observations and analysis of different aspects of model uncertainty under different rainfall resolution scenarios.

Such an extension of the study would allow giving reliable recommendations on what should be the model and rainfall resolution in order to prioritise either the improvement on rainfall estimation or catchment hydrological characterisation.

Appendix A: SOBEK software description

Sobek 212 is a semi-distributed hydrodynamic model from Deltares. It accounts for two modules: the rainfall–runoff module and the routing module. In the rainfall–runoff module four different types of surfaces are used depending on the runoff factor and slope: closed paved, open paved, flat roof and sloped roofs (with a slope greater than 4 %). These four categories show different runoff factor and storage coefficient. The resulting runoff is calculated based on the "rational method", where the runoff Q is given by

$$Q\left(\text{mmh}^{-1}\right) = c\left(h^{-1}\right) * p(\text{mm}), \tag{A1}$$

where p is the net rainfall and c is a runoff factor which accounts for the delay of the rainfall as overland flow to the entry point of the sewer system. The runoff factor is a function of the length, roughness and slope of the surface (Sobek, 2012). The runoff coefficient is defined as a number between 0 and 1. A coefficient of 0.5 will mean that 50 % of the runoff volume will reach the sewer entry point in 1 min. The runoff factor moves the centre of mass of the resulting hydrograph, thereby increasing the lag time. The runoff formula is applied to each one of the runoff areas connected to the node of the sewer. In semi-distributed models, the whole catchment is split into a number of subcatchments (runoff areas), each of which is treated as a lumped model (i.e. within each subcatchment rainfall input and hydrologic responses are assumed to be uniform; their spatial variability is not accounted for). Rainfall is input uniformly within each subcatchment and based on the subcatchment's characteristics; the total runoff is estimated and routed to the outlet point, which is a node of the sewer system.

Once the water enters the sewer, the routing is computed by means of the complete 1-D Saint Venant equations.

Acknowledgements. This work has been funded by the EU INTERREG IVB RainGain Project. The authors would like to thank the RainGain Project for supporting this research (www.raingain.eu). We also thank the reviewers and editor for their valuable comments and suggestions which greatly helped to improve the quality of the paper.

Edited by: P. Molnar

References

Balmforth, D. J. and Dibben, P.: A modelling tool for assessing flood risk, Water Practice & Technology, 1, 1, doi:10.2166/WPT.2006008, 2006.

Bell, V. A. and Moore, R. J.: The sensitivity of catchment runoff models to rainfall data at different spatial scales, Hydrol. Earth Syst. Sci., 4, 653–667, doi:10.5194/hess-4-653-2000, 2000.

Berne, A., Delrieu, G., Creutin, J.-D., and Obled, C.: Temporal and spatial resolution of rainfall measurements required for urban hydrology, J. Hydrol., 299, 166-179, doi:10.1016/j.jhydrol.2004.08.002, 2004.

Blanchet, F., Brunelle, D., and Guillon, A.: "Influence of the spatial heterogeneity of precipitation upon the hydrologic response of an urban catchment", 2nd Int. Symp. Hydrological Applications of Weather Radar, Hannover, p. 8, Preprint, 1992.

Dawdy, D. R. and Bergmann, J. M.: Effect of rainfall variability on streamflow simulation, Water Resour. Res., 5, 958–966, 1969.

Deltares: Sobek, Hydrodynamics, Rainfall Runoff and Real Time Control, User Manual. Version: 1.00.34157, 3 June 2014. Copyright© 2014 Deltares, Delft, NL, 2014.

Einfalt, T., Arnbjerg-Nielsen, K., Golz, C., Jensen, N.-E., Quirmbach, M., Vaes, G., and Vieux, B.: Towards a roadmap for use of radar rainfall data in urban drainage, J. Hydrol., 299, 186–202, doi:10.1016/j.jhydrol.2004.08.004, 2004.

Emmanuel, I., Andrieu, H., Leblois, E., and Flahaut, B.: Temporal and spatial variability of rainfall at the urban hydrological scale, J. Hydrol., 430–431, 162–172, doi:10.1016/j.jhydrol.2012.02.013, 2012.

Gires, A., Onof, C., Maksimovic, C., Schertzer, D., Tchiguirinskaia, I., and Simoes, N.: Quantifying the impact of small scale unmeasured rainfall variability on urban runoff through multifractal downscaling: A case study, J. Hydrol., 442–443, 117–128, doi:10.1016/j.jhydrol.2012.04.005, 2012.

Gires, A., Tchiguirinskaia, I., Schertzer, D., and Lovejoy, S.: Multifractal analysis of a semi-distributed urban hydrological model, Urban Water J., 10, 195–208, doi:10.1080/1573062X.2012.716447, 2013.

Goovaerts, P.: Geostatistical approaches for incorporating elevation into the spatial interpolation of rainfall, J. Hydrol., 228, 113–129, doi:10.1016/S0022-1694(00)00144-X, 2000.

Haberlandt, U.: Geostatistical interpolation of hourly precipitation from rain gauges and radar for a large-scale extreme rainfall event, J. Hydrol., 332, 144–157, 2007.

Jensen, N. E. and Pedersen, L.: Spatial variability of rainfall: Variations within a single radar pixel, Atmos. Res., 77, 269–277, doi:10.1016/j.atmosres.2004.10.029, 2005.

Johnson, J. T., MacKeen, P. L., Witt, A., Mitchell, E. D. W., Stumpf, G. J., Eilts, M. D., and Thomas, K. W.: The storm cell identification and tracking algorithm: An enhanced WSR-88D algorithm, Weather Forecast., 13, 263–276, 1998.

Koren, V., Finnerty, B., Schaake, J., Smith, M., Seo, D.-J., and Duan, Q.-Y.: Scale dependencies of hydrologic models to spatial variability of precipitation, J. Hydrol., 217, 285–302, doi:10.1016/S0022-1694(98)00231-5, 1999.

Krajewski, W. F., Lakshmi, V., Georgakakos, K. P., and Jain, S. C.: A Monte Carlo study of rainfall sampling effect on a distributed catchment model, Water Resour. Res., 27, 119–128, 1991.

Leijnse, H., Uijlenhoet, R., van de Beek, C. Z., Overeem, A., Otto, T., Unal, C. M. H., Dufournet, Y., Russchenberg, H. W. J., Figueras i Ventura, J., Klein Baltink, H., and Holleman, I.: Precipitation Measurement at CESAR, the Netherlands, J. Hydrometeorol., 11, 1322–1329, doi:10.1175/2010JHM1245.1, 2010.

Liguori, S., Rico-Ramirez, M. A., Schellart, A. N. A., and Saul, A. J.: Using probabilistic radar rainfall nowcasts and NWP forecasts for flow prediction in urban catchments, Atmos. Res., 103, 80–95, doi:10.1016/j.atmosres.2011.05.004, 2012.

Matheron, G.: Principles of geostatistics, Econom. Geol., 58, 1246–1266, 1963.

Neal, J., Villanueva, I., Wright, N., Willis, T., Fewtrell, T., and Bates, P.: How much physical complexity is needed to model flood inundation?, Hydrol. Process., 26, 2264–2282, doi:10.1002/hyp.8339, 2012.

Obled, C., Wendling, J., and Beven, K.: The sensitivity of hydrological models to spatial rainfall patterns: an evaluation using observed data, J. Hydrol., 159, 305–333, 1994.

Ogden, F. L. and Julien, P. Y.: Runoff model sensitivity to radar rainfall resolution, J. Hydrol., 158, 1–18, 1994.

Otto, T. and Russchenberg, H. W. J.: Estimation of Specific Differential Phase and Differential Backscatter Phase From Polarimetric Weather Radar Measurements of Rain, IEEE Geosci. Remote Sens. Lett., 8, 988–992, doi:10.1109/LGRS.2011.2145354, 2011.

Ozdemir, H., Sampson, C. C., de Almeida, G. A. M., and Bates, P. D.: Evaluating scale and roughness effects in urban flood modelling using terrestrial LIDAR data, Hydrol. Earth Syst. Sci., 17, 4015–4030, doi:10.5194/hess-17-4015-2013, 2013.

Parker, D. J., Priest, S. J., and McCarthy, S. S.: Surface water flood warnings requirements and potential in England and Wales, Appl. Geogr., 31, 891–900, doi:10.1016/j.apgeog.2011.01.002, 2011.

Pathirana, A., Tsegaye, S., Gersonius, B., and Vairavamoorthy, K.: A simple 2-D inundation model for incorporating flood damage in urban drainage planning, Hydrol. Earth Syst. Sci., 15, 2747–2761, doi:10.5194/hess-15-2747-2011, 2011.

Priest, S. J., Parker, D. J., Hurford, A. P., Walker, J., and Evans, K.: Assessing options for the development of surface water flood warning in England and Wales, J. Environ. Manage., 92, 3038–3048, doi:10.1016/j.jenvman.2011.06.041, 2011.

Schellart, A. N. A., Shepherd, W. J., and Saul, A. J.: Influence of rainfall estimation error and spatial variability on sewer flow prediction at a small urban scale, Adv. Water Resour., 45, 65–75, doi:10.1016/j.advwatres.2011.10.012, 2012.

Schmitt, T. G., Thomas, M., and Ettrich, N.: Analysis and modeling of flooding in urban drainage systems, J. Hydrol., 299, 300–311, doi:10.1016/j.jhydrol.2004.08.012, 2004.

Segond, M.-L., Wheater, H. S., and Onof, C.: The significance of spatial rainfall representation for flood runoff estimation: A numerical evaluation based on the Lee catchment, UK, J. Hydrol., 347, 116–131, doi:10.1016/j.jhydrol.2007.09.040, 2007.

Veldhuis ten, J. A. E. and Skovgård Olsen, A. Hydrological response times in lowland urban catchments characteristed by looped drainage systems, Urban Rain 14, 9th International Workshop on Precipitation in Urban Areas, 6–9 December, St. Moritz, Switzerland, 2012.

Vieux, B. E. and Imgarten, J. M.: On the scale-dependent propagation of hydrologic uncertainty using high-resolution X-band radar rainfall estimates, Atmos. Res., 103, 96–105, doi:10.1016/j.atmosres.2011.06.009, 2012.

Weisman, M. L. and Rotunno, R.: "A theory for strong long-lived squall lines" revisited, J. Atmos. Sci., 61, 361–382, 2004.

Winchell, M., Gupta, H. V., and Sorooshian, S.: On the simulation of infiltration and saturation excess runoff using radar based rainfall estimates: Effects of algorithm uncertainty and pixel aggregation, Water Resour. Res., 34, 2655–2670, 1998.

Wood, S. J., Jones, D. A., and Moore, R. J.: Static and dynamic calibration of radar data for hydrological use, Hydrol. Earth Syst. Sci., 4, 545–554, doi:10.5194/hess-4-545-2000, 2000.

15

On the efficiency of the hybrid and the exact second-order sampling formulations of the EnKF: a reality-inspired 3-D test case for estimating biodegradation rates of chlorinated hydrocarbons at the port of Rotterdam

Mohamad E. Gharamti[1,3], Johan Valstar[2], Gijs Janssen[2], Annemieke Marsman[2], and Ibrahim Hoteit[1]

[1]King Abdullah University of Science and Technology (KAUST), Thuwal 23955, Saudi Arabia
[2]Deltares, Princetonlaan 6, 3584 CB, Utrecht, the Netherlands
[3]Nansen Environmental and Remote Sensing Center (NERSC), Bergen 5006, Norway

Correspondence to: Mohamad E. Gharamti (gharamti@ucar.edu)

Abstract. This study considers the assimilation problem of subsurface contaminants at the port of Rotterdam in the Netherlands. It involves the estimation of solute concentrations and biodegradation rates of four different chlorinated solvents. We focus on assessing the efficiency of an adaptive hybrid ensemble Kalman filter and optimal interpolation (EnKF-OI) and the exact second-order sampling formulation ($\mathrm{EnKF_{ESOS}}$) for mitigating the undersampling of the estimation and observation errors covariances, respectively. A multi-dimensional and multi-species reactive transport model is coupled to simulate the migration of contaminants within a Pleistocene aquifer layer located around 25 m below mean sea level. The biodegradation chain of chlorinated hydrocarbons starting from tetrachloroethene and ending with vinyl chloride is modeled under anaerobic environmental conditions for 5 decades. Yearly pseudo-concentration data are used to condition the forecast concentration and degradation rates in the presence of model and observational errors. Assimilation results demonstrate the robustness of the hybrid EnKF-OI, for accurately calibrating the uncertain biodegradation rates. When implemented serially, the adaptive hybrid EnKF-OI scheme efficiently adjusts the weights of the involved covariances for each individual measurement. The $\mathrm{EnKF_{ESOS}}$ is shown to maintain the parameter ensemble spread much better leading to more robust estimates of the states and parameters. On average, a well tuned hybrid EnKF-OI and the $\mathrm{EnKF_{ESOS}}$ respectively suggest around 48 and 21 % improved concentration estimates, as well as around 70 and 23 % improved anaerobic degradation rates, over the standard EnKF. Incorporating large uncertainties in the flow model degrades the accuracy of the estimates of all schemes. Given that the performance of the hybrid EnKF-OI depends on the quality of the background statistics, satisfactory results were obtained only when the uncertainty imposed on the background information is relatively moderate.

1 Introduction

Subsurface contamination has received significant attention in the last few decades. Consequent cleanup costs have increased the awareness of environmental issues related to contaminated fields (Appelo and Postma, 1994; Drécourt et al., 2006). Historically, it was believed that subsurface contamination could be remediated to natural background contamination levels by digging in the soil and pumping out the contaminated groundwater. However, it was not too long before it was discovered that there were simply too many contaminated areas to completely remediate. In addition, all available cleaning technologies, including source removal, are economically not viable to fully resolve the problem (Cunningham and Berti, 1993; Starr and Cherry, 1994; Todd and Mays, 2005).

Governmental authorities are now considering another approach to remediation based on management of industrial groundwater contamination at regional scales. The idea is simply to prevent groundwater contamination from causing negative effects on humans or ecology, and to control any undesired spreading beyond the boundary of the contaminated site. In the European Water Framework Directive, an option was offered allowing groundwater aquifers to remain contaminated when remediation is too costly and when an adequate monitoring system of the contaminated area is set up (Chave, 2001; Mostert, 2003; Hering et al., 2010). This procedure relies mostly on natural attenuation of contaminant plumes without the need for a direct human intervention. This is often possible given that the size and concentration of dissolved contaminants are frequently subjected to considerable decline due to natural as well as, eventually, human induced biodegradation processes. The challenge is then to predict in a cost-effective way what type and when contaminants may cause a risk, so monitoring and, if needed, remediation may be undertaken to prevent any unacceptable spreading beyond specific planes of compliance. One efficient way to implement such a monitoring system at a regional scale is to use prediction models with monitoring data and combine them using advanced data assimilation techniques (McLaughlin, 2002; Reichle et al., 2002).

Various numerical groundwater contaminant models have been developed in the literature (e.g., Freeze and Cherry, 1979; Pollock, 1994; Dawson et al., 2004; Sun and Wheeler, 2006; Bear and Cheng, 2010). The idea behind forming such models is to simulate and predict the dynamic fluxes and energies, defined as state variables (e.g. groundwater pressure, contaminant concentration), as accurately as possible based on some selected parameters (e.g., porosity, permeability, sorption) that describe the subsurface geometry, fluid and rock properties, and surface–subsurface interactions (Moradkhani et al., 2005).

Groundwater contaminant models can be subject to several sources of uncertainties due to poorly known parameters, inputs and boundary conditions. For instance, we often know very little about the time at which contamination started, the amount of contaminant mass present in a pure phase source zone, the location of the pure phase and the rate at which biodegradation is taking place (Franssen and Kinzelbach, 2009; Gharamti et al., 2013). Other uncertain aspects are the heterogeneity of the parameters, such as the hydraulic conductivity, groundwater recharge and the redox state of the groundwater. Therefore, model predictions of where and when a contaminant crosses a plane of compliance, with what concentration and how long it takes before a pure phase source zone dries up can be quite uncertain.

One way to reduce uncertainty in model predictions and parameters is to assimilate data into the model. Data assimilation (DA) methods follow a Bayesian formulation by combining prior information of a dynamical system with available measurements to obtain an analysis of the system state and parameters (Hoteit et al., 2012; Gharamti et al., 2014a). Sequential DA techniques, such as the ensemble Kalman filter (EnKF), assimilate the data as they become available. The EnKF (Burgers et al., 1998; Evensen, 2003) is a popular DA method in hydrology, operating in consequent forecast and analysis steps. During the forecast, an ensemble of state realizations is run forward in time using the dynamical model. At the time of the update, a linear Kalman filter (KF) type analysis (Kalman, 1960; Gharamti et al., 2011) is applied to the ensemble members. The EnKF is relatively simple to implement, requiring only forward integrations of the dynamical and observational models. The EnKF has been proven useful in various subsurface hydrology applications (e.g., Chen and Zhang, 2006; Hendricks Franssen and Kinzelbach, 2008; Zhou et al., 2011; Li et al., 2012; Crestani et al., 2013; Panzeri et al., 2013; Gharamti and Hoteit, 2014). The parameters most often calibrated are those characterizing the flow and the general transport of the contaminants, such as permeability and porosity. Very few applications have tackled the estimation problem of reactive modeling parameters using sequential DA techniques. Bailey et al. (2013) used the EnKF to estimate spatially variable selenium and nitrate reaction rates in near-surface agricultural soil profiles. In another study, Bailey et al. (2012) used the ensemble smoother to infer the denitrification rate constants from synthetic observations of nitrate concentrations.

It is now widely recognized that the performance of the EnKF strongly depends on the ensemble size; a large enough ensemble is required to obtain good performances. Gharamti et al. (2014b) proposed an efficient hybrid EnKF assimilation scheme for state and parameters estimation, in which the predicted EnKF statistics are complemented with predefined static background covariance in order to mitigate for filter inbreeding and undersampling (Hamill and Snyder, 2000). The hybrid filter was applied to a small-scale synthetic reactive transport model and was found computationally efficient, providing reliable estimates using fairly small ensembles (50 members). In this study, we test the hybrid EnKF with a realistic large-scale contaminant model and further extend its formulation to allow for serial processing of the observations during the analysis step. For this, the objective function involved in the adaptive scheme is designed in such a way that the weighting between the background and the filter flow-dependent statistics is adjusted for each assimilated observation. Such an updating strategy could be more convenient given that observations from different sources and locations carry varying degrees of information to the system. This generalizes the adaptive scheme of Gharamti et al. (2014b), allowing for the weighting between the ensemble and the background covariances to change not only between different assimilation times but also for different observations at any update step.

The stochastic EnKF assimilates perturbed observations during the analysis step in order to (asymptotically) match the second moment of the KF (Burgers et al., 1998). This often introduces noise, which may become significant when the rank of the observational error covariance is larger than the ensemble size (Nerger et al., 2005). Ensemble square root filters, such as the ensemble transform Kalman filter (ETKF; Bishop et al., 2001), the singular evolutive interpolated Kalman filter (SEIK; Pham, 2001; Hoteit et al., 2002) and the deterministic ensemble Kalman filter (DEnKF; Sakov and Oke, 2008) do not require observations perturbations. Yet, the stochastic EnKF tends to "re-Gaussianize" the ensemble distribution, which improves the stability of the filter, unlike other deterministic schemes that follow the shape of the background distribution (Lawson and Hansen, 2004). In a recent study, Hoteit et al. (2015) proposed a serial EnKF algorithm to mitigate the observation sampling errors in the EnKF. The algorithm, referred to as EnKF with exact second-order observation perturbation sampling ($EnKF_{ESOS}$), is straightforward to implement in any existing serial EnKF code, requiring only removing a single rank from the sample forecast covariance matrix to exactly match the first two moments of the KF. Compared to the EnKF and the deterministic filters, the $EnKF_{ESOS}$ was shown to provide a more accurate estimates of the state of a 40-variable Lorenz-96 model. Here, we consider the $EnKF_{ESOS}$ in a realistic large-scale system and further study the impact of mitigating the observation undersampling errors in the EnKF's analysis not only on the state but also on the parameters estimates. This is the first study addressing the application of the $EnKF_{ESOS}$ scheme for parameter estimation. The idea is to investigate whether accounting for observation sampling errors can lead to a better tuning of the unknown parameters or not.

We focus on two aspects that are known to limit the efficiency of the EnKF, namely, the undersampling of the forecast errors in the forecast step and the observation errors in the analysis step. We consider an industrial groundwater contamination problem at the port of Rotterdam in the Netherlands. Many areas at the port site are contaminated due to various industrial activities. Contamination with chlorinated hydrocarbons (CH) have been detected at the port area. Reductive dechlorination process of four hazardous CH components, namely, tetrachloroethene (PCE), trichloroethene (TCE), 1,2-dichloroethene (DCE) and vinyl chloride (VC), is believed to be one of the main reactive processes taking place at the port site. We simulate this process using a coupled three-dimensional (3-D) flow–transport–reaction (3-D-FTR) model for a single plume. The contaminant data collected in 2012 by the municipality of Rotterdam is used for initializing the contaminant migration, which propagates to surface and deep aquifer layers (≈ 50 m below sea level). We use "synthetic" CH concentration data on a yearly basis, for a total of 50 years, to calibrate four biodegradation rates of the reaction chain. To the best of our knowledge, this is the first study in which biodegradation parameters of a reductive dechlorination process are estimated in a real-world system using a sequential DA procedure. Hydraulic parameters of the groundwater-flow model are not incorporated in the assimilation system as part of the unknown parameters. They are estimated in an offline procedure using borehole data and a probabilistic approach given the associated geologic lithofacies. On top of the biodegradation, the concentrations of the components are also constrained using the EnKF, the hybrid EnKF and the $EnKF_{ESOS}$ schemes. Sensitivity analyses are performed to study the efficiency and the accuracy of the assimilation schemes under different experimental settings. The filtering schemes are evaluated based on the accuracy of the estimated solute concentrations, the handling of the posterior distributions of the biodegradation rates, and computational complexity.

The rest of this paper is organized as follows. Section 2 presents the ensemble filtering schemes. Section 3 describes the large-scale subsurface reactive transport model and its numerical implementation. Section 4 presents the assimilation setup and the experimental scenarios. Results of the assimilation experiments are presented and analyzed in Sect. 5. Conclusions and further discussion are given in Sect. 6.

2 The data assimilation framework

The aim of DA is to combine measured observations and a dynamical model in order to compute the best possible estimates of the past, current and future states of the system, together with estimates of the associated uncertainties (Nichols, 2010). We follow the standard discrete nonlinear dynamical system:

$$\boldsymbol{x}_{k+1} = \mathcal{M}_k(\boldsymbol{x}_k, \boldsymbol{\Theta}_k) + \boldsymbol{\eta}_{k+1}, \tag{1}$$

where $\boldsymbol{x}_k \in \mathbb{R}^{N_x}$ denotes a state vector of N_x variables at time t_k, $\boldsymbol{\Theta}_k \in \mathbb{R}^{N_\Theta}$ is the vector of model parameters, $\mathcal{M}_k$: $\mathbb{R}^{N_x} \to \mathbb{R}^{N_x}$ is the nonlinear operator that propagates the model state from t_k to t_{k+1}. $\boldsymbol{\eta}_{k+1} \in \mathbb{R}^{N_x}$ is a model error accounting for model uncertainties, commonly assumed to follow a Gaussian distribution $\mathcal{N}(0, \mathbf{Q}_{k+1})$. The measurements obey the following observational system:

$$\boldsymbol{y}_{k+1} = \mathcal{H}_{k+1}(\boldsymbol{x}_{k+1}) + \boldsymbol{\varepsilon}_{k+1}, \tag{2}$$

where $\boldsymbol{y}_{k+1} \in \mathbb{R}^{N_y}$ is a vector of N_y observations at time t_{k+1}, $\mathcal{H}_{k+1}$: $\mathbb{R}^{N_y} \to \mathbb{R}^{N_x}$ is an observational map including grid interpolations, and could be nonlinear. The observation errors $\boldsymbol{\varepsilon}_{k+1} \in \mathbb{R}^{N_y}$ are assumed Gaussian with zero mean and covariance $\mathbf{R}_{k+1}$. We also assume independent model and observation errors.

Following the Bayesian filtering problem, the objective is to evaluate the joint probability density function (pdf), i.e., $p\left(\boldsymbol{x}_k, \boldsymbol{\Theta}_k \middle| \boldsymbol{y}_{0:k}\right)$, of the system state $\boldsymbol{x}_k$ and the parameters $\boldsymbol{\Theta}_k$ given all available observations $\boldsymbol{y}_{0:k}$. The observations, $\boldsymbol{y}_{0:k}$, are used to update the model forecast. The updated estimate is then used to compute a future prediction. Likewise, the estimation problem can be also tackled using variational approaches that involve minimization of a cost function (Dimet and Talagrand, 1986; Courtier et al., 1994; Hoteit et al., 2005; Altaf et al., 2013). Variational DA techniques, such as 3DVar and 4DVar, are widely used in geoscience applications. These methods look for an optimal state trajectory that best fits observational data over a time window, but do not offer an efficient framework for quantifying uncertainties in the solution. In this study, we will only consider the sequential Bayesian filtering problem.

2.1 The ensemble Kalman filter for state-parameter estimation

The computation of $p\left(\boldsymbol{x}_k, \boldsymbol{\Theta}_k \,|\, \boldsymbol{y}_{0:k}\right)$ is not feasible in real applications owing to the nonlinear character of the model and observation operators in addition to the very large dimension of the subsurface flow and transport system. The EnKF is an efficient Monte Carlo method that computes an approximation of the joint pdf, using the first two moments, at reasonable computational requirements. The EnKF represents the distribution of the system using a collection of state vectors, called ensemble. Generally, the true pdf of the system might not be accessible through this Monte Carlo approximation given the finite ensemble size. We follow the state-parameter augmentation procedure (Annan et al., 2005) and denote by ψ the jointly concatenated state and parameters vector. The parameters are time invariant so that their time-propagation function is simply the identity operator.

To illustrate, starting at time t_{k-1} from an analysis ensemble, $\{\boldsymbol{\psi}_{k-1}^{\mathrm{a},i}: \boldsymbol{x}_{k-1}^{\mathrm{a},i}, \boldsymbol{\Theta}_{k-1}^{\mathrm{a},i}\}_{i=1}^{N_\mathrm{e}}$, which represents $p(\boldsymbol{\psi}_{k-1} \,|\, \boldsymbol{y}_{0:k-1})$, the EnKF propagates the dynamical model (1) to compute the forecast ensemble at the time of the next available observation, t_k. Incoming measurements are then used to update the joint ensemble. The EnKF algorithm is summarized below.

- Forecast step: the analysis members are integrated forward in time to obtain the forecast ensemble from which we estimate the following:

$$\widehat{\boldsymbol{\psi}}_k^{\mathrm{f}} = \begin{bmatrix} \boldsymbol{x}_k^{\mathrm{f},i} \\ \widehat{\boldsymbol{\Theta}}_k^{\mathrm{f},i} \end{bmatrix} = \frac{1}{N_\mathrm{e}} \begin{bmatrix} \sum_{i=1}^{N_\mathrm{e}} \mathcal{M}\left(\boldsymbol{x}_{k-1}^{\mathrm{a},i}, \boldsymbol{\Theta}_{k-1}^{\mathrm{a},i}\right) \\ \sum_{i=1}^{N_\mathrm{e}} \boldsymbol{\Theta}_{k-1}^{\mathrm{a},i} \end{bmatrix}$$
$$\equiv \frac{1}{N_\mathrm{e}} \sum_{i=1}^{N_\mathrm{e}} \boldsymbol{\psi}_k^{\mathrm{f},i}, \tag{3}$$
$$\widehat{\mathbf{P}}_k^{\mathrm{f}} = \begin{bmatrix} \widehat{\mathbf{P}}_{xx}^{\mathrm{f}} & \widehat{\mathbf{P}}_{x\theta}^{\mathrm{f}} \\ \widehat{\mathbf{P}}_{\theta x}^{\mathrm{f}} & \widehat{\mathbf{P}}_{\theta\theta}^{\mathrm{f}} \end{bmatrix}$$
$$\equiv \frac{1}{N_\mathrm{e}-1} \sum_{i=1}^{N_\mathrm{e}} \left(\boldsymbol{\psi}_k^{\mathrm{f},i} - \widehat{\boldsymbol{\psi}}_k^{\mathrm{f}}\right)\left(\boldsymbol{\psi}_k^{\mathrm{f},i} - \widehat{\boldsymbol{\psi}}_k^{\mathrm{f}}\right)^T. \tag{4}$$

The joint sample covariance $\widehat{\mathbf{P}}_k^{\mathrm{f}}$, as shown in Eq. (4), consists of the sample state covariance $\widehat{\mathbf{P}}_{xx}^{\mathrm{f}}$, the state-parameter cross-correlation $\widehat{\mathbf{P}}_{\theta x}^{\mathrm{f}}$ and the sample parameter covariance $\widehat{\mathbf{P}}_{\theta\theta}^{\mathrm{f}}$ matrices. The joint state-parameter forecast estimate (mean) is denoted by $\widehat{\boldsymbol{\psi}}_k^{\mathrm{f}}$. The complexity of the forecast step grows with the ensemble size. If one supposes that $\mathcal{C}_\mathrm{M}$ is the cost for integrating the model to the next observation time, the computational requirement of the forecast step is $NN_e\mathcal{C}_\mathrm{M}$, where N is the final simulation time (Gharamti et al., 2014a). The superscripts a, f and i denote the analysis, forecast and ensemble number, respectively.

- Analysis step: when the observation $\boldsymbol{y}_k$ becomes available, the joint forecast members $\boldsymbol{\psi}_k^{\mathrm{f},i}$ are updated using the Kalman-update step; i.e.

$$\boldsymbol{\psi}_k^{\mathrm{a},i} = \boldsymbol{\psi}_k^{\mathrm{f},i} + \mathbf{K}\left(\boldsymbol{y}_k + \boldsymbol{\epsilon}_k^i - \widetilde{\mathbf{H}}_k \boldsymbol{\psi}_k^{\mathrm{f},i}\right), \tag{5}$$

where $\mathbf{K} = \widehat{\mathbf{P}}_k^{\mathrm{f}} \widetilde{\mathbf{H}}_k^T \left(\widetilde{\mathbf{H}}_k \widehat{\mathbf{P}}_k^{\mathrm{f}} \widetilde{\mathbf{H}}_k^T + \mathbf{R}_k\right)^{-1}$ is the Kalman gain and the analysis state is

$$\widehat{\boldsymbol{\psi}}_k^{\mathrm{a}} = \frac{1}{N_\mathrm{e}} \sum_{i=1}^{N_\mathrm{e}} \boldsymbol{\psi}_k^{\mathrm{a},i} \equiv \widehat{\boldsymbol{\psi}}_k^{\mathrm{f}} + \mathbf{K}\left(\boldsymbol{y}_k + \widehat{\boldsymbol{\epsilon}}_k - \widetilde{\mathbf{H}}_k \widehat{\boldsymbol{\psi}}_k^{\mathrm{f}}\right),$$
$$\text{with } \widehat{\boldsymbol{\epsilon}}_k = \frac{1}{N_\mathrm{e}} \sum_{i=1}^{N_\mathrm{e}} \boldsymbol{\epsilon}_k^i. \tag{6}$$

by $\boldsymbol{\epsilon}_k^i$, are sampled from a Gaussian distribution of zero mean and covariance $\mathbf{R}_k$. The observational operator $\widetilde{\mathbf{H}}_k = \left[\mathbf{H}_k, \mathbf{O}\right]$, acting on the augmented state-parameter vector, is assumed linear for simplicity, and the matrix $\mathbf{O}$ is a zero matrix. Computationally, the update step in hydrological applications is usually less demanding than the forecast step, with a complexity of $NN_\mathrm{e}N_yN_x + NN_\mathrm{e}^2(N_x + N_\Theta)$. The observations used in the update Eq. (5) are processed in one single batch. In our implementation, we will consider the serial EnKF update formulation in which the observations are assimilated one at a time. The reason for this will become clear in Sect. 2.3.

2.1.1 EnKF limitations

The performance of the EnKF strongly depends on the accuracy of the forecast error covariance matrix $\widehat{\mathbf{P}}^{\mathrm{f}}$. The errors in $\widehat{\mathbf{P}}^{\mathrm{f}}$ are essentially due to (1) model errors and the use of small ensemble sizes, and (2) propagation of errors in the sample covariance matrix $\widehat{\mathbf{P}}^{\mathrm{a}}$ at the previous step. The Gaussian assumption of the system's distribution is also a limiting factor but this was proven to be less problematic (Hoteit et al., 2008). The Gaussianity of the estimates often breaks when the parameters are also included as part of the state vector during assimilation (e.g., Liu et al., 2016).

The main advantage of the ensemble approximation (Eqs. 3 and 4) is that it does not involve any linearization and allows one to represent the first two moments of the states and parameters by an ensemble of vectors (Evensen, 2003). The use of large ensembles is practically not possible and thus the sample covariance $\widehat{\mathbf{P}}_k^{\mathrm{f}}$ may not well approximate the KF forecast covariance, $\mathbf{P}_k^{\mathrm{f}}$. As such, the joint forecast pdf of the system's state and parameters at any time t_k is only partially sampled, which means that there exists a null subspace in the error space that is not covered by the ensemble (Song et al., 2010; Mandel et al., 2011). To mitigate this, we will use a hybrid formulation of the forecast state and parameter statistics before performing the EnKF update (e.g., Wang et al., 2007). Further details are given in Sect. 2.2.

The limited ensemble size may also introduce noise in the update step of the EnKF when the rank of the observation error covariance is large (Hoteit et al., 2015). This is because the number of observation perturbations may not be enough to sample the observation error covariance matrix, $\mathbf{R}_k$. In addition, spurious correlations between the observation and the forecast perturbations may also introduce noise in the EnKF update (e.g., Bowler et al., 2013; Hoteit et al., 2015). To illustrate, the EnKF analysis assumes zero cross-correlations between the observation perturbations and the forecast ensemble; i.e.

$$\sum_{i=1}^{N_{\mathrm{e}}} \boldsymbol{\epsilon}_k^i \left(\boldsymbol{\psi}_k^{\mathrm{f},i} - \widehat{\boldsymbol{\psi}}_k^{\mathrm{f}}\right)^T = 0. \tag{7}$$

This can be easily seen by subtracting Eq. (5) from Eq. (6). After arranging the terms and using Eq. (4), one obtains

$$\boldsymbol{\Delta} = \left(\mathbf{I} - \mathbf{K}\widetilde{\mathbf{H}}_k\right) \frac{1}{N_{\mathrm{e}}-1} \sum_{i=1}^{N_{\mathrm{e}}} \left(\boldsymbol{\psi}_k^{\mathrm{f},i} - \widehat{\boldsymbol{\psi}}_k^{\mathrm{f}}\right) \boldsymbol{\epsilon}_k^{i\,T} \mathbf{K}^T + \mathbf{K} \frac{1}{N_{\mathrm{e}}-1} \sum_{i=1}^{N_{\mathrm{e}}} \boldsymbol{\epsilon}_k^i \left(\boldsymbol{\psi}_k^{\mathrm{f},i} - \widehat{\boldsymbol{\psi}}_k^{\mathrm{f}}\right)^T \left(\mathbf{I} - \mathbf{K}\widetilde{\mathbf{H}}_k\right)^T, \tag{8}$$

$$\widehat{\mathbf{P}}_k^{\mathrm{a}} = \left(\mathbf{I} - \mathbf{K}\widetilde{\mathbf{H}}_k\right) \widehat{\mathbf{P}}_k^{\mathrm{f}} \left(\mathbf{I} - \mathbf{K}\widetilde{\mathbf{H}}_k\right)^T + \mathbf{K}\mathbf{R}_k\mathbf{K}^T + \boldsymbol{\Delta}, \tag{9}$$

where $\boldsymbol{\Delta}$ is the sampling error term, not accounted for in the EnKF. Consequently, the ensemble analysis covariance matches the optimal KF covariance, $\mathbf{P}_k^{\mathrm{a}}$, only when the observational sampling errors and the cross-correlation terms in $\boldsymbol{\Delta}$ are indeed zero. This can be numerically achieved by assimilating the observations serially using the EnKF with exact second-order perturbations sampling, EnKF$_{\mathrm{ESOS}}$, as will be discussed in more detail in Sect. 2.3.

2.2 The hybrid EnKF

The hybrid EnKF and optimal interpolation (EnKF-OI) scheme was introduced as a way to mitigate for small ensemble sizes and model deficiencies in the EnKF (Hamill and Snyder, 2000). Using small ensembles results in rank deficient forecast covariance matrices, which strongly limit the fit to the observations (Song et al., 2010). Neglecting model errors might further produce small ensemble spread, and consequently unrealistic confidence in the forecast (Song et al., 2013). The standard solution for rank deficiency or covariance underestimation is to apply inflation and localization. Inflation artificially inflates the spread of the ensemble around the mean state (Hamill et al., 2001; Hoteit et al., 2002). It is also a simple way to account for neglected model errors (Pham et al., 1998). Covariance localization eliminates spurious correlations by a Schur product multiplication of the under-sampled covariance matrix with a function of local support (Houtekamer and Mitchell, 2001; Sakov and Bertino, 2011). Inflation and localization, although efficient and widely used (especially in atmosphere and ocean application), are generally model dependent and require important tuning efforts. They further do not introduce any new directions to diversify the ensemble, limiting the filter update to a small-dimensional ensemble subspace (Song et al., 2010, 2013). Moreover, global model parameters are not local quantities and therefore localization techniques might not be as straightforward (Devegowda et al., 2007). In addition, the parameters are dynamically constant quantities (static in time), and thus large ensembles are usually needed to well approximate the parameter distributions (Hendricks Franssen and Kinzelbach, 2008; Zhou et al., 2012).

The hybrid approach estimates the EnKF's forecast error covariance by a weighted sum of the ensemble covariance and a stationary covariance matrix, typically used in a variational or an optimal interpolation (OI) assimilation system. More specifically, the background state-state and state-parameter covariances are estimated as:

$$\widetilde{\mathbf{P}}_{xx}^{\mathrm{Hybrid}} = \alpha\widehat{\mathbf{P}}_{xx}^{\mathrm{EnKF}} + (1-\alpha)\mathbf{P}_{xx}^{\mathrm{b}}, \tag{10a}$$

$$\widetilde{\mathbf{P}}_{\theta x}^{\mathrm{Hybrid}} = \beta\widehat{\mathbf{P}}_{\theta x}^{\mathrm{EnKF}} + (1-\beta)\mathbf{P}_{\theta x}^{\mathrm{b}}, \tag{10b}$$

where $\widehat{\mathbf{P}}_{xx}^{\mathrm{EnKF}}$ and $\widehat{\mathbf{P}}_{\theta x}^{\mathrm{EnKF}}$ are the sample covariance and cross-correlation matrices of the EnKF ensemble, respectively. The background covariances are denoted by $\mathbf{P}_{xx}^{\mathrm{b}}$ and $\mathbf{P}_{\theta x}^{\mathrm{b}}$, respectively. Indeed, it was shown by Hamill and Snyder (2000) that this additional stationary background covariance may help represent part of the ensemble's null space that is not described by the limited ensemble. This procedure is based on physically reliable statistics, although flow independent, unlike inflation and localization (Wang et al., 2009). The scalar quantities α and β are weighting factors, taking values between 0 and 1.

2.2.1 Practical implementation

The static background covariance, $\mathbf{P}_{xx}^{\mathrm{b}}$, is often built on the basis of a long inventory of forecast errors (Wang et al., 2009). It is usually assumed to be of low-rank, r_x, and can be factorized into spectral modes using proper orthogonal decomposition (POD) as follows;

$$\mathbf{P}_{xx}^{\mathrm{b}} = \mathbf{S}\boldsymbol{\Omega}\mathbf{S}^T = \mathbf{S}\boldsymbol{\Omega}^{\frac{1}{2}}\left(\mathbf{S}\boldsymbol{\Omega}^{\frac{1}{2}}\right)^T = \widehat{\mathbf{S}}\widehat{\mathbf{S}}^T, \tag{11}$$

where $\mathbf{S}$ is a matrix of spectral coefficients, $\boldsymbol{\Omega}$ carries information about the associated spectral variances and $\boldsymbol{\Omega}^{\frac{1}{2}}$ is its Cholesky decomposition of $\boldsymbol{\Omega}$. The background perturbation matrix, $\widehat{\mathbf{S}}$, has r_x columns, with r_x much smaller than the number of state variables. The background state and parameter cross-covariance, $\mathbf{P}^{\mathrm{b}}_{\theta x}$, can also be approximated by a low-rank, r_θ, matrix using singular value decomposition (SVD) if the number of parameters is not equal to the number of state variables (and thus the matrix $\mathbf{P}^{\mathrm{b}}_{\theta x}$ is not square). This decomposition is useful in practice in order to reduce computational burden and memory storage. Accordingly, the complexity of the analysis step (referred to as $\mathcal{O}^{\mathrm{a}}$) in the hybrid EnKF-OI scheme becomes

$$\begin{aligned}\mathcal{O}^{\mathrm{a}}_{\mathrm{EnKF-OI}} &= NN_{\mathrm{e}}N_{y}N_{x} + NN_{\mathrm{e}}^{2}\left(N_{x}+N_{\Theta}\right)\\ &\quad + NN_{\mathrm{e}}\left(N_{x}r_{x}+N_{\Theta}r_{\theta}\right),\\ &= \mathcal{O}^{\mathrm{a}}_{\mathrm{EnKF}} + NN_{\mathrm{e}}\left(N_{x}r_{x}+N_{\Theta}r_{\theta}\right).\end{aligned} \tag{12}$$

Given that r_x and r_θ are usually small in subsurface flow and transport problems (Gharamti et al., 2014b), the complexity of the analysis step of the hybrid EnKF-OI is only marginally larger than that of the EnKF. The complexity of the forecast step of the EnKF and the hybrid EnKF-OI is the same when both are implemented with the same ensemble size.

The weighting factors α and β need to be specified in Eqs. (10a) and (10b). Careful tuning of α and β is very important (Hamill and Snyder, 2000). The simplest way is to select them based on trial and error but this can be computationally very intensive. A more efficient approach was introduced by Gharamti et al. (2014b) and consists of optimizing a 1-D objective function at every update step of the states and the parameters. Based on Kalman's update formulation, assimilating observations causes the uncertainties in the prior estimates to shrink. Thus, using the Kullback–Leibler (KL) divergence (Kullback and Leibler, 1951), one can choose the α and β that maximize the information gains at the analysis time t_k. In this study, we opt to assimilate the observations serially and thus one can adaptively compute optimal weighting factors as follows:

$$\begin{aligned}\arg\max_{\alpha}\mathcal{F}(\alpha) &= \arg\max_{\alpha}\operatorname{tr}\left[\widetilde{\mathbf{P}}^{\mathrm{f}}_{xx}-\widetilde{\mathbf{P}}^{\mathrm{a}}_{xx}\right],\\ &= \arg\max_{\alpha}\operatorname{tr}\left[\widetilde{\mathbf{P}}^{\mathrm{f}}_{xx}\mathbf{H}^{T}\left(\mathbf{H}_{k}\widetilde{\mathbf{P}}^{\mathrm{f}}_{xx}\mathbf{H}_{k}^{T}+\mathbf{R}_{k}\right)^{-1}\right.\\ &\qquad \left.\mathbf{H}_{k}\widetilde{\mathbf{P}}^{\mathrm{f}}_{xx}\right],\\ &\overset{\text{single observation}}{\equiv} \arg\max_{\alpha}\frac{1}{d}\sum_{m=1}^{N_x}\left(\boldsymbol{c}^{[m]}_{xx}\right)^{2},\end{aligned} \tag{13}$$

where $tr[\cdot]$ denotes the trace of a matrix and d is a scalar quantity equivalent to observation variance $\left(\mathbf{H}_{k}\widetilde{\mathbf{P}}^{\mathrm{f}}_{xx}\mathbf{H}_{k}^{T}+\mathbf{R}_{k}\right)$ when assimilating one observation. $\mathbf{c}^{[m]}_{xx}$ is the mth forecast variance-component corresponding to one observed variable. Similarly, one can define the objective function for the parameters' weighting factor as follows:

$$\begin{aligned}\arg\max_{\beta}\mathcal{G}(\beta) &= \arg\max_{\beta} tr\left[\widetilde{\mathbf{P}}^{\mathrm{f}}_{\theta\theta}-\widetilde{\mathbf{P}}^{\mathrm{a}}_{\theta\theta}\right],\\ &\overset{\text{single observation}}{\equiv} \arg\max_{\beta}\frac{1}{d}\sum_{m=1}^{N_\Theta}\left(\boldsymbol{c}^{[m]}_{\theta x}\right)^{2},\end{aligned} \tag{14}$$

where $\mathbf{c}^{[m]}_{\theta x}$ is the forecast cross-correlation component between the mth parameter and one observed variable. Such KL criterion describes the information gain from each individual observation as it reflects the difference between the prior and the posterior distributions. The interesting point here is that for each observation, different weights would be assigned to the background and the ensemble statistics. The maximization problems in Eqs. (13) and (14) are 1-D and bounded, yielding minimal forecast variance after the update. In terms of implementation, we perform the optimization, over the interval [0, 1], using a computationally efficient scheme combining both golden-section search and repeated parabolic interpolation (Forsythe et al., 1977).

2.3 Exact second-order observation perturbations sampling

The sampling error from neglecting the cross-correlation terms in Eq. (9) in the EnKF analysis is generally not globally small. It is often composed of a large number of elements that can add up after successive assimilation steps (Hoteit et al., 2015). This may degrade the filter's accuracy and increases the underestimation of the analysis error covariance (Whitaker and Hamill, 2002). Furthermore, such sampling errors can propagate to subsequent steps, eventually deteriorating the performance of the filter.

In a mathematical sense, for the condition in Eq. (7) to hold, the rank of the forecast perturbation matrix

$$\boldsymbol{\Psi}^{\mathrm{f}}_{k} = \left[\boldsymbol{\psi}^{\mathrm{f},1}_{k}-\widehat{\boldsymbol{\psi}}^{\mathrm{f}}_{k}, \boldsymbol{\psi}^{\mathrm{f},2}_{k}-\widehat{\boldsymbol{\psi}}^{\mathrm{f}}_{k}, \ldots, \boldsymbol{\psi}^{\mathrm{f},N_{\mathrm{e}}}_{k}-\widehat{\boldsymbol{\psi}}^{\mathrm{f}}_{k}\right] \tag{15}$$

plus the rank of $\mathbf{R}_k$ must not exceed $N_{\mathrm{e}}-1$, which is essentially the rank of $\boldsymbol{\Psi}^{\mathrm{f}}_{k}$ (Pham, 2001). Obviously, this is not possible given that $N_y+N_{\mathrm{e}}-1$ is always greater than $N_{\mathrm{e}}-1$. Yet, if we suppose that $\boldsymbol{\Psi}^{\mathrm{f}}_{k}$ has a rank $N_{\mathrm{e}}-2$, then when $\mathbf{R}_k$ is scalar, it is possible to draw the observation perturbations $\boldsymbol{\epsilon}^{i}_{k}$ such that the EnKF analysis first and second moments are exactly the same as those computed using the KF. Accordingly, Hoteit et al. (2015) proposed to remove one rank from $\boldsymbol{\Psi}^{\mathrm{f}}_{k}$ using an SVD decomposition:

$$\boldsymbol{\psi}^{\mathrm{f},i}_{k} \leftarrow \boldsymbol{\psi}^{\mathrm{f},i}_{k} - \left(\boldsymbol{\Psi}^{\mathrm{f}}_{k}\boldsymbol{w}_{k}\right)w^{i}_{k}, \tag{16}$$

where $\boldsymbol{w}_k$ is the normalized right singular vector of $\boldsymbol{\Psi}^{\mathrm{f}}_{k}$ associated with the smallest nonzero singular value. The ith component of $\boldsymbol{w}_k$ is denoted by w^{i}_{k} and the symbol $\leftarrow$ means "replaced by". Then, assimilating the observations serially

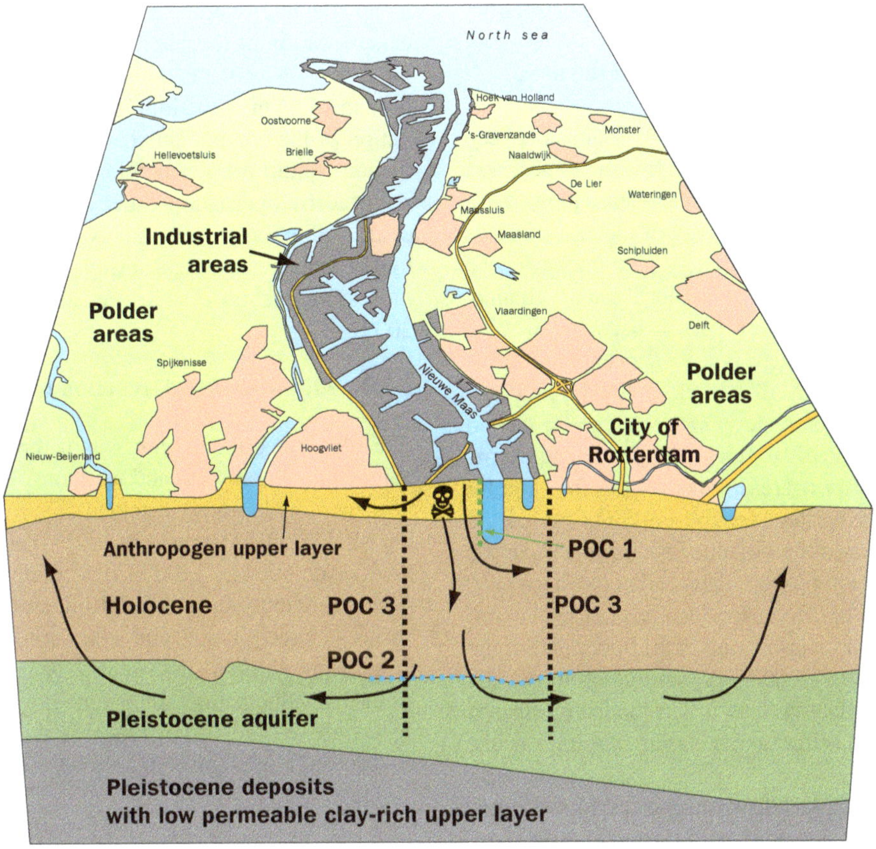

Figure 1. Schematic representation of the port of Rotterdam area with three main geologic layers: (i) Holocene clay and peat layer with sandy deposits ($\approx$ 20 m thick), (ii) Pleistocene layer with coarse sand ($\approx$ 10 m thick) and (iii) Pleistocene clay layer ($\approx$ 30 m thick). POC1, POC2 and POC3 refer to different planes of compliance at the port site.

and simply choosing $\boldsymbol{\epsilon}_k^i = \sqrt{(N_\mathrm{e}-1)\,\mathbf{R}_k}\, w_k^i$ would guarantee zero cross-correlations between the modified forecast perturbations and the observation perturbations. The algorithm, referred to as $\mathrm{EnKF_{ESOS}}$, involves a recursive update for $\boldsymbol{w}_k$ after each update (Hoteit et al., 2015). The serial analysis procedure of the $\mathrm{EnKF_{ESOS}}$ is summarized in the algorithm below:

- *While* $j \leq N_y$ *do*

1: $\widehat{z}_k = \mathbf{H}_k^{[j]} \widehat{\boldsymbol{\psi}}_k$

2: $z_k^{[j],i} = \mathbf{H}_k^{[j]} \widehat{\boldsymbol{\psi}}_k^{\mathrm{a},i}$

3: $\mathbf{K}^{[j]} = \sum_{i=1}^{N_\mathrm{e}} \left(\boldsymbol{\psi}_k^{\mathrm{f},i} - \widehat{\boldsymbol{\psi}}_k^{\mathrm{f}}\right)\left(z_k^{[j],i} - \widehat{z}_k\right) \left[\sum_{i=1}^{N_\mathrm{e}} \left(z_k^{[j],i} - \widehat{z}_k\right) + (N_\mathrm{e}-1)\mathbf{R}_k^{[j,j]}\right]^{-1}$

4: *For* i in $[1, 2, \ldots, N_\mathrm{e}]$ *do*

- $\boldsymbol{\psi}_k^{\mathrm{a},i} \leftarrow \boldsymbol{\psi}_k^{\mathrm{a},i} + \mathbf{K}^{[j]} \left(\boldsymbol{y}_k^{[j]} + s\sqrt{(N_\mathrm{e}-1)\,\mathbf{R}_k^{[j,j]}}\, w_k^i - z_k^{[j],i}\right)$

5: *EndFor*

6: $\widehat{\boldsymbol{\psi}}_k^{\mathrm{a}} \leftarrow \frac{1}{N_\mathrm{e}} \sum_{i=1}^{N_\mathrm{e}} \boldsymbol{\psi}_k^{\mathrm{a},i}$

7: *For* i in $[1, 2, \ldots, N_\mathrm{e}]$ *do*

- $w_k^i \leftarrow s\sqrt{(N_\mathrm{e}-1)\,\mathbf{R}_k^{[j,j]}}\, w_k^i - \left(z_k^{[j],i} - \widehat{z}_k\right) \left[\sum_{i=1}^{N_\mathrm{e}} \left(z_k^{[j],i} - \widehat{z}_k\right)^2 + (N_\mathrm{e}-1)\,\mathbf{R}_k^{[j,j]}\right]^{-\frac{1}{2}}$

8: *EndFor*

- *EndWhile*

where s is an independent plus or minus sign. The superscript $[j]$ denotes the jth element and row of the given vector and matrix, respectively. The superscript $[j, j]$ denotes the element in row and column j of the associated matrix. Note that unlike the EnKF, the observation perturbations cannot be Gaussian because of the constraint they satisfy in Eq. (7). In the experiments of Hoteit et al. (2015), these were shown to be almost Gaussian. In terms of complexity, the $\mathrm{EnKF_{ESOS}}$ algorithm has almost the same computational cost as that of the serial EnKF. Additional cost is required for iteratively updating the vector $\boldsymbol{w}_k$ and performing an SVD on $\boldsymbol{\Psi}_k$ to reduce its rank by one. Both operations are computationally almost negligible compared to the cost of integrating the subsurface model.

3 The subsurface model and assimilation experiments

3.1 The port of Rotterdam and geology of the area

The port of Rotterdam is located in the Netherlands between the city of Rotterdam and the North Sea. It is the largest port of Europe covering an area of 105 km^2 and stretching over a distance of 40 km. The original geology of the area consists of a top Holocene layer of approximately 20 m thick (Fig. 1). It is composed of clay and peat with local sandy channel deposits, but in the most western part, it becomes sandier. Under the Holocene layer, there is a Pleistocene aquifer of coarse sand of approximately 10 m thick. Below lays a Pleistocene clay layer of approximately 30 m thick and a second aquifer of approximately 140 m thick. The second aquifer is saline for most of the port areas, whereas the first aquifer is partly saline in the western section only. On top of the Holocene sediments, an anthropogenic layer of fine sand was added up to a level of 4 m (eastern part) to 6 m (western part) above the mean sea level. Moreover, locally a dense network of sand filled vertical drains was used in the upper part of the Holocene clay in order to speed up the settling of the clay. A large part of the industrial port area is surrounded by surface water, some of which continue to the bottom of the Holocene layer.

At the port site, more than 600 companies perform various activities such as the trans-shipment of containers (coal, oil, gas, etc), storage of oils and chemicals, building/repairing ships and oil/gas rigs, distribution and transport inland and disposal/treatment of chemical wastes. As a result of the long-term presence of these industrial activities, the soil and groundwater have become contaminated. This contamination is substantial, complex, and not limited to one particular site but affects the groundwater systems at a regional scale (Marsman et al., 2006; Ter Meer et al., 2007). Part of the contaminants are non-mobile such as heavy metals including Arsenic, cadmium, copper, mercury, lead and zinc. Other mobile contaminants are mineral oils, volatile aromatics, chlorinated solvents and pesticides.

3.2 Coupled 3-D subsurface model

3.2.1 Organic contaminants

Well monitoring and lab analysis have concluded that groundwater at the port area is contaminated, at different depth, with varying levels of pollutants (Marsman et al., 2006). One of the major contaminants are chlorinated hydrocarbons that had entered the subsurface as dense non-aqueous phase liquids (DNAPLs) and often have source zones of stagnant pure phases at considerable depth. Numerous industrial companies at the port manufacture or work with these organic molecules. Here, we simulate the degradation chain of four CH components, namely, PCE (a.k.a perchloroethene), TCE, DCE and VC. We use plume data from a real site, but for confidentiality reasons we do not show the exact location of the site. The horizontal area of the domain is equal to 1.5 km^2, extending 1 km in the transverse direction and 1.5 km in the longitudinal direction (Fig. 2). Degradation of the components happen, under anaerobic environmental conditions, when chlorine elements are subsequently replaced by hydrogen (Vogel and McCarty, 1985; Clement et al., 2000; Tobiszewski and Namieśnik, 2012). Chlorinated hydrocarbons can pose a serious threat to human health and the environmental (Ojajärvi et al., 2001; Lee et al., 2002, 2003).

3.2.2 Flow–transport–reaction model (FTR model)

The subsurface model consists of three major components, namely, flow, transport and reactions. First, the groundwater flow (assumed steady) is solved on a rectangular domain using MODFLOW (Harbaugh, 2005). The steady groundwater-flow assumption is valid at the current port location. Temporal variations, such as tidal influences and yearly fluctuations of precipitation and evapotranspiration are expected to happen, but on a small scale. Essentially, tidal influences and yearly fluctuations of precipitation and evapotranspiration are expected to be minor as the near-surface groundwater levels are controlled by the drainage levels of the drainage systems in the port area (3–4 m above sea level, a.s.l.). The deeper groundwater levels are predominantly influenced by surface water levels in the polders area (managed levels around or below sea level) and the large surface waters (approximately sea level). Temporal variations due to density-driven flow are also neglected as we would expect only minor changes in the most lower part of the model domain on the timescale of 50 years. MT3DMS (modular three-dimensional multi-species transport model) is used to solve the advection-dispersion-based transport of the components (Zheng and Wang, 1999), in which the degradation process of the components is added based on the module within the 3-D multi-species reactive package: RT3-D (Clement, 1997). The softwares are integrated in a sophisticated Fortran-based tool (with graphical interface) called iMOD (Vermeulen et al., 2013). In differential form, the fate and transport of the components is modeled following

$$\left(\phi+\rho_b k^\ell\right)\frac{\partial C^\ell}{\partial t}+\lambda\phi C^\ell=\nabla\cdot\left(\phi\mathcal{D}\nabla C^\ell\right)-\nabla\cdot\left(\nu C^\ell\right)+q_s C_s^\ell+r_C, \tag{17}$$

where ϕ is porosity, ρ_b is the bulk density of the soil, k is the distribution (sorption) coefficient, C is the solute concentration, λ is first-order reaction rate, $\mathcal{D}$ consists of hydrodynamic dispersion and molecular diffusion, ν denotes the Darcy velocity, q_s is the volumetric source/sink flow rate, C_s is the source/sink flux concentration and r_C refers to the reaction rates. The superscript ℓ corresponds to the component number taking values between 1 and 4 in this study. Along with the basic groundwater flow and transport equations as

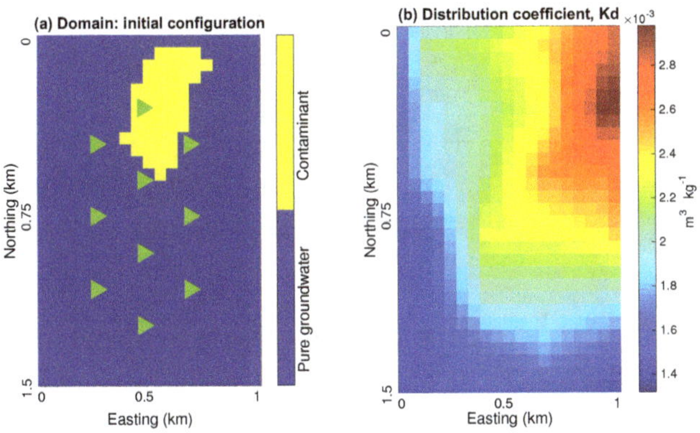

Figure 2. **(a)** Initial configuration and geometry of the study area, located at the port of Rotterdam. The blue part is the domain area ($1.5\,km^2$) of each layer and the yellow region is the plume of chlorinated hydrocarbon contaminants located in layer 60 at a depth of 22.5 m below the mean sea level. The green triangles indicate the measurement locations collected from layers 30, 50, 70 and 90; **(b)** 2-D spatial configuration of sorption (distribution coefficient) for trichloroethene (TCE) averaged over the first 10 layers of the domain.

well as using the reaction operator-split strategy (Clement et al., 1998), the biological reaction kinetics are assembled as a set of ordinary differential equations as follows:

$$\frac{\partial C_{\text{PCE}}}{\partial t} = -\frac{K_P C_{\text{PCE}}}{R_P}, \tag{18a}$$

$$\frac{\partial C_{\text{TCE}}}{\partial t} = -\frac{1}{R_T}\left(K_T \cdot C_{\text{TCE}} - S_{T/P} \cdot K_P \cdot C_{\text{PCE}}\right), \tag{18b}$$

$$\frac{\partial C_{\text{DCE}}}{\partial t} = -\frac{1}{R_D}\left(K_D \cdot C_{\text{DCE}} - S_{D/T} \cdot K_T \cdot C_{\text{TCE}}\right), \tag{18c}$$

$$\frac{\partial C_{\text{VC}}}{\partial t} = -\frac{1}{R_V}\left(K_V \cdot C_{\text{VC}} - S_{V/D} \cdot K_D \cdot C_{\text{DCE}}\right), \tag{18d}$$

where C_{PCE}, C_{TCE}, C_{DCE} and C_{VC} are the concentrations of the components, K_P, K_T, K_D and K_V are first-order anaerobic degradation rate constants, $S_{T/P}$, $S_{D/T}$ and $S_{V/D}$ are stoichiometric yield values, and R_P, R_T, R_D and R_V are retardation factors. Linear sorption conditions are assumed for all components.

The model domain as indicated by the blue region of Fig. 2a is discretized horizontally into grid cells of 20×30 from 50×50 m. In the vertical direction, we consider 120 layers for each 0.5 m of thickness. The discretization is based on the geological voxel model GeoTOP (Stafleu et al., 2011a). The top layer starts at 7.5 m a.s.l., whereas the lowest layer is located at around 52.5 m below sea level. Based on different simulations conducted as part of this study, the migration of the contaminants was found to be limited to a certain depth. We thus assume that only layers 21–100 are active. Figure 2a also shows the contaminant source (in yellow) consisting of four CH components with uniform concentration values. The plume data were obtained in January 2012 from a depth of 22.5 m below mean sea level (model layer 60), in which $C_{\text{PCE}} = 1083.0$, $C_{\text{TCE}} = 238.0$, $C_{\text{DCE}} = 633.0$ and $C_{\text{VC}} = 833.0\,\mu g\,l^{-1}$. This contaminant plume is considered as the initial condition of the transport simulations in this study. Furthermore, the PCE plume is used as a continuous contamination source and was included in the source/sink mixing (SSM) package of the MT3DMS simulator. Up to this date, other time series and well-contaminant data are not accessible due to confidentiality imposed by local companies. Modeling parameters required for running the coupled FTR model, such as porosity and hydraulic conductivity, are estimated in an offline procedure. To illustrate, the hydraulic conductivity is provided as a 3-D field in the database GeoTOP. The GeoTOP for the province of South Holland is constructed using 46 000 borehole data (Busschers et al., 2010). Using the borehole data, the most probable lithostratigraphy and lithofacies have been estimated in each voxel of $100 \times 100 \times 0.5$ m. The GeoTOP further uses relations between the lithostratigraphical units and the lithofacies with parameters, such as hydraulic conductivity, porosity and organic carbon content, in order to provide these parameters on the voxel scale. Further details about the GeoTOP methodology in addition to application to another province can be found in Stafleu et al. (2011b). Table 1 outlines the mean value (averaged over all layers) for some of these parameters. We further show in Fig. 2b the spatial map of the distribution coefficient of TCE averaged over the top 10 layers. The map shows larger sorption degrees in the northeast part of the domain. This gradually decreases towards the southern region.

3.3 Assimilation experiments

3.3.1 Reference run and pseudo-observations

In the scope of twin experiments, we first conduct a reference model run using some "true" (reference) parameters and initial condition. Next, we impose different uncertainties on the model and the initial conditions, and we assimilate pseudo-observations extracted from the reference run to recover the "true" trajectory of the model. The goal is to estimate the concentration of chlorinated hydrocarbons (i.e., state variables) and their associated degradation rates (i.e., parameters):

$$\boldsymbol{x} = \left[C_{\text{PCE}}^T, C_{\text{TCE}}^T, C_{\text{DCE}}^T, C_{\text{VC}}^T\right]^T, \tag{19a}$$

$$\boldsymbol{\Theta} = \left[K_P, K_T, K_D, K_V\right]^T. \tag{19b}$$

Based on the model's configuration, the dimension of the state is $N_x = 288\,000$ and the parameters $N_{\Theta} = 4$. The reference values of the anaerobic degradation rates are obtained through field and laboratory testing (Suarez and Rifai, 1999) and are given as $K_P = 0.068$, $K_T = 0.086$, $K_D = 0.004$ and $K_V = 0.153$ per day. We perform the reference run for a 50-year period using these rates along with the parameters of

Table 1. Different modeling parameters for the coupled flow–transport–reaction model. The values given for 3-D parametric fields, such as bulk density and distribution coefficients, are the mean values from the entire 121 layers.

Symbol	Parameter description	Value (unit)
ϕ	Porosity	0.30
ρ_b	Bulk density	1167 ($kg\,m^{-3}$)
k_{PCE}	Distribution coefficient of PCE	0.0012 ($m^3\,kg^{-1}$)
k_{TCE}	Distribution coefficient of TCE	0.0015 ($m^3\,kg^{-1}$)
k_{DCE}	Distribution coefficient of DCE	0.0014 ($m^3\,kg^{-1}$)
k_{VC}	Distribution coefficient of VC	0.0010 ($m^3\,kg^{-1}$)
κ_L	Longitudinal dispersivity	0.5 (m)
$\dot{\kappa}_T/\kappa_L$	Ratio of horizontal transverse dispersivity to longitudinal dispersivity	0.1
$\ddot{\kappa}_T/\kappa_L$	Ratio of vertical transverse dispersivity to longitudinal dispersivity	0.1
δ_m	Molecular diffusion	10^{-10} ($m^2\,s^{-1}$)

Table 1 and the initial conditions, $\boldsymbol{x}_0$, as defined above. We plot the resulting hydraulic head field at four different layers (i.e., 30, 50, 70 and 90) in Fig. 3. The maps clearly show the southward and downward flow direction of the groundwater. The hydraulic head varies between 1.5 m in the center of the domain and drops to around −1 m in the southern part. The top layers, on average, have larger hydraulic heads than the deeper ones. Overall, the flow configuration indicates that the contaminant plume would follow the behavior of the groundwater and predominantly moves vertically downwards and laterally in the southwards direction.

Based on this steady flow, we then simulated the reference transport of PCE, TCE, DCE and VC. The time step of the transport-reaction simulator was about 11 days. To visualize the migration process, in Fig. 4 we plot the contaminant evolution of PCE, TCE, DCE and VC in layers 40, 60, 80 and 100 after 50 years, respectively. As shown, the contaminant plume, which is originally present in layer 60, has moved into deeper Pleistocene layers. After 50 years, the maximum concentration of DCE in layer 80 reaches 650 $\mu g\,l^{-1}$. Careful assessment of the transport process shows that the four plumes have reached the last active layer in the second aquifer; i.e., layer 100. This is mostly due to the continuous PCE contamination source located in layer 60. Contaminant concentrations in the top Holocene layers are much smaller. Laterally, the contaminant plume is seen to expand from its initial location to a distance of 1.3 km southwards.

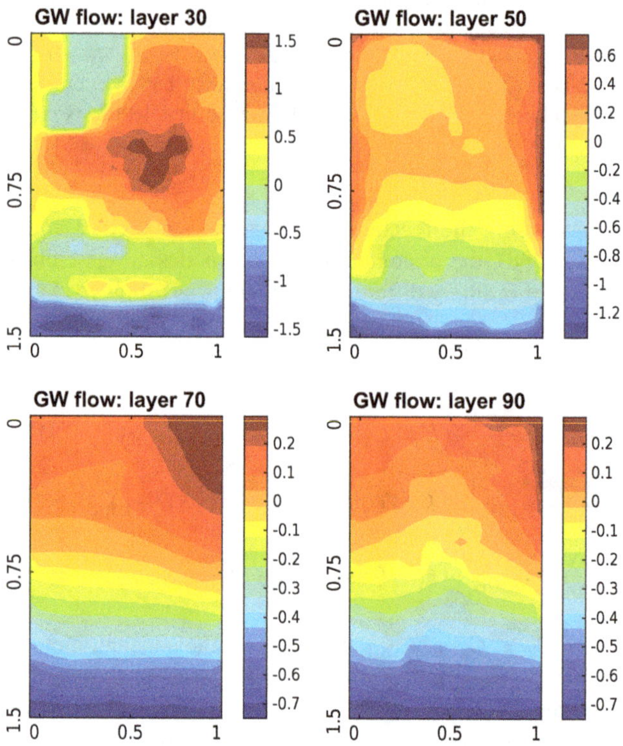

Figure 3. Groundwater (GW) hydraulic head configuration from four different active layers in the domain. The largest water head is located in the center of the domain and is equal to 1.5 m. The water head deceases in the southern part of the domain. The flow is computed using MODFLOW and plotted using iMOD's graphical interface utility.

From the reference run, we collect pseudo-observations of the concentration to use them later for assimilation. Observations are assumed available for all components from layers 30, 50, 70 and 90. From each of these four layers, 10 data points are collected and thus a total of $N_y = 160$. Note that N_y is much smaller than the number of state variables, N_x. This is usually the case in subsurface hydrology applications, given the significant and expensive cost incurred for preparing, drilling and completing wells. The observation points are uniformly distributed throughout the domain as denoted by green triangles in Fig. 2a. We assume that these 160 measurements are available for assimilation on a yearly basis. We also place a control well in layer 70 around the center of the domain, particularly at the local coordinates $x = 450$ and $y = 600$ m, to monitor the concentration evolution in time. We further assume that these observations are noisy, in order to mimic realistic settings. We thus perturb them with a Gaussian noise of mean 0 and standard devia-

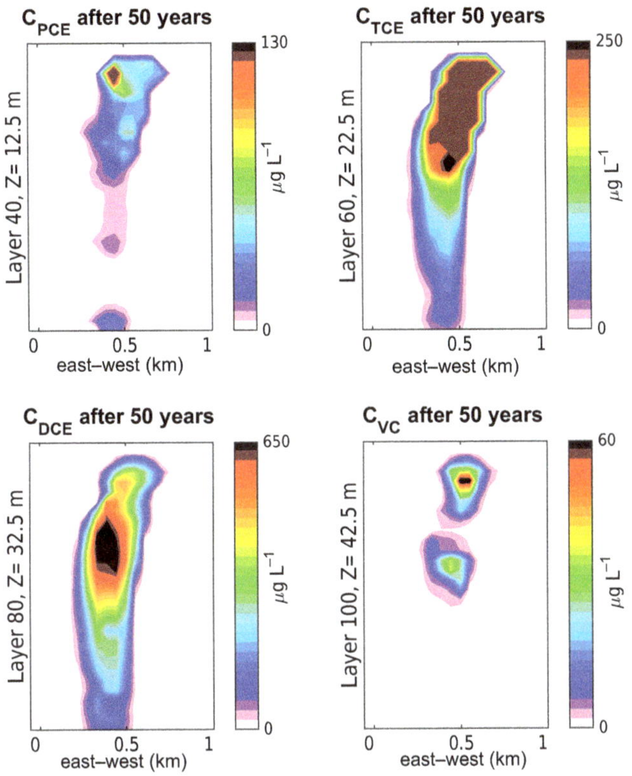

Figure 4. Contaminant plume after 50 years for PCE, TCE, DCE and VC in layers 40, 60, 80 and 100, respectively. Vertically, the contaminant plume tends to move downwards towards the Pleistocene clay layers and the second aquifer. In the lateral direction, displacement of the plume occurs southwards.

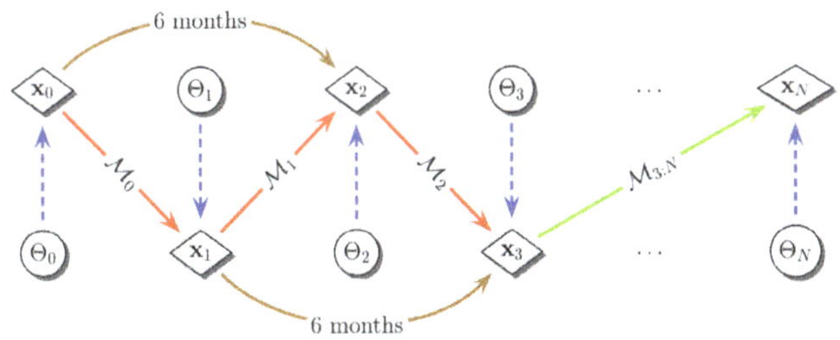

Figure 5. A sketch illustrating the procedure followed to construct the background statistics, $\mathbf{P}^{\mathrm{b}}_{xx}$ and $\mathbf{P}^{\mathrm{b}}_{\theta x}$; 3-month forecasts are performed starting from different initial conditions, $\boldsymbol{x}_{0,1,\ldots,N}$, and different degradations rate parameters, $\boldsymbol{\Theta}_{0,1,2,\ldots,N}$, where $N = 200$ steps summing up to 50 years. The background state covariance, $\mathbf{P}^{\mathrm{b}}_{xx}$, and state-parameter cross-correlations, $\mathbf{P}^{\mathrm{b}}_{\theta x}$, are then constructed using the first leading modes only.

tion equal to 15 % of the total observation mean (averaged over the entire 50 years). During assimilation, the updated concentration values are monitored to make sure that they are non-negative. Cell values that fall out of this constraint are set to zero to obtain a physically meaningful solution (Li et al., 2012; Gharamti et al., 2013).

3.4 Initial ensemble and background statistics

To initialize the filters, we perform an unconditional 50-year simulation run (referred to as free run) starting from the mean concentration of the reference model run. Thus, at time $t = 0$ the concentration of the four CH components is not only present in layer 60, but also spread-out in all layers. In this free model run, we use around 30 % larger degradation rates than the reference values. The concentration of the components was saved each 6 months. Next, we randomly select a set of N_{e} concentration snapshots from the free run outputs to form the state ensemble. This prior (initial) ensemble is quite far from the truth and further exhibits a relatively small spread. This is chosen for the purpose of testing the robustness of the assimilation schemes to challenge initial uncertainties. The initial parameter ensemble is sampled assuming a Gaussian distribution with mean equal to the reference rate values and variance of 40 %. The initial ensemble may be obtained in many different ways. One possible technique would typically involve deriving the initial concentration fields by simulating a warm-up period with historical data until the assimilation of measurements starts at a certain time t_0 (e.g., after a monitoring network was installed).

The background error statistics required for the hybrid EnKF-OI scheme are parameterized as follows. We form a set of 200 degradation realizations, as described above, and use these to perform 3-month forecasts starting from a series of initial concentrations distributed at 3-month intervals over a 50-year period, as outlined in Fig. 5. To illustrate, starting from the mean concentration of the reference run, one realization of the degradation rates $\boldsymbol{\Theta}_0$ is used to obtain a 3-month forecast of the concentration $\boldsymbol{x}_1$. Then, using $\boldsymbol{x}_1$ and $\boldsymbol{\Theta}_1$, the 3-D-FTR model is integrated forward to obtain $\boldsymbol{x}_2$ concentration after 6 months. We continue this process until the end of the 50-year period. Then, we collect the predicted contaminant states for all components and augment them with the corresponding degradation rates in a joint matrix form. POD and SVD are then performed on the augmented concentration–degradation forecast perturbations to summarize the correlations by a small number of orthogonal patterns (Hoteit et al., 2002; Skachko et al., 2009; Altaf et al., 2013). Consequently, the parameterization of the background covariance matrix, $\mathbf{P}^{\mathrm{b}}_{xx}$, is achieved using the leading 10 POD modes (i.e., $r_x = 10$) of this ensemble, which capture more than 98 % of the total variance. Concerning the background cross-correlations, we use the first 10 singular vectors (thus, $r_\theta = 10$) to parameterize $\mathbf{P}^{\mathrm{b}}_{\theta x}$ matrix. To visualize these correlations, we plot in Fig. 6, as an illustration, the spatial correlation between the rate at which PCE is degrading and the concentration of the ending product of the chain, VC. Clearly, the VC concentration in layer 60 exhibits the largest correlation values because of the continuous source term. Furthermore, since the groundwater flow is stronger in the downwards direction, as well as the contaminant migration, the cross-correlations in deeper layers are larger than those of the shallow layers. The background $\mathbf{P}^{b}_{\theta x}$ seem to vanish in the upper parts of the Holocene clay and peat layers. A consistent behavior is observed for the remaining three degradation rates, in which the largest correlations are those associated with C_{PCE} and smallest with C_{TCE}. PCE has the highest correlation because of the continuous source zone of PCE. Any removal of PCE due to biodegradation in the source zone is directly replenished in the next time step, and therefore K_P determines the total load of chlorinated hydrocarbons in the system. On the other hand, TCE has the lowest correlation because its value is high and that makes the parameter relatively insensitive to the amount of biodegradation taking place as compared to the other degradation rate constants.

State and parameter estimates of the EnKF, hybrid EnKF-OI and $\mathrm{EnKF_{ESOS}}$ schemes are evaluated using two metrics, namely, mean squared error (MSE) and average ensemble spread (AES):

$$\mathrm{MSE} = N_{x}^{-1} N_{\mathrm{e}}^{-1} \sum_{j=1}^{N_{\mathrm{e}}} \sum_{i=1}^{N_{x}} \left(x^{e}_{j,i} - x^{t}_{i} \right)^{2}, \tag{20a}$$

$$\mathrm{AES} = N_{x}^{-1} N_{\mathrm{e}}^{-1} \sum_{j=1}^{N_{\mathrm{e}}} \sum_{i=1}^{N_{x}} \left| x^{e}_{j,i} - \hat{x}^{\mathrm{e}}_{i} \right|, \tag{20b}$$

where x^{t}_{i} is the true value of the variable at location i, $x^{e}_{j,i}$ is the forecast ensemble value and $\hat{x}^{\mathrm{e}}_{i}$ is the corresponding ensemble mean. MSE measures the distance from the estimate to the truth and AES measures the spread or the uncertainty of the estimates (Hendricks Franssen and Kinzelbach, 2008).

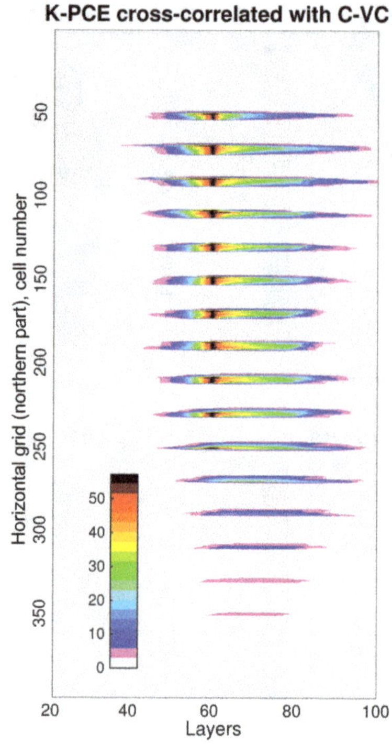

Figure 6. Individual cross-correlation terms of the background matrix $\mathbf{P}^{b}_{\theta x}$ associated with PCE biodegradation rate and VC concentration. The correlations are shown for all layers, assuming that the cells from each layer have been stretched in one vertical line (y axis). The largest correlation is present in layer 60 where the contaminant source is located. Biodegradation in shallow layers is not as strong as in deep layers because of the downwards groundwater-flow direction.

4 Results and discussion

In this section, we present and compare assimilation results with the Rotterdam port's 3-D-FTR model using the standard EnKF, the hybrid EnKF-OI and the $\mathrm{EnKF_{ESOS}}$ schemes. The observations are assimilated serially in all three schemes. Concentration and degradation rate estimates of the filters are compared in terms of accuracy and spread. Concentration data are assimilated every year for a total of 50 assimilation cycles. The ensemble size, N_{e}, is set to 48 in which batches of four members are run in parallel using Fortran's OpenMP library.

4.1 The hybrid EnKF-OI vs. the EnKF

4.1.1 Adjusting concentration statistics

To initiate the assimilation experiments, we first run the EnKF and the hybrid EnKF-OI, implemented using only state background statistics, i.e., using Eq. (10a) and $\beta = 1$. We carry out 10 different experiments by changing the weighting factor α, for each individual run, between 0 and 1 in increments of 0.1. To visualize the resulting estimates, we plot the average MSE of the contaminant concentrations, averaged over the four components and in time, as seen in Fig. 7. As shown in the left panel of Fig. 7, the most accurate concentration estimates are obtained using $\alpha = 0.7$. This indicates that out of the total forecast error variance, the best reconstruction of the reference contaminant solution is obtained when 30 % of this variance is traced from the background statistics. Increasing or decreasing this background contribution (i.e., 30 %) results in less accurate contaminant estimates. We also note that the least accurate estimates are those obtained when 90 % of the ensemble statistics are built based on the static background error covariance. On average, we found that when α takes values between 0.4 and 0.9, the hybrid EnKF-OI is 16 % more accurate than the EnKF.

We also study the effect of changing α on the resulting parameters' estimates. In Fig. 7b, we plot the average MSE of each individual degradation rate obtained using the EnKF and nine different hybrid EnKF-OIs using $\alpha = 0.1, 0.2, \ldots, 0.9$. We notice that the most influenced biodegradation rates are those associated with TCE and VC. In fact, K_T and K_V are the least identifiable parameters and therefore a small difference in the estimation algorithm (i.e., hybrid EnKF-OI and the EnKF) may lead to different estimates. In contrast, K_P and K_D are less sensitive to the weighting factor α. In accordance with the estimates of the contaminant concentrations, the best match for K_T is obtained using $\alpha = 0.7$. This is not the case for the other parameters, in which $\alpha = 0.3$, 1 and 0.8, which resulted in the best fit to the reference degradation rates of VC, PCE and DCE, respectively. On average, K_T and K_V estimates are 13 and 35 % more accurate than those of the EnKF, respectively. The key point is that complementing the state statistics, using a weighted error covariance as in Eq. (10a), does not only contribute to a better retrieval of the concentration but also helps adjusting the cross-correlations with the uncertain parameters. This is essentially the case when biodegradation is taking place at a higher rate, as in K_T and K_V, and thus the more information fed through observations, the better the state-parameter cross-correlations would be.

To better understand the performance of the hybrid EnKF-OI scheme, we further study how the uncertainties of these degradation rates are maintained in time. For this, we plot in Fig. 7c the overall ensemble spread of the four degradation rates obtained using the EnKF and the hybrid EnKF-OI (all tested α's). As shown, the EnKF's spread around these four parameters is quickly reduced after the first 2 or 3 years. This rapid reduction of the ensemble spread, which is due to the relatively small ensemble size and large initial uncertainties, limits the ability of the filter to impose larger corrections in the future, eventually degrading the accuracy of the estimated parameters. In contrast, the hybrid EnKF-OI maintains larger uncertainties in time for different weighting factor values. Therefore, as α decreases from 1 to 0.1 the performance approaches that of the EnKF. For instance, after the second year, the ensemble spread of the EnKF reaches 0.03 while it remains larger in the hybrid EnKF-OI ($\alpha = 0.3, 0.5, 0.7$) and equal to 0.09, 0.07 and 0.045, respectively. This allows the hybrid scheme to better exploit the concentration information from observations. This can be confirmed by noticing that the spread of the hybrid filter continues to decrease after 25 years of assimilation, unlike the EnKF that does not show signs of more corrections. All in all, selecting $\alpha = 0.7$ seems to maintain enough spread for both components' concentration and degradation rates and leads to the most accurate estimates, on average. For the sake of comparison, we refer hereafter to this scenario as EnKF-OI$_{\alpha=0.7}$.

4.1.2 Adjusting concentration and parameter statistics

In the following set of experiments, we fix α to 0.7 and focus on changing the weighting factor between the EnKF and background state and parameter cross-correlations; i.e., β. As in the previous section, we conduct nine experiments in which we change β between 0.1 and 0.9. Note that the larger β is, the closer the performance is to EnKF-OI$_{\alpha=0.7}$. To analyze the results, we plot in Fig. 8 the average MSE and AES of the chlorinated hydrocarbon concentrations. Compared to the previous runs that hybridize the state only, including background cross-correlations information slightly increases the spread around the ensemble mean of concentration, as observed for $0.1 < \beta < 0.3$. In terms of accuracy, varying β between 0.1 and 0.6 yields more accurate concentration estimates for all components. To illustrate, when $\beta = 0.1$ the average improvements over the EnKF and the EnKF-OI$_{\alpha=0.7}$ are around 50 and 32 %, respectively. This vigorous performance suggests that using only 10 % of the "flow-dependent" parameters' ensemble to characterize the pdf of the system is enough to outperform the EnKF. In essence, the background state and parameter cross-correlations seem to carry sufficient description of how the degradation rates and the concentration of each of the components are related. Consequently, only a small portion (i.e., 10 %) of the online parameters' ensemble is required to obtain an accurate biodegradation picture, while the rest of the information could come from the prescribed static background statistics. This could be due to the time-independent nature of the propagation step describing the evolution of the degradation rates, thereby manifesting a minimal dependence on the flow-dependent ensemble. This observation comes in accordance with the steady-state Kalman filter (El Serafy and Mynett, 2008) that assumes time-invariant error covariance matrix as long as accurate spatial correlations are used within the Kalman gain. Our experimental results suggest that the best parameter's hybrid covariance matrix is very close to a steady-state one. However, this is only for the parameters and this was not the case for the state as discussed in Sect. 4.1.1. Following the notation introduced earlier, we refer to this scenario, hereafter, as EnKF-OI$^{\beta=0.1}_{\alpha=0.7}$.

To have a better insight in the suggested performance, we plot in Fig. 9 the evolution of the concentration ensemble members for all components in time. For a fair comparison, we also plot the associated reference solution, the EnKF's and the EnKF-OI$_{\alpha=0.7}$'s ensemble members. As explained in Sect. 3.4, the initial ensemble spread is clearly far from the truth. When data are assimilated into the system, all schemes tend to move closer to the truth. By the end of the 50-year period, both EnKF and EnKF-OI$_{\alpha=0.7}$ underestimate the concentration of DCE and VC and end up with quite small ensemble spread. The EnKF-OI$^{\beta=0.1}_{\alpha=0.7}$, on the other hand, leads to the best performance, well matching the reference solution for all components. Moreover, this hybrid scheme is shown to better preserve the ensemble spread around the true final concentrations. In terms of the estimated degradation rates, we plot in Fig. 10 the temporal change of MSE for each individual degradation rate as they result from the EnKF-OI$^{\beta=1.0}_{\alpha=0.1...0.9}$ (top panels) and EnKF-OI$^{\beta=0.1...0.9}_{\alpha=0.7}$ (bottom panels). For all rates, the EnKF-OI$^{\beta=0.1...0.9}_{\alpha=0.7}$ performs much better during the first 10 years, especially for K_P and K_T. Averaging in time and over all cases, EnKF-OI$^{\beta=0.1...0.9}_{\alpha=0.7}$ is 33, 17, 33 and 15 % more accurate for retrieving K_P, K_T, K_D and K_V, respectively. From these results, one can see

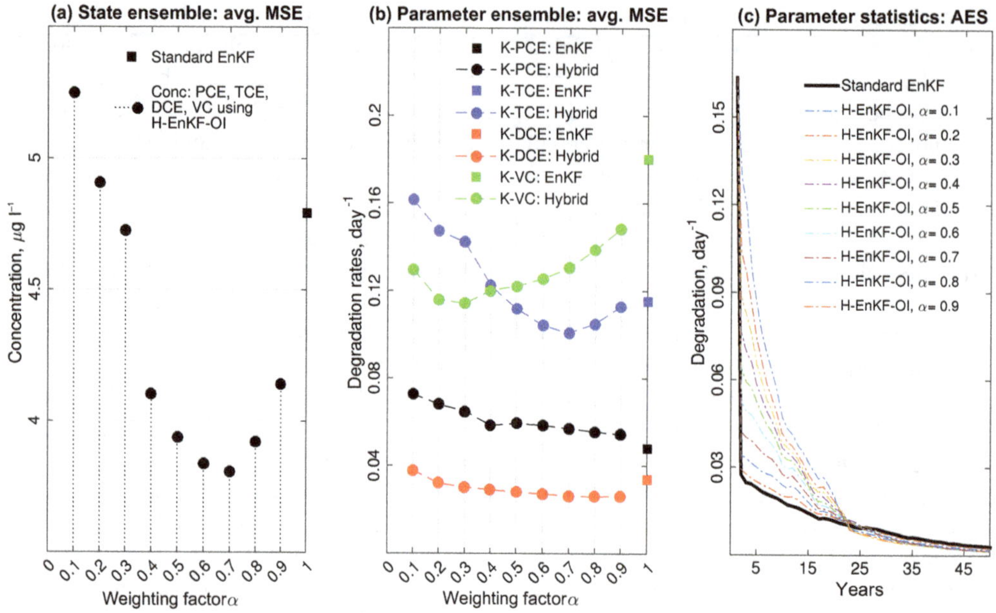

Figure 7. **(a)** Average mean square errors (MSE) of contaminant concentrations obtained using the EnKF and the hybrid EnKF-OI using $\alpha = 0.1, 0.2, \ldots, 0.9$. **(b)** Average MSE for PCE, TCE, DCE and VC biodegradation rates obtained using the EnKF and the hybrid EnKF-OI for different weight factor (α) values. **(c)** Time series change of average ensemble spread (AES) resulting from the EnKF and hybrid EnKF-OI using 48 members in which $\alpha = 0.1, 0.2, \ldots, 0.9$.

that the accuracy of the degradation rates tends to improve in time, except for K_D that is shown to degrade for small α and large β values. To interpret this behavior, one should recall that K_D (and also K_V) can only be estimated correctly as long as the concentration of the source component (C_{DCE} in this case) is accurately recovered. Before reaching this, the estimates of K_D are compensated for errors in K_P and K_T.

Next, instead of manually changing the weighting factors α and β, we follow Eqs. (13) and (14) and conduct a 1-D optimization problem prior to assimilating the observations serially. The idea is to get the maximum reduction in the prior uncertainties for both the concentration and the degradation rates as a way to "optimally" exploit the information in the assimilated observations. Therefore, different weights can be assigned to the background and the ensemble statistics. Based on this, we plot in Fig. 11 the resulting optimal α values at every assimilation cycle and for each observation. Recall that there is a total of 160 observations, such that each contaminant component is observed at 40 different locations. To better interpret the plot, we arrange these observation indices as follows: from top to bottom of the left y axis: PCE – 1 $\rightarrow$ 40, TCE – 41 $\rightarrow$ 80, DCE – 81 $\rightarrow$ 120 and VC: 121 $\rightarrow$ 160. As can be seen from the plot, the adaptive hybrid EnKF-OI algorithm selects either 0, and thus Eq. (10a) is purely based on $\mathbf{P}^{\mathrm{b}}_{xx}$, or 1 so that only the ensemble covariance, $\widehat{\mathbf{P}}^{\mathrm{EnKF}}_{xx}$, is included. When assimilating PCE, TCE and VC concentrations, the adaptive scheme tends to use the background covariance (i.e., $\alpha = 0$) for almost the first 25 years. Beyond this, the filter statistics are only based on the ensemble flow-dependent information (i.e., $\alpha = 1$). This is not surprising given the large initial uncertainties imposed on the contaminant concentrations. Once the statistics are adjusted towards the truth, the filter relies more on the correlations of the flow-dependent ensemble statistics. This performance changes when assimilating DCE observations in the sense that the filter builds its forecast error covariance mostly using static background statistics and less using the flow-dependent ensemble. This is in agreement with the results and conclusion drawn from Fig. 10 in which the background information of DCE are more useful than the ensemble statistics. Averaging over the entire optimal values of α, we obtain a global $\alpha^* = 0.64$, which is quite close to the 0.7 value that resulted in the best performance in Sect. 4.1.1. In terms of the adaptive β values, we found that maximizing the difference between the prior and the posterior parameters' covariance, $\widetilde{\mathbf{P}}_{\theta\theta}$, may not always be helpful. This is because doing such maximization can quickly diminish the ensemble spread, eventually paralyzing the filter's analysis. In fact, minimizing the difference[1] yielded more accurate degradation rates. To analyze this, we plot on the same figure the time-evolution of MSE of concentration when (1) max-

[1] Minimizing the difference between the prior and the posterior covariances does not mean that the filter does not apply any correction. Since the Kalman analysis equation always minimizes the variance, the adaptive algorithm now acts in a way such that only the lowest minimization possible is retrieved. Unlike standard Kalman filtering, this procedure moves at a slower pace towards the true trajectory.

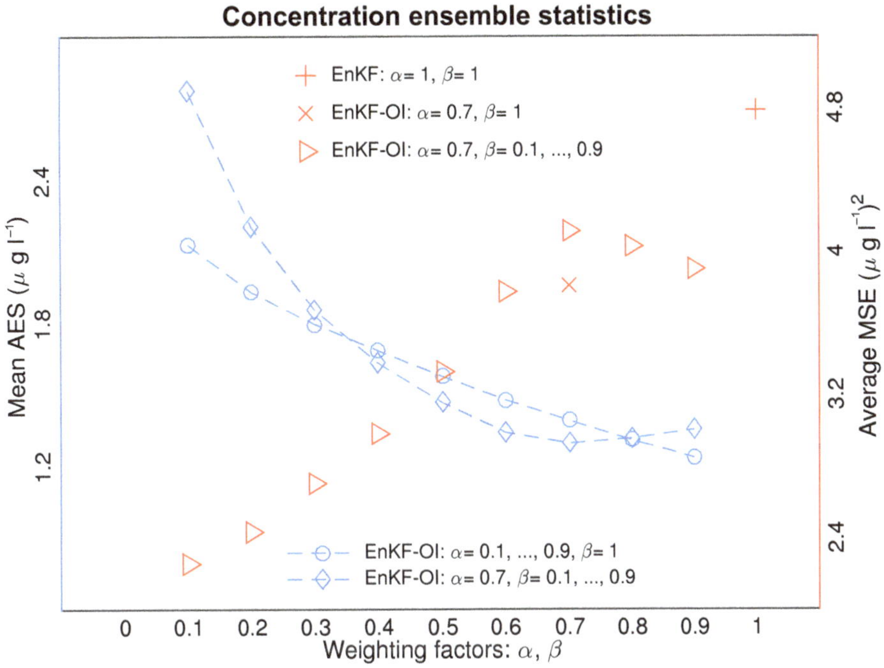

Figure 8. Left y axis: the AES of concentration ensemble obtained using the hybrid EnKF-OI by changing the individual weighting factors (α and β) between 0.1 and 0.9. Shown according to the right y axis the MSE obtained using the hybrid EnKF-$\mathrm{OI}_{\alpha=0.7}^{\beta=0.1,\ldots,0.9}$ (triangles), the hybrid EnKF-$\mathrm{OI}_{\alpha=0.7}$ (cross) and the EnKF (plus).

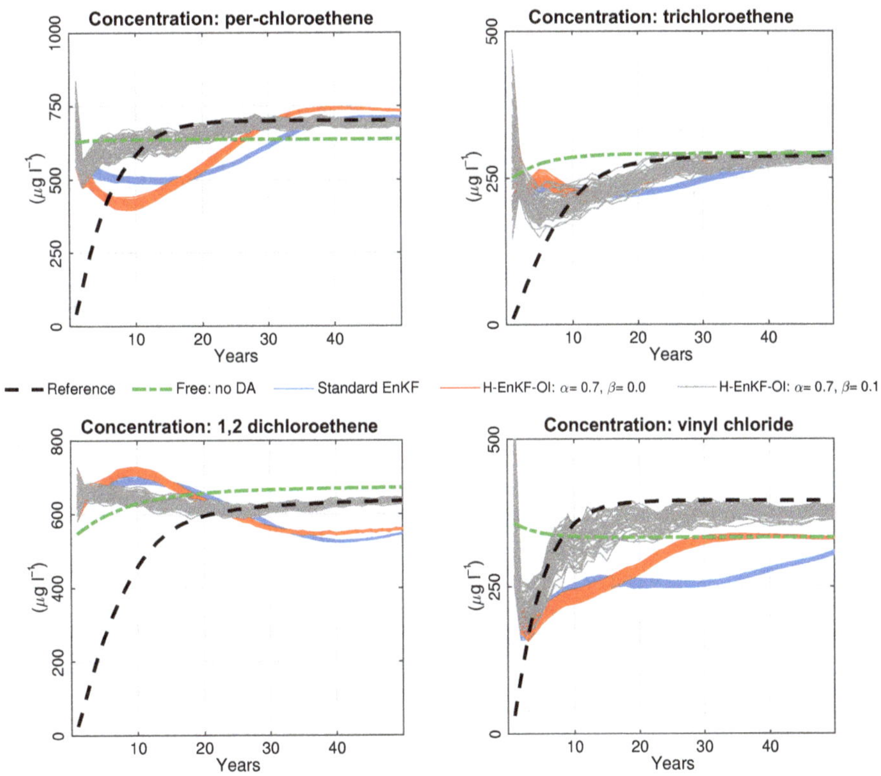

Figure 9. Forecast ensemble members of the PCE, TCE, DCE and VC concentrations. The evolution of these members (N_e = 48) is shown for the entire 50 years. Results are obtained using the standard EnKF, the EnKF-$\mathrm{OI}_{\alpha=0.7}$ and the EnKF-$\mathrm{OI}_{\alpha=0.7}^{\beta=0.1}$ schemes. Solid (black) dashed and dashed-dotted (green) lines correspond to the reference and free run concentration of each component.

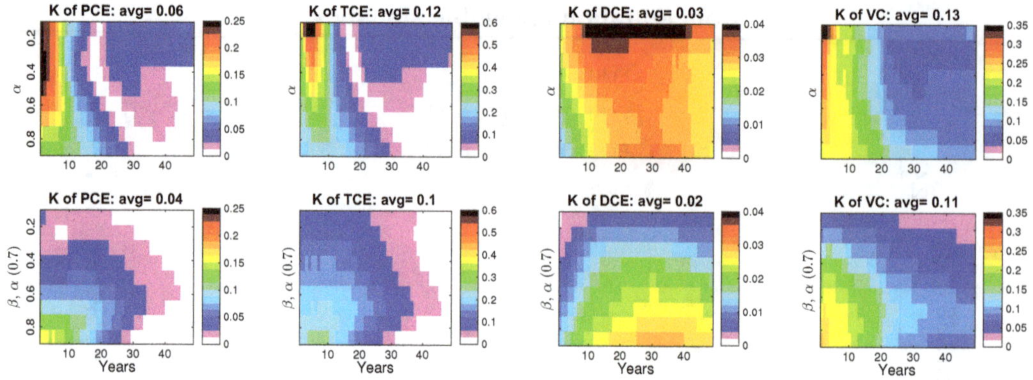

Figure 10. Images showing the MSE of PCE, TCE, DCE and VC degradation rates in time. These are obtained using the hybrid EnKF-OI scheme for (1) different α values (top panels) and (2) different β values keeping α fixed and equal to 0.7 (bottom panels).

imizing the information gain for both state and parameters, (2) minimizing the information gain for both and (3) maximizing the state's and minimizing the parameters' information gain. As seen from the three curves, the best performance is obtained when the information gain for concentration is maximized and the associated parameter's one is minimized. Compared to maximizing the information gain of both states and parameters, this mixed scheme now yields 37 % more accurate contaminant concentrations.

Motivated by this robust performance, we argue here that processing the observations serially in the adaptive hybrid scheme allows for a more efficient exploitation of the information content of the observations, as compared to the batch processing in the original algorithm proposed by Gharamti et al. (2014b). To prove our point, we plot in Fig. 12 the relative improvements suggested by the serial and the batch adaptive hybrid EnKF-OIs over the standard EnKF scheme. We also show the evolution of the value of α in time, as estimated by both filters. The relative improvements (RIs) are given in terms of MSE and averaged over all estimated concentration variables and degradation coefficients. The α values shown for the serial EnKF-OI are obtained by averaging over the entire 160 observations (i.e., each value corresponds to the mean of the columns shown in Fig. 11). As can be seen, the batch scheme imposes more weight on the static background covariance over the entire 50-year assimilation period. During the first 15 years, for instance, the optimal α values obtained using the batch and the serial schemes are 0.1 and 0.35, respectively. Therefore, given the large uncertainty in the initial ensemble, the batch scheme is shown to contribute to larger improvements over the EnKF estimates. However, this behavior changes after 15 years, beyond which the serial algorithm starts outperforming the associated batch EnKF-OI once the ensemble-based statistics have been properly adjusted. Towards the end of the assimilation, the performances of both schemes become comparable (RI $\simeq 0.7$), with a subtle advantage to the serial EnKF-OI. Overall, the mean RI suggested by the batch and the serial hybrid schemes over the EnKF is 0.52 and 0.57, respectively. Processing the observations serially leads to a more smooth selection of the weights between the ensemble and the static background covariance. In the batch scheme, the optimization is relatively more erratic, exhibiting stronger variations over time. From an algorithmic point of view, optimizing the parameters α and β in a serial formulation is computationally more efficient and does not involve any matrix inversion, in contrast with the batch processing that requires the inversion of the matrix $\left(\mathbf{H}_k\mathbf{P}_k^{\mathrm{f}}\mathbf{H}_k^T + \mathbf{R}_k\right)$ for every iteration of the optimization procedure.

4.2 The EnKF$_{\text{ESOS}}$ vs. the EnKF

In the previous section, all approaches and experimental results were intended to mitigate for the rank deficiency and the under-sampling of the ensemble's sample forecast error covariance. In the following experiment, we attempt to deal with the undersampling of the observation errors by implementing the EnKF$_{\text{ESOS}}$ algorithm presented in Sect. 2.3. We first note that the distribution of the new observation perturbations show reasonable deviations from the prescribed Gaussian errors in the original EnKF algorithm, as has been noticed by Hoteit et al. (2015). To assess the performance of the EnKF$_{\text{ESOS}}$ against the EnKF, we study at a closer glance the contaminant maps after 50 years as estimated by the ensemble means from both schemes. Thus, we plot in Fig. 13 the normalized errors for the components TCE and DCE at layers 70 and 80, respectively. These error maps are obtained by subtracting the ensemble mean concentration from the reference and then normalizing the result by the average of the reference solution. One common feature in these maps is the clear underestimation of TCE and DCE in the northern part of the domain. This is because the initial reference concentration is quite different from the one assigned to the initial ensemble using the free run setup as outlined in Sect. 3.4. In time, both filtering schemes try to push the contaminant plume, which has already moved towards the southern re-

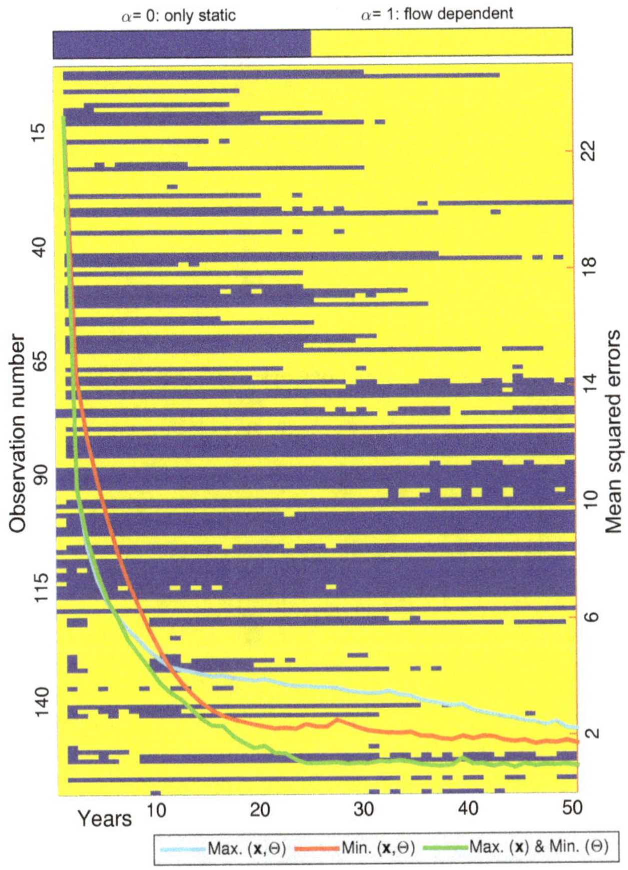

Figure 11. The colored image shows, according to the left y axis, the adaptive change in α values for each individual observation. The observation index $(1, \ldots, 160)$ is sorted such that the first 40 indices correspond to PCE measurements, the second 40 correspond to TCE, third 40 correspond to DCE and finally VC takes the last 40 indices. The yellow color indicates that no background covariance matrices have been used and the blue color suggests that only ensemble "flow-dependent" statistics are involved. The curves demonstrate the change in MSE, according to the right y axis, in time when maximizing the information gain (cyan), minimizing the information gain (red) and maximizing the formation for concentration and minimizing it for degradation rates (green).

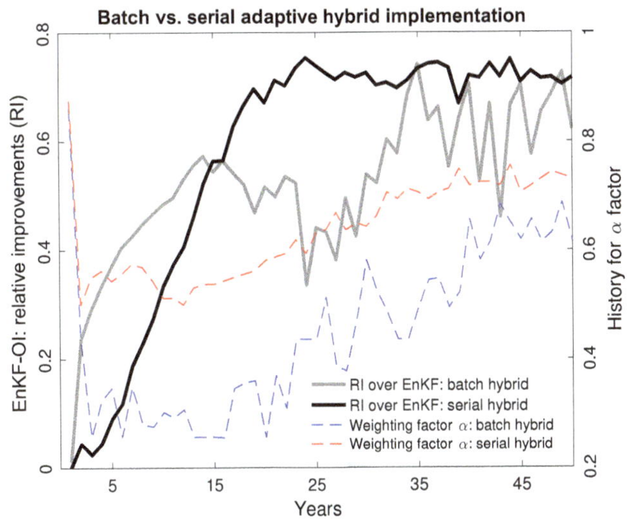

Figure 12. Left y axis: relative improvements (RIs) of the batch (i.e., observations processed all at once) and the serial (i.e, observations processed one at a time) adaptive EnKF-OIs over the standard EnKF scheme. RI is evaluated in time as follows: $\mathrm{RI}(t) = \frac{\mathrm{MSE}_{\mathrm{EnKF}}(t) - \mathrm{MSE}_{\mathrm{EnKF-OI}}(t)}{\mathrm{MSE}_{\mathrm{EnKF}}(t)}$. Right y axis: time series for the optimized α values resulting from the batch and the serial (optimal α's from all observations are averaged) EnKF-OI.

gion, upwards to match the truth. Moreover, as demonstrated in layer 70 and unlike the $\mathrm{EnKF_{ESOS}}$, the EnKF overestimates the TCE concentration in the center of the domain, which further continues to move southwards. In layer 80 (i.e., 5 m deeper), the EnKF tends to underestimate the concentration of DCE especially in the southern part of the domain. On the other hand, a slight overestimation of this DCE concentration towards the center is suggested by the $\mathrm{EnKF_{ESOS}}$. In general, and assessing similar patterns at other layers, the $\mathrm{EnKF_{ESOS}}$ exhibits higher accuracy in retrieving the contaminant concentration than the EnKF. This provides further evidence that ignoring the observation sampling errors within the EnKF can indeed deteriorate the quality of the state estimates.

To study the impact of the $\mathrm{EnKF_{ESOS}}$ on the estimates of the parameters, we examine the evolution of the approximate distribution of TCE degradation rate in time. We compare the resulting pdf's with those obtained using the EnKF after 5, 15, 30 and 45 years. On top of the pdf's, we also monitor the temporal evolution of K_T AES in Fig. 14. Starting from rather flat and uncertain pdf's of K_T, both EnKF and $\mathrm{EnKF_{ESOS}}$ correct the members of TCE degradation rate towards the truth, which is 0.086 per day. Notice that within the first 15 years, the pdf's seem to move in the wrong direction, away from the truth. This is due to the large concentrations at time 0, and thus the filter increases the degradation rates to fit the reference contaminant concentration. Beyond that and once the concentration is adjusted, the parameters from both filtering schemes begin moving closer toward the true degradation rate. However, the EnKF is seen to move faster towards the truth and further diminishes the uncertainty around K_T quite rapidly. Consequently, the resulting pdf of K_T after 45 years looks like a Kronecker delta function. This is, roughly speaking, not a very healthy assimilation system as the parameter updates become insignificant over the rest of the assimilation window. In contrast, the degradation rate obtained using $\mathrm{EnKF_{ESOS}}$ moves at a slower pace towards the true rate maintaining a large enough spread to fit the incoming observations. Compared to the EnKF, the AES suggested by the $\mathrm{EnKF_{ESOS}}$, as shown on the left y axis, is almost 40 to 50 % higher. As a matter of fact, this performance is more

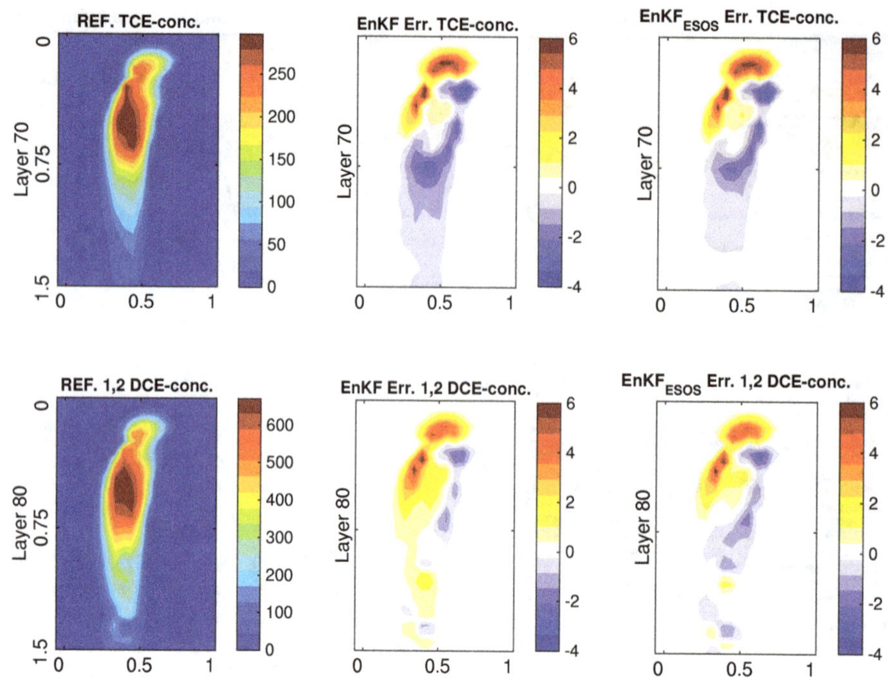

Figure 13. Top panels: TCE concentration and error maps in layer 70 obtained using the reference run (first column), the EnKF (second column) and the EnKF$_{ESOS}$ (third column). Bottom panels: same as top panels but for the concentration of DCE.

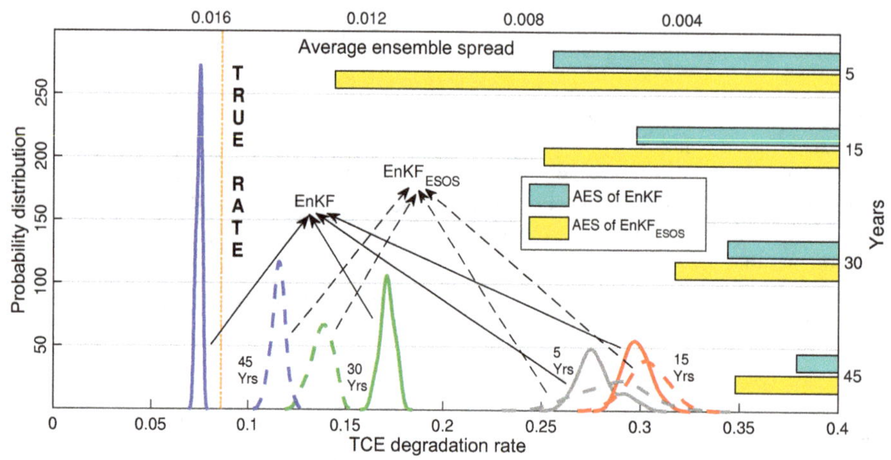

Figure 14. Left y axis: the time evolution of the prior probability density functions corresponding to TCE degradation rate obtained using the EnKF (solid lines) and the EnKF$_{ESOS}$ (dashed lines). The reference "true" rate is given at 0.086 day^{-1} in brown color. Right y axis: the AES of K_T suggested using the EnKF and the EnKF$_{ESOS}$.

trustworthy than that of the EnKF, indicating the essential need to account for observation sampling errors at the time of the analysis. Hoteit et al. (2015) found that the ensemble spread of the EnKF$_{ESOS}$ is larger than that of the EnKF for state estimation. In here, we experienced a similar, yet more pronounced behavior for the estimates of the parameters.

As a way to provide an overall assessment, we compare the best estimates obtained using all schemes considered earlier, i.e., the EnKF, the EnKF-OI$^{\beta=0.1}_{\alpha=0.7}$ and the EnKF$_{ESOS}$. The EnKF was found the least accurate. Accounting for observation sampling errors yields around 21 and 23 % more accurate state and parameters estimates, respectively. Tackling the rank deficiency of the EnKF results in 48 and 70 % more accurate state and parameters estimates, respectively. Accordingly, addressing the issues of observation sampling errors and rank deficient forecast ensemble matrices seems to be crucial and can highly improve the accuracy of the estimates. From our experimental results and for this particular

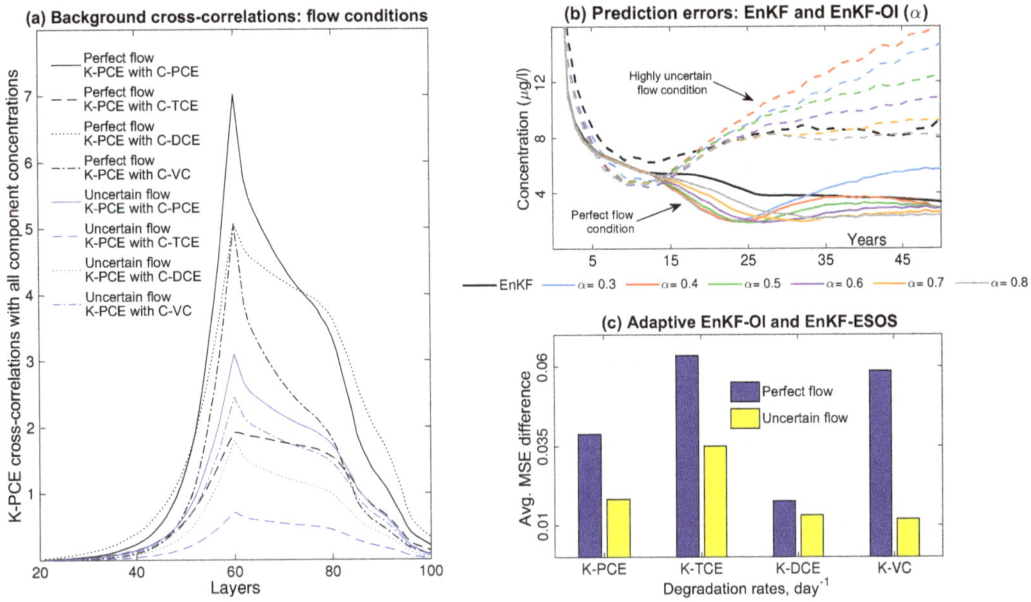

Figure 15. (a) Cross-correlation terms of the background matrix $\mathbf{P}^{\mathrm{b}}_{\theta x}$ associated with PCE biodegradation rate and all components' concentrations. The spatial correlations are averaged for all layers. The correlations in black are based on perfect groundwater-flow model. The correlations in blue are constructed after perturbing the hydraulic conductivity and porosity of the forecast model. The largest correlation is present in layer 60 where the contaminant source is located. **(b)** Performance of the EnKF and the hybrid EnKF-OI$^{\beta=1.0}_{\alpha=0.3,0.4,\ldots,0.8}$ before and after perturbing the hydraulic conductivity and porosity. The performance is assessed based on the temporal evolution of the MSE (prediction) errors. The imposed Gaussian uncertainty is high and obtained using standard deviation σ_{h}. **(c)** Average gain $\frac{1}{50}\sum_{i=1}^{50}\left(\mathrm{MSE}_{\mathrm{EnKF-OI}} - \mathrm{MSE}_{\mathrm{EnKF-ESOS}}\right)$ suggested for all degradation parameters. The improvements suggested by the adaptive EnKF-OI over EnKF$_{\mathrm{ESOS}}$ are shown for both perfect and perturbed (using σ_{m}) flow scenarios.

setting, resolving the rank deficiency issue appears to have the largest impact on the final estimates of the filter.

4.3 Incorporating uncertainties in the hydraulic parameters

In Sect. 4.1 and 4.2, the static background covariance matrices were derived on the basis of perfectly known hydraulic parameters. In this section, we test the impact of incorporating uncertainties in the groundwater-flow model on the performance of both the hybrid EnKF-OI and the EnKF$_{\mathrm{ESOS}}$ schemes. Generally, such procedure is expected to alter the precise description that the hybrid scheme utilizes to relate the biodegradation rates and the components' concentrations. To this end, the GeoTOP software package, described in Sect. 3.2.2, is used to obtain 48 different realizations for hydraulic conductivity and porosity. The realizations are built assuming Gaussian distributed hydraulic parameters with mean equal to the 3-D fields used in the reference model run (Sect. 3.2.2) and standard deviation parameterized in two different scenarios. We use $\sigma_m = 10\,\%$ of the mean in the first scenario (moderate uncertainty) and $\sigma_h = 30\,\%$ of the mean value in the second scenario (high uncertainty). The reason for this choice is to provide a realistic assessment of the filters under varying modeling uncertainty.

Before testing the performance of the hybrid EnKF-OI and the EnKF$_{\mathrm{ESOS}}$, we first construct the background covariances, i.e., $\mathbf{P}^{\mathrm{b}}_{xx}$ and $\mathbf{P}^{\mathrm{b}}_{\theta x}$, using a similar procedure to the one presented in Sect. 3.4 but based on perturbed conductivity and porosity realizations (here, σ_m is used). To interpret the influence of this modeling uncertainty, we plot in Fig. 15a the averaged cross-correlations of K_P rate with all four components. We observe that the dominant correlation patterns are similar to those obtained earlier using perfect flow conditions, especially in the shallow aquifer layers. In deeper layers, there are noticeable differences in the correlations of K_P and C_{TCE}. In addition, the magnitude of the new background correlations is considerably smaller. For instance, the estimated correlation value between K_P and C_{DCE} after imposing uncertainty on the hydraulic parameters has shrunk by 60 %. The fact that the major spatial patterns of the background correlations were preserved and their magnitude was influenced the most is related to the nature of the perturbed hydraulic parameters. Generally, porosity and conductivity affect the speed and the movement of groundwater in the aquifer and thus the degradation process would be expected to either slow down or accelerate. Compared to the previous parameterization, in this scenario 15 POD modes were required to capture around 97 % of the total variance.

With moderate uncertainty, both the hybrid EnKF-OI ($\alpha > 0.3$) and the $\mathrm{EnKF_{ESOS}}$ outperform the EnKF and recover reasonable estimates of the degradation parameters. However, compared to the ones obtained under perfect flow conditions, these were slightly less accurate. The largest gain, in terms of MSE, suggested by the EnKF-OI state estimates, is obtained with $\alpha = 0.7$ and is equal to 22 %. The $\mathrm{EnKF_{ESOS}}$ is a performant with an average gain of 20 %. Switching to the high uncertainty scenario (i.e., using σ_h), all three assimilation schemes failed to retrieve reliable estimates of the concentration and biodegradation. To illustrate, we show in Fig. 15b the temporal evolution of MSE obtained using the EnKF and $\text{EnKF-OI}^{\beta=1.0}_{\alpha=0.3,\ldots,0.8}$ with perfect and perturbed hydraulic parameters. The performance as shown starts degrading after 15 years, eventually leading to inaccurate concentration estimates and unreliable parameters. This suggests that quantifying the large uncertainty of the flow model alongside the reaction parameters, using any of the three schemes, might be necessary in such a challenging setup. On the other hand, under moderate uncertainty we compare the performance of the assimilation schemes to the ones obtained assuming perfect flow. For this, we plot in Fig. 15c the average gain (MSE difference), suggested by the adaptive hybrid scheme over $\mathrm{EnKF_{ESOS}}$ estimates of the degradation rates. The adaptive hybrid EnKF-OI remains more accurate even after perturbing the hydraulic parameters. However, the proposed accuracy is roughly halved for all parameters. This is indeed related to the less reliable static background covariances, which are now subjected to uncertainties in the groundwater-flow dynamics.

5 Conclusions

In this study, we examined and investigated the hybrid ensemble Kalman filter (EnKF-OI) and the second-order observation perturbations sampling ($\mathrm{EnKF_{ESOS}}$) schemes to estimate contaminant concentration and biodegradation rates of chlorinated hydrocarbons at the port of Rotterdam. We simulated the migration problem of a single plume consisting of tetrachloroethene (PCE), trichloroethene (TCE), cis-1,2-dichloroethene (DCE) and vinyl chloride (VC). Concentration data were used for yearly assimilation over a period of 50 years. The hybrid scheme complements the flow-dependent sample ensemble covariance of the EnKF with a prescribed static background covariance from an OI system to mitigate the undersampling of the ensembles and neglected model errors. The exact second-order sampling of the observation perturbations modifies the observation perturbations and assimilates the data one after the other, thus resolving the undersampling of the observation noise in the EnKF analysis. Challenging assimilation scenarios using a relatively small ensemble ($N_e = 48$) were presented, in which observations were processed serially. The key findings of this study and future research directions are summarized below:

1. Both the hybrid EnKF-OI and the $\mathrm{EnKF_{ESOS}}$ successfully provide accurate concentration and degradation rate estimates. On average, a tuned hybrid EnKF-OI (using $\alpha = 0.7$ and $\beta = 0.1$) suggests 48 and 70 % more accurate state and parameters estimates than those obtained using the EnKF. On the other hand, the $\mathrm{EnKF_{ESOS}}$'s state and parameters estimates are 21 and 23 % more accurate, respectively. In addition, the two schemes are easy to implement and computationally efficient requiring only a minimal change to an existing EnKF code.

2. Both filtering schemes demonstrated a better handling of the ensemble spread, for both states and parameters, avoiding collapse or false (unrealistic) confidence in the estimates, which enables better fit to the observations.

3. The hybrid scheme requires some effort to tune two weighting factors that adjust the background statistics for both state and parameters. The serial adaptive version of this scheme, which relies on maximizing the information gain between the forecast and analysis for each individual observation point, seems promising. From the experiments, we found that maximizing the information gain could however possibly deplete the uncertainty within the ensemble quite rapidly. Yet, this observation may vary between systems depending on the degree and the rate of uncertainty growth. One possible solution that we tested is to minimize the information gain, and thus decrease the update impact when fitting the observations. Further, one could also build the objective function in such a way that only a portion of the information gain is maximized. For instance, an example would be to enforce the ratio between the trace of the analysis and the forecast covariance matrices to be greater than 30 % meaning that at least 60 % of the ensemble uncertainty is preserved.

4. Failing to account for observation undersampling errors in the standard EnKF can impact not only the quality of the states but also, more importantly, the estimates of the parameters. In our experiments, the degradation rates obtained after assimilating the observations using the $\mathrm{EnKF_{ESOS}}$ scheme were more accurate, more reliable and more realistic.

5. Imposing large uncertainties on the hydraulic parameters of the flow model degrades the performance of all filtering schemes. Given that the performance of the hybrid EnKF-OI depends on the quality of the static background statistics, satisfactory results were obtained only when the uncertainty imposed on the background information is relatively moderate. The $\mathrm{EnKF_{ESOS}}$ further outperformed the standard EnKF with moderate flow uncertainty conditions.

6. Careful tuning of the hybrid EnKF-OI yields the best estimates of the concentration and the degradation rates as compared to the EnKF and the $\text{EnKF}_{\text{ESOS}}$. This manifests the importance of complementing the EnKF parameter cross-correlations with static ones.

Building a unified EnKF scheme, which tackles both the undersampling of the forecast covariance and the observation sampling errors simultaneously is an interesting line of research that we will consider in our future work. Further potential research includes generalizing the current assimilation system to allow for estimating various subsurface aquifer parameters (including spatial heterogeneity), boundary conditions and model structural errors.

6 Data availability

All data and codes used in this study can be obtained by emailing the first author (MG) at: gharamti@ucar.edu. More detailed and historical contaminant datasets are only accessible through the municipality of Rotterdam in the Netherlands.

Acknowledgements. The authors would like to thank three anonymous reviewers for their constructive comments and suggestions, which improved the quality of the manuscript. The authors would also like to thank the municipality of Rotterdam for their support and cooperation throughout this project. Research reported in this publication was supported by King Abdullah University of Science and Technology (KAUST).

Edited by: H.-J. Hendricks Franssen

References

Altaf, M., Gharamti, M., Heemink, A., and Hoteit, I.: A reduced adjoint approach to variational data assimilation, Comput. Method. Appl. M., 254, 1–13, doi:10.1016/j.cma.2012.10.003, 2013.

Annan, J., Hargreaves, J., Edwards, N., and Marsh, R.: Parameter estimation in an intermediate complexity earth system model using an ensemble Kalman filter, Ocean Model., 8, 135–154, 2005.

Appelo, C. and Postma, D.: Geochemistry Groundwater and Pollution. AA Balkema, Rotterdam, Brookfield, Tech. rep., 536 pp., doi:10.1017/s0016756800011523, 1994.

Bailey, R. T., Baù, D. A., and Gates, T. K.: Estimating spatially-variable rate constants of denitrification in irrigated agricultural groundwater systems using an Ensemble Smoother, J. Hydrol., 468, 188–202, 2012.

Bailey, R. T., Gates, T. K., and Halvorson, A. D.: Simulating variably-saturated reactive transport of selenium and nitrogen in agricultural groundwater systems, J. Contam. Hydrol., 149, 27–45, 2013.

Bear, J. and Cheng, A.-D.: Modeling groundwater flow and contaminant transport, vol. 23, Springer, doi:10.1007/978-1-4020-6682-5_7, 2010.

Bishop, C. H., Etherton, B. J., and Majumdar, S. J.: Adaptive sampling with the ensemble transform Kalman filter. Part I: Theoretical aspects, Mon. Weather Rev., 129, 420–436, 2001.

Bowler, N. E., Flowerdew, J., and Pring, S. R.: Tests of different flavours of EnKF on a simple model, Q. J. Roy. Meteor. Soc., 139, 1505–1519, 2013.

Burgers, G., Van Leeuwen, P. J., and Evensen, G.: Analysis scheme in the ensemble Kalman filter, Mon. Weather Rev., 126, 1719–1724, 1998.

Busschers, F. S., Dubelaar, C. W., Stafleu, J., and Maljers, D.: Lithological and sand grain-size variability in the three-dimensional GeoTOP model of Zuid-Holland, Tech. Rep. uGE 100 ENG, TNO Built Environment and Geosciences, Princetonlaan 6, 3508 TA Utrecht, The Netherlands, http://www2.dinoloket.nl/nl/about/modellen/uGE100A3%ENGFreekBusschersPrint.pdf (last access: 6 September 2011), 2010.

Chave, P.: The EU water framework directive: an introduction, IWA publishing, 2001.

Chen, Y. and Zhang, D.: Data assimilation for transient flow in geologic formations via ensemble Kalman filter, Adv. Water Resour., 29, 1107–1122, doi:10.1016/j.advwatres.2005.09.007, 2006.

Clement, T.: RT3D, A modular computer code for simulating Reactive multispecies Transport in 3-Dimensional groundwater aquifers, Pacific Northwest National Laboratory, Richland, Washington, 1–59, 1997.

Clement, T., Sun, Y., Hooker, B., and Petersen, J.: Modeling multispecies reactive transport in ground water, Groundwater Monitoring & Remediation, 18, 79–92, doi:10.1111/j.1745-6592.1998.tb00618.x, 1998.

Clement, T. P., Johnson, C. D., Sun, Y., Klecka, G. M., and Bartlett, C.: Natural attenuation of chlorinated ethene compounds: model development and field-scale application at the Dover site, J. Contam. Hydrol., 42, 113–140, doi:10.1016/s0169-7722(99)00098-4, 2000.

Courtier, P., Thépaut, J.-N., and Hollingsworth, A.: A strategy for operational implementation of 4D-Var, using an incremental approach, Q. J. Roy. Meteor. Soc., 120, 1367–1387, doi:10.1002/qj.49712051912, 1994.

Crestani, E., Camporese, M., Baú, D., and Salandin, P.: Ensemble Kalman filter versus ensemble smoother for assessing hydraulic conductivity via tracer test data assimilation, Hydrol. Earth Syst. Sci., 17, 1517–1531, doi:10.5194/hess-17-1517-2013, 2013.

Cunningham, S. D. and Berti, W. R.: Remediation of contaminated soils with green plants: an overview, In Vitro Cell. Dev.-Pl., 29, 207–212, doi:10.1007/bf02632036, 1993.

Dawson, C., Sun, S., and Wheeler, M. F.: Compatible algorithms for coupled flow and transport, Comput. Method. Appl. M., 193, 2565–2580, doi:10.1016/j.cma.2003.12.059, 2004.

Devegowda, D., Arroyo, E., Datta-Gupta, A., and Douma, S. G.: Efficient and robust reservoir model updating using ensemble Kalman filter with sensitivity-based covariance localization, in: SPE Reservoir Simulation Symposium, Society of Petroleum Engineers, doi:10.2523/106144-ms, 2007.

Dimet, F.-X. L. and Talagrand, O.: Variational algorithms for analysis and assimilation of meteorological observations: theoretical aspects, Tellus A, 38, 97–110, doi:10.3402/tellusa.v38i2.11706, 1986.

Drécourt, J.-P., Madsen, H., and Rosbjerg, D.: Calibration framework for a Kalman filter applied to a groundwater model, Adv. Water Resour., 29, 719–734, doi:10.1016/j.advwatres.2005.07.007, 2006.

El Serafy, G. Y. and Mynett, A. E.: Improving the operational forecasting system of the stratified flow in Osaka Bay using an ensemble Kalman filter–based steady state Kalman filter, Water Resour. Res., 44, doi:10.1029/2006wr005412, 2008.

Evensen, G.: The ensemble Kalman filter: Theoretical formulation and practical implementation, Ocean Dynam., 53, 343–367, doi:10.1007/s10236-003-0036-9, 2003.

Forsythe, G. E., Malcolm, M. A., and Moler, C. B.: Computer methods for mathematical computations, 270 pp., Prentice-Hall, Englewood Cliffs, NJ, 1977.

Franssen, H. and Kinzelbach, W.: Ensemble Kalman filtering versus sequential self-calibration for inverse modelling of dynamic groundwater flow systems, J. Hydrol., 365, 261–274, doi:10.1016/j.jhydrol.2008.11.033, 2009.

Freeze, R. A. and Cherry, J.: Groundwater, Prentice-Hall Englewood Cliffs, NJ, 604 pp., 1979.

Gharamti, M. and Hoteit, I.: Complex step-based low-rank extended Kalman filtering for state-parameter estimation in subsurface transport models, J. Hydrol., 509, 588–600, doi:10.1016/j.jhydrol.2013.12.004, 2014.

Gharamti, M., Hoteit, I., and Sun, S.: Low-rank Kalman filtering for efficient state estimation of subsurface advective contaminant transport models, J. Environ. Eng., 138, 446–457, doi:10.1061/(asce)ee.1943-7870.0000484, 2011.

Gharamti, M., Kadoura, A., Valstar, J., Sun, S., and Hoteit, I.: Constraining a compositional flow model with flow-chemical data using an ensemble-based Kalman filter, Water Resour. Res., 50, 2444–2467, doi:10.1002/2013wr014830, 2014a.

Gharamti, M., Valstar, J., and Hoteit, I.: An adaptive hybrid EnKF-OI scheme for efficient state-parameter estimation of reactive contaminant transport models, Adv. Water Resour., 71, 1–15, doi:10.1016/j.advwatres.2014.05.001, 2014b.

Gharamti, M. E., Hoteit, I., and Valstar, J.: Dual states estimation of a subsurface flow-transport coupled model using ensemble Kalman filtering, Adv. Water Resour., 60, 75–88, doi:10.1016/j.advwatres.2013.07.011, 2013.

Hamill, T. M. and Snyder, C.: A hybrid ensemble Kalman filter-3D variational analysis scheme, Mon. Weather Rev., 128, 2905–2919, 2000.

Hamill, T. M., Whitaker, J. S., and Snyder, C.: Distance-dependent filtering of background error covariance estimates in an ensemble Kalman filter, Mon. Weather Rev., 129, 2776–2790, 2001.

Harbaugh, A. W.: MODFLOW-2005, the US Geological Survey modular ground-water model: The ground-water flow process, US Department of the Interior, US Geological Survey Reston, VA, USA, 1–253, 2005.

Hendricks Franssen, H. and Kinzelbach, W.: Real-time groundwater flow modeling with the Ensemble Kalman Filter: Joint estimation of states and parameters and the filter inbreeding problem, Water Resour. Res., 44, 1–21, doi:10.1029/2007wr006505, 2008.

Hering, D., Borja, A., Carstensen, J., Carvalho, L., Elliott, M., Feld, C. K., Heiskanen, A. S., Johnson, R. K., Moe, J., Pont, D., and Solheim, A. L.: The European Water Framework Directive at the age of 10: a critical review of the achievements with recommendations for the future, Sci. Total Environ., 408, 4007–4019, doi:10.1016/j.scitotenv.2010.05.031, 2010.

Hoteit, I., Pham, D.-T., and Blum, J.: A simplified reduced order Kalman filtering and application to altimetric data assimilation in Tropical Pacific, J. Marine Syst., 36, 101–127, doi:10.1016/s0924-7963(02)00129-x, 2002.

Hoteit, I., Cornuelle, B., Köhl, A., and Stammer, D.: Treating strong adjoint sensitivities in tropical eddy-permitting variational data assimilation, Q. J. Roy. Meteor. Soc., 131, 3659–3682, doi:10.1256/qj.05.97, 2005.

Hoteit, I., Pham, D.-T., Triantafyllou, G., and Korres, G.: A new approximate solution of the optimal nonlinear filter for data assimilation in meteorology and oceanography, Mon. Weather Rev., 136, 317–334, 2008.

Hoteit, I., Xiaodong, L., and Pham, D.-T.: Particle Kalman Filtering: A Nonlinear Bayesian Framework for Ensemble Kalman Filters, Mon. Weather Rev., 140, 528–542, doi:10.1175/2011mwr3640.1, 2012.

Hoteit, I., Pham, D.-T., Gharamti, M., and Luo, X.: Mitigating observation perturbation sampling errors in the stochastic EnKF, Mon. Weather Rev., 143, 2918–2936, 2015.

Houtekamer, P. L. and Mitchell, H. L.: A sequential ensemble Kalman filter for atmospheric data assimilation, Mon. Weather Rev., 129, 123–137, 2001.

Kalman, R. E.: A new approach to linear filtering and prediction problems, J. Basic Eng., 82, 35–45, doi:10.1115/1.3662552, 1960.

Kullback, S. and Leibler, R. A.: On information and sufficiency, Ann. Math. Stat., 22, 79–86, doi:10.1214/aoms/1177729694, 1951.

Lawson, W. G. and Hansen, J. A.: Implications of stochastic and deterministic filters as ensemble-based data assimilation methods in varying regimes of error growth, Mon. Weather Rev., 132, 1966–1981, 2004.

Lee, L. J., Chung, C.-W., Ma, Y.-C., Wang, G.-S., Chen, P.-C., Hwang, Y.-H., and Wang, J.: Increased mortality odds ratio of male liver cancer in a community contaminated by chlorinated hydrocarbons in groundwater, Occup. Environ. Med., 60, 364–369, doi:10.1136/oem.60.5.364, 2003.

Lee, L. J.-H., Chan, C.-C., Chung, C.-W., Ma, Y.-C., Wang, G.-S., and Wang, J.-D.: Health risk assessment on residents exposed to chlorinated hydrocarbons contaminated in groundwater of a hazardous waste site, J. Tox. Environ. Health, 65, 219–235, doi:10.1080/15287390252800828, 2002.

Li, L., Zhou, H., Gómez-Hernández, J. J., and Hendricks Franssen, H.-J.: Jointly mapping hydraulic conductivity and porosity by assimilating concentration data via ensemble Kalman filter, J. Hydrol., 428, 152–169, doi:10.1016/j.jhydrol.2012.01.037, 2012.

Liu, B., Gharamti, M., and Hoteit, I.: Assessing clustering strategies for Gaussian mixture filtering a subsurface contaminant model, J. Hydrol., 535, 1–21, 2016.

Mandel, J., Cobb, L., and Beezley, J. D.: On the convergence of the ensemble Kalman filter, Appl. Math., 56, 533–541, 2011.

Marsman, A., Valstar, J., and ter Meer, J.: Risk analysis on groundwater contamination at the megasite Port of Rotterdam, IAHS-AISH publication, 309–314, 2006.

McLaughlin, D.: An integrated approach to hydrologic data assimilation: interpolation, smoothing, and filtering, Adv. Water Resour., 25, 1275–1286, 2002.

Moradkhani, H., Sorooshian, S., Gupta, H. V., and Houser, P. R.: Dual state–parameter estimation of hydrological models using ensemble Kalman filter, Adv. Water Resour., 28, 135–147, doi:10.1016/j.advwatres.2004.09.002, 2005.

Mostert, E.: The European water framework directive and water management research, Phys. Chem. Earth, 28, 523–527, doi:10.1016/s1474-7065(03)00089-5, 2003.

Nerger, L., Hiller, W., and Schröter, J.: A comparison of error subspace Kalman filters, Tellus A, 57, 715–735, 2005.

Nichols, N.: Mathematical concepts of data assimilation, in: Data Assimilation, 13–39, Springer, doi:10.1007/978-3-540-74703-1_2, 2010.

Ojajärvi, A., Partanen, T., Ahlbom, A., Boffetta, P., Hakulinen, T., Jourenkova, N., Kauppinen, T., Kogevinas, M., Vainio, H., Weiderpass, E., and Wesseling, C.: Risk of pancreatic cancer in workers exposed to chlorinated hydrocarbon solvents and related compounds: a meta-analysis, Am. J. Epidemiol., 153, 841–850, doi:10.1093/aje/153.9.841, 2001.

Panzeri, M., Riva, M., Guadagnini, A., and Neuman, S.: Data assimilation and parameter estimation via ensemble Kalman filter coupled with stochastic moment equations of transient groundwater flow, Water Resour. Res., 49, 1334–1344, 2013.

Pham, D. T.: Stochastic methods for sequential data assimilation in strongly nonlinear systems, Mon. Weather Rev., 129, 1194–1207, 2001.

Pham, D. T., Verron, J., and Roubaud, M. C.: A singular evolutive extended Kalman filter for data assimilation in oceanography, J. Marine Syst., 16, 323–340, 1998.

Pollock, D. W.: User's Guide for MODPATH/MODPATH-PLOT, Version 3: A Particle Tracking Post-processing Package for MODFLOW, the US: Geological Survey Finite-difference Ground-water Flow Model, US Department of Interior, 1994.

Reichle, R. H., McLaughlin, D. B., and Entekhabi, D.: Hydrologic data assimilation with the ensemble Kalman filter, Mon. Weather Rev., 130, 103–114, 2002.

Sakov, P. and Bertino, L.: Relation between two common localisation methods for the EnKF, Comp. Geosci., 15, 225–237, doi:10.1007/s10596-010-9202-6, 2011.

Sakov, P. and Oke, P. R.: A deterministic formulation of the ensemble Kalman filter: an alternative to ensemble square root filters, Tellus A, 60, 361–371, 2008.

Skachko, S., Brankart, J.-M., Castruccio, F., Brasseur, P., and Verron, J.: Improved turbulent air-sea flux bulk parameters for controlling the response of the ocean mixed layer: a sequential data assimilation approach, J. Atmos. Ocean. Technol., 26, 538–555, doi:10.1175/2008jtecho603.1, 2009.

Song, H., Hoteit, I., Cornuelle, B. D., and Subramanian, A. C.: An adaptive approach to mitigate background covariance limitations in the ensemble Kalman filter, Mon. Weather Rev., 138, 2825–2845, doi:10.1175/2010mwr2871.1, 2010.

Song, H., Hoteit, I., Cornuelle, B. D., Luo, X., and Subramanian, A. C.: An Adjoint-Based Adaptive Ensemble Kalman Filter, Mon. Weather Rev., 141, 3343–3359, doi:10.1175/mwr-d-12-00244.1, 2013.

Stafleu, J., Maljers, D., Gunnink, J., Menkovic, A., and Busschers, F.: 3D modelling of the shallow subsurface of Zeeland, the Netherlands, Neth. J. Geosci., 90, 293–310, 2011a.

Stafleu, J., Maljers, D., Gunnink, J., Menkovic, A., and Busschers, F.: 3D modelling of the shallow subsurface of Zeeland, the Netherlands, Neth. J. Geosci., 90, 293–310, 2011b.

Starr, R. C. and Cherry, J. A.: In Situ Remediation of Contaminated Ground Water: The Funnel-and-Gate System, Groundwater, 32, 465–476, doi:10.1111/j.1745-6584.1994.tb00664.x, 1994.

Suarez, M. P. and Rifai, H. S.: Biodegradation rates for fuel hydrocarbons and chlorinated solvents in groundwater, Biorem. J., 3, 337–362, doi:10.1080/10889869991219433, 1999.

Sun, S. and Wheeler, M. F.: A posteriori error estimation and dynamic adaptivity for symmetric discontinuous Galerkin approximations of reactive transport problems, Comp. Meth. Appl. Mech. Eng., 195, 632–652, doi:10.1016/j.cma.2005.02.021, 2006.

Ter Meer, J., Hartog, N., Valstar, J., Marsman, A., and Rijnaarts, H.: Contaminant sources, pathways and receptors at the Rotterdam Port Area – a non-site-specific approach, Grundwasser und Boden, p. 71, 2007.

Tobiszewski, M. and Namieśnik, J.: Abiotic degradation of chlorinated ethanes and ethenes in water, Environ. Sci. Poll. Res., 19, 1994–2006, doi:10.1007/s11356-012-0764-9, 2012.

Todd, D. K. and Mays, L. W.: Groundwater hydrology edition, Wiley, New Jersey, 1–625, 2005.

Vermeulen, P., Quan, N. H., Nam, N. D. G., Van Hungd, P., Tungd, N. T., Thanhe, T. V., and Dam, R.: Groundwater modeling for the Mekong Delta using iMOD, 20th International Congress on Modelling and Simulation, Adelaide, Australia, 2499–2505, 2013.

Vogel, T. M. and McCarty, P. L.: Biotransformation of tetrachloroethylene to trichloroethylene, dichloroethylene, vinyl chloride, and carbon dioxide under methanogenic conditions, Appl. Environ. Microbiol., 49, 1080–1083, 1985.

Wang, X., Hamill, T. M., Whitaker, J. S., and Bishop, C. H.: A comparison of hybrid ensemble transform Kalman filter-optimum interpolation and ensemble square root filter analysis schemes, Mon. Weather Rev., 135, 1055–1076, doi:10.1175/mwr3307.1, 2007.

Wang, X., Hamill, T. M., Whitaker, J. S., and Bishop, C. H.: A comparison of the hybrid and EnSRF analysis schemes in the presence of model errors due to unresolved scales, Mon. Weather Rev., 137, 3219–3232, doi:10.1175/2009mwr2923.1, 2009.

Whitaker, J. S. and Hamill, T. M.: Ensemble data assimilation without perturbed observations, Mon. Weather Rev., 130, 1913–1924, 2002.

Zheng, C. and Wang, P.: MT3DMS: a modular three-dimensional multispecies transport model for simulation of advection, dispersion, and chemical reactions of contaminants in groundwater systems; documentation and user's guide, Alabama Univ. University, 1999.

Zhou, H., Gómez-Hernández, J. J., Hendricks Franssen, H.-J., and Li, L.: An approach to handling non-Gaussianity of parameters and state variables in ensemble Kalman filtering, Adv. Water Resour., 34, 844–864, doi:10.1016/j.advwatres.2011.04.014, 2011.

Zhou, H., Li, L., Franssen, H.-J. H., and Gómez-Hernández, J. J.: Pattern recognition in a bimodal aquifer using the normal-score ensemble Kalman filter, Math. Geosci., 44, 169–185, doi:10.1007/s11004-011-9372-3, 2012.

PERMISSIONS

All chapters in this book were first published in HESS, by Copernicus Publications; hereby published with permission under the Creative Commons Attribution License or equivalent. Every chapter published in this book has been scrutinized by our experts. Their significance has been extensively debated. The topics covered herein carry significant findings which will fuel the growth of the discipline. They may even be implemented as practical applications or may be referred to as a beginning point for another development.

The contributors of this book come from diverse backgrounds, making this book a truly international effort. This book will bring forth new frontiers with its revolutionizing research information and detailed analysis of the nascent developments around the world.

We would like to thank all the contributing authors for lending their expertise to make the book truly unique. They have played a crucial role in the development of this book. Without their invaluable contributions this book wouldn't have been possible. They have made vital efforts to compile up to date information on the varied aspects of this subject to make this book a valuable addition to the collection of many professionals and students.

This book was conceptualized with the vision of imparting up-to-date information and advanced data in this field. To ensure the same, a matchless editorial board was set up. Every individual on the board went through rigorous rounds of assessment to prove their worth. After which they invested a large part of their time researching and compiling the most relevant data for our readers.

The editorial board has been involved in producing this book since its inception. They have spent rigorous hours researching and exploring the diverse topics which have resulted in the successful publishing of this book. They have passed on their knowledge of decades through this book. To expedite this challenging task, the publisher supported the team at every step. A small team of assistant editors was also appointed to further simplify the editing procedure and attain best results for the readers.

Apart from the editorial board, the designing team has also invested a significant amount of their time in understanding the subject and creating the most relevant covers. They scrutinized every image to scout for the most suitable representation of the subject and create an appropriate cover for the book.

The publishing team has been an ardent support to the editorial, designing and production team. Their endless efforts to recruit the best for this project, has resulted in the accomplishment of this book. They are a veteran in the field of academics and their pool of knowledge is as vast as their experience in printing. Their expertise and guidance has proved useful at every step. Their uncompromising quality standards have made this book an exceptional effort. Their encouragement from time to time has been an inspiration for everyone.

The publisher and the editorial board hope that this book will prove to be a valuable piece of knowledge for researchers, students, practitioners and scholars across the globe.

LIST OF CONTRIBUTORS

L. J. M. Peeters
CSIRO Land and Water, Water for a Healthy Country Flagship, Adelaide, Australia

G. M. Podger, T. Smith and S. M. Cuddy
CSIRO Land and Water, Water for a Healthy Country Flagship, Canberra, Australia

T. Pickett
CSIRO Land and Water, Water for a Healthy Country Flagship, Brisbane, Australia

R. H. Bark
CSIRO Ecosystem Sciences, Water for a Healthy Country Flagship, Brisbane, Australia

Léonard Santos, Guillaume Thirel and Charles Perrin
Irstea, HYCAR Research Unit, 1 rue Pierre-Gilles de Gennes, 92160 Antony, France

W. A. Timms, S. Bouzalakos and M. Whelan
School of Mining Engineering, University of New South Wales, Sydney, Australia
UNSW Connected Waters Initiative affiliated with the National Centre for Groundwater Research and Training, Sydney, Australia

R. Crane, D. McGeeney and R. I. Acworth
UNSW Connected Waters Initiative affiliated with the National Centre for Groundwater Research and Training, Sydney, Australia
Water Research Laboratory, School of Civil and Environmental Engineering, University of New South Wales, Sydney, Australia

D. J. Anderson and P. F. Rahman
Water Research Laboratory, School of Civil and Environmental Engineering, University of New South Wales, Sydney, Australia

María Carolina Rogelis
UNESCO-IHE, 2601 DA Delft, the Netherlands

Micha Werner
UNESCO-IHE, 2601 DA Delft, the Netherlands
Deltares, 2600 MH Delft, the Netherlands

Andrea Libertino, Daniele Ganora and Pierluigi Claps
Department of Environment, Land and Infrastructure Engineering (DIATI), Politecnico di Torino, Torino, Italy

A. Md Ali
Department of Integrated Water System and Knowledge Management, UNESCO-IHE Institute for Water Education, Delft, the Netherlands
Department of Irrigation and Drainage, Kuala Lumpur, Malaysia

D. P. Solomatine
Department of Integrated Water System and Knowledge Management, UNESCO-IHE Institute for Water Education, Delft, the Netherlands
Water Resource Section, Delft University of Technology, the Netherlands

G. Di Baldassarre
Department of Earth Sciences, Uppsala University, Sweden

Vijay P. Santikari and Lawrence C. Murdoch
Department of Environmental Engineering and Earth Sciences, Clemson University, Clemson, SC 29634, USA

Roger Moussa
INRA, UMR LISAH, 2 place Viala, 34060 Montpellier, France

Jean-Paul Lhomme
IRD, UMR LISAH, 2 place Viala, 34060 Montpellier, France

Xinjun Tu and Xiaohong Chen
Center of Water Resources and Environment, Sun Yat-sen University, Guangzhou, 510275, China
Center of Water Security Engineering and Technology in Southern China of Guangdong, Guangzhou, 510275, China

Yiliang Du, Kairong Lin and Haiou Wu
Center of Water Resources and Environment, Sun Yat-sen University, Guangzhou, 510275, China

Vijay P. Singh
Department of Biological and Agricultural Engineering, Texas A&M University, 2117 College Station, Texas 77843, USA
Zachry Department of Civil Engineering, Texas A&M University, 2117 College Station, Texas 77843, USA

Petr Máca, Petr Bašta, Radek Vlnas and Pavel Pech
Faculty of Environmental Sciences, Czech University of Life Sciences, Kamýcká 1176, Prague 6, Czech Republic

Martin Hanel
Faculty of Environmental Sciences, Czech University of Life Sciences, Kamýcká 1176, Prague 6, Czech Republic
T. G. Masaryk Water Research Institute, Podbabská 30, Prague 6, Czech Republic

G. Panthou, T. Vischel, T. Lebel, G. Quantin and G. Molinié
LTHE - UMR5564, Univ. Grenoble, IRD, CNRS, Grenoble, France

Claudia Cherubini
Department of Physics & Earth Sciences, University of Ferrara, Via Saragat 1-, 44122, Ferrara, Italy
School of Civil Engineering, University of Queensland, St Lucia, Brisbane 4072, Australia

Nicola Pastore and Concetta I. Giasi
DICATECh, Department of Civil, Environmental, Building Engineering, and Chemistry, Politecnico di Bari, Bari, Italy

Dimitra Rapti
New Energies and environment Company (NEA) Via Saragat 1-, 44122 Ferrara, Italy

Stanislaus J. Schymanski and Dani Or
Department of Environmental Sciences, ETH Zurich, 8092 Zurich, Switzerland

G. Bruni, N. C. van de Giesen and J. A. E. ten Veldhuis
Department of Water Management, Faculty of Civil Engineering and Geosciences, Delft University of Technology, Delft, the Netherlands

R. Reinoso
Department of Geoscience and Remote Sensing, Faculty of Civil Engineering and Geosciences, Delft University of Technology, Delft, the Netherlands

F. H. L. R. Clemens
Department of Water Management, Faculty of Civil Engineering and Geosciences, Delft University of Technology, Delft, the Netherlands
Deltares, Delft, The Netherlands

Johan Valstar, Gijs Janssen and Annemieke Marsman
Deltares, Princetonlaan 6, 3584 CB, Utrecht, the Netherlands

Index

R

S

T

U

V

W

www.ingramcontent.com/pod-product-compliance
Lightning Source LLC
LaVergne TN
LVHW060544210926
848485LV00005B/1252